METALLURGY AND CORROSION CONTROL IN OIL AND GAS PRODUCTION

WILEY SERIES IN CORROSION

R. Winston Revie, Series Editor

METALLURGY AND CORROSION CONTROL IN OIL AND GAS PRODUCTION

SECOND EDITION

ROBERT HEIDERSBACH

Edition History
Metallurgy and Corrosion Control in Oil and Gas Production, John Wiley & Sons, Inc. (1e, 2011)

Registered Office
John Wiley & Sons, Inc., 111 River Street, Hoboken, NJ 07030, USA

Editorial Office
111 River Street, Hoboken, NJ 07030, USA

For details of our global editorial offices, customer services, and more information about Wiley products visit us at www.wiley.com.

Wiley also publishes its books in a variety of electronic formats and by print-on-demand. Some content that appears in standard print versions of this book may not be available in other formats.

Library of Congress Cataloging-in-Publication Data

Names: Heidersbach, R. (Robert), author.
Title: Metallurgy and corrosion control in oil and gas production / by Robert Heidersbach.
Description: Second edition. | Hoboken, NJ : Wiley, 2018. | Series: Wiley series in corrosion | Includes index. |
Identifiers: LCCN 2018012501 (print) | LCCN 2018013527 (ebook) | ISBN 9781119252382 (pdf) | ISBN 9781119252375 (epub) |
    ISBN 9781119252054 (cloth)
Subjects: LCSH: Oil fields–Equipment and supplies. | Metallurgy. | Corrosion and anti-corrosives.
Classification: LCC TN871.5 (ebook) | LCC TN871.5 .H45 2018 (print) | DDC 622/.3382–dc23
LC record available at https://lccn.loc.gov/2018012501

Cover design by Wiley
Cover image: © Wan Fahmy Redzuan/Shutterstock

Set in 10/12pt TimesTen by SPi Global, Pondicherry, India

Printed in the United States of America

V10003867_082218

To Dianne Heidersbach, my wife of over 50 years

CONTENTS

5 Forms of Corrosion

PREFACE

"Were I to wait perfection, my book would never be finished." This quote from Tai K'ung was in the preface to the first edition, and it is worth repeating.

I had no idea that the second edition of a book would be so much work, but it is now time to submit my efforts to the publisher and thank those who have helped me get to this point. The first edition was dedicated to my daughter, Krista Heidersbach, who also helped me with this update. Over the years her advice and insights have been invaluable. In addition to the items that have been included in this book, she also advised on those subjects that were best omitted. No father could be prouder than I am of her, and I thank her for all the time she spent helping me on this book.

Most engineers can handle the routine; it's the details that have caused problems in recent years. The Pareto principle, often called the 80/20 rule, suggests that most problems are associated with only some of the equipment in any installation or oilfield. In recent years I have been involved in working on both corrosion under insulation (CUI), which can cause surprising leaks in piping systems that are hard to inspect, and failure analysis on high-strength steel fasteners. The problem with broken bolts of subsea equipment in the Gulf of Mexico is an ongoing technical challenge that has resulted in a number of new or revised industrial standards. These relatively small components on large structures can lead to catastrophic consequences. This second edition has increased discussion on these two subjects that reflects increased concern by industry, both in North America and worldwide. It is likely that updates and changes on these and other topics will continue in future years, and I have attempted to suggest where additional information on these, and other problem areas, may be found.

I suggest the reader should pay particular attention to the differences between corrosion monitoring, which is routine in many organizations, and the need for better inspection. Monitoring cannot replace inspection. No organization has the resources to inspect everything that can corrode or break, and the developments and implementation of risk-based inspection, concentrating on equipment most likely to fail and have high consequences, cannot be overemphasized. It is unfortunate that the oil and gas industry continues to experience equipment failures that would have been prevented if management, and the technical experts who advise them, had understood that monitoring *cannot* replace, or substitute for, inspection.

In recent years the LinkedIn online discussion groups started by Riky Bernardo, currently with RasGas in Qatar but originally from Indonesia where I met him, have grown to over 50 000 members. Steve Jones, whom I have never met, often contributes ideas to this discussion group and is one of the international group of leaders that I have relied upon to help me develop this revised manuscript. I urge the reader to monitor the discussions at the LinkedIn Oil and Gas Corrosion and Materials Selection site. On the day that I am writing this, there is informed discussion of erosion corrosion comparing API RP14E and DNV RP-0501, hydrogen embrittlement of high-strength steels, post-weld heat treatment of clad piping, and a variety of coatings topics. The reader can learn a lot from these discussions, because no international standard, let alone this book, can cover the entire range of technical questions that come up.

I am a metallurgical engineer by training, but materials selection and corrosion control is so multidisciplinary that many other fields must also be considered for safe and reliable oilfield operations. In addition to the people identified in the preface to the first edition of this book, I would also like to acknowledge the following people who have guided me on the preparation of this revised second edition:

Steve Jones, Ivan Gutierrez, Arun Soman, Reza Javaherdashti, and Roger Francis have provided many helpful discussions on Riky Bernardo's LinkedIn sites and have taught me valuable lessons. I have never met these people, and probably never will, but their advice has been invaluable to me and to many others.

Herb Townsend, Tom Goin, Ian MacMoy, and Candi Hudson have been leaders in redefining how high-strength fasteners can be used in marine and other environments. I urge the reader to keep up discussions on this still evolving technology.

Bob Gummow reviewed the 2011 discussions on cathodic protection and offered valuable advice on how it could be improved.

Travis Tonge, Ramesh Bapat, Chelsea LeHaye, and Juan Imamoto are examples of experts I met while teaching corrosion classes. I have learned far more from them than the little I was able to teach them.

Juan Imamoto and Tom Kuckertz offered me the opportunity to work on CUI projects in widely different situations. I learned a lot from them and from the work we did together. Bob Guise and Kash Prakash worked with me on one of these projects, and they taught me a lot.

Tim Bieri and coworkers at BP have remained valuable resources. I talked to Tim on several occasions, and he is one of my "go-to" experts. The 2006 NACE BP article on inspection and monitoring is one of the best explanations of the advantages and limitations of both inspection and monitoring that I have found, and I have relied heavily on this and their other advice.

Chemists look at things differently than engineers, and both Mark Kolody and Luz Marina Calle are two of my favorite chemists. When I am confused in my understanding, they have the unique ability to explain things in ways that I can understand.

All writers need to have critical reviewers. Gurudas Saha, whom I have only met through the Internet, is one of these experts whose advice I value highly.

BOB HEIDERSBACH
Cape Canaveral, Florida, USA

1

INTRODUCTION TO OILFIELD METALLURGY AND CORROSION CONTROL

The American Petroleum Institute (API) divides the petroleum industry into the following categories:

- Upstream
- Downstream
- Pipelines

Other organizations use terms like production, pipelining, transportation, and refining. This book will discuss upstream operations, with an emphasis on production, and pipelines, which are closely tied to upstream operations. Many "pipelines" could also be termed gathering lines or flowlines, and the technologies involved in materials selection and corrosion control are similar for all three categories of equipment.

Until the 1980s metals used in upstream production operations were primarily carbon steels. Developments of deep hot gas wells in the 1980s led to the use of corrosion-resistant alloys (CRAs), and this trend continues as the industry becomes involved in deeper and more aggressive environments [1, 2]. Nonetheless, most metal used in oil and gas production is carbon or low-alloy steel, and nonmetallic materials are used much less than metals.

Increased emphasis on reliability also contributes to the use of newer or more corrosion-resistant materials. Many oilfields that were designed with anticipated operating lives of 20–30 years are still economically viable after more than 50 years. This life extension of oilfields is the result of increases in the market value of petroleum products and the development of enhanced recovery techniques that make possible the recovery of

larger fractions of the hydrocarbons in downhole formations. Unfortunately, this tendency to prolong the life of oilfields creates corrosion and reliability problems in older fields when reductions in production and return on investment cause management to become reluctant to spend additional resources on maintenance and inspection.

These trends have all led to an industry that tends to design for much longer production lives and tries to use more reliable designs and materials. The previous tendency to rely on maintenance is being replaced by the trend to design more robust and reliable systems instead of relying on inspection and maintenance. The reduction in available trained labor for maintenance also drives this trend.

COSTS

A US government report estimated that the cost of corrosion in upstream operations and pipelines was $1372 billion per year, with the largest expenses associated with pipelines followed by downhole tubing and increased capital expenditures (primarily the use of CRAs). The most important opportunity for savings is the prevention of failures that lead to lost production. The same report suggested that the lack of corrosion problems in existing systems does not justify reduced maintenance budgets, which is a recognition that, as oilfields age, they become more corrosive at times when reduced returns on investment are occurring [3]. The 2013 environmental cracking problems with offshore

Metallurgy and Corrosion Control in Oil and Gas Production, Second Edition. Robert Heidersbach.
© 2018 John Wiley & Sons, Inc. Published 2018 by John Wiley & Sons, Inc.

pipelines in the Caspian Sea Kashagan oilfield are estimated to have cost billions of dollars for pipeline replacement costs plus lost production [4]. It is estimated that corrosion costs are approximately equal to mechanical breakdowns in maintenance costs.

SAFETY

While proper equipment design, materials selection, and corrosion control can result in monetary savings, a perhaps more important reason for corrosion control is safety. Hydrogen sulfide, H_2S, is a common component of many produced fluids. It is poisonous to humans, and it also causes a variety of environmental cracking problems. The proper selection of H_2S-resistant materials is a subject of continuing efforts, and new industrial standards related to defining metals and other materials that can safely be used in H_2S-containing (often called "sour") environments are being developed and revised due to research and field investigations [2].

Pipelines and other oilfield equipment frequently operate at high fluid pressures. Crude oil pipelines can leak and cause environmental damage, but natural gas pipeline leaks, like the corrosion-related rupture in Carlsbad, New Mexico, shown in Figure 1.1, can lead to explosions and are sometimes fatal [5].

High-pressure gas releases can also cause expansive cooling leading to brittle behavior on otherwise ductile pipelines. API standards for line pipe were revised in 2000 to recognize this possibility. Older pipelines, constructed before implementation of these revised standards, are usually made from steel with no controls on low-temperature brittle behavior and may develop brittle problems if they leak. Gas pipelines are more dangerous than liquid pipelines, because of the stored energy associated with compression of enclosed fluid.

ENVIRONMENTAL DAMAGE

Environmental concerns are also a reason for corrosion control [6]. Figure 1.2 shows oil leaking from a pipeline that suffered internal corrosion followed by subsequent splitting along a longitudinal weld seam. The damages due to this leak are minimal compared with the environmental damages that would have resulted if the leak had been on a submerged pipeline. Figure 1.3 shows an oil containment boom on a river where a submerged crude oil pipeline was leaking due to external corrosion caused by nonadherent protective coatings that shielded the exposed metal surfaces from protective cathodic protection currents.

In the 1990s, the entire downtown area of Avila Beach, California, was closed because of leaking underground oil pipelines. The cleanup from these corroded pipelines took years and cost millions of dollars.

Figure 1.1   Natural gas pipeline rupture near Carlsbad, New Mexico, in 2000. Source: From Pipeline Accident Report [5].

Figure 1.2   Aboveground leak from an internally corroded crude oil pipeline.

Figure 1.3 An oil containment boom to minimize the spread of crude oil from an external corrosion leak on a submerged pipeline.

Figure 1.4 Offshore platform leg in Cook Inlet, Alaska. The extra metal for the corrosion allowance is submerged twice a day during high tides.

CORROSION CONTROL

The environmental factors that influence corrosion are:

- CO_2 partial pressure
- H_2S partial pressure
- Fluid temperature
- Water salinity
- Water cut
- Fluid dynamics
- pH

Corrosion is normally controlled by one or more of the following:

- Material choice
- Protective coatings
- Cathodic protection
- Inhibition
- Treatment of environment
- Structural design including corrosion allowances
- Scheduled maintenance and inspection

Figure 1.4 shows an offshore platform leg in relatively shallow water, approximately 30 m (100 ft) deep, in Cook Inlet, Alaska. The leg is made from carbon steel, which would corrode in this service. Corrosion control is provided by an impressed current cathodic protection system. The bottom of the leg is 2½ cm (1 in.) thicker than the rest of the leg, and this is intended as a corrosion allowance for the submerged portions of the platform legs. Note that the water level goes above the corrosion allowance twice a day during high tides, because the platform is located in water 3 m (10 ft) deeper than was intended during design and construction. Fortunately the cathodic protection system was able to provide enough current, even in the fast-flowing abrasive tidal waters of Cook Inlet, to control corrosion. This platform was obsolete when the picture was taken, but it was less expensive to operate and maintain the platform than it was to remove it. Thirty-five years later oil prices had increased, recovery methods had improved, and the platform was economically profitable. Robust designs, adequate safety margins, and continuous reevaluation of corrosion control methods are important, not just for marine structures but for all oilfield equipment.

While it might seem desirable to stop all corrosion, this is not necessarily cost effective. An 80 : 20 Pareto-type rule probably applies: 80% of corrosion can be prevented for relatively modest cost, but the increased cost of the remaining corrosion would not be justified [7]. The British ALARP (as low as reasonably practicable) terminology is a similar concept discussed in many recent corrosion-related documents and standards [8].

REFERENCES

1 Kane, R. (2006). Corrosion in petroleum production operations. In: *Metals Handbook, Volume 13C – Corrosion: Corrosion in Specific Industries*, 922–966. Materials Park, OH: ASM International.

2 Iannuzzi, M. (2011). Chapter 15: Environmentally-assisted cracking in oil and gas production. In: *Stress Corrosion Cracking: Theory and Practice* (ed. V. Raja and T. Shoji), 570–607. Oxford: Woodhead Publishing, Ltd.

3 Ruschau, G. and Al-Anezi, M. (September 2001). Appendix S: Oil and gas exploration and production. In: *Corrosion Costs and Preventive Strategies in the United States*, Report FHWA-RD-01-156. Washington, DC: US Government Federal Highway Administration.

4 Nurshayeva, R. (2014). Update 1 – new pipelines to cost Kashagan oil project up to $3.6 bn. *Reuters* (10 October). http://www.reuters.com/article/oil-kashagan-idUSL6 N0S52P420141010 (accessed 17 May 2017).

5 National Transportation Safety Board (2003). *Pipeline Accident Report, Natural Gas Pipeline Rupture and Fire Near Carlsbad, New Mexico, 19 August 2000*, NTSB/PAR-03/01 (11 February 2003). Washington, DC: National Transportation Safety Board.

6 Javaherdashti, R. and Nikraz, H. (2010). *A Global Warning on Corrosion and Environment: A New Look at Existing Technical and Managerial Strategies and Tactics*. Saarbrucken, Germany: VDM Verlag.

7 Palmer, A.C. and King, R.A. (2008). *Subsea Pipeline Engineering*, 2e. Tulsa, OK: Pennwell.

8 Health and Safety Executive (HSE-UK). ALARP "at a glance". http://www.hse.gov.uk/risk/theory/alarpglance.htm (accessed 22 May 2017).

2

CHEMISTRY OF CORROSION

Corrosion, the degradation of a material due to reaction(s) with the environment, is usually, but not always, electrochemical in nature. For this reason, an understanding of basic electrochemistry is necessary to the understanding of corrosion. More detailed descriptions of all phenomena discussed in this chapter are available in many general corrosion textbooks [1–8].

ELECTROCHEMISTRY OF CORROSION

Most corrosion involves the oxidation of a metal that is accompanied by equivalent reduction reactions, which consume the electrons associated with the corrosion reaction. The overall corrosion reactions are often referred to separately as "half-cell" reactions, but both oxidation and reduction are interrelated, and the electrical current of both anodes, where oxidation is prevalent, and cathodes, where reduction predominates, must be equal in order to conserve electrical charges in the overall system.

Electrochemical Reactions

A typical oxidation reaction for carbon steel would be

$$Fe \rightarrow Fe^{+2} + 2e^- \tag{2.1}$$

Common reduction reactions associated with corrosion include

Hydrogen evolution $\quad 2H^+ + 2e^- \rightarrow H_2 \quad$ (2.2)

Oxygen reduction

In acid solutions $\quad O_2 + 4H^+ + 4e^- \rightarrow 2H_2O \quad$ (2.3)

In neutral or basic solutions $\quad O_2 + 2H_2O + 4e^- \rightarrow 4OH^-$

$$\tag{2.4}$$

Metal ion reduction or deposition is also possible:

$$Fe^{+3} + e^- \rightarrow Fe^{+2} \tag{2.5}$$

$$Fe^{+2} + 2e^- \rightarrow Fe \tag{2.6}$$

The reduction reaction is usually corrosion rate controlling, because of the low concentrations of the reducible species in most environments compared with the high concentration (essentially 100%) of the metal. As one example, the dissolved oxygen concentration in most air-exposed surface waters is slightly lower than 10 ppm (parts per million). This relatively low dissolved oxygen concentration is usually much higher than the concentration of any other reducible species, and the control of air leakage into surface facilities is a primary means of controlling internal corrosion in topside equipment and piping.

More than one oxidation or reduction reaction may be occurring on a metal surface, e.g. if an alloy is corroding or if an aerated acid has high levels of dissolved oxygen in addition to the hydrogen ions of the acid.

Electrochemical reactions occur at anodes, locations of net oxidation reactions, and at cathodes, locations of net reduction reactions. These anodes and cathodes can be very close, for example, different metallurgical phases

Metallurgy and Corrosion Control in Oil and Gas Production, Second Edition. Robert Heidersbach.
© 2018 John Wiley & Sons, Inc. Published 2018 by John Wiley & Sons, Inc.

on a metal surface, or they can have wide separations, e.g. in electrochemical cells caused by differences in environment or galvanic cells between anodes and cathodes made of different materials.

Electrolyte Conductivity

The electrical conductivity of an environment is determined by the concentration of ions in the environment, and the resulting changes in corrosivity can be understood by considering Ohm's law:

$$E = IR \qquad (2.7)$$

where

E = the potential difference between anode and cathode, measured in volts.
I = the electrical current, measured in amperes.
R = the resistance of the electrical circuit, determined by the distances between anode and cathode and by ρ, the resistivity of the electrolyte, which is usually expressed in ohm-centimeters (Ω-cm). In most cases the distance between anode and cathode is not known, but the changes in the corrosion rate can be monitored and correlated in changes in resistivity, e.g. the changes in resistivity of soils caused by changes in moisture content, which alter the ionic content of the soil electrolyte.

The resistivity (inverse of conductivity) of liquids and solids is determined by the ions dissolved in the bulk solution. Hydrocarbons such as crude oil, natural gas, and natural gas condensates are covalent in nature and very poor electrolytes, because they have very high resistivities. Oilfield corrosion is usually caused by chemicals in the water phase that, among other things, lower the natural resistivity of water, which is also mostly covalent. Water is a very efficient solvent for many chemicals, and most oilfield corrosion occurs when metal surfaces become wetted by continuous water phases having dissolved chemicals, which lower the natural high resistivity (low conductivity) of water.

Faraday's Law of Electrolysis

The mass of metal lost due to anodic corrosion currents can be determined from Faraday's law for electrolysis, Equation (2.8), which is also used by the electroplating industry:

$$W_{\text{corroded}} = \frac{F(it)M}{n} \qquad (2.8)$$

where

W_{corroded} = mass (weight) of corroded/electrodeposited metal.
F = Faraday's constant.
i = current in amps.
t = time of current passage.
M = molar mass of the element in question.
n = ionic charge of the metal in question.

The amount of a substance consumed or produced at one of the electrodes in an electrolytic cell is directly proportional to the amount of electricity that passes through the cell. Methods of measuring the corrosion current are difficult and are discussed in Chapter 7.

Electrode Potentials and Current

The electromotive force series (EMF series) is an orderly arrangement of the relative standard potentials for pure metals in standard, unit activity (one normal, 1 N), solutions of their own ions (Table 2.1). The more active metals on this list tend to be corrosion susceptible, and the less active, or noble metals, will resist corrosion in many environments.

It should be noted that two sign conventions are followed in publishing the EMF series. This can cause confusion, which can be avoided if the reader understands that active metals like magnesium and aluminum will always be anodic to carbon steel and corrosion-resistant metals like silver and palladium will be cathodic.

The EMF series shows equilibrium potentials for pure metals in 1 N (one normal or unit activity of ions) solutions of their own ions. While this is the basis for much theoretical work in corrosion and other areas of electrochemistry, pure metals are seldom used in industry, and oilfield corrosive environments never have 1 N metal ion concentrations. The more practical galvanic series (Figure 2.1), which shows the relative corrosion

TABLE 2.1 The Electromotive Force Series for Selected Metals

	Metal	Ion Formed	Potential	
Anodic	Magnesium	Mg^{+2}	+2.96	Active
	Aluminum	Al^{+3}	+1.70	
	Zinc	Zn^{+2}	+0.76	
	Iron	Fe^{+2}	+0.44	
	Nickel	Ni^{+2}	+0.23	
	Tin	Sn^{+2}	+0.14	
	Lead	Pb^{+2}	+0.12	
	Hydrogen	*H^{+1}*	*0.00*	
	Copper	Cu^{+2}	−0.34	
	Silver	Ag^{+1}	−0.80	
Cathodic	Palladium	Pd^{+2}	−0.82	Noble

Source: Adapted from Parker and Peattie [9].

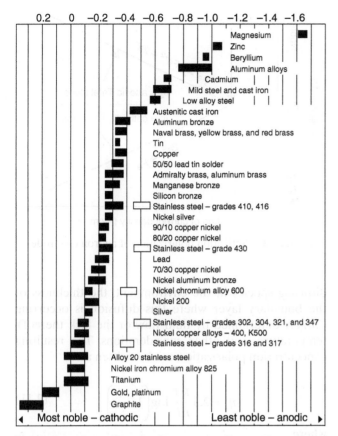

Figure 2.1 Galvanic series in seawater. *Source*: Reproduced with permission of John Wiley & Sons.

potentials of many practical metals, is often used in corrosion control. This is based on experimental work in seawater and serves as the basis for many corrosion-related designs [10].

The galvanic series in seawater shown in Figure 2.1 is widely used for engineering designs. Some authorities claim that the relationships between various alloys must be determined for each environment, but this is seldom done. The reason for this precaution is that zinc and carbon steel undergo a polarity reversal in some fresh waters at approximately 60 °C (140 °F). The only other polarity reversal that has been reported is when tin, which would normally be cathodic to carbon steel, becomes anodic to carbon steel in deaerated organic acids, such as are found in the common tin cans used for food storage. It is unlikely that any other polarity reversals will be found in oilfield environments, and designers should assume that the relationships shown in Figure 2.1 are valid. Revie and Uhlig offer a brief review of polarity reversals [2].

The Nernst equation, first published in 1888 by the German chemist who later won the 1920 Nobel Prize in chemistry, explains how potentials of both anodic and cathodic reactions can be influenced by changes in the temperature and chemical compositions of the environment. The reduction potential can be expressed as

$$E = E° - \frac{RT}{nF} \ln \frac{[\text{Products}]}{[\text{Reactants}]} \qquad (2.9)$$

where

E = the electrochemical potential of the reaction in question.

$E°$ = the standard electrode potential at 25 °C in a 1 N (normal) solution of the ion formed by oxidation of the reactants in question.

R = the Boltzmann distribution constant, normally referred to as the universal gas constant = 8.31(15) J K^{-1} mol^{-1}.

T = the absolute temperature, °K.

n = the charge on the ion being reduced.

F = Faraday's constant, the number of coulombs per gram-mole of electrons = 9.63 × 10^4 C mol^{-1}.

At standard temperature conditions this equation can be simplified to

$$E = E° - \frac{0.059}{n} \log \frac{[\text{products}]}{[\text{reactants}]}$$

The details of this relationship are described in many general corrosion textbooks [1–7]. What is important to understand for oilfield corrosion control is that electrochemical cells (corrosion cells) can be caused by changes in:

- Temperature
- Chemical concentrations in the environment

Both types of electrochemical cells are important in oilfield corrosion and will be discussed further in Chapter 5.

It is simplistic to describe a chemical reaction as either oxidation or reduction. In actuality the reversible chemical reactions are happening in both directions simultaneously. The equilibrium potential, determined by the Nernst equation, is the potential where the oxidation and reduction currents, measured in current density on an electrode surface, are equal. The current density at this point is called the exchange current density. Some metals, e.g. the platinum and palladium used in impressed current anodes, have very high exchange current densities. This means that a small surface area of these materials can support much higher anodic currents than other anode materials such as high-silicon cast iron or graphite. Figure 2.2 shows the idea of exchange current densities for hydrogen oxidation/reduction reactions. A platinum surface can support 10000 times the current

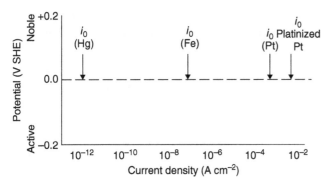

Figure 2.2 Hydrogen–hydrogen ion exchange current densities.

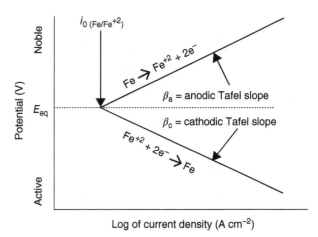

Figure 2.3 Activation polarization of an iron electrode.

density of an iron anode for the same reaction. This increase in efficiency is used in the cathodic protection industry to justify the use of relatively expensive precious metal surfaces to replace much heavier, and therefore harder to install, high-silicon cast iron anodes.

As potentials change from the equilibrium potential, the electrode surface becomes either an anode or a cathode. It is common to plot the shifts in potential on linear-logarithmic plots, because in many cases these plots show a region of activation-controlled electrode behavior where the voltage of anodes and cathodes follows a log-linear pattern, called the Tafel slope after the German scientist who first explained this behavior in 1905.

On an anode, the Tafel equation can be stated as

$$\eta_a = \beta_a \ln\left(\frac{i}{i_0}\right) \tag{2.10}$$

where

η_a = the overpotential or change between the measured potential and the potential at the current density of interest. The subscript "a" indicates that this polarization is "activation polarization," which occurs at low current densities near the equilibrium potential.
β_a = the so-called Tafel slope.
i = the current density, $A\,m^{-2}$.
i_0 = the exchange current density, $A\,m^{-2}$.

At low electrode current densities, the change in potential can be plotted as shown in Figure 2.3. These plots of potential vs. logarithm of current are often termed Evans diagrams, after Professor U.R. Evans, of Cambridge University, who popularized their use [5].

As stated above, most oilfield corrosion rates are controlled by the low concentrations of reducible species in the environment. These species must migrate, or diffuse, to the metal surface in order to react. The rate of this diffusion is controlled by the concentration of the

diffusing species in the environment, the thickness of the boundary layer where this diffusion is occurring (largely determined by fluid flow or the lack thereof), temperature, and other considerations. The resulting concentration polarization can be written as

$$\eta_c = 2.3\frac{RT}{nF}\log\left(1-\frac{i}{i_L}\right) \tag{2.11}$$

where

η_c is the overpotential, or polarization, caused by the diffusion of reducible species to the metal surface.
F is the Faraday's constant.
i is the current on the electrode.
i_L is the limiting current density determined by the diffusivity of the reducible species; this is the maximum rate of reduction possible for a given corrosion system.

The other terms are the same as described above in discussions of the Nernst equation and activation polarization (Tafel slope) behavior.

Concentration polarization is shown in Figure 2.4. In corrosion, the limited concentrations of reducible species produce concentration polarization only at cathodes. At low current densities, the concentration polarization is negligible, and as the reduction current density approaches the limiting current, the slope quickly becomes a vertical downward line.

The total polarization of an electrode is the sum of both the activation and concentration polarization. The combined polarization for a reduction reaction on a cathode is

$$\eta_{red} = -\beta_c\frac{\log i}{i_0} + 2.3\frac{RT}{nF}\log\left(1-\frac{i}{i_L}\right) \tag{2.12}$$

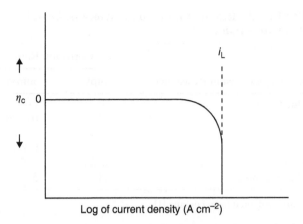

Figure 2.4 Concentration polarization curve for a reduction reaction.

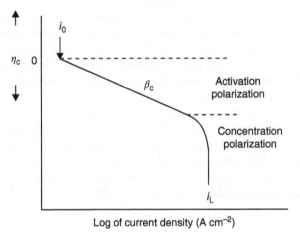

Figure 2.5 Combined polarization curve for activation and concentration polarization on a cathode.

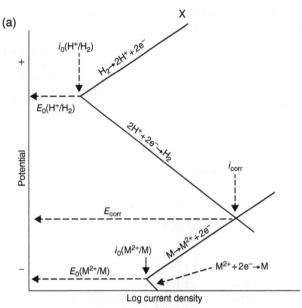

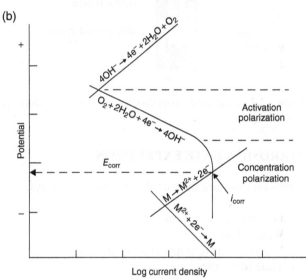

Figure 2.6 (a) Corrosion current and potential of iron determined by the polarization of iron and the hydrogen reduction reaction. (b) Corrosion current and potential of iron determined by the concentration polarization of oxygen. *Source*: Beavers [11]. Reproduced with permission of NACE International.

This is shown in Figure 2.5.

As stated earlier, most oilfield corrosion rates are determined by the concentration of the reducible chemicals in the environment. Figure 2.6a shows how the polarization of both the oxidation of a metal and the reduction of hydrogen ions determines the corrosion rate, i_{corr}, and the corrosion potential, E_{corr}, for a generic metal.

For surface equipment, most corrosion rates are determined by the concentration of dissolved oxygen in whatever water is available. This is shown in Figure 2.6b, where the oxidation line showing Tafel behavior intersects the vertical (concentration limited) portion of the reduction reaction.

The importance of potential in determining corrosion rates is apparent from the above discussions. Academic chemistry reports tend to describe potentials relative to the standard hydrogen electrode (SHE), which has been arbitrarily set to a potential of zero. In field applications, it is common to use other reference electrodes. The most common reference electrodes used in oilfield work are the saturated copper–copper sulfate electrode (CSE), used in onshore applications, and the silver–silver chloride electrode used for offshore measurements, where contamination of the CSE electrode would produce variable readings. Table 2.2 shows conversion factors for these electrodes and other commonly used reference electrodes compared with the SHE. As an example, an electrode that measures −0.300V vs. CSE would measure +0.018V vs. SHE. Figure 2.7 shows a standard copper–CSE.

TABLE 2.2 Potential Values for Common Reference Electrodes

Name	Potential V vs. SHE
Copper–copper sulfate	+0.318
Saturated calomel	+0.241
Silver–silver chloride	+0.222
Standard hydrogen	+0.000

Source: Adapted from Jones [3].

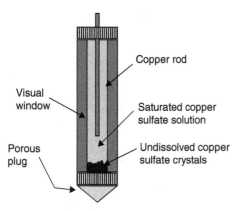

Figure 2.7 Saturated copper–copper sulfate reference electrode.

CORROSION RATE EXPRESSIONS

Corrosion rates are measured in a number of ways:

- Depth of penetration
- Weight loss
- Electrical current associated with corrosion
- Time to failure

The simplest of these concepts to understand is depth of penetration. It can be expressed in mm yr^{-1} or mpy (mils or thousandths of an inch per year). The loss of wall thickness is often used to determine remaining equipment life or safe operating pressures for piping systems, storage tanks, etc. Table 2.3 shows a commonly used classification of relative corrosion rates. The US standard units, mpy, produce small numbers that are easy to understand, and corrosion rates in mpy are commonly used worldwide, although other expressions are also common [1].

Weight loss measurements are commonly used on exposure samples used to monitor corrosion rates in oil and gas production. It is a simple matter to convert these weight loss measurements into average depths of penetration, although this can be very misleading, because most corrosion is localized in nature and the average

TABLE 2.3 Relative Corrosion Resistance vs. Annual Penetration Rates

Relative Corrosion Resistance	Corrosion Rate	
	mpy	mm yr^{-1}
Outstanding	<1	<0.02
Excellent	1–5	0.02–0.1
Good	5–20	0.1–0.5
Fair	20–50	0.5–1
Poor	50–200	1–5
Unacceptable	200+	5+

Source: Adapted from Fontana [1].

penetration rate seldom gives an indication of the true condition of oilfield equipment.

The electrical current associated with anodic dissolution of a metal can be used to determine the corrosion rate using Faraday's law. This calculation of mass loss can be converted into remaining thickness. Once again, the reader is cautioned that most corrosion is localized in nature and calculations assuming uniform loss of cross section are frequently misleading.

The time to failure, however defined, is the most common concern of managers and operators of equipment. For some forms of corrosion testing, e.g. stress corrosion cracking, the time to failure is used to screen alloys, environments, or other variables.

pH

The pH of an environment is one of the major factors determining if corrosion will occur. It also influences the type of corrosion that is experienced.

pH is defined as

$$pH = -\log\left[H^+\right] \quad (2.13)$$

where the [H$^+$] expression shows the hydrogen ion activity of the environment. The [H$^+$] depends on the ionization of water and varies with temperature. The pH of neutral water at standard temperature (25°C) is 7, but neutrality varies with temperature as shown in Figure 2.8. Downhole oilfield temperatures are usually elevated, and it is common to calculate the *in situ* pH of any fluids that might affect corrosion or scale deposition. There are many software packages available for this purpose. Figure 2.9 shows the effects of pH on the corrosion rates of iron in water. At low pH bare metal is exposed to the environment, and acid reduction on the surface controls corrosion rates. For intermediate pH a partially protective film of iron oxide reduces the corrosion rate and the

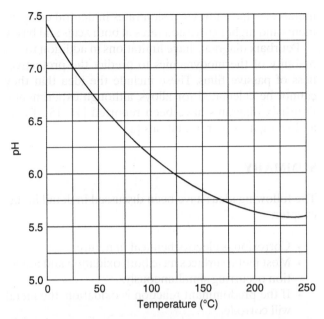

Figure 2.8 pH values of pure water at various temperatures [12]. *Source*: Reproduced with permission of John Wiley & Sons.

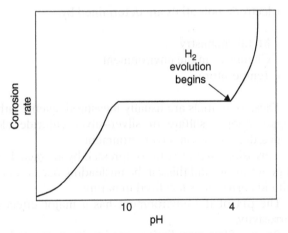

Figure 2.9 The effect of pH on the corrosion rate of iron in water at room temperature. *Source*: Adapted from Revie and Uhlig [2].

diffusion of oxygen to cathodic locations on the metal surface controls. As the pH increases to even higher values, the surface becomes covered with mineral scales, and corrosion is reduced.

PASSIVITY

Passivity is a phenomenon that is frequently misunderstood. Most metals form oxide films in most corrosive environments. These passive films can be protective and retard or even effectively stop corrosion, but they can

also lead to fairly deep localized corrosion in situations where the protective films are removed or defective. Except in rare circumstances, the oxide films formed on carbon steel are not adequately protective, and other means of corrosion control are necessary. This is in contrast to stainless steels, titanium, and aluminum – oilfield metals that form protective passive films that are commonly the primary means of corrosion control for these alloys. On many corrosion-resistant alloys such as stainless steels, the passive films may be only dozens of atoms thick. This means that they are very weak and subject to mechanical damage, and this can lead to localized corrosion at the damaged locations.

Potential-pH (Pourbaix) Diagrams

Marcel Pourbaix developed a means of explaining the thermodynamics of corrosion systems by plotting regions of thermodynamic stability of metals and their reaction plots on potential vs. pH plots [12–14]. The regions of a Pourbaix diagram can be described as:

- **Immunity** The metal cannot oxide or corrode (although it may still be subject to hydrogen embrittlement).
- **Corrosion** Ions of the metal are thermodynamically stable and the metal will corrode.
- **Passivity** Compounds of the metal and chemicals from the environment are thermodynamically stable, and the metal may be protected from corrosion if the passive film is adherent and protective.

Many users of Pourbaix diagrams miss the final point above. Thermodynamics alone cannot predict if passive films will be protective or not [2, 13–15].

Figure 2.10 shows the Pourbaix diagram for water. Water is thermodynamically stable over a potential region of 1.23 V, and the potentials at which oxidation and evolution (bubbling off) of oxygen from water at the top of the diagram or the evolution of hydrogen at the bottom of the diagram depend on the pH of the environment.

The Pourbaix diagram of iron is superimposed on the diagram for water in Figure 2.11. Similar diagrams are available for most structural metals for which thermodynamic data are available [2, 13–15].

These diagrams make a number of important points useful for oilfield corrosion control:

- Water is only stable over a potential range of slightly more than one volt. This is very important in cathodic protection.
- Iron (carbon steel) is covered with iron oxides (passive films) in most aqueous environments.

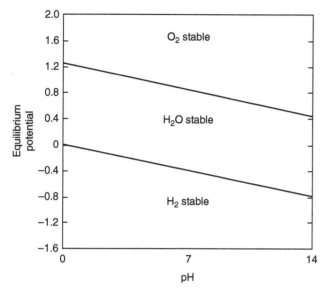

Figure 2.10 Potential-pH (Pourbaix) diagram for water [13]. *Source*: Reproduced with permission of National Association of Corrosion Engineers.

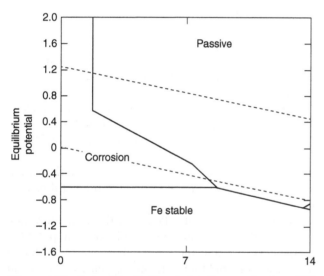

Figure 2.11 Potential-pH (Pourbaix) diagram for iron [13]. *Source*: Reproduced with permission of National Association of Corrosion Engineers.

Unfortunately, these passive films are usually not sufficiently protective, and other means of corrosion control are necessary.

- The potentials at which iron (carbon steel) is protected from corrosion do not coincide with the immunity regions on the Pourbaix diagram. This point is discussed in greater detail in Chapter 6.

The diagrams for zinc, aluminum, and cadmium, commonly referred to as the amphoteric coating metals, have passive regions in neutral environments. These

metals also have low corrosion rates in neutral environments and higher corrosion rates in both acids and bases.

Pourbaix diagrams have limitations in addition to the inability of thermodynamics to predict the protectiveness of passive films. These include the idea that they cannot be calculated for alloys, although experimental Pourbaix diagrams have been reported [14, 15]. Revie and Uhlig list other limitations [2].

SUMMARY

The following ideas have been discussed in detail in this chapter:

- Corrosion is electrochemical in nature.
- Most metal surfaces have both oxidation and reduction occurring simultaneously.
- If the predominant reaction is oxidation, the metal will corrode.
- The most important reduction reaction is oxygen reduction for many oilfield systems. If no oxygen is available, the corrosion rate will often be very low.

Electrode potentials are determined by:

- Metal chemistry
- Chemicals in the environment
- Temperature

These potentials are usually measured against either copper–copper sulfate or silver–silver chloride electrodes, depending on the environment.

Corrosion rates are often expressed by average depth of penetration, and this can be misleading because most oilfield corrosion is localized in nature.

The pH of the environment has a major effect on corrosivity.

Passive films may limit corrosion in many environments, but carbon steel, the most common oilfield metal, seldom forms adequately protective passive films, and other means of corrosion are often necessary.

REFERENCES

1 Fontana, M. (1986). *Corrosion Engineering*. New York: McGraw-Hill.
2 Revie, W.R. and Uhlig, H.H. (2008). *Corrosion and Corrosion Control*. Hoboken, NJ: Wiley-Interscience.
3 Jones, D.A. (1996). *Principles and Prevention of Corrosion*, 2. Upper Saddle River, NJ: Prentice-Hall.
4 Perez, N. (2004). *Electrochemistry and Corrosion Science*. Dordrecht, The Netherlands: Kluwer Academic Publishers.
5 Ahmad, Z. (2006). *Principles of Corrosion Engineering and Corrosion Control*. Boston, MA: Elsevier.

6   Bradford, S.A. (2001). *Corrosion Control*, 2. Edmonton, Alberta, Canada: CASTI Publishing Inc.

7   Trethewey, K.R. and Chamberlin, J. (1995). *Corrosion for Science and Engineering*, 2. London, UK: Longman Scientific and Technical.

8   McCafferty, E. (2009). *Introduction to Corrosion Science*. Berlin: Springer.

9   Parker, M. and Peattie, E. (1984 [1995 reprinting]). *Pipeline Corrosion and Cathodic Protection*, 3. Houston, TX: Gulf Publishing.

10   LaQue, F.L. (1975). *Marine Corrosion, Causes and Prevention*. New York: Wiley.

11   Beavers, J. (2000). Chapter 15: Fundamentals of corrosion. In: *Peabody's Control of Pipeline Corrosion*, 2 (ed. R. Bianchetti), 297–317. Houston, TX: NACE International.

12   Baboian, R. (ed.) (2002). *Corrosion Engineer's Reference Book*, 3, 78. Houston, TX: NACE International.

13   Pourbaix, M. (1974). *Atlas of Electrochemical Equilibria in Aqueous Solutions*, 2e English. Houston, TX: NACE International, and Brussells: Centre Belge d'Etude de La Corrosion (CELBECOR).

14   Verink, E.D. Jr. (2000). Chapter 6: Simplified procedure for constructing Pourbaix diagrams. In: *Uhlig's Corrosion Handbook*, 2 (ed. R.W. Revie), 111–124. New York: Wiley-Interscience.

15   Thompson, W.T., Kaye, M.H., Bale, C.W., and Pelton, A.D. (2000). Chapter 7: Pourbaix diagrams for multielement systems. In: *Uhlig's Corrosion Handbook*, 2 (ed. R.W. Revie), 125–136. New York: Wiley-Interscience.

3

CORROSIVE ENVIRONMENTS

A very limited amount of oilfield corrosion is associated with very high-temperature atmospheric exposures, common in flares, and with liquid metals, usually mercury found in natural gas and some crude oils. The great majority of oilfield corrosion requires liquid water. Downhole formation water that comes to the surface with oil and gas production can also include the following impurities that can affect corrosion rates [1–3]:

- Oxygen – this is normally a problem only with surface equipment, because oxygen is unlikely to occur naturally in downhole formations.
- Sulfur-containing species.
- Naturally occurring radioactive materials (NORM).
- CO_2.
- H_2S.

Fresh surface water is generally considered less corrosive than seawater or produced formation water. Designers will often assume that fresh water will require less stringent corrosion control efforts, but this can be a mistake, because surface water will usually have dissolved oxygen contents high enough to promote corrosion. Uncontaminated formation water, while usually very high in mineral content and very salty, becomes more corrosive if air (and oxygen) is allowed to enter. Oxygen scavengers are often used in produced water systems to limit the corrosion rates before produced water is reinjected downhole to maintain formation pressure.

For this reason, as an oilfield ages and the water cut increases, corrosion also increases. This is shown in Figure 3.1, which shows the effect of water cut on corrosion rates [4]. Many operators use rules of thumb such as the idea that corrosion is not a problem until the water cut reaches 40 or 50%. For some oilfields, this may take several years before corrosion becomes a problem. Unfortunately, this means that corrosion and other maintenance problems become more important at a time when maintenance funds, often related to production rates, decrease.

Water has very limited solubility in hydrocarbons, and the presence of a separated water phase is necessary for corrosion. The low corrosion region in Figure 3.1 is where most of the metal surface is in contact with a water-in-oil emulsion. The small water droplets are not continuous, and most of the metal surface is in contact with nonconductive hydrocarbons. As the water cut increases, the amount of the metal surface in contact with water gradually increases until the emulsion reverses, and the liquid becomes continuous water with entrained hydrocarbon droplets. Production and fluid flow rates, along with temperature and pressure considerations, determine when this will happen. Figure 3.2 shows how water separates out on production tubing.

In contrast to oil wells, natural gas wells are corrosive from the beginning. This is due to the fact that all natural gas reservoirs will produce some water, and minor components of the natural gas, which condense from the gas stream as temperatures and pressures are reduced, dissolve in this water and make it corrosive. Condensed water lacks dissolved minerals, which could lower corrosion rates by buffering pH changes. Rainwater has enough dissolved CO_2 to lower the pH to between 5 and

Metallurgy and Corrosion Control in Oil and Gas Production, Second Edition. Robert Heidersbach.
© 2018 John Wiley & Sons, Inc. Published 2018 by John Wiley & Sons, Inc.

5.6, and the increased pressures in pipelines and downhole systems can lower the pH to even more acidic levels.

Most downhole hydrocarbon reservoirs have virtually no dissolved oxygen in the fluids, and this is fortunate, because the presence of oxygen at the parts-per-billion (ppb) level has been shown to promote corrosion. This is in contrast to carbon dioxide (CO_2) and hydrogen sulfide (H_2S), which may be present in varying quantities in both oil and gas fields. The relative effects of these three gases are shown in Figure 3.3. Oxygen is approximately 50 times more corrosive than CO_2 and more than a hundred times more corrosive than H_2S.

Downhole corrosion, in the absence of oxygen, is largely determined by the concentrations of CO_2 or H_2S in the produced fluids. The terms "sweet corrosion" to describe corrosion caused by CO_2 and "sour corrosion" to describe problems with H_2S have been used for many years to differentiate which of these two

gases is likely to predominate in a given field [1–3, 6–9]. Other considerations that affect corrosion rates include temperature and pressure, which determine the nature of the fluid (gas, liquid, etc.) on the metal surface, and minor constituents in the liquid water phase. Figure 3.4 shows how complex the determination of corrosivity can be.

EXTERNAL ENVIRONMENTS

The external environments discussed in this section are not unique to oil and gas production, but much of the information comes from oilfield experience with production platforms, buried or subsea pipelines, and similar equipment. External corrosion can affect all equipment, from the bottom of the well to the surface.

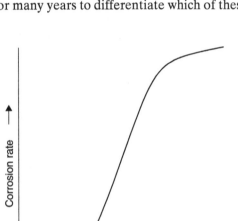

Figure 3.1 The effects of water cut on the corrosion rate of oil well tubing. *Source*: Craig [4]. Reproduced with permission of NACE International.

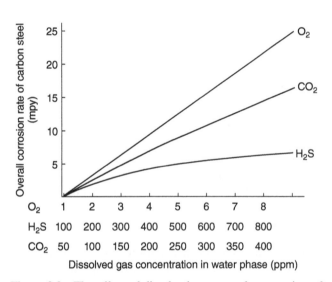

Figure 3.3 The effect of dissolved gases on the corrosion of carbon steel. *Source*: From Shankardass [6].

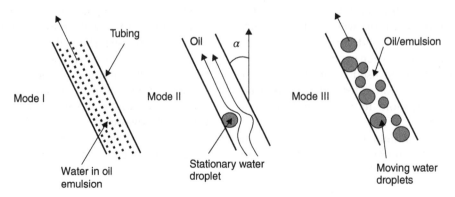

Figure 3.2 Water wetting producing corrosion on deviated oil wells. *Source*: de Waard et al. [5]. Reproduced with permission of NACE International.

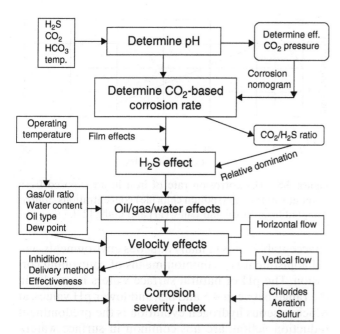

Figure 3.4 A flowchart indicating the factors that determine the corrosion severity to be expected in an oil or gas field [10]. *Source*: Reproduced with permission of Corrosionsource.

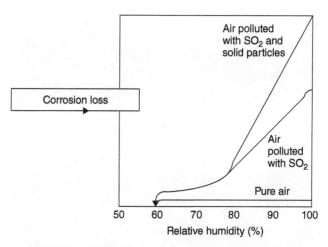

Figure 3.5 Simplified diagram showing the effect of relative humidity and pollution on the corrosion of carbon steel. *Source*: Revie and Uhlig [11]. Reproduced with permission of John Wiley & Sons.

Atmospheric Corrosion

Like all other types of corrosion discussed in this book, this form of corrosion requires the presence of condensed water on the metal surface in order for corrosion to occur. The only exception to this general rule is at very elevated temperatures, e.g. those associated with flares and other combustion processes, where corrosion can occur without liquid water. Even here most corrosion occurs below the dew point, because these high-temperature applications require special alloys to withstand corrosion at high temperatures, and this equipment often suffers the worst corrosion during shutdown, when acidic moisture can condense on rough surfaces and cause corrosion similar to that on automotive mufflers during times when the system is cold enough for condensation.

It would seem logical that atmospheric corrosion would not occur until the relative humidity is 100%, but this is not the case. Research dating back to the 1920s has shown that corrosion can occur once the humidity reaches a "critical humidity" of approximately 60–70% [11–13]. Many structures, especially on the away-from-the-sun side (the north side in northern latitudes), stay above this critical humidity virtually all the time, at least whenever the temperature is above freezing [13]. It is important to realize that heat sinks, e.g. large structural members on offshore structures, can remain above the critical humidity long after the sun comes up and corrosion has diminished elsewhere on

the same structure. The presence of deliquescent salts means that many surfaces remain wetted even in sunlight. Salt-contaminated surfaces have been found to be wet below 20% relative humidity [14]. This is a very important consideration when painting structures, because "flash rusting" due to surface moisture can quickly form and severely degrade the adherence of primary coatings to painted structures (Figure 3.5).

Most oilfield metal exposed to atmospheric corrosion is carbon steel, and the most common method of corrosion control is by the use of protective coatings (painting) [11]. Some process equipment, storage tanks, and electronic control systems are protected by the use of inerting gases, heaters, deliquescing agents, or vapor phase inhibitors. Control lines, conduit, and similar tubing are often stainless steel on offshore structures.

The atmospheric corrosion exposure procedures described in international standards are of limited use for operating personnel, and they should only be used for potential coating systems evaluation [15–20]. Atmospheric corrosion is most severe in local areas on structures. The atmospheric exposure test panels shown in Figure 3.6 were intended to identify portions of locations a seaside petrochemical processing facility where corrosion damage inspections should be concentrated. The actual atmospheric corrosion on nearby equipment, as shown in Figure 3.7, was determined by the equipment and structure geometries and drainage patterns. The boldly exposed simple geometry exposure samples shown in Figure 3.6 cannot identify where the in-plant corrosion inspection and maintenance should be concentrated. Accelerated tests in artificial atmospheres are useful for comparing prospective coating systems, but they cannot predict long-term performance [21].

Figure 3.6 Atmospheric exposure panels at a large seaside petrochemical facility.

Figure 3.7 Localized atmospheric corrosion at the same facility shown in Figure 3.6.

NORSOK and other materials selection guidelines suggest that atmospheric corrosion of carbon steels should have protective coatings, but corrosion-resistant alloys (CRAs) usually do not require protective coatings except under insulation or when submerged in seawater [21]. This NORSOK guidance is questioned by some operators having concerns with stress corrosion cracking (SCC) of stainless steels. The alternative to this practice is to use more alloys having better corrosion resistance than standard stainless steels, e.g. nickel-based alloys, for atmospheric exposure temperatures above 60 °C.

Water as a Corrosive Environment

The effect of pH on corrosion of carbon steel was discussed in Chapter 2, Figure 2.9. Carbon steel, the most common metal used in oilfield systems, corrodes at

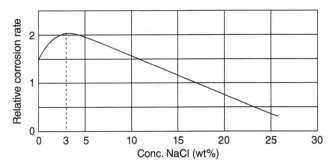

Figure 3.8 The corrosion rate of iron in air-exposed fresh water at varying salt (sodium chloride) concentrations. *Source:* Adapted from Revie and Uhlig [11] and Uhlig [23].

unacceptable rates in many aqueous environments, and pH adjustment is a common means of controlling corrosion. The pH of natural surface waters is usually in the range between 4.5 and 8.5, and lower pH values, at which gaseous hydrogen evolution is the predominant reduction action, are not common in surface waters. Pure water is not corrosive in the absence of dissolved gases [22].

Most readers are familiar with the idea that salt water is more corrosive than fresh water. The combined effects of dissolved oxygen and salt concentration on the corrosivity of water are shown in Figure 3.8. As increasing amounts of salt are added to water, the electrical conductivity of the electrolyte increases and so does the corrosion rate. At the same time, the oxygen solubility decreases continuously with additional concentrations of salt, and this limits the corrosion rate because oxygen reduction is the rate-controlling chemical (reduction) reaction [11]. The same phenomenon happens with all other salts. The maximum corrosion rate is at approximately 3% salt – the exact concentration depends on temperature and the salt involved [11, 23]. This explains why highly concentrated brines, such as those used in packer fluids, are noncorrosive, provided they are properly pH adjusted and have little or no dissolved oxygen.

Figure 3.8 shows that fresh water, low in salt, is less corrosive than salt water, but the most important point to be learned from this picture is that, even at its most corrosive, only about one-third of the corrosion in salt water is due to salt – most of the corrosion would occur anyway due to the presence of oxygen. It should be noted that even deionized water can be corrosive if it is exposed to air [11, 24, 25]. Fresh surface water is generally considered less corrosive than seawater or produced formation water. Designers will often assume that fresh water will require less stringent corrosion control efforts, and one possible definition of fresh water will be water with less than 100 ppm chloride ions, although other definitions range from 50 to 500 ppm, dependent upon temperature.

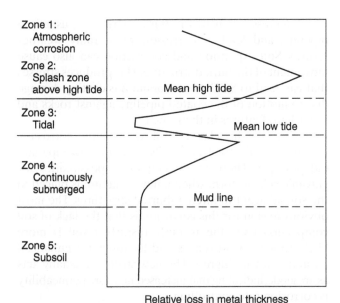

Figure 3.9 Zones of corrosion for steel piling in seawater. *Source*: Adapted from Baboian and Treseder [26] and LaQue [27].

Figure 3.10 Waterline corrosion.

Figure 3.11 Leaking seawater filter vessel on an offshore platform due to waterline corrosion.

Many reports on corrosion ascribe corrosion damage to the presence of chlorides, the most common anions found in seawater and often found in fresh water as well. This dates back to analytical chemistry practices in the early twentieth century, when qualitative analysis techniques (methods of determining the presence of various chemicals in the environment) were relatively new. The field methods for identifying chloride were relatively easy, and many authors started blaming chlorides for damage caused by salts. It was unnecessary to identify the other components of the salt, as there will always be cations (positively charged ions) present to balance the charge of the negatively charged anions. This practice continues. Any highly ionic salt would result in similar damage, but chloride salts are the most common in most natural environments. It is important to remember, as shown in Figure 3.8, that most of the corrosion in any location is due to the presence of dissolved oxygen or some other chemically reducible species (oxidizer). Salt *cannot* cause corrosion – it can only increase the corrosion rate by increasing the conductivity of the electrolyte.

Figure 3.9 shows the corrosion rates of piling in seawater at various elevations. The highest corrosion rates are in the splash zone, where the metal is frequently covered with air-saturated water. The relatively low corrosion rates in the tidal region are due to the oxygen concentration cells between the highly aerated tidal zone and the fully submerged zone just below. The tidal zone, having high oxygen concentrations, is cathodic to the fully submerged zone just below, which is anodic. As the water deepens, the oxygen concentrations lower and

corrosion decreases. NORSOK suggests an additional corrosion allowance (thicker metal) and the use of thick-film protective coatings in the splash zone for both carbon steel and for martensitic (13Cr) stainless steels [21]. The same guidelines suggest Alloy 625 (*UNS.* N06625) and other nickel alloys with equal or higher pitting resistance, titanium alloys, or glass-reinforced polymer composites for submerged service [21].

In locations with no tidal flows, the most corrosive location is at the air–water interface. This problem occurs on pilings (Figure 3.10) and in storage tanks and other equipment (Figure 3.11).

Produced water can vary from very salty, which is common in oil wells, to almost pure, the condensate associated with some gas wells. This "pure" water can become very corrosive, because the dissolved gases,

CO_2 and H_2S, plus acidic hydrocarbons can drastically lower the pH, especially at downhole temperatures and pressures.

Most oilfield metal exposed to corrosive waters is carbon steel, and the most common method of corrosion control is by the use of protective coatings (painting), which is often supplemented by cathodic protection. Corrosion inhibitors and CRAs are also used, especially in downhole environments.

Soils as Corrosive Environments

Much has been written on corrosion in soils. The definitive work on this subject was published by M. Romanoff of the National Bureau of Standards (now the National Institute of Standards and Technology) in 1957, and, when the government report went out of print, it was republished by NACE [28]. Many advances have been made in the understanding of corrosion and cathodic protection since the original publication, but the data in this report represents one of the most extensive sources of corrosion in soil data that is available.

Water and gas occupy much of the space between the solid particles of soil, and these are very important in determining the corrosivity of soils. The air–water interface, wherever located, is the most corrosive location for buried structures, and this location often varies with seasonal rainfall patterns. The minerals in soil dissolve in water and affect the soil resistivity. This directly affects corrosivity, as shown in Table 3.1.

Sandy soils drain well and tend to have the highest resistivities and lowest corrosion rates. Clays, that can swell when wetted, sometimes produce situations where drainage is prevented and buried structures remain wet and corrode.

Soil pH can also affect corrosion. Table 3.2 shows the effect of soil pH on corrosivity. Acidic soils are encountered in swampy locations, volcanic regions, and areas with silicate rocks and high moisture.

Some dry soil, especially clay-rich soil, contracts during dry seasons as shown in Figure 3.12. This can lead

to air ingress down to the buried structure, usually a pipeline, and lead to corrosion when rainy weather returns. Soil expansion and contraction can also cause movement of the buried structure. This produces stresses that can lead to SCC. More common is coating damage due to motion of the coated pipeline against rocks and other hard features in the trench.

Peabody, in his classic book on pipeline corrosion, cautioned against galvanic cells between new pipe and old pipe [31]. There are a number of possibilities for galvanic cells to form when a new structure is placed in the soil adjacent to already-buried structures. The most obvious reason for this corrosion is that the lack of soil compaction over the recently disturbed soil is more likely to leave void spaces and locations for enhanced air and moisture ingress. The new structure usually acts as an anode, indicating that increased moisture permeability is corrosion rate controlling.

Buried pipelines are in disturbed soil near soil that has been in place for many years. The differences in aeration and moisture are evident in the vegetation patterns over many pipelines. This is shown in Figure 3.13, which shows two obvious right-of-way locations, each of which has several parallel buried pipelines.

TABLE 3.2 Soil pH vs. Corrosivity

pH	Degree of Corrosivity
<5.5	Severe
5.5–6.5	Moderate
6.5–7.5	Neutral
>7.5	None (alkaline)

Source: Roberge [30]. Reproduced with permission of NACE International.

TABLE 3.1 Corrosivity Ratings Based on Soil Resistivity

Soil Resistivity (Ω-cm)	Corrosivity Rating
>20000	Essentially noncorrosive
10000–20000	Mildly corrosive
5000–10000	Moderately corrosive
3000–5000	Corrosive
1000–3000	High corrosive
<1000	Extremely corrosive

Source: Bianchetti [29]. Reproduced with permission of NACE International.

Figure 3.12 Cracked soil due to drying after the rainy season.

Figure 3.14 shows how the corrosion rate of buried steel decreases with time. This is due to soil compaction and other poorly documented factors.

The most corrosive location in any buried structure is usually where the structure crosses the air-to-soil interface. This is shown in Figure 3.15. It is important to concentrate inspections in these locations, because cathodic protection, which protects buried structures, cannot be effective in the loosely compacted soil at these locations. Abrasion, motion due to solar-induced expansion and contraction, and a variety of other factors are likely to cause coating damage at these locations.

Virtually all oilfield equipment buried in soil is protected by a combination of protective coatings supplemented by cathodic protection. The exceptions are in those rare locations where the resistivity of the soil is so high that corrosion is unlikely and cathodic protection would be difficult to achieve. Even in these locations, it is common to use protective coatings. Cathodic protection also cannot work at elevated temperatures where any moisture in the soil will quickly evaporate and no electrolyte is available. Many locations where this occurs, e.g. in acidic volcanic soils, require specialized protective coatings, and the lack of widely accepted coatings for these applications is a continuing problem for many operators.

Corrosion Under Insulation

Corrosion under insulation (CUI) is an increasingly important problem. Most air-exposed insulated piping and vessels are covered with porous insulation that is

Figure 3.13 Differences in vegetation over two parallel pipeline rights of way.

Figure 3.15 Corroded pipeline at the air-to-soil interface.

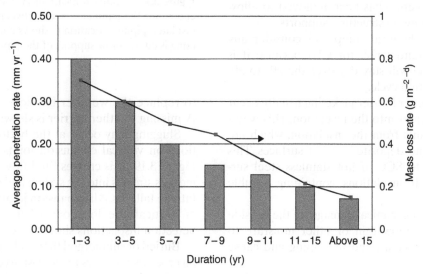

Figure 3.14 Variation of corrosion penetration rate as a function of buried steel exposure duration. *Source*: Adapted from Ricker [32] and Logan [33].

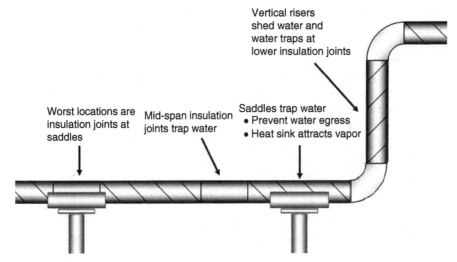

Figure 3.16 Problem locations for insulated aboveground pipelines.

protected from moisture by sheet metal covering. Figure 3.16 shows the corrosion locations identified by a major operator. They are virtually identical to the locations discussed in NACE SP0198, The Control of Corrosion Under Thermal Insulation and Fireproofing Materials – A Systems Approach, which contains detailed discussions on how this corrosion occurs and approaches to minimize it [34]. American Petroleum Institute (API) and other international standards and guidelines for CUI also exist [35–38].

Figure 3.17 shows seams in the jacketing around insulated piping. The seams should be located near the bottom (6 o'clock location) so that any moisture can drain from inside the jacketing. The two seams shown in this figure are at 9 and 11 o'clock locations where moisture can enter but cannot drain out of the system. Location C shows where the jacketing has been removed to allow placement of the piping on structural supports.

Many large petrochemical companies consider any insulated pipe a potential risk for CUI because it is unlikely that the pipe will stay dry over the 20- to 30-year required inspection cycle.

Breaks in the weather barrier jacketing usually occur at joints and allow water into the insulation. This water can leach contaminants from the insulation, which are then concentrated and deposited on hot surfaces. This can eventually lead to SCC of hot stainless steel surfaces, especially with insulations containing soluble chloride salts.

Figure 3.18 shows mechanical damage on the outside jacketing, which will allow rain or other moisture to enter the enclosed space around the piping and cause corrosion.

CUI is hard to inspect, and it is important to conduct these inspections on a frequent basis. It is also important

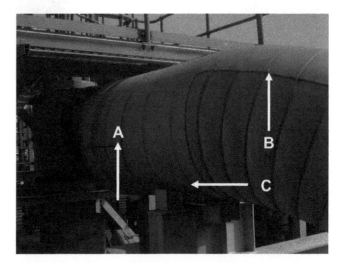

Figure 3.17 Seams at locations A and B where moisture can enter into the annular spaces between the jacketing and the insulated piping. Location C shows a location where the jacketing is cut to allow support of the piping system.

to replace the weather barriers after the inspection. A missing weather barrier is shown in Figure 3.19.

Slugging may occur at the bottom of upstream locations on vertical expansion joints like those shown in Figure 3.19. This creates fluid flow problems, but it also induces stresses into the system and may lead to corrosion fatigue failures. All expansion joints, whether horizontal or vertical, are locations where corrosion and crack detection inspections should be concentrated.

Buried or submerged thermally insulated piping relies on protective coatings for corrosion control. The application of cathodic protection to thermally insulated structures has not worked in most instances [38, 39].

Figure 3.18 Mechanical damage to insulation jacketing.

Figure 3.19 Vertical expansion loops in piping for secondary recovery steam injection. The top arrow indicates where a repair has been covered with painted sheet metal. The middle arrow indicates a location where the sheet metal outer covering is missing, but, because the system operates at elevated temperatures, only superficial tarnishing has occurred. The bottom arrow indicates the bottom of a vertical expansion loop that may cause internal slugging problems, especially with low-quality steam.

In addition to the proper design of moisture shields and draining for insulated piping, many operators have adopted the practice of always coating the carrier piping (even if it is made from a CRA) with an immersion grade coating [35–37].

It is also important to provide inspection capabilities for insulated piping. This is becoming an increasingly important consideration as more and more subsea pipelines are insulated to maintain noncorrosive temperatures on the inside of pipelines and gathering lines that

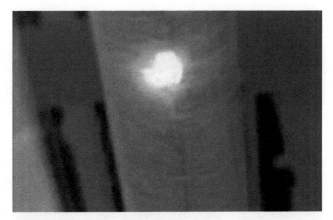

Figure 3.20 Wet insulation identified by infrared (IR) thermography.

Figure 3.21 Insulation removed for maintenance and not replaced.

would otherwise quickly cool to the temperatures of the ambient seawater environments.

CUI is difficult to detect, and the use of infrared inspection cameras is one means of trying to locate places where moisture is accumulating and temperatures are different than elsewhere on nearby piping (Figure 3.20) [40].

It is important that all personnel working around insulated piping should be encouraged to report indications of leaks. Figure 3.21 shows insulated piping that was exposed for maintenance, and the jacketing was not replaced or repaired. Figure 3.22 shows that a piping system was installed with no protective coating near the welds. Repairs in this location were completed, but no one was notified that the system seemed to have been constructed contrary to design drawings and general industry practice at the time of construction. Figure 3.23 shows moss growing where moisture is leaking out of

Figure 3.22 Missing protective coating near welds on insulated crude oil piping.

Figure 3.23 Moss growing at a seam on insulated piping. This moss indicates that excess moisture is leaking from within the protective jacketing.

the insulation jacketing. If this moisture indication had been reported, CUI on the crude oil piping might have been detected before it became serious.

Figure 3.24 shows the corrosion that resulted downstream from the piping shown in Figures 3.22 and 3.23. If the indications of problems apparent in these earlier figures had been noted, the corrosion in Figure 3.24 might have been detected much earlier.

International standards recommend against the use of insulation on topside applications to the extent possible, and insulation should only be used for safety or processing reasons. Many times insulation is installed to protect personnel from hot surfaces. An alternative to insulation is the installation of wire "standoff" cages. These cages are simple and inexpensive, and they eliminate CUI concerns.

INTERNAL ENVIRONMENTS

The environments discussed in this section all relate to internal corrosion in oilfield production and piping systems. These internal environments will vary depending on production fields under consideration. Figure 3.25 shows a typical production field where oil and gas wells may be located.

Crude Oil

Crude oil is not generally corrosive, but the minor constituents found with crude oil can cause corrosion if they separate into the water phase. The viscosity of crude oil also affects water dropout and surface wetting, which affect corrosion.

Oil is a generally benign environment for most metals. Corrosion of oil-wetted metals is rare, but the impurities that are present in oil, especially those that make any water that may be present acidic, can cause severe

(a)   (b)

Figure 3.24 Corrosion at uncoated weld on the crude oil piping system shown in Figures 3.22 and 3.23. (a) Corrosion exposed when jacketing and insulation was removed. (b) Ultrasonic inspection markings indicating that greater than 75% wall loss had occurred in some locations.

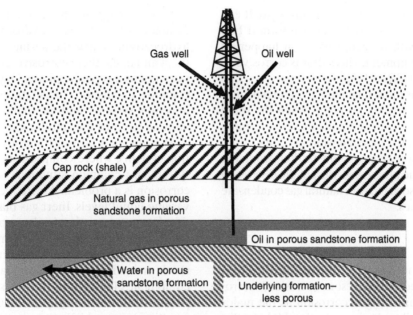

Figure 3.25 A typical oil and gas field.

corrosion [1, 41]. For this reason it is common practice to try and keep walls of containers oil wet whenever possible. This is done by means such as ensuring that the oil is flowing fast enough to cause turbulence and create an emulsion, a mixture of the immiscible water and oil fluids, which will keep the surfaces from becoming water wet. Certain crude oils, e.g. the heavy crude associated with Canadian tar sands, may be corrosive due to a tendency to deposit sludge on the bottom of equipment [1], and methods for testing this tendency are available [41–44].

Natural Gas

Unlike oil wells, which may produce noncorrosive fluids for many years before corrosion becomes a problem, natural gas wells are corrosive from the beginning. This is due to the fact that all natural gases will have at least some associated water, and this water, unlike the brines associated with oil wells, is usually very pure. This pure water has no natural buffering agents, and it becomes acidic due to the presence of dissolved gases, CO_2 and/or H_2S. Most of the discussion on CO_2 corrosion is associated with natural gas condensate in gas wells or pipelines. Table 3.3 shows typical compositions for natural gas wells and for natural gas from wells that also produce oil.

Natural gas deposits also produce small amounts of mercury. Offshore gas processing systems remove the mercury before the gas enters aluminum heat exchangers used to cool the gas before sending it to shore. This is less of a problem onshore, because aluminum heat exchangers, used offshore for weight savings, are not common in onshore gas processing plants.

TABLE 3.3 Compositions of Typical Natural Gas Wells

Natural Gas	
Hydrocarbon	
Methane	70–98%
Ethane	1–10%
Propane	Trace–5%
Butanes	Trace–2%
Pentanes	Trace–1%
Hexanes	Trace–1/2%
Heptanes +	Trace–1/2%
Nonhydrocarbon	
Nitrogen	Trace–15%
Carbon dioxide[a]	Trace–5%
Hydrogen sulfide[a]	Trace–3%
Helium	Up to 5%, usually trace or none

Gas from a Well That Also Produces Petroleum Liquid	
Hydrocarbon	
Methane	45–92%
Ethane	4–21%
Propane	1–15%
Butanes	1/2–7%
Pentanes	Trace–3%
Hexanes	Trace–2%
Heptanes +	None–1½%
Nonhydrocarbon	
Nitrogen	Trace–up to 10%
Carbon dioxide[a]	Trace–4%
Hydrogen sulfide[a]	None–trace–6%
Helium	None

[a] Occasionally natural gases are found which are predominantly carbon dioxide or hydrogen sulfide.

Elemental sulfur can be present in natural gas. It can precipitate onto metal surfaces. It can also form if H_2S comes into contact with oxygen, which can happen in topside piping and equipment. This sulfur is corrosive to both carbon steel and CRAs [45].

Natural gas contains more than methane, and pipelines frequently develop corrosion problems when the small amounts of water, CO_2 and/or H_2S, and other acid-forming compounds collect on the pipeline walls as temperature and pressure conditions change the farther from the compressor the pipeline progresses. Most of the discussion on CO_2 corrosion is associated with natural gas condensate in gas wells or pipelines.

Oxygen

The API guidelines on corrosion discuss "sweet corrosion," corrosion caused by CO_2; "sour corrosion," corrosion caused by H_2S; and oxygen corrosion as the major corrodents found in oil and gas production [46]. As discussed in Chapter 2, corrosion cannot occur unless a chemical is available to be reduced at the same time that metal is oxidized or corroded. The most common chemical that serves this purpose is oxygen, which makes up approximately 20% of the air that we breathe and is found in most soil and liquid environments. The maximum solubility level of oxygen in surface waters is only approximately 10 ppm (8 ppm in seawater, as high as 11 ppm in fresh water), but, as shown in Figure 3.3, oxygen is so corrosive that the effects of oxygen can overwhelm the effects of CO_2 and H_2S, which are less corrosive.

Figure 3.26 shows how the corrosion rates for steel vary with water temperature. The corrosion rate increases with temperature in closed systems, but if air is allowed to escape, the corrosion rate decreases as outgassing of the water increases with temperature [36].

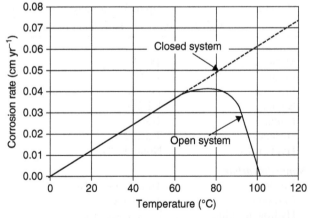

Figure 3.26 The effect of temperature on the corrosion rate of steel in water. *Source*: Adapted from Baboian and Treseder [26] and Speller [47].

Gaseous oxygen is not naturally present in geological formations that produce hydrocarbons, but once these hydrocarbons reach the surface, oxygen can leak into them and make them corrosive, especially if water separates out as a separate phase. Oxygen corrosion is most common in offshore installations, brine handling, and injection systems and in shallow wells where air is allowed to enter the annular spaces [46].

Rod-pumped oil wells and low-pressure gas production systems, where compressors can suck air into the system, are common situations where oxygen-induced corrosion is a problem. Air frequently enters tanks with varying liquid levels. Inert gas blanketing can minimize this. Oxygen corrosion is also a problem in waterflood injection equipment [46]. Oxygen attack can be either general overall corrosion or pitting corrosion, which is much more likely to produce leaks in fluid-containing equipment [48]. One of the ways oxygen can promote pitting corrosion is by providing oxygen to bacteria, resulting in microbially influenced corrosion (MIC).

As the ability to detect dissolved oxygen improved, it became apparent to the oil and gas industry that corrosion problems start if oxygen levels as low as 50–100 ppb are exceeded. This requires careful control of surface fluids and the use of chemical oxygen scavengers.

Oxygen also reacts with dissolved metal ions, usually iron or manganese in water, and produces suspended solids, which can be erosive and clog filters.

Any dissolved chlorine gas can act in conjunction with dissolved oxygen with an oxygen equivalent of 0.3 times the concentration of chlorine [21].

Most operators have found that the best way to prevent oxygen corrosion is by keeping oxygen out of their systems. This is cheaper than treating the problems that oxygen causes. This requires careful maintenance of seals, valves, and other locations where air and oxygen can leak into the system. Keeping a positive pressure on equipment is considered by some authorities to be another effective means of limiting oxygen, but some oxygen can enter due to diffusion and still cause corrosion. Inert gas blankets are used over storage tanks. Other methods include corrosion inhibitors, to include oxygen scavengers, protective coating, and cathodic protection of water-wetted interior surfaces of process vessels and storage tanks. CRAs are also used, but this is more common for other corrosive environments and is not often necessary for oxygen corrosion [46].

Carbon Dioxide

Carbon dioxide is not corrosive provided it stays dry. Mixing CO_2 with water produces carbonic acid that can be very aggressive in some circumstances. Steel in CO_2-containing environments forms scales, which can be protective, but this scale often breaks down. Figure 3.27 shows how the carbonate–bicarbonate–carbonic acid equilibrium shifts with pH. This picture is for standard temperature (25 °C [77 °F]), but the principle is valid for downhole and pipeline conditions as well. At high pHs carbonate films will form on metal surfaces, and at low pHs the bare metal will be exposed to liquid acids.

The solubility of carbonate scales and films varies with temperature and pH. This is shown in Figure 3.28, which is produced by one of the numerous software packages used to predict corrosive conditions in oilfield equipment.

Software packages, like the one that produced Figure 3.28, are available that predict the stability of carbonates under a variety of conditions. Corrosion rates are strongly affected by temperature, pressure, scaling tendency, and dissolved gases.

Figure 3.29 shows a sucker rod made of an iron–chrome alloy that corroded when the "protective" scale broke down locally and allowed the underlying metal to be exposed to aggressive carbonic acid. Relatively clean circular pits surrounded by unattacked metal characterize CO_2 corrosion. The clean pits and sharp edges of the pits have earned this corrosion term "mesa corrosion," because a cross section of the metal looks like the flat-topped mesa (table) mountains common in the southwestern United States.

Figure 3.30 shows CO_2 channeling corrosion when CO_2-rich water accumulated along the bottom of a horizontal gathering line. The same kind of corrosion is also common on the bottom of steam condensate lines in power plants. This form of corrosion is discussed in more detail in Chapter 5.

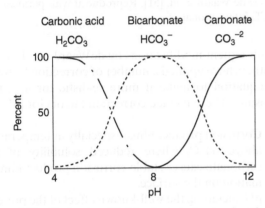

Figure 3.27 Carbonic acid, bicarbonate ion, and carbonate distribution as a function of pH.

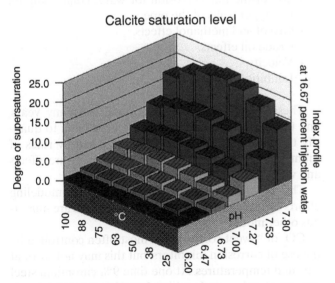

Figure 3.28 Effect of pH and temperature on the scaling tendency of calcium carbonate [49]. *Source*: Figure courtesy FrenchCreek Software – Downhole SAT®.

Figure 3.29 Mesa corrosion on a sucker rod.

Figure 3.30 CO_2 channeling along the bottom of a horizontal gathering line.

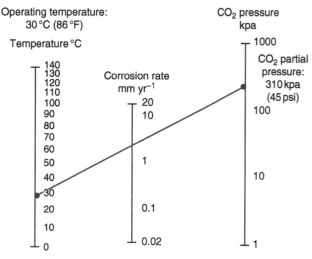

Maximum unmitigated corrosion rate: 5.0 mm yr⁻¹ (200 mpy)

Figure 3.31 Nomogram used to predict corrosion rates in CO_2-containing environments. *Source*: de Waard and Lotz [50]. Reproduced with permission of NACE International.

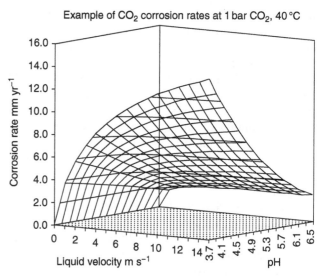

Figure 3.32 The effect of liquid velocity on CO_2 corrosion. *Source*: de Waard et al. [51]. Reproduced with permission of NACE International.

CO_2 corrosion has become increasingly important in recent years because many deep hot gas wells have high concentrations of CO_2. Figure 3.31 is a nomograph developed by deWaard and Milliams to predict corrosion rates in the presence of CO_2. The same research group published the information in Figure 3.32, which shows the effects of liquid velocity on CO_2 corrosion.

The deWaard–Milliams model used to generate Figure 3.31 is available as an algorithm and is particularly easy to use:

$$\text{Log}(V_{cor}) = 5.8 - \frac{1710}{(273+T)} + 0.67 \log\left(P_{CO_2}\right) \left(\text{mm yr}^{-1}\right)$$

(3.1)

where

V_{cor} = corrosion rate, mm yr⁻¹
T = temperature, °C
P_{CO_2} = carbon dioxide partial pressure, bar

This equation is intended for the initial corrosion rate of carbon steel in 5% brine, before the development of protective carbonate corrosion product films that passivate the surface. Nevertheless, the deWaard and Milliams equation is considered a landmark in the estimation of carbon dioxide corrosion rates and is widely used to judge the potential severity of CO_2 corrosion. Note that this equation takes into account both temperature and carbon dioxide partial pressure. It is much better than the old rules of thumb based only on carbon dioxide pressure.

Subsequent modifications by deWaard and Lotz and by others have applied a number of correction factors to the equation to make it more realistic for long-term exposures. These include correction factors for [51–53]:

- Corrosion product films, especially at temperatures above 60°C, where reduced solubility of iron carbonate causes stable corrosion product film formation on the surface.
- pH, including the well-known effect of the presence of organic acids in gas condensate.
- Effects of system pressure on the fugacity of CO_2.
- Top-of-the-line corrosion for water condensing on the upper walls of the pipe.
- Glycol and methanol effects.
- Crude oil effects.
- Velocity.
- Inhibition.

These effects have been discussed for many years, and the results of modeling efforts have been to quantify the information that had been applied by rules of thumb in the past. Figure 3.33 shows temperature effects and why they occur.

This discussion has concentrated on the modeling efforts by deWaard and coworkers, but there are numerous other models available [55–59].

CO_2 or "sweet corrosion" is most often controlled by the use of corrosion inhibitors, but this may not work at elevated temperatures. At one time 9% chromium steel was used for downhole tubing, but SCC problems developed, and this use has been discontinued. Martensitic stainless steels (12% chromium and higher) and other

CRAs have been successfully used in recent years, especially as downhole temperatures have increased to levels where organic chemical-based corrosion inhibitors cannot be used [44, 58–61]. Drilling fluid corrosion is often controlled by pH control with caustic soda (sodium hydroxide).

Hydrogen Sulfide

Oil and gas that contain sulfur are termed sour gas or sour crude, and the most common form of sulfur is H_2S gas [4, 46, 62]. Sour conditions can come from H_2S in the producing formation or from surface sources (injection water, lift gases, etc.). Hydrogen sulfide is toxic, and releases of H_2S can cause death within seconds. H_2S is more soluble in crude oil than in water, with a ratio of 1.7/1 at 32 °C (90 °F). The saturation level is 5000 ppm, and concentrations in the range of 100–200 ppm are common [1]. It also forms a weak mineral acid that can lower pH and make the environment acidic, similar to the effect of CO_2. Sulfur is also involved in the metabolism of some microorganisms, and the presence of H_2S can be associated with MIC.

H_2S can also lead to several forms of metal cracking, variously termed sulfide stress cracking (SSC), hydrogen stress cracking (HSC), hydrogen-induced cracking (HIC), SCC, stress-oriented hydrogen-induced cracking (SOHIC), etc. [9, 58]. This is generally a more serious problem, as all forms of environmental cracking can produce sudden gas releases and their associated safety problems. H_2S serves as a hydrogen entry promoter, and

steels that are subject to any form of hydrogen-related cracking are more likely to do so in the presence of H_2S.

Long-life oilfield equipment should be designed for sour conditions even if the production starts out noncorrosive. The souring of many fields is sometimes attributed to surface water injection or reinjection, but there are other causes as well [8, 9, 58, 62, 63].

ANSI/NACE MR0175/ISO 15156 – Materials for Use in H_2S-Containing Environments in Oil and Gas Production Problems with cracking of various types led to the development of NACE MR0175, which was extensively modified to become NACE MR0175/ISO 15156 in 2003. It has also become an ANSI standard in recent years.

The original 1975 document, based on work by various NACE and other working groups, covered only valves and wellhead equipment, but the scope was expanded in later revisions. It described various environments where H_2S cracking was considered to be a problem and placed restrictions, based on temperature and pressure, on where carbon steel could be used in these environments [9, 64]. At temperatures and pressures where carbon steel was deemed unsuitable, the use of CRAs was indicated. The document underwent various changes over the years driven, in part, by the increased temperatures and H_2S partial pressures encountered.

Early versions of the MR0175 had figures like Figures 3.34 and 3.35. These figures clearly showed where MR0175-qualified alloys were required and where the H_2S and total pressures were low enough that H_2S-related cracking was considered unlikely.

The original 1975 version of MR075 also restricted the hardness of metals to no greater than HRC 22. This was based on a series of experiments conducted by a consortium of oil companies that showed that the time to cracking in H_2S environments was very short for

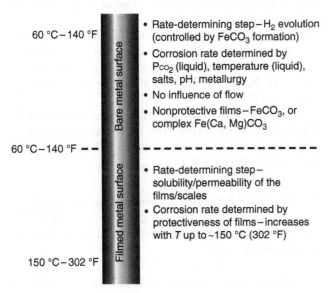

Figure 3.33 CO_2 rate-determining factors at various temperatures. *Source*: Schmidt [54]. Reproduced with permission of NACE International.

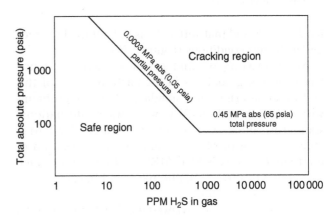

Figure 3.34 Sulfide stress cracking in sour gas environments. Simplified version of figure A-1, NACE MR0175-03.

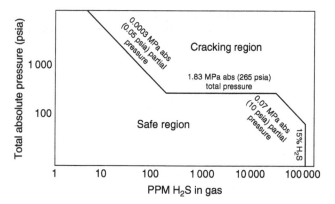

Figure 3.35 Sulfide stress cracking in sour multiphase systems. Simplified version of figure A-2, NACE MR0175-03.

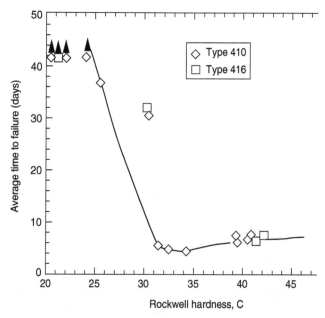

Figure 3.36 The effect of hardness on time to failure in H_2S environments. *Source*: Phelps [65]. Reproduced with permission of NACE International.

harder steels and that softer steels did not tend to crack under the test conditions (Figure 3.36).

Many existing oil and gas fields were designed based on the guidelines in the early versions of this and other international standards. Various problems with MR0175 and conflicts with other international standards, along with ongoing research, lead to a major revision of MR0175 and to the publication of joint standards as NACE MR0175/ISO 15156, which was issued in three separate parts and became effective at the end of 2003.

Milliams and Tuttle reviewed the development of the original NACE MR0175 and other H_2S-related materials standards [64]. They also explained how the new

MR0175/ISO15156 documents were organized into the following parts:

- **Part 1** – General principles for the selection of cracking-resistant materials [66].
- **Part 2** – Cracking-resistant carbon and low-alloy steels [67].
- **Part 3** – CRAs and other alloys [68].

The revised standard gives requirements and recommendations for the selection and use of cracking-resistant alloys for equipment used in oil and gas production and in natural gas treatment plants. It does not cover refinery equipment. It supplements, but does not replace, other codes, standards, or regulations, and it does not address corrosion except if it is related to H_2S-assisted cracking. Changes introduced in 2003 incorporated information obtained on H_2S cracking by the European Federation of Corrosion [69, 70].

The standard addresses only environmental cracking in H_2S-containing environments, and it does not address whether the alloys are immune to cracking under other service conditions (e.g. SCC due to chloride-containing environments). It also does not address mass-loss (weight-loss) corrosion, e.g. pitting or crevice corrosion, even though it suggests the use of a pitting resistance equivalent number formula for ranking potential CRAs in Part 3 of the standard.

Industrial standards of all types undergo periodic review and updating, and specifiers should designate the current version of standards when ordering equipment. It is not appropriate to state "current version"; the exact year of the modification needs to be in all specification and purchasing documents.

NACE MR0103 was introduced in 2003 and covers similar questions related to downstream (refining) operations [71]. Until then it was common to use the advice in NACE MR0175 for H_2S environments in refineries [72].

Table 3.4 shows the types of equipment covered by ANSI/MR0175/ISO15156. The same table appears in both Part 2, which covers carbon steel and low-alloy steels, and in Part 3, CRAs.

Part 1: General Principles for the Selection of Cracking-Resistant Materials The scope of ANSI/MR0175/ISO15156 is applicable to materials used for the following equipment [66]:

- Selection of materials based on the guidance in Parts 2 and 3.
- Qualification and selection of materials for specific H_2S environments not covered in Parts 2 and 3.

TABLE 3.4   Equipment Covered by ANSI/MR0175/ISO 15156 Parts 2 and 3 [66, 67]

ANSI/NACE MR0175/ISO 15156 Is Applicable to Materials Used for the Following Equipment	Permitted Exclusions
Drilling, well construction, and well-servicing equipment	Equipment exposed only to drilling fluids of controlled composition[a] Drill bits Blowout-preventer (BOP) shear blades[b] Drilling riser systems Work strings Wireline and wireline equipment[c] Surface and intermediate casing
Wells, including subsurface equipment, gas lift equipment, wellheads, and Christmas trees	Sucker rod pumps and sucker rods[d] Electric submersible pumps Other artificial lift equipment Slips
Flowlines, gathering lines, field facilities, and field processing plants	Crude oil storage and handling facilities operating at a total absolute pressure below 0.45 MPa (65 psi)
Water-handling equipment	Water-handling facilities operating at a total absolute pressure below 0.45 MPa (65 psi) Water injection and water disposal equipment
Natural gas treatment plants	—
Transportation pipelines for liquids, gases, and multiphase fluids	Lines handling gas prepared for general commercial and domestic use
For all equipment above	Components loaded only in compression

[a] See ANSI/NACE MR0175/ISO 15156-2:2009, A.2.3.2.3 for more information.
[b] See ANSI/NACE MR0175/ISO 15156-2:2009, A.2.3.2.1 for more information.
[c] Wireline lubricators and lubricator connecting devices are not permitted exclusions.
[d] For sucker rod pumps and sucker rods, reference can be made to NACE MR0176.
Source: Courtesy of NACE International.

- Determination of qualifications for existing equipment that is to be exposed to an increased level of H_2S.
- Qualification for service may be based on laboratory testing or field experience.

Part 2: Cracking-Resistant Carbon and Low-Alloy Steels [67]   The severity of the environment is determined in accordance for carbon, and low-alloy steel will be assessed using Figure 3.37.

The key to this figure is:

$X = H_2S$ partial pressure, kPa
$Y = in situ$ pH

Appendix D of Part 2 has a detailed discussion of how *in situ* pH is determined, including a number of tables to aid in this determination.

In region 0 ($P_{H_2S} < 0.3$ kPa [0.05 psi]), no precautions are normally necessary, but factors that can affect steel performance should be considered:

- Steels susceptible to SCC and HSC may crack due to other mechanisms, e.g. chloride SCC and liquid

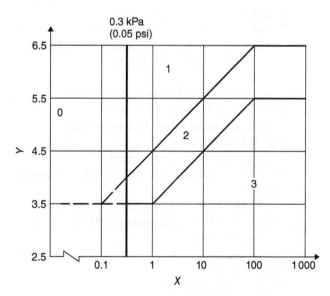

Figure 3.37   Figure showing H_2S environment classifications for selection of carbon and low-alloy steels. *Source*: ANSI/NACE MR0175/ISO 15156-2 [67]. Reproduced with permission of NACE International.

metal embrittlement. These forms of cracking and mass-loss (weight-loss) corrosion are not covered by ANSI/NACE MR0175/ISO 15156.

- Very high strength steels can suffer HSC without the presence of H_2S. Yield strengths above 965 MPa (140 ksi) may need special processing to ensure that these steels do not suffer HSC or SCC in Region 0 environments.

Regions 1, 2, and 3 define conditions of increasing severity. The types of materials likely to be suitable in these environments are discussed.

Part 3: CRAs and Other Alloys [68] Susceptibility to cracking produces restrictions on what alloys can be used in H_2S environments. Other important variables in H_2S environments include salt (expressed as chlorides), temperature, and P (total and H_2S).

Part 3 classifies CRAs into the following categories [68]:

- Austenitic stainless steels.
- Highly alloyed austenitic stainless steels.
- Solid-solution nickel-based alloys.
- Ferritic stainless steels.
- Martensitic stainless steels.
- Duplex stainless steels.
- Precipitation-hardened stainless steels.
- Precipitation-hardened nickel-based alloys.
- Cobalt-based alloys.
- Titanium and tantalum.

Copper and aluminum can be used without restrictions on P_{H_2S}, Cl^-, or *in situ* pH.

Then there are general restrictions based on the alloys, usually by group, and also additional materials selection tables for these applications:

- Casing, tubing, and downhole equipment:
  - Downhole tubular components.
  - Packer and other subsurface equipment.
  - Gas lift equipment.
  - Injection tubing and equipment.
  - Downhole control line tubing and downhole screens.
- Wellheads, Christmas trees, valves, and chokes:
  - Wellhead and tree components (with various specified exclusions).
  - Valves and choke components (with various specified exclusions).
  - Shafts, stems, and pins.
  - Nonpressure-containing internal valve, pressure regulator, and level controller components.
- Process plant:
  - Compressor components
- Materials selection tables for other equipment:

- Instrument tubing and associated compression fittings, surface control line tubing, and surface screens.
- Springs.
- Diaphragms, pressure measuring devices, and pressure seals.
- Seal rings and gaskets.
- Snap rings.
- Bearing pins.
- Miscellaneous equipment as named in the tables (including hardware [e.g. set screws], downhole, and surface temporary service tool applications).

Then the various alloys have individual tables, e.g. Table 3.5, which shows the environmental and materials limits for martensitic stainless steels used for equipment. Other tables are more complex and have different limits for various components.

These general limits are then supplemented by tables showing limits for specific applications, such as those shown in Table 3.6.

In the past it has been a common practice for operating companies to require that equipment suppliers deliver equipment and supplies in compliance with NACE MR0175 or other similar documents. Older equipment may not meet the more stringent requirements of current versions of ANSI/MR0175/ISO 15156.

Control of H_2S Corrosion and Cracking The amount of H_2S in drilling fluid environments is usually controlled by pH control using caustic soda (sodium hydroxide) and the use of H_2S scavengers. At one time these scavengers were zinc-based minerals, but concern with environmental pollution has led to nonmetallic scavengers in recent years. In production tubing the use of corrosion inhibitors is the normal means of corrosion control. Flowlines and pipelines use a combination of corrosion inhibitors and H_2S scavengers. Some flowlines and pipelines use internal coatings – either organic or cementitious [67].

Cracking control in H_2S requires the selection of the appropriate alloys in accordance with the guidance of ANSI/NACE MR0175/ISO 15156 [67, 68]. NACE TM0177 describes how metals should be tested for resistance to H_2S environments [73].

Organic Acids

Oilfield organic acids sometimes cause oilfield corrosion [74–76]. Like most organic compounds, they tend to be covalent and form weak acids. Their presence is most important at high pressures in gas condensate environments. Many operators have software packages that include organic acids in their downhole pH calculations.

TABLE 3.5   Environmental and Materials Limits for Martensitic Stainless Steels Used for Any Equipment or Components

Individual Alloy UNS Number	Temperature Max. °C (°F)	Partial Pressure H_2S pH_2S Max. kPa (psi)	Chloride Concentration Max. mgl^{-1}	pH	Sulfur Resistant	Remarks
S41000	See "Remarks" column	10 (1.5)	See "Remarks" column	≥3.5	NDS[a]	
S41500						
S42000						
J91150						Any combination of temperature
J91151						and chloride concentration
J91540						occurring in production
S42400						environments is acceptable
S41425	See "Remarks" column	10 (1.5)	See "Remarks" column	≥3.5	No	

These materials shall also comply with the following:
(a) Cast or wrought alloys UNS S41000, J91150 (CA15), and J91151 (CA15M) shall have a maximum hardness of 22 HRC and shall be:
  (i)   Austenitize and quenched or air cooled.
  (ii) Tempered at 621 °C (1150 °F) minimum, than coded to ambient temperature.
  (iii) Tempered at 621 °C (1150 °F) minimum, but lower than the first tempering temperature, than cooled to ambient temperature.
(b) Low-carbon, martensitic stainless steels, either card J915-40 (CA6NM) or wrought S42400 or S41500 (F6NM), shall have a maximum hardness of 23 HRC and shall be:
  (i)   Austenitize at 1010 °C (1850°) minimum, than air- or oil quenched to ambient temperature.
  (ii) Tempered at 649–691 °C (1200–1275 °F), than air cooled to ambient temperature.
  (iii) Tempered at 593–521 °C (1100–1150 °F), than air cooled to ambient temperature.
(c) Cast or wrought alloy UN5 S42000 shall have a maximum hardness of 22 HRC and shall be in the quenched and tempered head-treatment condition.
(d) Wrought low-carbon UNS S41425 martensitic stainless steel in the austenitize, quenched, and tempered condition shall have a maximum hardness of 26 HRC.

[a] No data submitted to ascertain whether these materials are acceptable for service in the presence of elemental sulfur in the environment.
Source: From table A.18, Ref. [68]. Courtesy of NACE International.

Scale

Solids in oil and gas production can include both produced solids that are entrained in the oil and gas and scales such as calcium carbonate, calcium sulfate, barium sulfate, and silicates that form on tubing walls and surface equipment as a result of changes in chemistry, temperature, and pressure as fluids are produced, separated, and transported. These scales tend to form on any surface, to include production tubing, sand, and topside piping wherever the temperature and pressure conditions are such that dissolved mineral solubility limits are exceeded [2].

Figure 3.38 shows scales formed on the inside of production tubing. As can be seen, these scales can substantially reduce the flow cross section, and eventually they could plug the pipes. Similar problems occur within producing formations and are the reason why acidizing treatments are sometimes necessary to reopen plugged formations.

The software package that produced Figure 3.28 is one of the many calculation methods used to determine the downhole conditions where this scaling will occur. Similar packages are also used to predict scaling tendencies within formations and to predict the effectiveness of secondary and tertiary (e.g. CO_2 injection) recovery methods used on aging fields.

Scale generally keeps metals dry and reduces general, or uniform, corrosion [8, 77–82]. Unfortunately, scale can be imperfect, and corrosion, such as that shown in Figure 3.29, occurs at breaks in the scale. Scale or produced solids can also be abrasive and cause erosion–corrosion, especially at wellheads and on surface equipment.

Many operators generate steam for various purposes. The same minerals that can plug production tubing can also plug steam piping if the feedwater treatment procedures are inadequate. Figures 3.39 and 3.40 show boiler scale, which is a major problem in steam generating equipment.

TABLE 3.6   Environmental and Materials Limits for Martensitic Stainless Steels Used as Downhole Tubular Components and for Packers and Other Subsurface Equipment

Specification/Individual Alloy UNS Number	Temperature Max. °C (°F)	Partial Pressure H₂S pH₂S Max. kPa (psi)	Chloride Concentration Max. mg l⁻¹	pH	Sulfur Resistant	Remarks
ISO 11960 L-80 Type 13 Cr, S41426, S42500	See "Remarks" column	10 (1.5)	See "Remarks" column	≥3.5	NDS[a]	Any combination of temperature and chloride concentration occurring in production environments is acceptable
S41429	See "Remarks" column	10 (1.5)	See "Remarks" column	≥4.5	NDS[a]	

For these applications, these materials shall also comply with the following:

(a) UNS S41426 tubular components shall be quenched and tempered to maximum 27 HRC and maximum yield strength 724 MPa (105 ksi).

(b) UNS S425DD (15Cr) tubing and casing is acceptable as Grade 80 (SMYS 556 MPa [80 ksi] only and shall be in the quenched and double-tempered condition, with a maximum hardness of 22 HRC. The quench and double-temper process shall be as follows:

  (i) Austenitize art minimum 900 °C (1652 °F), then air- or oil quench.

  (ii) Temper at minimum 730 °C (1346 °F), then cool to ambient temperature.

  (iii) Temper at minimum 620 °C (1148 °F), then cool to ambient temperature.

(c) UNS S41429 tubular components shall be quenched and tempered or normalized and tempered to a maximum hardness of 27 HRC and a maximum yield strength of 827 MPa (120 ksi).

[a] No data submitted to ascertain whether these materials are acceptable for service in the presence of elemental sulfur in the environment.

Source: From table A.19, Ref. [68]. Courtesy of NACE International.

The scale on steam generator piping (Figure 3.39) has a similar effect to the scales shown in Figure 3.38: it plugs the piping. Scales on boiler tubes can be more serious, because the loss of heat transfer through the low heat conductivity scale can cause hot spots that lead to overheating and stress ruptures producing leaks and sometimes explosions. Safety concerns related to boiler explosions led the US Bureau of Mines to conduct the research, which led to boiler water treatment procedures used worldwide today [78–80].

Water piping is prevented from corrosion by the presence of thin scales of calcium- or magnesium-containing minerals on the metal surfaces [23, 24].

The carbonate minerals shown in Figures 3.38–3.40 are soluble in mineral acids, and the same mineral acids used for descaling downhole formations can be used to remove scale from the inside of piping. If the acid is left in the equipment too long, or if the corrosion inhibitors added to the cleaning acid are inadequate, then rapid corrosion can occur, resulting in perforated tubing within a matter of hours (Figure 3.41).

NACE and other organizations provide guidance on how to evaluate scale inhibitors intended to prevent scale formation [81].

The scales shown in Figures 3.38–3.40 tend to be calcium minerals, but other metallic ions can be incorporated as impurities in these scales. Some of these impurities are radioactive strontium and other isotopes that make some scales and produced fluids radioactive

Figure 3.38   Calcite (left) and gypsum (right) scale in production tubing.

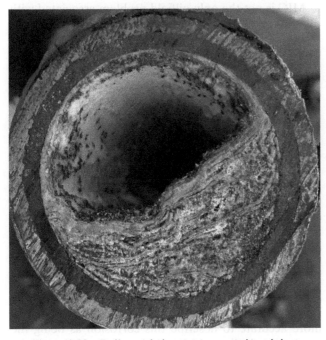

Figure 3.39   Boiler scale in steam generating piping.

Figure 3.40   Scale formed inside a heat exchanger [24]. *Source*: Reproduced with permission of Springer.

Figure 3.41 Perforated production tubing caused by uncontrolled acidizing treatment.

and a health and waste disposal hazard. As long as the scales remain downhole, there is no health hazard, but scrap tubing and the fluids recovered from scale removal acidizing operations require special handling and disposal [82].

Microbially Influenced Corrosion (MIC)

MIC is a phenomenon that has been recognized for many years and is the subject of a number of books, as well as numerous reports on the subject [83–111]. It is a growing problem in the oil and gas industry, and, unfortunately, many of the problems are introduced by improper water handling of surface waters. One of the problems associated with MIC is that most oilfield engineers and technicians have very little understanding of biology and, therefore, are likely to believe "experts" or "rules of thumb" whether or not they have validity.

A number of terms have been used for MIC including "microbiologically induced corrosion," "biocorrosion," and others. NACE standardized on the term "microbially influenced corrosion" in the early 1990s, and this term emphasizes that microbes can increase or decrease corrosion. Both phenomena have been reported, although increased corrosion is obviously of more technical and economic interest.

The following observations apply to MIC [71, 72]:

- MIC can occur in environments where corrosion is not expected, e.g. in downhole pumping equipment

TABLE 3.7 Examples of Operational Problems That May Be Caused by Bacteria [92]

Increased frequency of corrosion failures
Increasing H_2S concentrations
Reservoir souring
Rapid production decline
Metal sulfide scales
Failure of downhole equipment due to metal sulfide deposits
Inefficient oil/water separation
Inefficient heat exchange
Black water
Black powder in gas transmission lines
Filter plugging
Loss of injectivity

Source: Reproduced with permission of NACE International.

removed from any sources of oxygen or other apparent corrodents.
- MIC corrosion rates can be very rapid.
- Liquid culture techniques, the long-standing standard method of identifying the biological sources of MIC, do not provide accurate assessments of the numbers and types of organisms involved in field situations.
- Mitigation and control strategies have shifted from the widespread use of biocides to manipulation of the environment, e.g. introduction of smooth surfaces where biofilm attachment is difficult and MIC is less likely.

MIC is not the only oilfield problem associated with biofilms and bacteria. Table 3.7 shows some of the other oilfield problems caused or accelerated by bacteria [109]. The reader is cautioned that this list is not inclusive. While this discussion of MIC has concentrated on internal problems with MIC, exteriors of pipelines and other equipment have also been reported to have similar problems.

There are a number of possible mechanisms involved in MIC, and some have been better described than others [83]. Many advocates of MIC claim that whenever high microbe populations are found in the presence of corrosion, this is evidence that MIC caused the corrosion. This is not always the case, and several authors have discussed how MIC can be identified. MIC can only be confirmed when all other possible explanations for the observed corrosion have been eliminated [85].

MIC occurs within and underneath biofilms that form on metal surfaces. These films start out microscopically thin but can become much thicker. It might be more accurate to describe this corrosion as being the result of biofouling, which includes macroscopic growths, e.g. mussels, barnacles, etc. One definition of MIC is "an electrochemical type of corrosion in which certain micro-organisms have a role, either enhancing or inhibiting" [72].

Like most other forms of corrosion, MIC is an electrochemical process. Microorganisms can affect the extent and severity of corrosion, and, like all organism-related phenomena, water is necessary for microbial life, and therefore water is necessary for MIC.

Bacteria attach to metallic surfaces and start to form thin biofilms consisting of cells, living or dead. These films also incorporate water and debris from the environment. Growth of these films can change the chemical concentrations of the water at the biofilm/metal substrate interface. Films as thin as 12 μm can prevent diffusion of oxygen and produce localized areas that are anaerobic enough to promote the growth of SRB. One result of biofilm formation is the creation of concentration gradients that produce electrochemical cells that can be explained by the Nernst equation discussed in Chapter 2.

Biofilms can form in minutes to hours, and MIC can be detected within 10–20 days in stagnant waters, e.g. improperly drained and dried equipment that has been hydrotested with microbe-containing water [107–112].

There are many ways of classifying bacteria, but two distinctions are important to understand for control of oilfield corrosion:

- Sessile bacteria are attached to surfaces and become motionless.
- Planktonic bacteria freely float or swim in a body of water.

It is relatively easy to sample a fluid for planktonic (free-floating) bacteria, but the actions of sessile bacteria are more important in determining corrosion rates. This usually requires insertion of coupons or probes into the fluid at the elevation where the sessile bacteria are most likely to be forming. Sessile bacteria lead to most MIC-related corrosion problems in oilfield equipment [83–85]. Planktonic bacteria can also produce corrosive chemicals, which can lead to corrosion and H_2S-related cracking.

The two best known bacteria classifications for oilfield corrosion are sulfate-reducing bacteria (SRBs) and acid-producing bacteria (APB). NACE classifies the most important types of oilfield bacteria as [93]:

- Sulfate-reducing bacteria
- Iron-oxidizing bacteria (IOB)
- Acid-producing bacteria (APB)
- Sulfur-oxidizing bacteria (SOB)
- Slime-forming bacteria

Another widely used classification is the distinction between aerobic (air-breathing, more correctly oxygen-

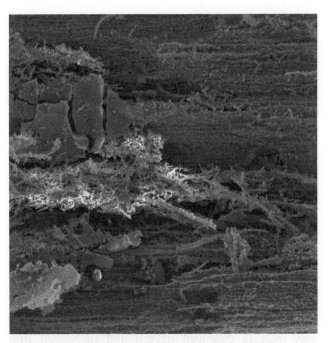

Figure 3.42 Biofilm inside corrosion pit. *Source*: Photo courtesy K. Pytlewski, Anamet, Hayward, California.

breathing) bacteria and anaerobic bacteria, which do not require air or oxygen for their respiration [83–85].

All of these bacteria contribute to the formation of biofilms, where bacteria can grow (Figure 3.42). These biofilms typically contain from 40 to 60% pore space by volume, and most of this pore space is full of water where the bacteria can find nutrients and reproduce. Biofilms also serve as collection points for larger microbes, fungi, debris, scale precipitates, and other solids that both increase the volume of the biomass and also help to shield the sessile bacteria within the biomass from chemical treatment biocides. Thick biofilms may have different characteristics from the outside, where slime may attract inorganic debris, to the inside, which may become highly acidic anaerobic environments conducive to growth of SRB populations that lead to further corrosion. These biofilms can be microscopic or macroscopic in nature.

Many authorities claim that SRB are the most important bacteria associated with oilfield corrosion, but this is disputed [84, 85]. Nonetheless SRB are important, widely studied, and widely reported. SRB are anaerobic; they do not require dissolved oxygen in their metabolism, but they can tolerate oxygen, usually by becoming dormant and not reproducing in oxygen-rich environments. SRB are typically found in dead legs and other quiescent locations, but planktonic SRB can survive in turbulent waters and then settle and become sessile when fluid flow diminishes. Most SRB strains thrive at 25–35 °C (77–95 °F), but some can thrive up to 90 °C (195 °F).

SRB cause corrosion by oxidizing organic compounds to CO_2 or organic acids. They reduce sulfates and other sulfur compounds to sulfide ions, which can then become gaseous or dissolved H_2S. Steel corrosion in acids produces monatomic hydrogen atoms at the cathodes. This can polarize the surface and lower corrosion rates, but SRB combine with the hydrogen forming H_2S and thus depolarize surfaces, increasing corrosion rates, especially in corrosion pits. The reaction of H_2S with dissolved iron ions can also produce iron sulfide scales on nearby surfaces. These scales can alter corrosion rates, either increasing or decreasing corrosion depending on the circumstances.

IOB are also known as iron-depositing bacteria or iron-related bacteria (IRB). They are usually found in mounds, called tubercles, over pits on metal surfaces. The presence of these tubercles is not always due to IOB, but it has been claimed that if the tubercles are shiny, rather than dull, this is a strong indication of IOB-related MIC [108]. Rust-colored water or yellow slime may indicate dissolved oxygen and suspended iron oxides, but it may also indicate the presence of IOB. IOB are found in open ponds, supply wells, filters, piping, equipment, and injection wells [108].

IOB oxidize dissolved Fe^{+2} ions to Fe^{+3}. They can also oxidize Mn^{+2} to Mn^{+3}. Depending on the pH, these ions may then combine with dissolved anions to form $Fe(OH)_3$ or $FeCl_3$, which can then deposit on tubercles, sealing them and creating anaerobic conditions conducive to the growth of SRB [108].

There are many types of APB that can become trapped under biofilms where they create acids leading to underfilm corrosion.

SOB are aerobic. Some produce sulfuric acid, H_2SO_4, which is usually more corrosive than the relatively nonionic H_2S acid formed by SRB.

Some forms of SOB require sunlight for photosynthesis [108]. SOB can exist over a range of pHs from 0 to 4 but survive best at pHs around 2.5 [108].

Slime-forming bacteria can form capsules over biofilms. These slime capsules can protect the underlying biomass from biocides. They can also create differential aeration cells, and slime-producing bacteria can be either aerobic or anaerobic [108].

Table 3.8 shows a number of classes of bacteria, their oxygen requirement, the metals they corrode, and the mechanism whereby they are thought to operate.

No unique form of corrosion is associated with MIC; this makes identification difficult. The presence of large amounts of planktonic or sessile bacteria does not necessarily mean that MIC is present. At one time it was common to analyze corrosion pit morphology (shape, depth, etc.) in an attempt to confirm MIC and identify the types of microbe likely to be involved. This approach

TABLE 3.8   Classes of Bacteria Considered to Be Important in Oilfield MIC

Bacteria	Oxygen Requirement	Metals Affected
Desulfovibrio	Anaerobic	Carbon steel, stainless steel, zinc, copper, aluminum
Desulfomonas	Anaerobic	Carbon steel
Thiobacillus thiooxidans	Aerobic	Carbon steel, copper
Thiobacillus ferrooxidans	Aerobic	Carbon steel
Gallionella	Aerobic	Carbon steel, stainless steel
Sphaerotilus	Aerobic	Carbon steel, stainless steel
Pseudomonas	Aerobic	Carbon steel, stainless steel

Source: From Wheeler and Adams [110].

has been discredited in recent years [84, 85]. One authority claims that MIC identification requires that all other possible explanations for the observed corrosion must be eliminated before the problem can be attributed to MIC [85].

Field personnel often rely on collecting water samples for planktonic bacteria or probes to collect sessile bacteria. They then follow prescribed procedures to allow the growth and identification of the collected bacteria [102].

Culture methods are commonly applied for detecting and estimating bacterial numbers in the petroleum industry. These methods rely on supplying all the requirements for microbial growth for the bacteria and then use visual means to determine the presence of these bacteria. The media must supply an energy source, a carbon source, and trace elements, as well as maintain the proper pH, temperature, and oxygen requirements for the microorganisms. The method generally used in the field involves the use of small glass bottles containing media sealed with a rubber septum. The sample from the system is injected using a hypodermic needle into the first bottle. After mixing, a diluted sample is withdrawn from this bottle and injected into the next bottle in the series [95].

Subsequent dilutions are made and the bottles are then incubated and observed at intervals for clouding which would indicate microbial growth. This method is known as the *"Serial Dilution or Dilution-to-Extinction Method"* since the first clear bottle indicates how many dilutions are required to achieve an absence of microorganisms in the sample.

The ideal procedure for estimating bacterial numbers is to use a triplicate dilution series then incorporate statistically valid Most Probable Number (MPN) techniques to estimate bacterial numbers. At least a duplicate series of dilutions should be used in order to avoid the possibility of error created by contaminated samples. In practice, however, most oilfield personnel use a single dilution series.

A large number of media are available for performing Serial Dilution Tests. These allow the estimation of numbers of total heterotrophic populations, anaerobic populations, various types of SRB, iron bacteria, and other organisms of interest.

However, it should be kept in mind that culture media detect only the number of bacteria in a sample, which will grow in the culture media. They do not represent an unequivocal estimation of total or specific bacterial numbers.

This method is a powerful tool for investigation of oil field problems, but should not be used as the sole source of information for assessing system characteristics [103].

The same reference that supplied the above quote found that field test kits were unreliable, at least under the conditions tested.

The collection and identification of sessile bacteria is much more difficult than for planktonic bacteria, because the locations where sessile bacteria are located can vary, and the types of bacteria under different biofilms within a piping system may also vary.

The following situations and locations are likely to be associated with MIC corrosion problems:

- **Stagnant conditions**: Hydrotesting and dead legs are two common sources of MIC problems [85]. Improperly drained hydrotest water can produce measurable corrosion within two weeks. Consumption of biocides in dead legs or in low spots of hydrotested equipment cannot be expected to last for extended periods.
- **Welds** are locations of surface roughness conducive to the growth of bacterial colonies protected from the action of biocides. Why welds are preferentially corroded by MIC is not completely understood, but there are many reports to indicate that this phenomenon is real [85, 104–106].
- **Particulates** are places for bacteria to attach. They also settle in low-velocity locations providing deposits under which biofilms can form

and grow. The increased surface area associated with particulate production can also remove biocides from fluids and increase the demand for biocides.

There are many types of equipment that develop MIC problems, but pipelines and piping systems predominate in reports on MIC problems. Any stagnant water location is likely to develop this problem. Improperly treated injection waters are known to produce reservoir souring when surface bacteria proliferate in the downhole or produced water environment.

MIC can be controlled through a number of methods such as:

- Regular mechanical cleaning if possible.
- Chemical treatment of the water with biocides to control the population of bacteria.
- Complete drainage and dry storage.
- Use of higher alloyed stainless steels, although this may be the most expensive route.
- Filtration and ultraviolet irradiation has been demonstrated, but it is unlikely to become an economically viable means of MIC control.

The most common means of control is through the use of a combination of mechanical cleaning and biocides. Without the necessary mechanical cleaning, biocides are unlikely to reach sessile bacteria, which are shielded by scale, sludge, and biofilm deposits.

It is easy for a chemical biocide to kill planktonic bacteria as they are unprotected by scale, debris, or biofilms. In the sessile environment abrasive pigging as well as chemical biocide treatment may be required to disinfect a system.

Pigging or other mechanical cleaning methods are a prime means of controlling MIC. Biofilms and other deposits on metal surfaces can shield microbes from chemical treatments, and their removal is necessary to ensure that piping and similar structures are efficiently treated by biocide injection.

Biocides are used to kill or render harmless biological organisms. Biocides can be either oxidizing or nonoxidizing chemicals. Hypochlorite, an oxidizing biocide that is often used as a biocide in drinking water, is also used in oilfield waters. It can be released from gaseous chlorine generators, from chemical injection, or by electrolytic chlorine generation. It often produces faster action than nonoxidizing biocides [79]. Unfortunately, chlorine and other oxidizing agents from any source also oxidize other species in oilfield brines – organic acids, soluble iron, and H_2S, and this oxidation can produce solids, which must be removed from treated water before injection. Chlorine

TABLE 3.9   Compatibility of Common Biocides with Various Metals and Elastomers [93]

Biocide	Compatible[a]	Incompatible
Quaternary amines	UNS S31603 (Type 316L SS) Polyvinylchloride Polyolefin PTFE Polyvinylfluoroethylene Perfluoroelastomer Vinyl ester	Carbon steel (CS) Natural rubber Neoprene Acrylonitrile-butadiene rubber (NBR)
Glutaraldehyde	Stainless steel (SS) Polyethylene Reinforced plastics	CS Galvanized iron Aluminum (Al) Tin (Sn) Zinc (Zn)
Acrolein	SS Butyl rubber Perfluoroelastomer PTFE Polyethylene Polypropylene	Neoprene Fluoroelastomer Acrylonitrile-butadiene rubber (NBR) Polyvinylchloride Polyurethane Galvanized metals
Isothiazolone	UNS S31603 Fiberglass-reinforced epoxy Polyester Vinyl ester Polyethylene Polypropylene PTFE Hydrocarbon rubber Fluoroelastomer Polyphenylene sulfide (PPS)	
THPS	SS Al Polyvinylchloride Nylon PTFE Polyethylene Polypropylene Polyurethane Silicone Fluoroelastomer Nitrile rubber Natural rubber	Copper Brass Mild steel Cast iron Zn
DBNPA	Fluoroelastomer PTFE Polyethylene Polypropylene Polyvinylfluoroethylene Fiberglass-reinforced plastic	Mild steel UNS S30400 (Type 304 SS) Al Nickel (Ni)

[a] Compatible with field strength product at ambient temperature. Compatibilities are typically verified under use concentration and conditions.
Source: Copy of table 3 NACE Publication 31205. Courtesy of NACE International.

also responds to the oxygen scavengers used in injection water treatment systems and can accelerate corrosion of metals and degradation of gaskets and seals.

Nonoxidizing inhibitors are usually more cost effective in systems with low hydrocarbon contents (injection water and freshwater makeup systems). Table 3.9 lists several nonoxidizing inhibitors and their compatibility with various materials used in their handling and injection equipment.

Many commercial biocides are blends of various chemicals. The reason for this is that different microbes

Figure 3.43 MIC on an ocean-bottom mooring chain [37]. *Source*: Reproduced with permission of Elsevier.

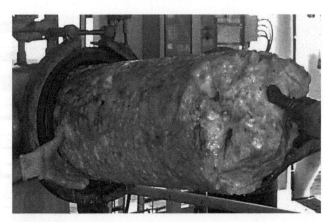

Figure 3.44 Hydrate plug removed from a subsea pipeline. *Source*: Photo courtesy Dendy Sloan, Colorado School of Mines.

respond to different biocides, and it is virtually impossible to identify all of the microbes present in a given water. These biocides are marketed as "broad-spectrum" biocides and are often quite effective. An alternative to broad-spectrum biocides is alternating biocides with the intention of preventing the buildup of biomasses that are resistant to the biocide being used.

The major problems associated with MIC are in low-flow areas and places where deposits of any type are allowed to form. Effective treatment for MIC requires mechanical cleaning as well as biocide treatment.

While this discussion has focused on MIC as an internal corrosion problem, it can also damage external equipment surfaces on pipelines, storage tanks, and similar equipment exposed to soil or water immersion [37, 94]. Figure 3.43 shows a mooring chain that suffered MIC near bacteria- and deposit-supporting ocean-bottom sediments.

Mercury

Mercury is an element found in trace amounts in most hydrocarbon formations [113, 114]. It can cause liquid metal cracking problems in brazed aluminum heat exchangers used for offshore gas processing, and this is the reason why mercury removal systems are commonly placed before the aluminum heat exchangers on offshore processing platforms [114]. Mercury forms very inefficient cathodic surfaces when condensed on steel and is not a serious galvanic corrosion problem [115]. The principle threat of mercury, or any liquid metal, is liquid metal embrittlement, a form of SCC.

Hydrates

Hydrates are icelike deposits that can form in natural gas systems [116]. Methane networks "trap" water and form very hard plugs like those shown in Figure 3.44. Hydrates

are a major flow assurance problem in subsea gas and multiphase pipelines. Thermodynamic inhibitors, which lower the temperatures at which hydrates can form, are a common method of preventing hydrate formation in subsea pipelines. The most common thermodynamic inhibitors are methanol, monoethylene glycol (MEG), and diethylene glycol (DEG). Corrosion inhibitors must be tested for compatibility with these hydrate inhibitors.

Fluid Flow Effects on Corrosion

Figure 3.45 shows various fluid flow regimes common in oilfield tubing, pipelines, and process equipment. Fluid flow regimes determine where erosion–corrosion, underdeposit corrosion, or other forms of corrosion are likely to occur. The flow regimes also determine the type of wetting that occurs, e.g. in pipelines and piping where top-of-the-line corrosion may occur due to the presence of corrosion-causing condensation in locations where corrosion inhibitors have not been applied. It would not be unusual for a gas well to progress from single-phase liquid flow at the bottom of a well to single-phase vapor flow as the tubing progressed from temperature and pressure conditions at the bottom of the well to the lower temperatures and pressures near the surface [117]. Additional discussions on fluid flow aspects of corrosion are presented in Chapters 5 and 8.

SUMMARY

Oilfield environments vary by location within upstream oil and gas operations. Liquid water wetting is necessary for corrosion, and the solids and gases dissolved in water strongly affect corrosivity. The three primary corrosion problems in oilfield waters are

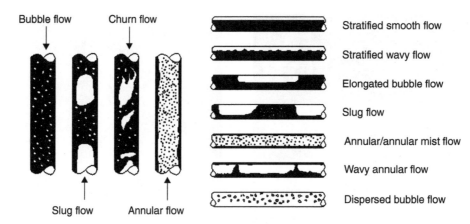

Figure 3.45 Two-phase fluid flow regimes common in oilfield tubing and piping. The dark areas represent liquid and the light areas are gas.

oxygen in surface equipment and CO_2 and H_2S from production fluids. Crude oil generally produces a benign environment with relatively few corrosion problems until the field ages and the water cut increases to the point that liquid water wetting starts to occur. In contrast to crude oil, gas wells and piping tend to be corrosive from the beginning. This is due to the relatively low mineral content of condensed waters that allows any dissolved acid gases, usually CO_2 and/ or H_2S, to lower the pH and cause corrosion. The presence of scales (mineral deposits on surfaces) and microbes influence corrosion but are less important than the dissolved gases – O_2, CO_2, and H_2S.

REFERENCES

1 Papavinasam, S. (2014). *Corrosion Control in the Oil and Gas Industry*. Houston, TX: Gulf Professional Publishing.

2 Chillingar, G., Mourhatch, B., and Al-Qahtani, G. (2008). *The Fundamentals of Corrosion and Scaling for Petroleum and Environmental Engineers*. Houston, TX: Gulf Publishing.

3 Becker, J. (1998). *Corrosion and Scale Handbook*. Tulsa, OK: PennWell.

4 Craig, B. (August 1996). Corrosion in oil/water systems. *Materials Performance* 39 (8): 61–62.

5 de Waard, C., Smith, L., and Craig, B. (2003). *The Influence of Crude Oils on Well Tubing Corrosion Rates*, NACE 03629. Houston, TX: NACE International.

6 Shankardass, A. (2004). *Corrosion Control in Pipelines Using Oxygen Stripping. Oilsands Water Usage Workshop 2004*. Edmonton, Alberta, Canada: CONRAD: Canadian Oil Sands Network for Research and Development. https:// www.scribd.com/document/241667826/Corrosion-Control-in-Pipelines-Using-Oxygen-Stripping-Shankardass (accessed 12 April 2008).

7 Anonymous (October 1990). *Corrosion of Oil and Gas-Well Equipment*, API Book 2 of the Vocational Training Series, 2. Washington, DC: API.

8 Kane, R. (2006). Corrosion in petroleum production operations. In: *Metals Handbook, Volume 13C – Corrosion in Specific Industries*, 922–966. Materials Park, OH: ASM International.

9 Iannuzzi, M. (2011). Chapter 15: Environmentally-assisted cracking in oil and gas production. In: *Stress Corrosion Cracking: Theory and Practice* (ed. V. Raja and T. Shoji), 570–607. Cambridge, UK: Woodhead Publishing, Ltd.

10 Jangama, V. and Srinivasan, S. (2008). Calibration of an integrated model for prediction of corrosivity of CO_2/H_2S environments. http://www.corrosionsource.com/events/intercorr/techsess/papers/session7/abstracts/vamshi.html (accessed 10 November 2008).

11 Revie, W.R. and Uhlig, H.H. (2008). *Corrosion and Corrosion Control*, 4. Hoboken, NJ: Wiley-Interscience.

12 Vernon, J. (1935). A laboratory study of the atmospheric corrosion of metals: part II, iron – the primary oxide film. Part III, the secondary product of rust influence of sulphur dioxide, carbon dioxide, and suspended particles on the rusting of iron. *Transactions of the Faraday Society* 31: 1668–1700.

13 Bayliss, D.A. and Deacon, D.H. (2002). *Steelwork Corrosion Control*, 10–11. London: CRC Press/Taylor & Francis Group.

14 Schindelholtz, E., Kelly, R.G., Cole, I.S. et al. (2013). Comparability and accuracy of time of wetness sensing methods relevant for atmospheric corrosion. *Corrosion Science* 67: 233–241.

15 ASTM G50. *Atmospheric Corrosion Tests on Metals*. West Conshohocken, PA: ASTM International.

16 ASTM G84. *Time-of-Wetness on Surfaces Exposed to Wetting Conditions as in Atmospheric Testing*. West Conshohocken, PA: ASTM International.

17 ASTM G92. *Characterization of Atmospheric Test Sites*. West Conshohocken, PA: ASTM International.

18 ISO 9223. *Corrosivity of Atmospheres – Classification, Determination and Estimation*. Geneva: ISO.

19 ISO 9225. *Corrosivity of Atmospheres – Measurement of Environmental Parameters Affecting Corrosivity of Atmospheres*. Geneva: ISO.

20 ISO 9227. *Corrosion Tests in Artificial Atmospheres – Salt Spray Tests*. Geneva: ISO.

21 NORSOK (2002). *Standard M-001, Materials Selection*. Lysaker, Norway: Standards Norway.

22 Roberge, P. (2008). *Corrosion Engineering – Principles and Practice*. New York: McGraw-Hill.

23 Uhlig, H.H. (1948). *The Corrosion Handbook*, 131. New York: Wiley.

24 Groysman, A. (2010). *Corrosion for Everybody*. New York: Springer.

25 Fredj, N., Burleigh, T.D., Heidersbach, K.L., and Crowder, B.R. (2012). *Corrosion of Carbon Steel in Waters of Varying Purity and Velocity*, NACE C2012-0001461. Houston, TX: NACE International.

26 Baboian, R. and Treseder, R. (eds.) (2002). *NACE Corrosion Engineer's Reference Book*, 3. Houston, TX: NACE International.

27 LaQue, F. (1975). *Marine Corrosion: Cause and Prevention*, 116. New York: Wiley.

28 Romanoff, M. (1989). *Underground Corrosion*. Houston, TX: NACE International.

29 Bianchetti, R. (2001). Chapter 5: Survey methods and evaluation techniques. In: *Peabody's Control of Pipeline Corrosion*, 2 (ed. R. Bianchetti), 49–64. Houston, TX: NACE International.

30 Roberge, P. (2006). *Corrosion Basics: An Introduction*, 2. Houston, TX: NACE International.

31 Peabody, A.W. (1967). *Control of Pipeline Corrosion*, 7. Houston, TX: NACE International.

32 Ricker, R.E. (2007). *Analysis of Pipeline Steel Corrosion Data from NBS (NIST) Studies Conducted Between 1922–1940 and Relevance to Pipeline Management*, NISTIR 7415 (2 May 2007). Gaithersburg, MD: National Institute of Standards and Technology.

33 Logan, K.H. (1945). *Underground Corrosion*. Washington, DC: National Bureau of Standards.

34 NACE RP0198. *The Control of Corrosion Under Thermal Insulation and Fireproofing Materials – A Systems Approach*. Houston, TX: NACE International.

35 API RP 583. *Corrosion Under Insulation and Fireproofing*. Washington, DC: API.

36 ISO 12736. *Wet Thermal Insulation Coatings for Pipelines, Flow Lines, Equipment and Subsea Structures*. Geneva: ISO.

37 Winnick, S. (ed.) (2015). *EFC 55 – Corrosion Under-Insulation (CUI) Guidelines*, 2. Cambridge, UK: Woodhead Publishing.

38 NACE (2006). *Effectiveness of Cathodic Protection on Thermally Insulated Underground Metallic Structures*, NACE Publication 10A392. Houston, TX: NACE International.

39 Gibson, S., Hogarth, M., and Crone, L. (2017). *Challenges in Providing Effective Cathodic Protection to Thermally Insulated Pipeline Risers*, NACE C2017, Product Number 51317-9550-SG. Houston, TX: NACE International.

40 Soman, A.K. (2016). CUI detection techniques for process pipelines (Part 2) (11 March 2016). https://www.corrosionpedia.com/2/5364/corrosion-under-insulation-cui/cui-detection-techniques-for-process-pipelines-part-2 (accessed 23 May 2017).

41 ASTM G205. *Guide for Determining Corrosivity of Crude Oils*. West Conshohocken, PA: ASTM International.

42 ASTM D665. *Test Method for Rust-Preventing Characteristics of Inhibited Mineral Oil in the Presence of Water*. West Conshohocken, PA: ASTM International.

43 NACE TM0172. *Determining Corrosive Properties of Cargoes in Petroleum Product Pipelines*. Houston, TX: NACE International.

44 Yari, M. (2017). The 6 corrosive components that can be found in crude oil (1 May 2017). https://www.corrosionpedia.com/2/1424/corrosion/the-6-corrosive-components-that-can-be-found-in-crude-oil (accessed 23 May 2017).

45 ISO 21457. *Materials Selection and Corrosion Control for Oil and Gas Production Systems*. Geneva: ISO.

46 American Petroleum Institute (1990). *Corrosion of Oil-and Gas-Well Equipment*, 2. Dallas: API.

47 Speller, F.N. (1951). *Corrosion: Causes and Prevention: an Engineering Problem*, 2, 168. New York: McGraw-Hill.

48 DNV Report No. 2006-3496 (2006). *Material Risk-Ageing Offshore Installations*. Oslo, Norway: DNV GL.

49 *DownHole SAT and DownHole Rx Series Product Information*. http://www.frenchcreeksoftware.com (accessed 12 April 2018).

50 de Waard, C. and Lotz, U. (1993). *Prediction of CO_2 Corrosion of Carbon Steel*, NACE 93069. Houston, TX: NACE International.

51 de Waard, C., Lotz, U., and Dugstad, A. (1995). *Influence of Liquid Flow Velocity on CO_2 Corrosion: A Semi-Empirical Approach*, NACE 95128. Houston, TX: NACE International.

52 Pots, B., John, R., Rippon, I. et al. (2002). *Improvements on de Waard-Milliams Corrosion Prediction and Applications to Corrosion Management*, NACE 02235. Houston, TX: NACE International.

53 Smart, J. (September 2001). A method for calculating the corrosion allowance for deepwater pipelines and risers. *Journal of Pipeline Integrity* 1 (1): 73.

54 Schmidt, G. (1984). CO_2 corrosion of steels: an attempt to range parameters and their effects. In: *Advances in CO_2 Corrosion* (ed. R.H. Hausler and H. Godard), 1–9. Houston, TX: NACE International.

55 NORSOK. *Standard M-506, CO_2 Corrosion Rate Calculation Model*. Lysaker, Norway: Standards Norway.

56 Nyborg, R. (2010). *CO_2 Corrosion Models for Oil and Gas Production Systems*, NACE2010-10371. Houston, TX: NACE International.

57 Wang, H., Cai, J.-Y., and Jepson, W.P. (2002). *CO_2 Corrosion Mechanistic Modeling and Prediction in Horizontal Slug Flow*, NACE 02238. Houston, TX: NACE International.

58 Brondel, D., Edwards, R., Hayman, A. et al. (1994). *Corrosion in the oil industry*. *Oilfield Review* (April): 4–18.

59 Jackman, P.S. and Smith, L.M. (1999). *Advances in Corrosion Control and Materials in Oil and, CO$_2$ Corrosion in Oil and Gas Production*, EFC Report 26. London: The Institute of Materials.

60 ISO 17348. *Materials Selection for High Content CO$_2$ for Casing, Tubing and Downhole Equipment*. Geneva: ISO.

61 ISO 17349. *Offshore Platforms Handling Streams with High Content of CO$_2$ at High Pressures*. Geneva: ISO.

62 Smith, L. and Craig, B. (2005). *Practical Corrosion Control Measures for Elemental Sulfur Containing Environments*, NACE 05646. Houston, TX: NACE International.

63 El-Raghy, S.M., Wood, B., Abuleil, H. et al. (1998). *Microbiologically Influenced Corrosion in Mature Oil Fields – A Case Study in El-Morgan Field in the Gulf of Suez*, NACE 98279. Houston, TX: NACE International.

64 Milliams, D.E. and Tuttle, R.N. (2003). *ISO 15156/NACE MR0175 – A New International Standard for Metallic Materials for Use in Oil and Gas Production in Sour Environments*, NACE 03090. Houston, TX: NACE International.

65 Phelps, E.H. (1981 and 1994). Stress corrosion of ferritic-martensitic stainless steels. In: *H$_2$S Corrosion in Oil & Gas Production* (ed. R. N. Tuttle and R. D. Kane), 352–357. Houston, TX: NACE International.

66 ANSI/NACE MR0175/ISO 15156-1. *Petroleum and Natural Gas Industries – Materials for Use in H$_2$S-Containing Environments in Oil and Gas Production – Part 1: General Principles for Selection of Cracking-Resistant Materials*. Geneva: ISO.

67 ANSI/NACE MR0175/ISO 15156-2. *Petroleum and Natural Gas Industries – Materials for Use in H$_2$S-Containing Environments in Oil and Gas Production – Part 2: Cracking-Resistant Carbon and Low-Alloy Steels, and the Use of Cast Irons*. Geneva: ISO.

68 ANSI/NACE MR0175/ISO 15156-3. *Petroleum and Natural Gas Industries – Materials for Use in H$_2$S-Containing Environments in Oil and Gas Production – Part 3: Cracking-Resistant CRAs (Corrosion-Resistant Alloys) and Other Alloys*. Geneva: ISO.

69 Eliassen, S. and Smith, L. (2009). *Guidelines on Materials Requirements for Carbon and Low Alloy Steels for H$_2$S-Containing Environments in Oil and Gas Production*, EFC Report 16, 3. Leeds, UK: Maney Publishing.

70 Anonymous (European Federation of Corrosion Publications) (2002). *Corrosion Resistant Alloys for Oil and Gas Production: Guidance on General Requirements and Test Methods for H$_2$S Service*, EFC Report 17. Leeds, UK: Maney Publishing.

71 NACE MR0103 (2003). *Materials Resistant to Sulfide Stress Cracking in Corrosive Petroleum Refining Environments*. Houston, TX: NACE International.

72 Bush, D.R., Brown, J.C., and Lewis, K.R. (2004). Introduction to NACE Standard MR0103. *Hydrocarbon Processing* (November): 73–77.

73 NACE TM0177. *Laboratory Testing of Metals for Resistance to Sulfide Stress Cracking and Stress Corrosion Cracking in H$_2$S Environments*. Houston, TX: NACE International.

74 Joosten, M., Kolts, J., Hembree, J., and Achour, M. (2002). *Organic Acid in Oil and Gas Production*, NACE 02294. Houston, TX: NACE International.

75 Fajardo, V., Canto, C., Brown, B., and Nesic, S. (2007). *Effect of Organic Acids in CO$_2$ Corrosion*, NACE 07319. Houston, TX: NACE International.

76 Andersen, T.R., Halvorsen, A.M.K., Valle, A. et al. (2007). *The Influence of Condensation Rate and Acetic Acid Concentration on TOL-Corrosion in Multiphase Pipelines*, NACE 07312. Houston, TX: NACE International.

77 Davies, M. and Scott, P.J.B. (2006). *Oilfield Water Technology*. Houston, TX: NACE Press.

78 GE. *Handbook of Industrial Water Treatment*. http://www.gewater.com/handbook/index.jsp (accessed 13 June 2016).

79 Flynn, D. (2009). *Nalco Water Handbook*. New York: McGraw-Hill.

80 Lane, R. (1993). *Control of Scale and Corrosion in Building Water Systems*. New York: McGraw-Hill.

81 NACE Publication 31105. *Dynamic Scale Inhibitor Evaluation Apparatus and Procedures in Oil and Gas Production*. Houston, TX: NACE International.

82 Smith, K.P. (December 1993). *An Overview of Naturally Occurring Radioactive Materials (NORM) in the Petroleum Industry*, Argonne National Laboratory Report ANL/EAIS-7. Argonne, IL: Argonne National Laboratory.

83 Skovhus, T., Enning, D., and Lee, J. (eds.) (2017). *Microbiologically Influenced Corrosion in the Upstream Oil and Gas, Industry*. Boca Raton, FL: CRC Press.

84 Little, B.J. and Lee, J. (2007). *Microbiologically Influenced Corrosion*. New York: Wiley-Interscience.

85 Javaherdashti, R. (2017). *Microbiologically Influenced Corrosion: An Engineering Insight*, 2. London: Springer-Verlag.

86 Videla, H.A. (1996). *Manual of Biocorrosion*. Boca Raton, FL: CRC Press.

87 Dexter, S.C. (ed.) (1986). *Biologically Influenced Corrosion*. NACE Reference Book No. 8. Houston, TX: NACE International.

88 Kearns, J.R. and Little, B.J. (eds.) (1994). *Microbiologically Influenced Corrosion Testing, ASTM STP 1232*. Philadelphia, PA: ASTM.

89 Borenstein, S.W. (1994). *Microbiologically Influenced Corrosion Handbook*. Cambridge, UK: Woodhead Publishing, Ltd.

90 Kobrin, G. (ed.) (1993). *A Practical Manual on Microbiologically Influenced Corrosion*. Houston, TX: NACE International.

91 Stoecker, J.G. (ed.) (2001). *A Practical Manual on Microbiologically Influenced Corrosion*, vol. 2. Houston, TX: NACE International.

92 NACE TPC 3 (1990). *Microbiologically Influenced Corrosion and Biofouling in Oilfield Equipment*. Houston, TX: NACE International.

93 NACE TR 31205 (February 2006). *Selection, Application, and Evaluation of Biocides in the Oil and Gas Industry*. Houston, TX: NACE International.

94 NACE TM0106-2016 (2016). *Detection, Testing and Evaluation of Microbiologically Influenced Corrosion*

(MIC) on External Surfaces of Buried Pipelines. Houston, TX: NACE International.

95   NACE TM0194-2014 (2014). *Field Monitoring of Bacterial Growth in Oil and Gas Systems.* Houston, TX: NACE International.

96   NACE TM0212-2012 (2012). *Detection, Testing, and Evaluation of Microbiologically Influenced Corrosion on Internal Surfaces of Pipelines.* Houston, TX: NACE International.

97   NACE SP0499. *Corrosion Control and Monitoring in Seawater Injection Systems.* Houston, TX: NACE International.

98   Tiller, A.K. and Sequeira, C.A.C. (eds.) (1994). *Microbial Corrosion, EFC Report 15.* Leeds, UK: Maney Publishing.

99   Eckert, R. (2016). *Introduction to Corrosion Management of Microbiologically Influenced Corrosion.* Houston, TX: NACE International.

100 Little, B.J., Lee, J.S., and Ray, R.I. (2006). Diagnosing microbiologically influenced corrosion: a state of the art review. *Corrosion* 62 (11): 1006–1017.

101 Herro, H.M. (1998). *MIC Myths – Does Pitting Cause MIC?*, NACE 98278. Houston, TX: NACE International.

102 NACE TM0194. *Field Monitoring of Bacterial Growth in Oil and Gas Systems.* Houston, TX: NACE International.

103 Al-Sulaiman, S., Al-Mithin, A.W., Murray, G. et al. (2008). *Advantages and Limitations of Using Field Test Kits for Determining Bacterial Proliferation in Oil Field Waters*, NACE 08655. Houston, TX: NACE International.

104 Sreekumari, K.R., Nandakumar, K., and Kiuchi, Y. (2004). *Effect of Metal Microstructure on Bacterial Attachment: A Contributing Factor for Preferential MIC Attack of Welds*, NACE 04597. Houston, TX: NACE International.

105 Walsh, D.W. and Willis, E. (1995). The effect of weld thermal cycling on microbial interaction in low alloy steels. *Trends in Welding Research, Proceedings of the 5th International Conference*, Gatlinburg, TN, pp. 579–587.

106 Walsh, D. (1999). *The Implications of Thermomechanical Processing for Microbially Influenced Corrosion*, NACE 99188. Houston, TX: NACE International.

107 Penkala, J., Fichter, J., and Ramachandran, S. (2010). *Protection Against Microbiologically Influenced Corrosion by Effective Treatment and Monitoring During Hydrotest Shut-in*, NACE 2010-10404. Houston, TX: NACE International.

108 Lewandowski, Z. and Beyenal, H. (2009). Mechanisms of microbially influenced corrosion. In: *Marine and Industrial Biofouling* (ed. H.C. Flemming, P.S. Murthy, R. Venkatesan and K.E. Cooksey), 35–65. Berlin: Springer.

109 Little, B.J., Wagner, P., and Mansfeld, F. (1997). *Microbiologically Influenced Corrosion, Volume 5, Corrosion Testing Made Easy.* Houston, TX: NACE International.

110 Wheeler, C.L. and Adams, D.L. (2009). Failure analysis of microbiologically influenced corrosion in middle eastern applications. *Proceedings, 5th Middle East Artificial Lift Forum (MEALF)*, Bahrain (16–19 February 2009).

111 Tatnall, R.E. and Pope, D.H. (1993). Chapter 8: Identification of MIC. In: *A Practical Manual on Microbiologically Influenced Corrosion* (ed. G. Kobrin). Houston, TX: NACE International.

112 Darwin, A., Annadorai, K., and Heidersbach, K. (2010). *Prevention of Corrosion in Carbon Steel Pipelines Containing Hydrotest Water – An Overview*, NACE 10401. Houston, TX: NACE International.

113 Wilhelm, S.M. and Bloom, N.S. (2000). Mercury in petroleum. *Fuel Processing Technology* 63 (1): 1–27.

114 Anonymous (2014). *Gas Conditioning and Processing, Volume 1 – The Basic Principles*, 9. Norman, OK: John M. Campbell Company.

115 Wilhelm, S.M. and Hill, D.M. (2008). Galvanic corrosion of steel coupled to liquid elemental mercury in pipelines. *Journal of Corrosion Science and Engineering* 11 (Preprint 6) (26 August). http://www.jsce.org (accessed 4 June 2009).

116 Dendy Sloan, E. and Koh, C.A. (1998). *Clathrate Hydrates of Natural Gases*, 2. Boca Raton, FL: CRC Press.

117 Lyons, W.C. and Plisga, G.J. (2005). *Standard Handbook of Petroleum & Natural Gas Engineering*, 2, 6-44–6-46. Boston: Gulf Publishing.

4

MATERIALS

The most important materials used in oil and gas production are carbon steels. Oilfield corrosion control of carbon steels has traditionally used corrosion inhibitors for internal corrosion and a combination of protective coatings and cathodic protection for external corrosion. Starting in the 1980s many new production environments became too aggressive for this approach, and the use of corrosion-resistant alloys (CRAs) has increased [1, 2]. Polymers and flexible polymers, called elastomers, are included in this chapter, but most of the discussion will be concerned with metals.

The discussions of mechanical properties and other properties of metals concentrate on carbon steels, and it is estimated that approximately 90% of all materials used in oilfield applications are carbon steels. This situation is unlikely to change in the foreseeable future. The principles of heat treatment, welding, and other operations for carbon steels are similar to those for other alloys.

The uses of metals and polymers for cathodic protection anodes and for protective coatings are discussed in Chapter 6.

METALLURGY FUNDAMENTALS

Virtually all metals used by industry are alloys. The exceptions are coatings, which sometimes involve commercially pure metals, and conductors. Pure metals are better electrical and thermal conductors than alloys. Strength considerations (stiffness) of heat transfer tubing usually mean that alloys are used in heat exchangers and similar devices, but pure copper, and sometimes other metals, is used for electrical conductors. These conductors are not unique to oilfield applications and will not be discussed in this book.

Crystal Structure

Most solids, with the exception of glasses and organic materials, are crystalline. This means that the atoms in the crystal are arranged in one of seven possible arrangements, only three of which are common in metals [3]. Figure 4.1 shows the most common crystal arrangements: body-centered cubic (BCC), found in low-temperature iron; face-centered cubic (FCC), found in high-temperature iron, aluminum, and austenitic stainless steel; and hexagonal close-packed (HCP), found in zinc and titanium.

The type of crystal structure, defects in the crystal structure, and size of the crystals combine to determine the mechanical, and to some extent corrosion resistance, properties of oilfield alloys. Note how the FCC crystal has 50% more close-packed directions (directions where the atoms "touch" their nearest neighbors) than the BCC crystals. This is why FCC metals, to include high-temperature iron and carbon steel, are both weaker and more ductile than BCC metals.

A solid metal consists of many crystals containing numerous defects. The combination of alloying additions, crystal size, and different crystal structures determines the mechanical and corrosion resistance properties of the alloy. Figure 4.2 shows how three crystals, with different orientations but the same chemistry, join and

Metallurgy and Corrosion Control in Oil and Gas Production, Second Edition. Robert Heidersbach.
© 2018 John Wiley & Sons, Inc. Published 2018 by John Wiley & Sons, Inc.

(a)   (b)   (c)

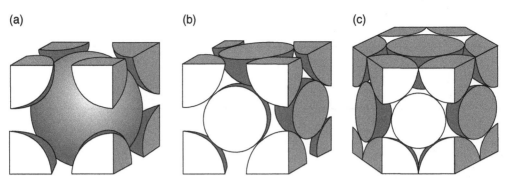

Figure 4.1   Crystal structures found in metals: (a) body-centered cubic, (b) face-centered cubic, (c) hexagonal close-packed.

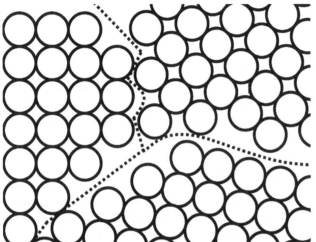

Figure 4.2   Grain boundaries in a metal.

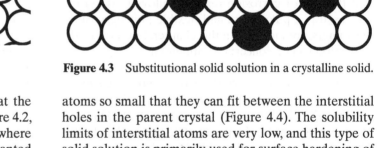

Figure 4.3   Substitutional solid solution in a crystalline solid.

form grain (crystal) boundaries. It is obvious that the grain boundaries, indicated by dotted lines in Figure 4.2, have larger spaces between the atoms, and this is where impurity atoms are most likely to be located. Unwanted segregation of impurity atoms to grain boundaries can cause major embrittlement and corrosion problems.

A typical crystal has millions of atoms, and the defects in the individual crystals determine some of their mechanical and corrosion resistance properties. Defects in crystalline solids include vacancies (locations where atoms are missing from the crystal), impurities (atoms of different elements than the base metal), and grain boundaries between crystals. These defects affect the mechanical properties and corrosion resistance of metals. Detailed discussions of defects, and how they affect the strength and other properties of metals, are available in metallurgical texts [4].

Most metals are alloys, deliberate combinations of two or more elements to improve achieve the desired properties. Alloying additions can produce substitutional solid solutions when the atomic sizes of the solute and solvent atoms are similar (Figure 4.3) or interstitial solid solutions when the alloying addition involves

atoms so small that they can fit between the interstitial holes in the parent crystal (Figure 4.4). The solubility limits of interstitial atoms are very low, and this type of solid solution is primarily used for surface hardening of steel using carbon, oxygen, boron, or nitrogen.

If the solubility limits of the secondary atoms are exceeded, a different crystal structure is formed, and this produces a stronger alloy, because atomic motion caused by stresses is impeded whenever the atoms reach a grain boundary and must change direction.

Material Defects, Inclusions, and Precipitates

Oilfield materials are produced and used in tonnage quantities. The economics of production for high-volume materials means that all oilfield materials and equipment will have defects that may affect their performance. Unlike the electronics industry, where ultrapure materials are necessary, the oil and gas industries have learned to recognize and tolerate some imperfections in the materials they use. If a defect becomes critical to the performance of the metal in question, then means of processing to remove the

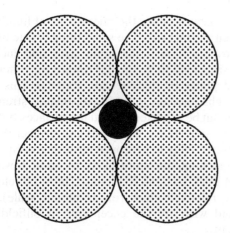

Figure 4.4 Interstitial solid solution in a crystalline solid.

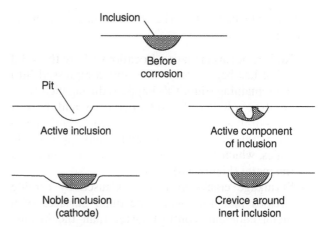

Figure 4.5 Effects of surface inclusions on corrosion. *Source*: Davis [5]. Reprinted with permission of ASM international.

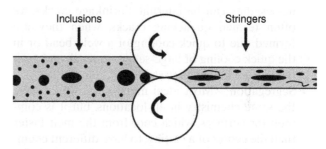

Figure 4.6 Formation of stringers parallel to the rolling direction in metal plate.

defects are necessary. This section discusses some of these processing defects and how they affect performance in the field.

Alloys are intentional mixtures of elements to gain desired properties. The microstructure of an alloy can contain multiple phases (different crystals having either different crystal structures or different chemistries, often both), and the distribution and amount of the second phases or precipitates are controlled to develop desired properties, for example, increased strength or toughness. Other second-phase particles can be undesirable. Examples of undesirable phases are oxides and sulfides, which precipitate in the metal from dissolved oxygen and sulfur in the metal-producing process. This results in a distribution of inclusions (small particles of oxide, sulfide, etc.) throughout the alloy.

When these inclusions are exposed at the metal surface to a corrosive environment, they can affect corrosion behavior. The effects of inclusions at the metal surface are shown schematically in Figure 4.5. The uppermost image represents an inclusion exposed at the metal surface prior to corrosion, and the lower images indicate the behavior under different conditions. If the inclusion is active, that is, less corrosion resistant than the matrix, then the inclusion dissolves, leaving a hole or pit in the metal surface. If only portions of the inclusion are active, then the exposed portions are attacked, leaving the other portions intact. If the inclusion is noble (more corrosion resistant than the matrix), then accelerated attack of the matrix adjacent to the noble inclusion can be observed. In other cases where the surface inclusion is inert to attack, accelerated corrosion adjacent to the inclusion can still occur because of a crevice generated between the inclusion and the matrix.

Subsurface inclusions can affect mechanical properties, and silicate inclusions in the form of stringers (long inclusions parallel to the forming direction) can develop parallel to the rolling or forming direction (Figure 4.6).

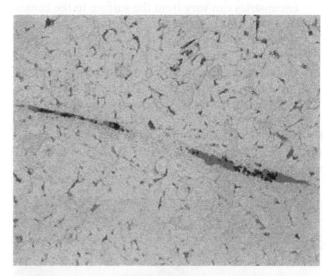

Figure 4.7 Sulfide stringer inclusion in a lap-welded carbon steel natural gas pipeline [6]. *Source*: Reproduced with permission of Corrosion Testing Laboratories, Inc.

These subsurface inclusions can strongly affect fatigue resistance and other properties. Stringers can be sources of weakness on surfaces, as shown on the lap-welded surface shown in Figure 4.7.

Other defects, which can be on the surface or internal, include:

- Rolling or forging laps – Locations where the solid metal has been folded over and compressed into the remaining solid. This happens during high-temperature rolling or forging operations, and the surfaces of these laps, even if they become buried into the metal, are covered with high-temperature oxide scales, which produce weakness discontinuities in the metal (Figure 4.8).
- Shrinkage cracks or tears – Caused by cooling stresses that arise when the outer surfaces of a metal cool and contract faster than the insides, which remain hotter and larger. These are usually surface defects, especially in rolled plate products, but in forgings with different cross-section thicknesses, they can be internal. Shrinkage cracks are often termed quenching cracks, when they are formed due to quick cooling of a weld bead or in the quick cooling of large-section quench and tempered (Q&T) low-alloy steels.
- Segregation – Most ideal metal parts would have the same chemistry in all locations, but it is common for surfaces, which cool from the melt faster than the center of a casting, to have different chemistries. This can affect both mechanical properties and corrosion resistance.
- Grain-size distribution – In the same way that chemistries can vary from the surface to the center of a part, it is possible for the grain size and grain distribution to be different. This is especially important when considering the effects of hydrogen embrittlement on thick section parts. Note that the inclusions and stringers in Figure 4.6 are smaller near the surface in this idealized drawing.

Of course the size of inclusions is greatly exaggerated in Figure 4.6. It is normally necessary to view these defects under the microscope, and Figures 4.7 and 4.8 were

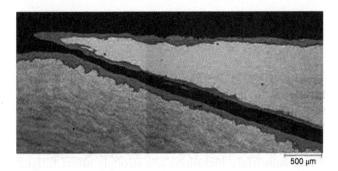

Figure 4.8 Forging lap with high-temperature oxidation on the lap surface [7]. *Source*: Reproduced with permission of Cambridge University Press.

originally viewed under the microscope at hundreds of times their normal size. The same thing is true in a real metal cross section – the grain sizes and inclusion-stringer sizes tend to be smaller toward the outer edges of rolled plates or thick-section forgings. This means that a slight increase in grain-size-related strength or hardness can be expected toward the surfaces, but it also means that defects, e.g. inclusions and stringer that may cause susceptibility to hydrogen blistering, or hydrogen-related cracking, are more likely to occur toward the center of thick sections. Remember, most oilfield metal parts are relatively thick (on an atomic scale). Sheet metal and wire are the exceptions in oilfield metal equipment.

Hydrogen blistering and environmental cracking are discussed in detail in Chapter 5. The blistering and cracking is often associated with microscopic defects within the metal.

Strengthening Methods

Metals are strengthened by one of the following strengthening mechanisms.

Work Hardening   Low-temperature deformation of metals introduces atomic motion that produces crystalline defects and dislocations that strengthen the metal. This works for very thin materials, wire, and sheet, but is not practical for thicker metals found in oil country tubular goods (OCTGs), plate, and structural steels.

Grain-size Refinement in Fully Killed Steels   Modern steel OCTGs are often strengthened by the addition of aluminum or other alloying elements that cause finer grain sizes in the finished product.

Oilfield steels will often be fully killed, which means any dissolved oxygen in the steel has been removed. At one time this was done by adding silicon to the melt, but modern steel-making practice often uses aluminum for this purpose. The aluminum reacts with the oxygen forming small aluminum oxide particles. The presence of these particles also serves to limit grain growth producing fine-grained steels, which are stronger and more ductile at ambient and lower temperatures than steels with larger grains. Silicon-killed steels are more common in pressure vessels, which often operate at high temperatures where the low-temperature strength advantages of aluminum killing are lost. Many American Petroleum Institute (API) and American Society of Mechanical Engineers (ASME) codes and standards will specify silicon-killed steels, even if aluminum-killed fine grain practice might be a better choice.

Silicon-killed steel can produce stringers, as discussed above.

Bolting materials should be fully killed unless otherwise specified [8].

Alloying   All metals used for strength are alloys of two or more different elements.

Second-phase Hardening   The presence of a second phase having different chemistry and/or crystal structure (usually both) is a major means of strengthening alloys. Carbon steel, the most common structural alloy, is a combination of almost pure iron crystals and other crystals containing iron carbide particles called cementite because of their hardness.

Precipitation Hardening   Second phases formed by heat treatment alter the microstructure and produce controlled microstructures with different grain sizes and/or crystals than would occur from chemistry alone.

Thermomechanical Processing   This is a term that has been applied to oilfield metals in recent years. This is a combination of plastic deformation of the metal and associated heat treatments to develop the optimum properties. It usually involves work hardening, but it can also involve deformation at temperatures so high that work hardening does not occur.

Mechanical Properties

It would often be desirable to make equipment out of hard, high-strength materials. Unfortunately, most high-strength/hardness materials are also relatively brittle. This is a major limitation for many oilfield applications. Most operators assume that as their fields age they may become more aggressive, and it is common to require that all produced-fluid equipment be made from materials that are compatible with H_2S, which causes environmentally assisted cracking in susceptible alloys [9–11].

Strength   For most applications, the strength of a metal is the most important property. Strength can be defined in a number of ways, but most industrial specifications set targets for yield strength and ultimate tensile strength. The hardness, ductility, and related properties of toughness are also important.

The tensile strength (resistance to elongation) of metals is measured by pulling a standard metal sample like the one shown in Figure 4.9 in tension and recording the elongation of the gauge length in the smallest diameter section, where most of the elongation occurs.

The results are plotted on a stress (load per unit area) versus strain (elongation, usually measured in percent) diagram shown in Figure 4.10.

Figure 4.10 shows the stress–strain plot decreasing and then increasing several times in the vicinity of the proportionality limit and yield stress. This is due to the initial unlocking of internal defects called dislocations that are the start of the work-hardening process. The dislocations move by slip at approximately 45° to the tensile axis and may form a phenomenon called Lüders bands on the metal surface [12].

Many engineering design codes require the structure to be loaded to only a fraction of the yield stress determined by a safety factor. This is the stress that cannot be

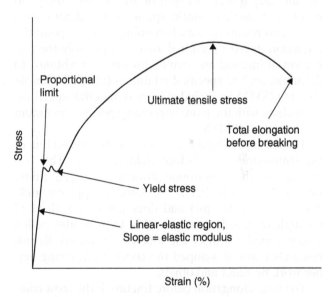

Figure 4.10   Stress–strain curve for low-strength carbon steel showing upper and lower yield stresses.

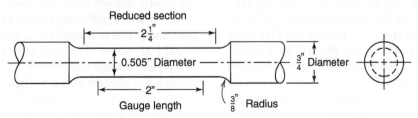

Figure 4.9   Standard sample used for determining the tensile strength of a metal [4]. *Source*: Reproduced with permission of John Wiley & Sons.

exceeding during operation of the structure or equipment in question. Safety factors depend on the type of structure under discussion.

The proportionality limit is almost the same as the yield stress. It is the point on the stress–strain plot where the deformation ceases to be in direct proportion to the stress. Up to this point, the stress–strain curve is considered to follow Hooke's Law:

$$\varepsilon = \sigma E \qquad (4.1)$$

where

ε = strain = deformation/original length, normally shown in percent

σ = load/(original cross-sectional area)

E = elastic modulus (Young's modulus)

Beyond the proportionality limit the stress–strain diagram becomes curved, and Hooke's law is no longer obeyed.

The highest point on a stress–strain plot determines the ultimate tensile strength or stress. Once the yield point is exceeded plastic (permanent), deformation starts. This results in work hardening. At some point the reduction in cross-sectional area, and possibly the formation of internal microvoids, lowers the resistance to deformation. The specified minimum (ultimate) tensile strength (SMTS) is included in most materials specifications along with the more important specified minimum yield strength (SMYS).

The elastic modulus is the slope of the straight line of the stress–strain plot before yielding occurs. This is a measure of the interatomic attraction between atoms. For steels this number is the same at approximately 200 GPa (30×10^6 psi) and does not vary with yield strength or tensile strength. This term is also called Young's modulus after the nineteenth-century British researcher who developed the concept advancing earlier work by Euler and others.

The total elongation before fracture is the most common measure of the ductility of a metal. This elongation will depend on the size of the tensile sample as well as the material being tested [12].

The above concepts were developed and incorporated into design codes before the development of modern measuring devices. It is now recognized that most metals, to include the carbon steels that are the primary alloys used in the oilfield, will exhibit some deviation from linear-elastic (Hooke's Law) behavior even at relatively low stresses. Modern methods of determining the yield stress are based on determining the load at which only 0.2% permanent offset (elongation) occurs after the load is released or by measuring the stress necessary

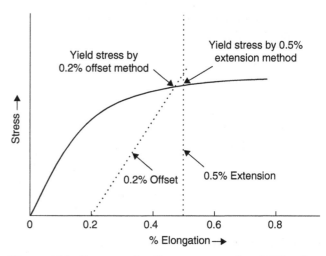

Figure 4.11 Stress–strain diagram comparing 0.2% offset and 0.5% extension methods of determining yield stress.

to produce 0.5% deformation while the load is applied. The two methods produce similar results, as shown in Figure 4.11, and both are accepted for oilfield materials [2, 13–16]. The term "proof stress" is sometimes applied to yield stresses determined by the permanent offset method.

Most materials specifications will define the SMYS and SMTS. Minimum elongation before breaking, which depends on sample size, may also be specified in addition to properties described below.

While most downhole and piping standards rely on yield strengths, it is common to use ultimate tensile strength as the basis for design calculations on pressure vessels and similar process equipment. The reason for this is that the primary concern on pressure vessels is to prevent catastrophic explosion. If the pressure on a vessel were to exceed yield strength, it may be acceptable, because slight changes in geometry are acceptable on process equipment. Downhole metallurgy is based on yield strength, because deformed pipe is unacceptable in downhole environments.

Modern steels used in oil and gas production have higher strengths than those used in previous decades. This has led to a steady rise in the yield stress/ultimate tensile stress ratio, the so-called Y/T ratio. A low Y/T ratio has been considered as providing a high capacity for plastic deformation and a safe margin against fracture, but the significance of the Y/T ratio is more complex and involves several other considerations [17].

API 5L states: "For cold expanded pipe, the ratio of body yield strength and body ultimate tensile strength of each test pipe on which body yield strength and body tensile strength are determined shall not exceed 0.93" [18]. Table 4.1 compares the allowable maximum ratios of yield stress to tensile stress according to several different international organizations [17].

TABLE 4.1   Yield Strength/Tensile Strength Ratios

Limits to Y/T Ratio in Accordance with Various Design Codes

Code and Application	Country	Maximum Allowed Y/T Ratio	Comments
API 5L (pipeline)	USA	0.93	All line pipe steel grades
HSE offshore (guidance note)	UK	0.7	Tubular joints
NS3472 (npd) (offshore)	N	0.83	All steels up to S460 (higher strength not permitted)
EPRG (European pipeline working group)	EU	0.88–0.93	Dependent on wall thickness

Source: From Bannister [17].

Hardness   Hardness is a material property that is often important, e.g. for wearing surfaces. In oilfield practice it is also used as a convenient method for field inspection of carbon steel products to determine if the metal in question has the necessary strength for the application in question. The hardness of a metal increases as the strength increases, and tables to convert measured hardness into approximate tensile strength are available.

The principle behind hardness testing is very simple. A penetrator of a known hardness, greater than the material being tested, is forced into the sample with a predetermined load. The larger the indentation produced, the softer the sample being tested. A number of hardness testers have been developed, but the Rockwell hardness test, developed in the United States in the early twentieth century, has been the most popular test for steels in North America. Other tests, differing primarily in the shape of the indenter, have gained use in other locations.

National Association of Corrosion Engineers (NACE) standards for hardness testing originally specified the use of Rockwell hardness testing, and, for a long time, HRC – Rockwell hardness using a 120° diamond cone penetrator (Figure 4.12) – was the standard used for oilfield testing in North America [9–11, 19, 20]. Other penetrators utilize round spheres (Brinell or Rockwell B) or different-shaped diamond pyramids (Vickers and Knoop) [21–25].

Recent revisions of NACE standards now cite Vickers hardness values, but API standards continue to refer to Rockwell hardness test. Conversion charts to compare the results of the various testing methods are available, but the user is cautioned that these conversions may be inaccurate, and the material in question should be tested with the hardness tester stated in the appropriate materials specification [10–12, 26]. Table 4.2 shows some conversions between different hardness tests and the approximate tensile strengths of carbon steels associated with these hardnesses.

Figure 4.12   Side view of a diamond cone penetrator for Rockwell hardness testing.

Because no exact mathematical relation has been demonstrated between any two methods of measuring hardness, it is important to identify how the hardness in question has been tested or approximated. Conversions from one measurement to another scale should carry notations like "_____ converted from _____," for example, "248 Vickers converted from HRC 22."

ANSI/NACE MR0175/ISO 15156 and other standards have used hardness values, which are easily confirmed in the field and are usually considered to be nondestructive tests, to determine if metals can be used in H_2S-containing oilfield environments [9–11, 26]. Part 2 of this three-part standard states:

> For ferritic steels, EFC Publication 16 shows graphs for the conversion of hardness readings, from Vickers (HV) to Rockwell (HRC) and from Vickers (HV) to Brinell (HBW), derived from the tables of ASTM E140 and ISO 18265. Other conversion tables also exist. Users may establish correlations for individual materials.

It is important to follow the hardness and other materials properties specifications appropriate for the equipment in question.

It is unfortunate that yield strengths, which are much more widely used instead of tensile strengths in oilfield materials specifications, are not readily available in tables like Table 4.2. Craig has suggested a "rule of thumb" that for carbon and low-alloy steels, the yield strength is approximately 75–90% of the tensile strength [2]. A publication by researchers at the Colorado School of Mines has suggested some yield strength correlations, but this data has not yet met with widespread acceptance for oilfield applications and standards [25].

Ductility Ductility is usually considered to be the ability of a metal to be stretched in tension before fracture. API Specification 5CT for oilfield tubing and casing requires a minimum elongation, depending on sample thickness, of between 8 and 30%. The minimum elongation depends on the sample size and the strength of the metal, with stronger metals having less ductility [12]. In addition to elongation before breaking, some definitions of ductility specify the reduction in cross-sectional area at fracture as a measure of ductility. The opposite of ductility is brittleness. Some authorities consider any

metal to be brittle that has less than 5% elongation before breaking [4].

Toughness Toughness is a measure of the resistance of a material to impact loading. This is an important materials property that has been gradually recognized by the petroleum industry and has been added to many materials specifications and design procedures. The mechanical properties described in previous paragraphs are measured at relatively low strain rates. Materials also need to withstand shock loading, and this can be measured by a number of different techniques. The most common technique is the Charpy impact test. Figure 4.13 shows a typical Charpy impact specimen. The specimen, with a premachined notch so that it will break at the desired location, is loaded into a low-friction pendulum apparatus (Figure 4.14) and struck with a known

TABLE 4.2   Carbon Steel Hardness Values and Approximate Tensile Strengths

Rockwell				Brinell		Vickers or Firth Diamond Hardness Number	Tensile Strength	
Diamond Brale			1/16″ Ball	10 mm⁻¹ Ball				
				3000 kg Load				
150 kg C Scale	60 kg A Scale	100 kg D Scale	100 kg B Scale	Diameter of Ball Impression in mm	Hardness Number		ksi	mPa
30	65	48	105	3.6	285	302	142	979
29	65	47	104	3.65	277	294	138	951
28	64	46	103	3.7	269	286	134	923
27	64	45	103	3.75	262	279	131	903
26	63	45	102	3.8	255	272	126	869
25	63	44	101	3.8	255	266	124	855
24	62	43	100	3.85	248	260	122	841
23	62	42	99	3.9	241	254	118	813
22	62	42	99	3.95	235	248	116	800
21	61	41	98	4	229	243	113	780
20	61	40	97	4.05	223	238	111	756

Source: Adapted from Material Hardness Conversion Table, http://www.corrosionsource.com/handbook/mat_hard.htm, July 2009.

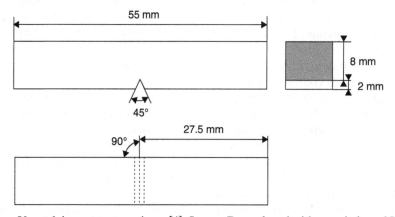

Figure 4.13 Charpy V-notch impact test specimen [4]. *Source*: Reproduced with permission of John Wiley & Sons.

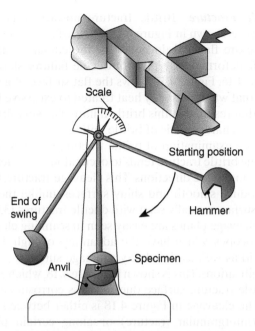

Figure 4.14 Charpy V-notch impact tester [4]. *Source*: Reproduced with permission of John Wiley & Sons.

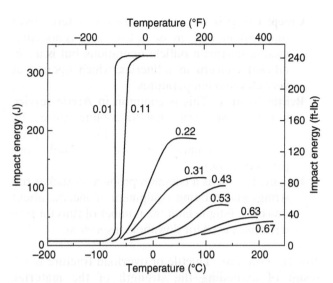

Figure 4.15 Influence of carbon content on the Charpy V-notch energy-versus-temperature behavior for steel [4]. *Source*: Reproduced with permission of John Wiley & Sons.

TABLE 4.3 Impact Energy Requirements for Carbon Steel Drill Pipe [28]

Specimen Size	Minimum Average Charpy V-Notch Impact Energy of Each Set of Three Specimens		Minimum Charpy V-Notch Impact Energy of Any Specimen of a Set	
mm × mm	Ft-lb	Joules	Ft-lb	Joules
10 × 10	40	54	35	47
10 × 7.5	32	43	28	38
10 × 5.0	22	30	19	26

Source: Reproduced with permission of API Publishing.

standard impact energy. The energy absorbed when the sample is impacted and breaks is then measured by determining the difference between the potential energy before releasing the pendulum and comparing it with the potential energy at the end of the swing. The difference in elevation is directly proportional to the energy absorbed in breaking the sample [27].

BCC metals, including carbon steels, become brittle when cold, and this can become a major problem for some applications or locations. FCC metals tend to be ductile at all temperatures used in oil and gas production, and this is the reason why aluminum, stainless steel, and iron–nickel alloys (e.g. 9Ni) are used for liquefied natural gas (LNG) storage and piping. Figure 4.15 shows a plot of ductile–brittle transition temperatures (DBTTs) for a series of carbon steels. As the carbon content, and the strength and hardness, increases, the transition temperatures decrease.

Table 4.3 shows ambient temperature (21 °C, 70 °F) impact resistance requirements for drill pipe. Similar requirements have been introduced for other OCTGs in recent years.

The NORSOK materials selection standard cautions about low-temperature effects on offshore structures. It also states that free-machining steels, which have lower ductility, are not suitable for pressure containing purposes [29].

While ambient and operating temperatures are obvious concerns for low-temperature brittle behavior of carbon steels and other materials, the expansive cooling of gases (the Joule–Thompson effect or autorefrigeration) is equally important [30, 31]. This is further discussed later on "Brittle Fracture" section of this chapter.

Carbon steel DBTTs are affected by many parameters, but grain size is probably one of the most important [2]. Grain-size refinement for line pipe and other OCTGs has been introduced in recent years, and this has improved several properties to include toughness and DBTTs.

Fracture Materials fracture when they are overloaded. The forms of fracture for many metals are:

Overload (ductile) fracture or deformation: This is relatively uncommon in upstream operations. Conservative safety factors predominate in most designs and pressure relief systems also help. Drill pipe and sucker rod strings may have this problem on occasion.

Creep: Creep is the elongation of a material over time without an increase in loading. It is not common in upstream oilfield operations, but is a significant concern in refineries, which operate at very elevated temperatures.

Brittle fracture: This is common in Arctic service but can also occur due to cooling caused by expansion of released gases (Joule–Thompson cooling) from natural gas pipelines and other pressure vessels.

Fatigue: Fatigue is a common problem in sucker rod strings and rotating equipment. Concerns about low-cycle fatigue limit the number of runs for tubing strings used for downhole inspection.

Ductile Fracture   Ductile or overload fracture is the result of exceeding the strength of the material. Bulging or bending is a frequent warning that overload failures are about to occur as the load exceeds the material's yield stress and plastic deformation begins. Water hammer is one example where this warning may not be present before the final overload failure. This can be a problem in piping systems with slug flow.

Ductile fracture is accompanied by plastic deformation. This can be seen microscopically, under the scanning electron microscope, as shown in Figure 4.16. Notice the curved surfaces where plastic deformation produced microscopic voids that formed and grew together before the final overload failure.

Brittle Fracture   Brittle fracture absorbs very little energy, as shown in Figure 4.15. The surfaces of a brittle failure are flat and do not show the curvature due to plastic deformation typical of ductile failures shown in Figure 4.16. Figure 4.17 shows the flat surface of a valve stem that was incorrectly heat treated to excessive hardness that resulted in this brittle failure. The very flat surface is a characteristic of brittle failures.

Close examination of brittle fracture surfaces shows that the brittle fracture tends to occur along well-defined crystallographic directions. This cleavage fracture tends to produce smooth and shiny surfaces, unlike the dull gray surfaces usually seen with ductile fractures [33, 34]. The cleavage planes are easily seen in scanning electron microscopes, which have the advantage of high depths of field as well as the ability to image surfaces at high magnifications. This is shown in Figure 4.18, which shows a brittle fracture surface due to stress corrosion cracking. The cleavage in Figure 4.18 is either between crystals (intergranular fracture) or along certain planes within the individual crystals (transgranular fracture). This transition from intergranular to transgranular fracture is common on many brittle fracture surfaces [33].

A major problem with brittle fractures is that, once they reach a critical flaw size, they spread at the speed of sound. This means that inspection for the defects that start brittle fractures is very important. Once the cracks start to run, it is too late to prevent major damage. Figure 4.19 shows a typical brittle fracture at a joint as a result of a hydrostatic pressure test on a welded pressure vessel. Note the arrows indicating the shiny fast-fracture surface. The features on this shiny surface are too fine to be seen in Figure 4.19. Figure 4.20, from another fast fracture, shows the chevron patterns characteristic of brittle fracture propagation surfaces. These markings, which are left on the surface of brittle fractures in both

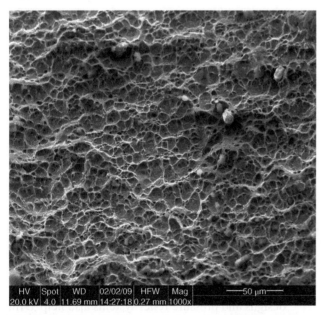

Figure 4.16   Ductile fracture surface. *Source*: Photo courtesy A. Michaels, Forensic Materials, San Jose, California.

Figure 4.17   Brittle flat fracture surface on incorrectly heat-treated valve stem [32]. *Source*: Reproduced with permission of NACE International.

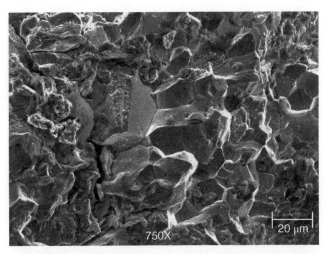

Figure 4.18 Brittle fracture surface showing flat and angular surfaces. *Source*: Photo courtesy J. Ribble, Materials Evaluation & Technology Corporation, www.metco-ndt.com.

Chevron marks

Figure 4.20 Chevron marks on the surface of a brittle fracture. *Source*: Photo courtesy R. Craig Jerner, PhD, PE, J.E.I. Metallurgical, Inc., Dallas, Texas, www.metallurgist.com.

Figure 4.19 Brittle fracture of a pressure vessel that failed during hydrotesting. The arrows indicate the shiny fast-fracture surface.

metals and polymers, point to the origin of the crack and are useful in failure analysis. Identifying why the crack started in the first place is often necessary in order to ensure that the repair or replacement does not suffer the same fate.

Brittle fracture can result from stress corrosion cracking and hydrogen embrittlement from improper welding procedures or heat treatment, from low temperatures that cause metals to become brittle, from the presence of sharp defects (stress risers), and from other causes. Figure 4.15 shows how carbon steels become brittle, less able to absorb impact loads, at different temperatures depending on carbon contents, which affect the strength and hardness levels.

Engineers must take ductile–brittle transformation temperatures into account when designing equipment. This means that the equipment must be designed in accordance with the following concepts.

Minimum Design Metal Temperature (MDMT) This is the lowest temperature expected in service including consideration for operating temperature, operational upsets, autorefrigeration, atmospheric temperature, and any other sources of cooling.

Design Minimum Temperature (DMT) This is the API term for the ASME MDMT. They are the same, but specifiers should be careful which code (API or ASME) a pressure vessel or other equipment must meet.

Critical Exposure Temperature (CET) This is the lowest temperature the equipment will see under "significant stress," which in most cases is assumed to be 8 ksi (55 MPa) [35].

Minimum Allowable Temperature (MAT) The lowest permissible temperature limit for a material at a specified thickness based on the material's resistance to fracture. This is the lowest safe temperature for the equipment in question.

The lowest temperature expected in service is the CET, and it must always be above the MAT, which is the lowest temperature the equipment in question can safely handle.

Detailed discussions on the above-listed temperature concepts are available in pressure vessel and piping reference books and design codes [35–43].

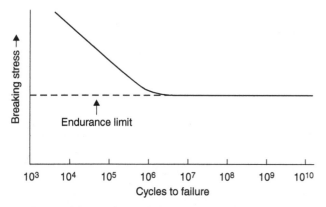

Figure 4.21   Fatigue curve showing endurance limit.

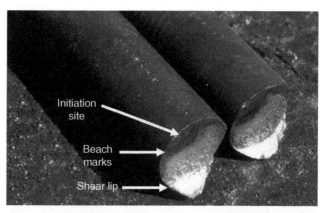

Figure 4.22   Fatigue fracture in a sucker rod.

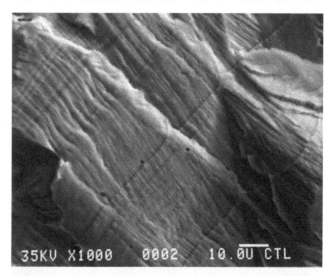

Figure 4.23   Striations caused by fatigue. The horizontal magnification is 1000×. *Source*: Photo courtesy Corrosion Testing Laboratories, Inc.

Design codes prior to the 1960s did not require toughness testing for equipment unless the equipment was to be operated below −20 °F (−29 °C). Thus older equipment may be susceptible to unexpected brittle fracture. This is a concern for any equipment, especially high-pressure gas pipelines that may undergo rapid cooling due to Joule–Thompson expansive cooling. Rapidly expanding gas can cause the pipeline to become brittle and lead to brittle crack propagation in the pipeline [44]. The requirement for toughness testing was added to API specification 5L for line pipe in 2000, and many gas pipelines constructed prior to this time may be subject to unexpected brittle behavior.

Fatigue   Fatigue fracture is the failure of metal or equipment due to repeated loading and stress cycles. Figure 4.21 shows a typical fatigue curve for carbon steel, which is assumed to have an endurance, or fatigue, limit below which failure will not occur even after many loading cycles. The endurance limit is an important concept in the designs of sucker rod strings, other pump components, and rotating equipment. Fatigue is also very important to welded and tubular structures, to include offshore platform structural components and piping and pipelines.

Fatigue crack initiation sites are usually surface flaws that act as stress risers that concentrate or magnify the applied stress. These can be corrosion pits on offshore structures, tong marks on drill pipe, machining grooves, or metallurgical defects. Once the crack starts to grow, the surface will frequently have concentric markings on the surface known as clamshell marks or beach marks. These concentric half-oval marks are the result of differential weathering as the crack progresses. Once the crack progresses to a certain level, the stresses are too high, and the part fails by normal overload. This is shown in Figure 4.22, where the final failure produced shear lips as the sucker rod pulled apart by tensile overload.

Fatigue fracture surfaces will often produce a pattern showing individual crack propagation markings, called striations. This is shown in Figure 4.23, where the striations progress from the lower left to the upper right. These striations can only be seen at very high magnifications using electron microscopes. The spacing between striations is so small that they can only be seen using electron microscopes at very high magnifications.

Most fatigue failures can be classified as due to high-cycle fatigue, where failure, if it occurs, is after 10^6 cycles or more. Some oilfield equipment, e.g. coiled tubing used for downhole inspections, is considered to be subject to low-cycle fatigue, which can occur after many fewer cycles, usually in the hundreds or less. Low-cycle fatigue is due to loading beyond the yield stress, whereas high-cycle fatigue is due to loading below the yield stress [45].

It is important to remember that submicroscopic defects, undetectable by modern inspection techniques, lead to eventual crack growth and propagation into fatigue failures. Even if no cracking is observed, it does not mean that the material has not been damaged by repeated loading cycles. This is the reason why drill pipe, coiled tubing for downhole inspections, and wireline are retired after too many "trips" or uses downhole.

Stress Risers Modern engineering practice has come to recognize that sharp defects, present in welds, fatigue cracks, corrosion pits, machined notches in fasteners, etc., can raise the effective stress level above the stress that would be calculated using a simple load per cross-sectional area calculation.

The recognition of this problem has led to the development of the field of engineering known as fracture mechanics. Modern computer programs using numerical techniques (finite element analysis, boundary integral analysis, etc.) enable engineers to predict the effect of defects of different sizes and geometries on the strength of various structures [12]. The effects of various stress risers on oilfield structures have been recognized for decades, and several commercial software packages are in widespread use in oilfield applications [30, 45–53].

Figure 4.24 shows an example of a location likely to produce fatigue cracking – hot spots – identified by DNVGL-RP-C203 [45]. Most of the hot spots identified in this standard/report are associated with welds, and the ring stiffener shown in Figure 4.24 could also be a welded stiffener instead of the bolted flange shown in the figure. Stiffeners shown in Figure 4.24 are common on risers for oil and gas production. Figure 4.25 shows how the critical hot spot on a welded connection is at the weld toe, and the standard provides guidance on how to minimize the fatigue problems associated with these welds. Numerous other examples of hot spots, and the means to calculate their fatigue loading, are contained in the standard.

Creep Creep is the time-dependent permanent (plastic) deformation of a material due to loading below the yield stress. In metals it is caused by atomic diffusion parallel to the stress axis. It is usually considered a high-temperature problem ($T > 0.4$ of the absolute melting temperature) and not a problem in upstream operations, but creep can cause elongation and bending in large continuous structures, e.g. tubing in deviated deep wells.

Figure 4.26 shows how creep elongation progresses with time. Most creep-susceptible equipment, e.g. flare stacks on offshore platforms, are monitored for changes in dimension. When the creep rate accelerates (tertiary creep in Figure 4.26), it becomes necessary to replace the equipment in service before stress rupture or creep cracking occurs.

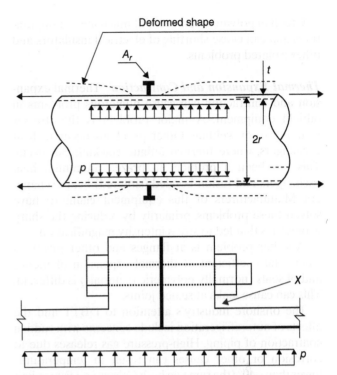

Figure 4.24 Upper figure shows how the stiffener produces distortion around the stiffener. Lower figure identifies the hot spot where fatigue crack initiation is most likely to occur.

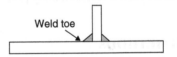

Figure 4.25 Critical hot spot at the weld toe on a stiffener.

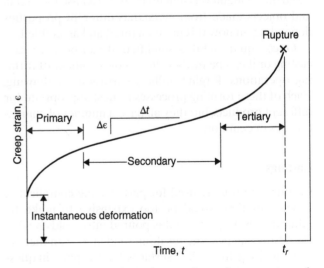

Figure 4.26 Creep elongation at elevated temperatures in metals.

Creep in polymers can occur at much lower temperatures and can cause shorting of electrical insulators and other isolated problems.

Thermal Expansion and Contraction   Thermal expansion and contraction can cause numerous problems in oilfield equipment. A major problem is the stresses resulting from welding. Other problems occur in heat exchangers, where thermal fatigue cracking can occur. This has been a problem in brazed aluminum heat exchangers and fusion-bonded (sintered) heat exchangers. Manufacturers of this equipment claim to have solved these problems, primarily by reducing the sharp geometries that led to stress intensity magnification.

Another problem is at flanges and other locations where the coefficient of thermal expansion of metals and of seals (normally polymeric materials) is different. This can cause leaks at sealed joints.

The offshore industry's attention to DBTT and the effects of autorefrigeration include concerns with sudden contraction of piping. High-pressure gas releases due to corrosion or other causes can result in temperatures lower than −40° (the same on both Celsius and Fahrenheit scales). The contraction of piping due to this cooling can create problems if the piping has too many mechanical constraints. This can even happen in piping leading to flare systems. The piping leading to flares is often carbon steel (e.g. ASTM A333), even though the flare tips require some form of corrosion resistant alloy [54].

FORMING METHODS

Oilfield metals and equipment can be fabricated and formed in a number of ways. It is important to understand how a metal was manufactured, because the metal, and objects made from metal, has different properties depending on how it is manufactured and assembled.

Once liquid metal is solidified, it can be used as a casting or it can be further shaped by a number of forming operations: forging, rolling, extrusion, or drawing. Each of these forming processes is most appropriate for different products, but they all have a number of characteristics in common.

Castings

Castings are usually used for parts having complicated geometries that would be too expensive to make by other means. Liquid metal is poured into molds having the desired shapes. The resulting solid metal can then be machined, e.g. for sealing surfaces, but the parts in question are usually used in approximately the same shape as they assumed upon solidification. The mechanical

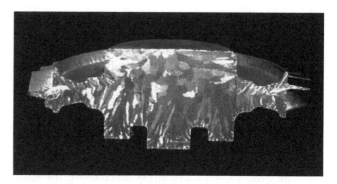

Figure 4.27   A casting that has been cut and polished to reveal the crystal structure.

properties of castings are usually lower than for wrought (deformed after solidification) products, because of porosity, large crystal structures (which tend to be weaker than finer-grained products), and a variety of other considerations. Their corrosion resistance is often somewhat lower, but this is offset by the fact that most castings are relatively thick and corrosion tolerant.

Figure 4.27 shows a casting that has been cut in two. The cut surface has been polished so that the crystal structure can be seen. Note the long slender columnar crystals at the bottom and edges of Figure 4.27 and the larger equiaxed crystals near center and the top. This image shows how the metal was oriented in the mold. Liquid metal was poured into the top of the mold, and the first crystals to form were at the perimeter of the mold where the liquid first cooled. Cooling toward the center and top of the mold was slower, and this allowed larger crystals to form. The crystals in this casting are so large that they can be seen without magnification. This is in contrast to most wrought metals, where the crystals are smaller and can only be seen at high magnification. The casting in Figure 4.27 is approximately 7.5 cm (3 in.) high.

Typical oilfield uses for castings include pumps, valves, and other fluid-control devices. Castings are not common in downhole applications because of their relatively low mechanical properties – strength, hardness, etc.

Wrought Metal Products

All wrought (deformed after solidification) metals have crystal structures that reflect their forming process. Rolled plate, which is used to manufacture large-diameter pipe and process equipment, will have different crystal structures and different mechanical properties in all of the three principal directions shown in Figure 4.28. The same principles apply to wrought products formed by forging, drawing, and extrusion. This directionality does not only apply to mechanical

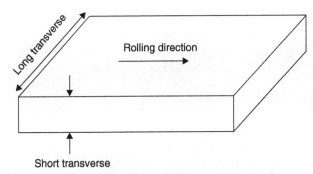

Figure 4.28 Principal directions of rolled plate.

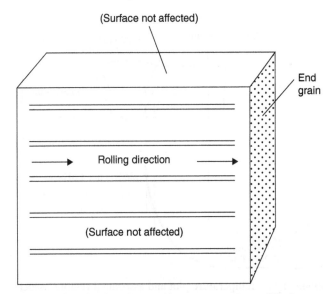

Figure 4.29 End grain attack on cut surfaces [5]. *Source*: Reprinted with permission of ASM international.

properties. The corrosion resistance of metals will be different depending on the orientation of the exposed surface to forming directions. For rolled products, to include pipelines made from rolled plate, the most corrosion-susceptible direction is the short transverse direction. This is because there are more grain boundaries exposed on these surfaces. End grain attack can be a problem on cut or drilled surfaces because of these grain boundaries (Figure 4.29) [5].

Defects in castings include inclusions, impurities from the melted metal or from the mold, and porosity due to entrained gases in the liquid metal. Wrought products are formed from castings and will have the same chemical compositions and volumes of defect inclusions as the castings from which they are formed. The inclusions are generally ceramic materials due to material impurities or from the mold. The forming process breaks these inclusions, which tend to be very brittle, and spreads them out parallel to the primary forming direction (Figures 4.6 and 4.7). Metal crystals in wrought products are usually microscopic in size, whereas many crystals in castings can be large enough to be seen on polished surfaces with the naked eye. Grain boundaries in metals are a primary source of strengthening, so wrought products, with much finer grain sizes, tend to be stronger and more ductile than castings. Any porosity in the cast metal is likely to have been removed by the compression of the forming process. For all of these reasons, wrought metal products are stronger and more reliable than castings.

Welding

Welding is the preferred joining method for most oilfield piping systems and process equipment. With most welding processes some of the metal is heated beyond the melting point, while some of the structure is not heated at all. In between these two extremes is a wide variety of temperatures and times at different temperatures. This produces at least three distinct regions in the metallic structure:

- **Weld bead or fusion zone** This is a combination of filler metal and melted metal from the base metal being joined. It solidifies as a casting with the problems of castings plus added stresses due to the thermal contraction caused by solidification and cooling to the ambient temperature at different rates than the surrounding metal.
- **Heat-affected zone** This is a region where the base metal has been affected by the heating associated with the weld. Phase changes in the metal can occur and produce different microstructures than the weld bead and the base metal. The results can be differences in mechanical and corrosion resistance properties.
- **Base metal** Metal that has not been heated enough to alter the metallic structure or corrosion resistance.

The three different heat-related regions are shown in Figure 4.30, and some of the defects associated with welds are shown in Figure 4.31.

Common Weld-Related Defects

- **Porosity** Internal porosity can be caused by trapped gases that have inadequate time to escape from the weld pool prior to solidification. Porosity is the cause of approximately 50% of all weld repairs [40]. Porosity can also be a problem in cast objects, but these objects are usually thicker than many welded structures and less subject to the loss of strength associated with porosity.

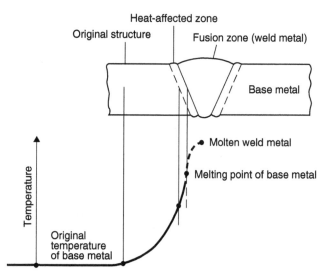

Figure 4.30 Temperature plot and associated regions associated with oxyfuel and arc welding [55].

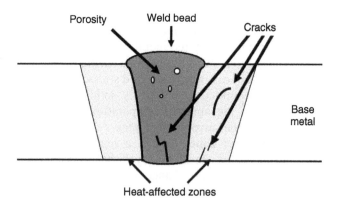

Figure 4.31 Defects associated with welding.

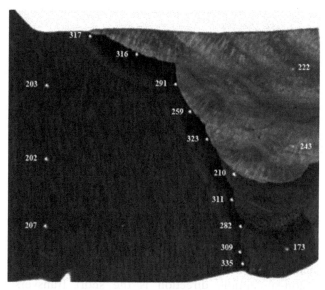

Figure 4.32 Hardness traverses on improperly welded low-alloy steel. *Source*: Photo courtesy J. Ribble, Materials Evaluation & Technology Corporation, www.metco-ndt.com.

- **Cold cracking** Hydrogen cracking is the principle cause of this problem. Quick cooling prevents the escape of hydrogen from the weld pool before solidification. Trapped hydrogen can cause hydrogen embrittlement and a variety of other problems that are discussed in detail under the "Environmentally Assisted Cracking" section in Chapter 5. Cold cracking is controlled by keeping filler metal electrodes dry and by avoiding hydrocarbon contamination of filler metal surfaces.
- **Hot cracking** This is also called sulfur cracking. It is restricted to low-grade carbon steel and some forgings with appreciable sulfur contents. Iron sulfides have low melting points and usually concentrate near the center of weld beads, where the metal solidifies last. Sulfide inclusions are weak and produce cracks as the weld cools.

- **Slag inclusions** This is usually due to poor welding procedures including inadequate cleaning of surfaces before welding.
- **Lack of fusion** Sharp cracks form where the weld bead does not bond to the base metal. This may be due to inadequate surface cleaning prior to welding or due to insufficient shielding gas allowing a surface oxide to form on the weld bead. A lack of fusion usually produces sharp, crack-like defects.
- **Incomplete penetration** When the root pass (first pass) weld bead leaves crevices where the molten metal has not penetrated, this produces relatively sharp defects that may be detectable by radiography. Unlike porosity, these relatively sharp defects must be analyzed using a fracture mechanics approach [56].
- **Hard spots** These are the result of rapid cooling that causes steel to transform from high-temperature austenite to either bainite or martensite upon uncontrolled cooling. Most authorities claim that welding-related hard spots produce martensite by uncontrolled quick cooling from high-temperature austenite to the martensite stability temperature (approximately 250–400 °C [400–750 °F] depending on alloy content). Figure 4.32 shows hardness traverses near an improperly welded low-alloy steel structure showing how much harder the steel is immediately adjacent to the parent metal-weld bead interface. The hardness readings are using the Vickers hardness (HV) scale, which, because it uses a fine-tipped indenter, is considered more suitable for microhardness testing than the more commonly

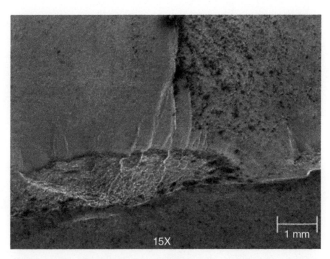

Figure 4.34 Angular distortion of a butt weld.

Figure 4.33 Fatigue crack initiation site on the improperly welded low-alloy steel shown in Figure 4.32. *Source*: Photo courtesy J. Ribble, Materials Evaluation & Technology Corporation, www.metco-ndt.com.

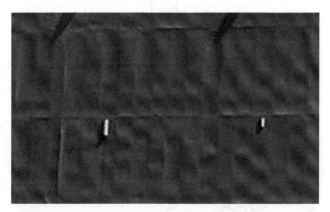

Figure 4.35 Distortion of exterior plates due to weld shrinkage.

used Rockwell HRC testing. Figure 4.33 shows the fatigue cracks starting at the surface of this weld bead interface.

- **Striking marks** These are the result of welding electrodes touching the metal surface and causing a spark. The heat generated by the strike can produce "hard spots" that can lead to subsequent cracking [57].
- **Other possible causes of cracks** A partial list of other causes of cracking includes movement before the weld sets, excessive delay between weld passes leading to internal hard spots caused by excessive cooling rates, and lamellar tearing due to inadequate inspection for laminations before starting the weld process.

Other Welding-related Problems The above listing shows relatively small defects associated with the welding process. Stresses caused by shrinkage of high-temperature welds adjacent to relatively low-temperature base metal can also cause distortions on pipelines, tank walls and floors, and other structures. This is shown in Figure 4.34, which shows how a fillet weld that is larger at the top will cause distortion on welded plates. If the pieces of metal that have been welded are restrained by other portions of the structure, the distortion may take years to become visible, as shown in Figure 4.35, which shows weld-shrinkage distortion on the side of a floating vessel. Similar distortions on the bottom of large storage tanks can lead to areas where water collects and is prevented from draining and can lead to corrosion.

Post-weld heat treating is sometimes specified to restore mechanical properties, reduce hardness and susceptibility to embrittlement, and relieve residual stresses. This adds to the cost, especially for field welds in piping systems.

The constraints associated with welding two structural members together can produce "hot spots" where fatigue is likely to produce cracking [45].

Figure 4.36 shows partial penetration welds. It is harder to inspect for fatigue cracks that start at the weld root than at the weld toe, where surface inspection methods, e.g. dye penetrant or magnetic particle inspection, can be used. DNV states that "it is difficult to detect internal defects by NDE in fillet/partial penetration welds. Such connections should therefore not be used in structural connections of significant importance for the integrity" [45].

Guidelines on where weld cracking is most likely to occur on structures are available. These guidelines also suggest structural details less likely to produce uninspectable fatigue cracks. Figure 4.37 shows how a backing strip is used to ensure that the most likely surface to develop fatigue cracks on a welded structural joint will be on the easiest to inspect surface [45].

Welds are also subject to testing for a number of properties including:

- Charpy impact tests at temperatures depending on the materials and the service environment.
- Microhardness tests to ensure that the welded structures are in accordance with appropriate standards, e.g. ANSI/NACE MR0175/ISO 15156 [9–11].

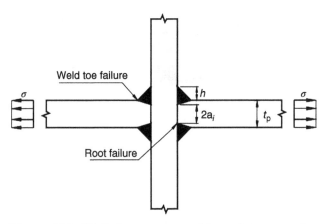

Figure 4.36 Welded connection with partial penetration welds [45]. *Source*: Reproduced with permission of DNVGL Group.

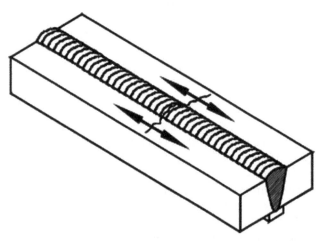

Figure 4.37 Automatic butt welds made from one side only, with a backing bar, but without start–stop positions [45]. *Source*: Reproduced with permission of DNVGL Group.

- Corrosion testing of stainless steels to avoid weld zone sensitization.
- Microstructural examination of duplex stainless steel welds to ensure no unwanted precipitates [57].

Lamellar tearing is cracking within plates due to shrinkage forces from weld bead cooling. It happens parallel to the rolling plane in locations where stresses perpendicular to the rolling plane tend to pull the plate apart forming cracks and tears in the metal (Figure 4.38). The tears nucleate at inclusions, so metal processing that produces cleaner, low inclusion content steels is less likely to have this problem. The problem is most common in plate 3 cm (1½ in.) and thicker, and hydrogen from the welding process is thought to contribute to the problem. Recent advances in steelmaking and reduction in sulfur-containing inclusions have lowered the

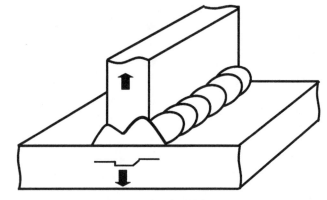

Figure 4.38 Lamellar tearing in thick plate steels caused by weld bead shrinkage.

incidence of this problem, which, if it happens, can only be detected by ultrasonic testing.

Weld Inspection The following inspection methods are routinely specified for weld inspection:

- **Radiography** This technique is mostly used to detect porosity, the problem associated with approximately 50% of all weld defect repairs. It can also detect incomplete penetration.
- **Ultrasonics** This is another technique that is largely associated with detecting porosity. It can be automated and rapid, but usually does not leave the visual images provided by radiography. Ultrasonics can identify the depth of porosity, whereas radiography can only locate the defects in two dimensions.
- **Surface detection of cracks** Both radiography and ultrasonics are primarily useful in detecting porosity. Crack-like defects, if they reach or approach the surface, are detected by magnetic particle and dye penetrant inspection.

Preferential Weld Corrosion Preferential weld corrosion can be caused by the differences in microstructure between weld beads, heat-affected zones (HAZs), and base metals. This is minimized by filler metal chemistry specifications, which call for the filler metals to be cathodic to the base metals being joined. The small galvanic effect making the weld bead cathodic to the base metal is often achieved by the same alloying additions that are used to strengthen welds. Figure 4.39 shows a case of corrosion caused by improper welding. The corrosion in Figure 4.39 was due to a combination of factors including the use of weld filler metal that was anodic, instead of cathodic, to the base metal. The recommended solution to this corrosion on already assembled equipment with dozens of welds is to use increased corrosion inhibitor dosages. The problem would have

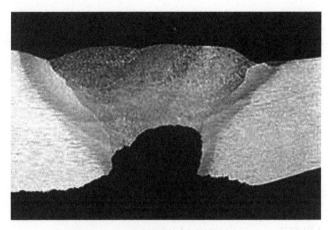

Figure 4.39 Preferential weld corrosion.

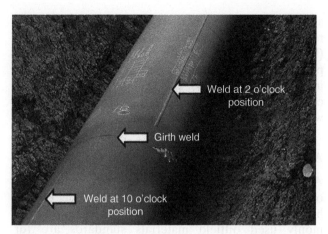

Figure 4.40 Pipeline with alternating longitudinal seam welds at 10 and 2 o'clock.

been minimized if a filler metal cathodic to the carbon steel base metal had been used [53].

Weld Fracture and Defects In-service repairs of piping and other equipment pose special problems not associated with welding for new construction. High cooling rates can be caused by liquid on the other side of the repair weld, and this can cause unwanted hard spots and cracking [44].

The possibility of defects in welds is much higher than in wrought metal, e.g. the plate from which welded pipelines are assembled. These welds can also be locations where slightly different microstructures lead to corrosion, either in the weld bead or in the HAZ. For these reasons it is common to specify that longitudinally welded pipe joints be placed in pipelines with their longitudinal seam welds alternating at the 10 and 2 o'clock positions [58]. This puts the welds in the upper quadrant of the pipeline, where corrosion is least likely to occur, and it decreases the chance that a rupture that may be associated with one longitudinal seam weld will run beyond the girth weld and into the next joint. This construction practice is shown in Figure 4.40.

Corrosion can also be concentrated at welds due to their rougher surfaces compared to the as-formed nearby metal. Roughness leads to increases in fluid turbulence, scale imperfections, and microbially influenced corrosion.

Clad Metals

It is common practice in oilfields to use clad metals to control corrosion. Less-expensive carbon steel is used for the primary structure (pipe, wellhead equipment, etc.), and CRAs are added to surfaces exposed to corrosive fluids. These CRAs can be applied by a number of

different methods depending on the size and complexity of the equipment involved. Detailed discussions on this topic are available [59].

Additive Manufacturing

Additive manufacturing, which is also known as 3D printing, is a relatively new method of manufacturing that is primarily used for parts with complicated geometries – the kind of part traditionally made with castings or forgings. Virtually all major oil and gas developments require that any production fluid-exposed equipment must meet the international standards for H_2S exposure, because, even if oil fields originally produce with low H_2S levels, as fields age they tend to "sour" – become enriched in H_2S in the production fluids.

The uncertainties about the quality of additive manufactured 3D printed parts mean that they must be tested in accordance with acceptance procedures used for cast alloy parts outlined in Part 3 of the ANSI/NACE/ISO standard [11, 60]. It remains to be seen if conventional CRA surfacing, e.g. cladding by weld overlays or similar processes, will result in equipment acceptable by this standard.

In addition to concerns about H_2S compatibility, questions remain on the mechanical properties of 3D printed parts. These parts may have similar properties to powder metal parts that are consolidated by hot isostatic pressing (HIP) processing.

MATERIALS SPECIFICATIONS

There are many organizations that issue materials standards used in the oilfield. This section explains some of the organizations that issue standards or specifications followed by the worldwide industry. It is not comprehensive, and many other standards are also used. The relatively recent development of the Universal Numbering System (UNS) for Metals and Alloys has helped designers and purchasing organizations to compare various alloys. Most international suppliers of oilfield equipment and supplies can identify which standards their equipment or supplies will meet.

API – The American Petroleum Institute

API standards tend to be performance standards, in other words the manufacturer is given wide leeway on how to produce the intended product. The most commonly used oilfield materials standards are for OCTGs. At one time, these products were described in one API standard, API 5, but in recent years this has been subdivided into API 5D for drill pipe, API 5CT for casing and tubing, and API 5L for linepipe [13, 18, 28].

A typical API specification will require well-defined strength levels and also specify mandatory limits on impact resistance, on carbon equivalents for weldability, and on a large number of other mechanical property and dimension requirements. The chemistry specifications will not translate to other common alloy designation systems, such as the UNS.

The above discussion has concentrated on OCTGs, the most common application for API-designated steels. It should be noted that other applications may also have API specifications. As one example, API 2H is a standard for plate steel used on offshore structures, whereas API 2B is a specification for structural steel pipe. It should be noted that pipeline steel is ordered as an OCTG to API 5L – line pipe steel. Both API 2B and 2H are for carbon–manganese steels. Many European standards not listed in this book also specify carbon–manganese steels, but this is not common in North America, where carbon steels are more commonly used. It should be noted that most steels will have residual manganese contents, because manganese is used in steelmaking for impurity controls as well as for improving mechanical properties.

API standards may also specify different product specification levels (PSLs). API 5L for line pipe lists two PSLs levels, with higher quality levels for more severe service or when regulatory agencies require higher reliability. API 5L PSL1 does not require Charpy impact testing, but PSL 2 includes this and other mandatory testing not included in PSL 1. API 6A for wellhead and Christmas tree equipment describes five PSLs, and other API standards may also list different PSL requirements. Table 4.4 lists API standards for materials.

TABLE 4.4 Selected API Standards for Materials

API 2B	Structural Steel Pipe
API 2H	Carbon Manganese Steel Plate for Offshore Structures
API 2MT1	Carbon Manganese Steel Plate with Improved Toughness for Offshore Structures
API 2MT2	Rolled Shapes with Improved Notch Toughness
API 2W	Steel Plates for Offshore Structures Produced by Thermo-Mechanical Control Processing (TMCP)
API 2Y	Steel Plates, Quenched and Tempered for Offshore Structures
API 5CT	Casing and Tubing
API 5D	Drill Pipe
API 5L	Line Pipe
API 5LC	CRA Line Pipe
API 5LCP	Coiled Line Pipe
API 5LD	CRA Clad or Lined Steel Pipe
API 6A	Wellhead and Christmas Tree Equipment
API 15HR	High-pressure Fiberglass Line Pipe
API 15LE	Polyethylene (PE) Line Pipe
API 15LR	Low-pressure Fiberglass Line Pipe
API 20A	Castings
API 20B	Open Die Forgings
API 20C	Closed Die Forgings
API 20D	Nondestructive Examination
API 20E	Alloy and Carbon Steel Bolting for Use in the Petroleum and Natural Gas Industries, 1st Edition, August 2012
API 20F	Corrosion-resistant Bolting for Use in the Petroleum and Natural Gas Industries, June 2015
API 20G	Welding Services (under development)
API 20H	Heat-treating Services, October 2015

AISI – The American Iron and Steel Institute

The American Iron and Steel Institute (AISI) steel specifications refer to chemical composition ranges and limits on steels, in the same way as Society of Automotive Engineers (SAE) steel designations, and they were often stated together, e.g. AISI/SAE 4340. While this practice continued for several decades, in 1995 AISI, which never wrote any of the specifications, turned the maintenance of the system over to SAE.

ASTM International (Formerly the American Society for Testing and Materials)

The American Society for Testing and Materials (ASTM) has a long history of supplying alloy designation standards. Metal standards are listed by the sponsoring committee. Standards starting with the letter

A pertain to ferrous metals, e.g. A36 ("Carbon Structural Steel"), ASTM A333 ("Seamless and Welded Steel Pipe for Low-Temperature Service and Other Applications with Required Notch Toughness"), and ASTM A353 ("Pressure Vessel Plates, 9 Percent Nickel, Double-Normalized and Tempered").

Standards starting with the letter B pertain to nonferrous metals, e.g. B209 ("Aluminum and Aluminum Alloys") and B337 ("Titanium and Titanium Alloy Pipe").

Other letters are used for nonmetallic materials, test methods, etc. [61].

In addition to ASTM standards for alloys, ASTM has cooperated with other organizations to develop a Universal Numbering System described in ASTM E527 – Numbering Metals and Alloys in the Universal Numbering System [58].

ASME – The American Society of Mechanical Engineers

ASME materials specifications are derived from ASTM standards and have an additional S in the prefix. As an example, ASTM A106 becomes ASTM A106/ASME SA 106 – Seamless Carbon Steel Pipe for High Temperature Service.

Most pressure vessel design is covered by international codes, the most common of which are ASME B31.3 – Process Piping and the ASME Boiler and Pressure Vessel Code Sections. In general, only materials recognized by ASME can be used in these designs. Because ASTM has jointly developed the UNS with SAE (see below), it is possible that metals produced to other standards can be cross-referenced to become acceptable for ASME design codes.

ASME materials standards are commonly prescribed for pressure vessels both onshore and offshore. The ASME Boiler and Pressure Vessel Code Section II – Materials includes the following parts:

- Part A – Ferrous Material
- Part B – Nonferrous Material
- Part C – Welding Rods, Electrodes, and Filler Metals
- Part D – Material Properties in Both Customary and Metric Units of Measure

Most custom carbon steel pressure vessels manufactured for the oil and gas industry are made from SA-516-70 carbon steel or its equivalent. It is also possible to find topside piping specified to ASME codes. Note that the ASME Parts A–D above use the term "material" where elsewhere in this book and in most other materials-related documents the terms "metal" or "alloy" would be used. This is presumably because the ASME pressure vessel codes discuss equipment where metallic materials are the obvious choice for construction – other materials would be too brittle or could not withstand the temperatures found in many pressure vessels.

SAE International (Formerly the Society of Automotive Engineers)

For many years SAE standards were usually listed as AISI/SAE standards, but in recent years the practice has been discontinued, because AISI does not issue chemical requirement specifications. Common SAE chemistry designations are listed in Table 4.5. The first two numbers indicate the alloy grouping, and the last two digits list the nominal percentage carbon expressed in 0.01% increments, e.g. 1040, 4140, and 4340 alloys all have 0.40% C as indicated by the last two digits of the alloy designation. Note that Table 4.5 lists "manganese steels" having 1.75% Mn. These are often referred to as carbon–manganese steels in European literature. API 2H plate steels, used for offshore structures, are also carbon–manganese steels.

UNS – The Universal Numbering System

Many different standards exist for materials, and the UNS for Metals and Alloys is a joint effort by ASTM and SAE to list metals in a uniform manner instead of using proprietary or local standard designations. UNS designations are not specifications (Table 4.6), because they establish no requirements for form, condition, property, or quality. UNS numbers are identifiers of a metal or alloy having controlling chemical composition limits in specifications published by some other standards organization. Whenever possible, identification

TABLE 4.5   SAE Carbon and Alloy Steel Grades Used in Oilfield Equipment

SAE Designation	Type
Carbon Steels	
10xx	Plain carbon steel (Mn 1.00% max)
15xx	Plain carbon steel (Mn 1.00–1.65%)
Manganese Steels	
13xx	Mn 1.75%
Chromium–molybdenum (Chromoly) Steels	
41xx	Cr 0.5–0.8%, Cr 0.5%, Mo 0.12–0.30%
Nickel–Chromium–Molybdenum Steels	
43xx	Ni 1.8%, Cr 0.5%, Mo 0.25%

TABLE 4.6   UNS Numbering for Alloys

Designation	Alloy System
Axxxxx	Aluminum alloys
Cxxxxx	Copper alloys, including brass and bronze
Fxxxxx	Cast iron
Gxxxxx	Carbon and alloy steels
Hxxxxx	Steels – AISI H steels
Jxxxxx	Steels – cast
Kxxxxx	Steels, including maraging, stainless steel, HSLA, iron-base superalloys
M1xxxx	Magnesium alloys
Nxxxxx	Nickel alloys
Rxxxxx	Refractory alloys
R03xxx	Molybdenum alloys
R04xxx	Niobium (columbium) alloys
R05xxx	Tantalum alloys
R3xxxx	Cobalt alloys
R5xxxx	Titanium alloys
R6xxxx	Zirconium alloys
Sxxxxx	Stainless steels, including precipitation hardening stainless steel and iron-based superalloys
Txxxxx	Tool steels
Zxxxxx	Zinc alloys

numbers from existing systems were incorporated into UNS designations, e.g. AISI/SAE 304 stainless steel is UNS S30400 [61–63].

API specifications are performance specifications and frequently do not have controlling limits, e.g. on chemistry. For this reason, API materials, e.g. 5L line pipe, do not have UNS numbers. They can be manufactured from a variety of UNS-designated alloys.

NACE – The Corrosion Society (Formerly the National Association of Corrosion Engineers)

NACE does not classify alloys, but ANSI/MR0175/ISO 15156 is a commonly used guideline into oilfield alloy selection. This standard, and all recent NACE publications, refers to alloys by UNS number whenever such numbers are available. The alloys discussed in various parts of the standard are [9–11]:

- Carbon steels and cast irons – note that while carbon–manganese steels are mentioned in Paragraph A.2.1.4 of this standard, this is the only paragraph where the term "carbon–manganese" steel is used [10].
- Austenitic stainless steels.
- Highly alloyed austenitic stainless steels.
- Solid-solution nickel-based alloys.
- Martensitic stainless steels.
- Duplex stainless steels.
- Precipitation-hardened stainless steels.

- Precipitation-hardened nickel-based alloys.
- Cobalt-based alloys.
- Titanium and tantalum.
- Copper and aluminum.

Other Organizations

The International Standards Organization, American National Standards Institute, Det Norske Veritas, and other organizations supply standards for materials. In many cases recent standards have reflected efforts to make existing standards compliant with the larger, non-materials organizations' standards, e.g. in 2009 API 5L for line pipe became ANSI/API 5L – Specification for Line Pipe.

ISO 21457, first published in 2009, is intended to provide general guidance on materials selection to both equipment fabricators and purchasing/specifying operators [64, 65]. It grew out of efforts to develop NORSOK M-001 as a means of standardizing, whenever possible, materials selection choices for upstream oil and gas operations [29].

Use of Materials Specifications

This book has been careful to avoid listing specifications and standards by specific date, and new equipment should be specified according to current versions of applicable codes and standards. Codes and standards are written based on the best available information at the time of writing, and they are updated. When specifying a standard in design or purchasing documents, it is not appropriate to use a general statement referring to "the latest version" of a code or standard. The specific edition of the standard in question should be clearly stated in the specification or purchasing document to avoid any disputes after the order is placed [66].

Many materials will meet several industry standards, but it is always important to specifically state which specification, of several possibilities, is the governing standard for purchasing and design documents.

As materials producers have improved process control and reduced inventories in recent years, it has become common for some suppliers to offer equipment or supplies that will exceed the specified mechanical properties of a given alloy. Unfortunately, stronger materials are frequently brittle, and the user must be sure to specify unacceptable maximum mechanical properties in addition to the commonly specified minimums. Some materials standards do not have specified maximum properties. Even when standards, e.g. the API 5L line pipe standard, have clearly stated maximums, it is common for suppliers to offer stronger materials if the specified material is not in stock. This concern with exceeding maximum properties is especially important for any application where embrittlement due to cold temperatures or H₂S is likely to be encountered [44].

For large projects, e.g. subsea pipelines, it is important to collect the statistical distribution data on the delivered material properties and retain it in a central location. Once the project is finished, it is difficult to obtain this information if it is necessary due to any one of a number of circumstances [44].

CARBON STEELS, CAST IRONS, AND LOW-ALLOY STEELS

Carbon steel is the most commonly used metal in the oilfield. Depending on the chemistry, primarily carbon content, and the heat treatment history, carbon steel can vary in yield strength from about 250 mPa (36 ksi) for structural steel to over 1380 Mpa (200 ksi) for wireline. Most carbon steels used in the oilfield are specified based on yield strength, and yield strengths of over 690 mPa (100 ksi) are common for many OCTG steels. Cast iron is much less commonly used in oilfield applications.

Figure 4.41 is the iron–iron carbide phase diagram showing the phases that form in iron–carbon alloys used for making steel and cast irons. Steels can have up to 2% carbon, but most oilfield alloys have only a small fraction of that, up to 0.4% in the case of drill pipe and other high-strength applications, but much lower for most other applications. Cast irons have 2% carbon or more, which makes them less expensive due to their lower melting points, but they are so brittle that their applications in the oilfield are very limited.

The diagram shows that α-ferrite has a maximum carbon content of only 0.02% at 723 °C (1333 °F). This means that carbon steel at room temperature will consist of a mixture of essentially pure α-ferrite and a compound ceramic material, Fe₃C, known as cementite. This two-phase structure serves to greatly strengthen carbon steel, which is much stronger than pure iron.

Figure 4.42 shows a typical microstructure for carbon steel. The light areas are α-ferrite, almost pure iron. The dark areas are alternating bands of α-ferrite and cementite, iron carbide Fe₃C. These alternating bands are a composite structure known as pearlite. Approximately 25% of the surface area shown in this picture is pearlite, indicating that the carbon steel had approximately 0.2% carbon. Most carbon steels used in oilfield applications will have similar microstructures and carbon contents.

The following phases are commonly found in carbon steels:

α-Ferrite: This phase, which is usually called ferrite, is the BCC form of iron thermodynamically stable at room temperature. It is reasonably ductile and can be work hardened by low-temperature plastic deformation.

Cementite: Cementite is a complex ceramic material having the approximate composition Fe₃C. Cementite was named by early metallographers, who noted how hard it is. Any plastic deformation in the ferrite phase tends to be stopped when it reaches a cementite grain boundary, and the hardness of cementite accounts for the strength of carbon steels.

Austenite: Above 723 °C (1333 °F) carbon steel starts to form the γ phase of iron called austenite. This phase is FCC, very ductile, and relatively soft. All of the carbon in carbon steels is also soluble in austen-

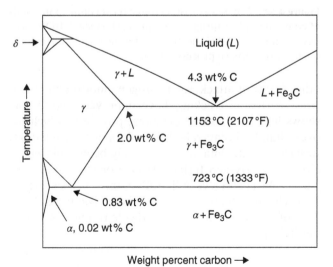

Figure 4.41   The iron–iron carbide phase diagram.

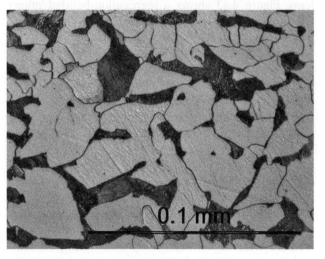

Figure 4.42   Microstructure of a typical carbon steel. *Source:* Photo courtesy J. Ribble, Materials Evaluation & Technology Corporation, www.metco-ndt.com.

ite, and this is very useful, because it makes hot steel ductile and amenable to forging, rolling, and other forming methods even in relatively thick sections.

Pearlite: The microscopes available to nineteenth-century steel researchers did not show the alternating plates of ferrite and cementite shown in Figure 4.42. They named the gray areas on the steel surface pearlite thinking it was a separate phase. Pearlite forms at the austenite grain boundaries as steel is cooled from higher temperatures. There is insufficient energy to allow diffusion into separated ferrite and cementite phases, so the alternating plate structure forms from the last austenite to transform. The composition of pearlite is 88% ferrite and 12% cementite.

Martensite: Martensite is a body-centered tetragonal form of iron that forms at lower temperatures (approximately 250 °C [480 °F]) from austenite that has not transformed into ferrite at higher temperatures during the cooling process. It is considered undesirable in pipeline steels and many other oilfield structures. The section on heat treatment of carbon steels discusses this in greater detail.

δ-Ferrite: The first solid metal to solidify from liquid steel has a BCC crystal structure. It is called ferrite, because it has the same high iron crystal structure as low-temperature α-ferrite. The presence of δ-ferrites is generally considered to be undesirable and an indication of improper heat treatment or welding.

Spheroidized and bainitic steels are specialized terms applied to carbon steel to indicate the shape of the cementite in the two-phase structure. They are generally not important in oilfield steels.

Segregation of chemicals in steels and other alloys can lead to mechanical and corrosion resistance problems. Figure 4.43 shows a segregation-banded microstructure in a medium-carbon steel (0.4% carbon), where the essentially pure iron α-ferrite bands run parallel to the darker pearlite colonies, which provide much of the strength in carbon steels. The bands run parallel to the rolling axis of the plate steel and are often reported to be a result of manganese segregation. This banding is generally undesirable, and some specifications suggest that banded steels should be rejected, e.g. for high strength bolting or for pressure vessel or pipeline steels [61–67].

Advances in steelmaking processes in recent decades have produced much more consistent, as well as stronger, carbon and low-alloy steels. The introduction of fine-grain practices and thermomechanical processing has resulted in steels having better mechanical properties and, to some extent, better environmental resistance. While carbon steels and low-alloy steels remain susceptible to weight-loss corrosion, their susceptibility to

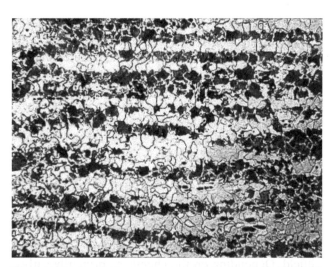

Figure 4.43   Banded structure in a medium-carbon (0.4% C) steel. The lighter bands are mostly α-ferrite, and the darker bands are pearlite that is not resolved at this magnification into separate α-ferrite and Fe_3C cementite.

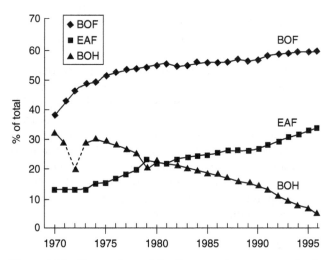

Figure 4.44   Comparison of the three most common methods of converting blast-furnace iron into carbon steel. BOF, basic oxygen furnace; EAF, electric arc furnace; and BOH, basic open hearth furnace processes [67].

other forms of attack, e.g. hydrogen embrittlement and stress corrosion cracking, have improved. Figure 4.44 shows how the basic steelmaking processes, which for over a hundred years relied on basic open hearth (BOH) furnaces for steelmaking, has largely been replaced by either basic oxygen furnace (BOF) or electric arc furnace (EAF) processes [67]. Both of these processes produce cleaner steels and have other advantages. Over 70% of all steel produced worldwide is now made with the BOF process.

All these improvements in mechanical properties have not resulted in major changes in weight-loss corrosion resistance. Figure 4.45 shows riveted carbon steel

Figure 4.45 Riveted carbon steel water pipeline that remained in service for over 100 years. *Source*: Photo courtesy K. Rickvalsky, McWayne Ductile, www.mcwaneductile.com.

water line that remained in service for over 100 years, and it is logical to expect that some modern pipelines will last just as long.

Classifications of Carbon Steels

The most common oilfield uses for carbon steel are for structures, OCTGs, and piping and process equipment. The term "carbon steel" is applied to any alloy consisting of iron plus carbon. Other elements may be added for deoxidation or machinability purposes. The term "killed steel" is applied to steels where the dissolved gases have been removed by the addition of either silicon or aluminum. It is common to have minimum residual content requirements for these elements to ensure that most of the dissolved gases have been removed. Most carbon steels used in the oilfield are killed steels.

Alloying additions are sometimes added to steels to make them easier to machine in high-speed production processes. These additions introduce small inclusions of softer materials (typically sulfur, selenium, or lead). Metals with these additions cannot be used in H_2S environments or for pressurized-fluid retaining purposes [9–11, 26].

Low-carbon Steels These steels can contain up to 0.30% C. Line pipe, tubing, and casing are usually made from low-carbon steels [13, 18].

Medium-carbon Steels Medium-carbon steels have carbon contents from 0.30 to 0.60% C and have increasing manganese contents intended to minimize the effects of impurities such as sulfur. The somewhat higher carbon content allows them to be heat treated and used in the quenched and tempered (Q&T) condition.

High-carbon Steels These steels have from 0.60 to 1.00% C and from 0.30 to 0.90% Mn. They can be hardened by the quenching and tempering process and are

very strong and hard. Their primary applications are for relatively small equipment needing maximum hardness. They are seldom specified for oilfield applications because of their brittle behavior. One notable exception is springs for control devices, although these are usually made from alloys other than carbon steel.

High-strength Low-alloy (HSLA) Steels These are also called microalloyed steels and have from approximately 0.25 to 0.50% C and Mn contents up to 2.0%. Small quantities of chromium, nickel, molybdenum, copper, nitrogen, vanadium, titanium, and zirconium are added in various combinations.

Several attempts have been made to distinguish "low-" and "high-" alloy steels, but the definitions vary between countries and between standard-setting organizations. As a general indication, low-alloy steel can be regarded as alloy steels (by the ISO definition) containing between 1% and less than 5% of elements deliberately added for the purpose of modifying properties.

Carbon–Manganese Steel The manganese content in carbon steels is sometimes increased for the purpose of increasing depth of hardening and improving strength and toughness. Carbon steels containing over 1.2% up to approximately 1.8% manganese are referred to as carbon–manganese steels. Some published books and standards discuss the use of carbon–manganese steels for pipelines [44, 68], but a review of the API 5L standard for line pipe steel shows only maximum manganese contents (from 0.60 to 1.85 depending on specified strength levels) [18]. With no minimum standards for manganese, purchases according to API 5L must specify the manganese level if it is desired. ANSI/NACE RP0175/AISI15156-2 for carbon and low-alloy steels in H_2S service recognizes carbon–manganese steels for downhole service [10].

Low-alloy Steel A common definition of low-alloy steels is steels with total alloying additions of 8% or less. Examples of low-alloy steels are listed in Table 4.3. Most authorities consider them as alternatives to carbon steels, and they are usually chosen because of their improved mechanical properties. They are often cathodic to carbon steels, but they must be protected from corrosion in the same ways as carbon steels. The only exceptions to the previous sentence are weathering steels, a form of structural steel intended for use in atmospheric exposure without protective coatings.

High-alloy Steel These alloys have 8% or more alloying additions and are considered to be CRAs. The most common examples of high-alloy steels are stainless steels. It should be noted that the term "high-alloy steel" is not synonymous with "stainless steel," because most authorities consider stainless steels to be ferrous alloys with a minimum of 11% or more chromium.

Alloying Elements and Their Influence on Properties of Steel

When analyzing a steel specification or product analysis, it is important to understand which elements are required (present for a reason) and which elements are impurities that must be controlled. In most materials specifications the necessary elements will have minimum and maximum contents, whereas impurities will have only maximum specified contents. This is not always true, as one example most OCTG specifications from API may not list minimums for most elements, even though it is recognized that *some* of the element in question is necessary. Recent versions of API 5CT for OCTG tubing and casing show only maximum contents for both carbon and manganese, even though it is recognized that both of these elements are necessary in carbon steels. This is shown in Table 4.7, from API Specification 5CT/ISO 11960 for OCTG casing and tubing [13].

The following list describes the effects of some alloying or impurity elements on oilfield steel properties:

- **Aluminum (Al)** – Aluminum is used as a deoxidizer in aluminum-killed steels and for grain-size refinement.
- **Calcium (Ca)** – Calcium is added to steel for sulfide shape control, where it promotes the formation of rounded, brittle inclusions. It is also a strong deoxidizer.
- **Carbon (C)** – Carbon has a major effect on the properties of steel and is the major hardening element. In carbon steels carbon forms iron carbides (Fe_3C-cementite), and the amount of carbon and the dispersion of the iron–carbide phases determine the strength and hardness of carbon steels. Most oilfield carbon steels will have 0.4% carbon or less. Tool steels and springs will often have higher carbon contents. In stainless steels, which rely on chromium for corrosion resistance, it is usually necessary to restrict the carbon content, because of a tendency to preferentially form chromium carbides, thus lowering the corrosion resistance of stainless steels.
- **Chromium (Cr)** – Chromium is added to steel to improve corrosion resistance in stainless steels in levels above approximately 10%. At lower levels, usually in combination with nickel and other elements, it is used to improve hardenability.

- **Copper (Cu)** – Copper in significant amounts is detrimental to hot working and welding of steels. Copper is beneficial to atmospheric corrosion resistance when present in amounts more than 0.30%. Weathering steels are made with copper content greater than 0.30%.
- **Lead (Pb)** – Lead is almost insoluble in liquid or solid steel. However, lead is sometimes added to carbon and alloy steels to form small lead-rich inclusions used to improve machinability. Sulfur, which is usually an impurity in steels, is also used for this purpose. These free-machining steels cannot be used for pressure-retaining purposes. Free-machining steels are unacceptable for pressure-retaining purposes [25].
- **Manganese (Mn)** – Manganese is the second most important alloying element in carbon steels, after carbon. The principal function of manganese is as a deoxidizer, and it is less likely to segregate than many other alloying elements. Manganese also affects the hardenability of steels. Manganese also combines with sulfur forming manganese sulfide inclusions, which can lower mechanical properties and increase susceptibility to hydrogen-related damage mechanisms (stress corrosion cracking, hydrogen embrittlement, etc.).
- **Molybdenum (Mo)** – Molybdenum is used in OCTGs needing high strength, especially if they must be heat treated (usually Q&T) to achieve the desired mechanical properties, examples include 4140 steel-UNS G41400 (Cr, Mo steel) and 4340 steel-UNS G43400 (Cr, Ni, Mo steel). Recent efforts to increase environmental cracking resistance have lead to the use of up to 0.8% Mo in some low-alloy steels with subsequent reductions in Cr and Mn.
- **Nickel (Ni)** – Nickel is a ferrite strengthening addition and is a useful alloying addition for low-temperature applications, e.g. for cryogenic storage. Nickel is also used in combination with Cr in the 43xx (UNS G43XXX) alloys, e.g. the 4340 steel mentioned above in the discussion of Mo additions. These alloys are used when thermal processing (quenching and tempering) are used to achieve the desired hardness/strength. Quenching and tempering is discussed later in this chapter.
- **Niobium (Columbium) (Nb)** – Niobium increases yield strength and to a lesser degree the tensile strength of carbon steels by promoting fine-grained microstructures with improved strength and toughness.
- **Phosphorus (P)** – Phosphorus is usually considered an impurity in oilfield steels. Phosphorous levels are normally controlled to low levels (Table 4.7 above).

TABLE 4.7 Chemical Compositions for Carbon Steel Tubing and Casing

Chemical Composition, Mass Fraction (%)

Group	Grade	Type	C Min.	C Max.	Mn Min.	Mn Max.	Mo Min.	Mo Max.	Cr Min.	Cr Max.	Ni Max.	Cu Max.	P Max.	S Max.	Si Max.
1	2	3	4	5	6	7	8	9	10	11	12	13	14	15	16
1	H40	—	—	—	—	—	—	—	—	—	—	—	0.030	0.030	—
	J55	—	—	—	—	—	—	—	—	—	—	—	0.030	0.030	—
	K55	—	—	—	—	—	—	—	—	—	—	—	0.030	0.030	—
	N80	1	—	—	—	—	—	—	—	—	—	—	0.030	0.030	—
	N80	Q	—	—	—	—	—	—	—	—	—	—	0.030	0.030	—
2	M65	—	—	—	—	—	—	—	—	—	—	—	0.030	0.030	—
	L80	1	—	0.43[a]	—	1.90	—	—	—	—	0.25	0.35	0.030	0.030	0.45
	L80	9Cr	—	0.15	0.30	0.60	0.90	1.10	8.00	10.0	0.50	0.25	0.020	0.010	1.00
	L80	13Cr	0.15	0.22	0.25	1.00	—	—	12.0	14.0	0.50	0.25	0.020	0.010	1.00
	C90	1	—	0.35	—	1.20	0.25[b]	0.85	—	1.50	0.99	—	0.020	0.010	—
	C90	2	—	0.50	—	1.90	—	NL	—	NL	0.99	—	0.030	0.010	—
	C95	—	—	0.45[c]	—	1.90	—	—	—	—	—	—	0.030	0.030	0.45
	T95	1	—	0.35	—	1.20	0.25[d]	0.85	0.40	1.50	0.99	—	0.020	0.010	—
	T95	2	—	0.50	—	1.90	—	—	—	—	0.99	—	0.030	0.010	—
3	P110	[e]	—	—	—	—	—	—	—	—	—	—	0.030[e]	0.030[e]	—
4	Q125	1	—	0.35	—	1.35	—	0.85	—	1.50	0.99	—	0.020	0.010	—
	Q125	2	—	0.35	—	1.00	—	NL	—	NL	0.99	—	0.020	0.020	—
	Q125	3	—	0.50	—	1.90	—	NL	—	NL	0.99	—	0.030	0.010	—
	Q125	4	—	0.50	—	1.90	—	NL	—	NL	0.99	—	0.030	0.020	—

[a] The carbon content for L80 may be increased up to 0.50% max. If the product is oil quenched.
[b] The molybdenum content for Grade C90 Type 1 has no minimum tolerance if the wall thickness is less than 0.700 in.
[c] The carbon content for C95 may be increased up to 0.55% max. if the product is oil quenched.
[d] The molybdenum content for T95 Type 1 may be decreased to 0.15% min. if the wall thickness is less than 0.700 in.
[e] For EW Grade P110, the phosphorus content shall be 0.020% max. and the sulfur content 0.010% max.
NL, no limit. Elements shown shall be reported in product analysis.
Source: Reproduced with permission of API Publishing, Table E.5 in Ref. [13].

- **Silicon (Si)** – Silicon is the main deoxidizer used in most oilfield steels (silicon-killed steels).
- **Sulfur (S)** – Sulfur is normally present as an impurity in steels, and this is controlled by manganese additions that form manganese sulfide inclusions. It decreases ductility and notch impact toughness especially in the transverse direction. Weldability of steels decreases with increasing sulfur content. Sulfur is found primarily in the form of sulfide inclusions. Sulfur levels in steels are controlled to low levels except where sulfur is added to improve machinability.
- **Vanadium (V)** – Vanadium is used in microalloyed steels to improve strength and hardness by grain-size refinement.

Strengthening Methods for Carbon Steels

The following discussion covers the most common strengthening methods for carbon steels. Similar methods are applicable for some CRAs, but many CRAs cannot be strengthened by some of the methods applicable to carbon steels.

Work Hardening   Ductile metals, including carbon steels, can be work hardened by deformation. The only common use of this practice in oilfield applications is wire, e.g. for downhole wireline applications. It is not practical to work harden plate and similar thick shapes due to the excessive pressures necessary to do so.

Alloying   The most important alloying addition for carbon steels is carbon. Unfortunately, brittleness can result in high-carbon steels, so oilfield alloys typically have approximately 0.2% for most applications and approximately 0.4% for drill pipe, erosion-resistant wellhead equipment, and other high-hardness applications. The strengthening effects of carbon additions are limited by metal thickness, which slows cooling and limits this kind of hardening in thicker structures. Common alloying additions for through-thickness hardening include the chromium-molybdenum (chromoly) and nickel–chromium–molybdenum alloys shown in Table 4.5. This is discussed further in the section on heat treatment.

Grain-size Refinement   Small-grained alloys are stronger than larger-grained alloys having the same chemistry. Steels can be specified for fine-grained practice, which may cost an additional 10–40% for typical carbon steel OCTGs. Steelmakers add nucleation agents such as aluminum to ensure that desired small-grain products are produced with practical cooling rates. It is common to normalize steel after solidification to further refine the grain size. Normalization (a form of heat treatment discussed below) can also be used on conventional steels and will have grain-size refinement effects, though not as much as with deliberately microalloyed steels.

Second-phase Hardening   All carbon steels have multiple phases. Carbon is insoluble in low-temperature α-ferrite, and the presence of cementite in various forms is the primary strengthening mechanism for most oilfield carbon steels.

Heat Treatment of Carbon Steels   The iron–iron carbide phase diagram, Figure 4.41, is an equilibrium diagram. It shows the phases that will be present if the metal is exposed for relatively long times at the indicated temperatures. Quick cooling, or quenching, can result in residual high-temperature phases at lower temperatures. This is why welding with insufficient preheating of the adjacent parent metal can produce δ-ferrites in the weld bead even though most of the metal has a different microstructure.

Most oilfield steel is too thick to be cold worked, so it is common that the final shape of a forging or the final thickness of plate to be formed into line pipe or process equipment was accomplished while the steel was at elevated temperatures in the relatively ductile austenite condition. Cooling from this temperature can produce different microstructures and properties depending on the cooling rate. The centers of thick sections will cool at slower rates and have different microstructures.

The following terms are commonly used in heat treating steels:

Annealing   This is a general metallurgical term for a heating process used to soften a metal. In steels it often refers to slow cooling from the austenite temperature region, but stress-relief annealing at lower temperatures is also common, e.g. in welded structures.

Quenching   Quenching refers to the quick cooling of a metal. The rapid removal of energy from the metal tends to limit atomic diffusion and may "freeze in" a higher-temperature crystal structure.

Tempering   Steels are tempered to remove unwanted martensite, a brittle metal phase formed by quickly quenching austenite so fast that the thermodynamically stable a-ferrite plus cementite structure cannot form. This is discussed in detail in the section on quenching and tempering.

Normalizing   The "normal" way of cooling steel is to let it air cool. Some manufacturing processes, e.g.

welding or upsetting the heated end of a tube (done with the upset region at austenite-stable temperatures), produce different microstructures, and different corrosion resistance, in the fully heated region, in the HAZ near the hottest region, and in those sections of the part that were not heated. The entire part is then reheated to form austenite and then allowed to air cool producing a uniform "normalized" microstructure throughout the entire part. This is called "full-length normalizing."

Hardenability The hardenability of steel is a measure of the hardness and the depth to which a steel may be hardened during quenching. Low-carbon steels lack insufficient carbon to be hardened. Medium- and high-carbon steels often have additional alloying additions to improve their hardenability.

Quench and Tempered (Q&T) Steels

Quenching is the quick cooling of a metal, usually to "freeze" in a high-temperature microstructure or to prevent transformation into an unwanted microstructure. It is applied to steels by first heating the steel to the austenite-stable region and converting all of the metal to a single-phased alloy with a uniform distribution of alloying and impurity elements. The steel is then quickly cooled to minimize the formation to α-iron ferrite plus cementite. Metal that remains as austenite is then converted to martensite, a body-centered tetragonal crystal structure. This conversion to martensite starts at temperatures between 250 and 350 °C (440–660 °F) depending on alloy composition. The temperatures at which martensite forms are so low that diffusion of carbon to form cementite is practically nonexistent [4].

Martensite is very hard and brittle and is, for all practical purposes, useless. The reason for this hardness is the fact that the carbon atoms, which were soluble in the high-temperature austenite, are trapped in interstitial sites in the martensite crystal. These trapped atoms are too big for the interstitial sites and put tremendous pressure on the surrounding iron atoms.

Quenched steels are then "tempered" by raising the metal to temperatures below the austenite transformation temperature and allowing the carbon to diffuse and form a fine-grained microstructure called "tempered martensite." This tempered martensite is a mixture of α-ferrite and cementite, and, due to the fine-grained structure, it is much stronger and tougher than pearlitic steel formed by normalizing or other heat treatment processes. Because the quenching and tempering process involves quenching (removing energy for diffusion and phase transformation) and tempering (reintroducing energy), the final product frequently contains a mixture of retained austenite, untransformed martensite,

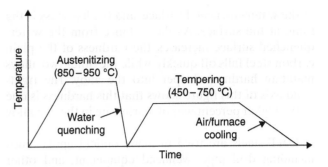

Figure 4.46 The quenching and tempering process for carbon- and low-alloy steels.

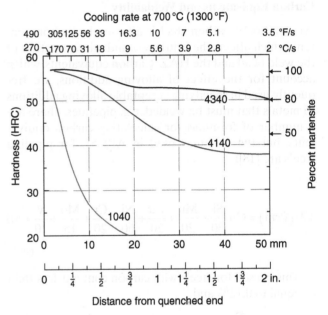

Figure 4.47 Hardness after quenching for three different steels having the same (0.40%) carbon content [4]. *Source*: Modified from figure 14.8, p. 580 in Callister and Rethwisch.

and tempered martensite. The amount of each phase depends on the time at annealing temperature, the thickness of the metal in question, and the cooling medium – oil, water, or salt water [4]. Figure 4.46 shows how quenching and tempering is accomplished.

The elimination of brittle martensite during the tempering process brings back ductility lost due to quenching, and the very fine microstructure means that the steel is very strong. The final properties produced by the Q&T process depend on the amount of carbon in the steel, so this process cannot be applied to low-carbon steels.

Alloying is used to increase the thick-section hardenability of Q&T steels. Figure 4.47 compares the hardenability of three steels, all of which have the same carbon content. Because they all have the same carbon content, they all form the same amount of martensite at the water-quenched surface, and the hardness is the same at the surface. As the distance from the water-quenched surface increases, the hardness of the plain carbon steel falls off quickly, while the other two alloys maintain hardness deeper into the alloy. The right-hand axis of this figure shows that this hardness is due to the higher percentages of martensite in the two alloy steels.

Q&T steels are used in many oilfield applications including drill pipe, wellhead equipment, and other equipment that must be very strong, hard, and ductile.

Carbon Equivalents and Weldability

Weldability of carbon steel is affected by carbon content, which affects the brittleness and microstructure of the weld bead and the HAZ. Carbon equivalents, which account for the effect of alloying additions, are frequently used in specifying acceptable alloying additions to metals that must be welded, e.g. pipelines. There are a number of formulas for calculating carbon equivalents, but the formulas used in API 5L for pipeline steels are [18]:

$$CE(Pcm) = C + \frac{Si}{30} + \frac{Mn}{20} + \frac{Cu}{20} + \frac{Ni}{60} + \frac{Cr}{20} + \frac{Mo}{15} + \frac{V}{10} + 5B$$

$$(4.2)$$

This is used for steels with carbon contents less than or equal to 0.12% and

$$CE(IIW) = C + \frac{Mn}{6} + \frac{Cr+Mo+V}{5} + \frac{Ni+Cu}{15} \quad (4.3)$$

This is used when the carbon content is greater than 0.12%.

Maximum carbon equivalents are intended to avoid brittleness in welds and HAZs.

Hard Spots Welding can leave "hard spots" in carbon steel structures, such as pipelines. This is often caused by quickly cooling (quenching) steels from high temperatures to form martensite. If post-weld heat treatment is not practical, then the martensite remains in the metal and can become a potential source of problems and brittle fractures, including those associated with H_2S cracking, can occur. The normal way to avoid this problem is to heat the surrounding metal prior to welding. This prevents the quick cooling that allows martensite to form. One of the purposes of carbon equivalents in metal specifications is to minimize the likelihood of hard spot formation [69].

Cleanliness of Steel

Steel purity strongly affects many properties. As steel first solidifies from the melt and is formed into plate or similar products, the cooling and forming processes tend to concentrate impurities, especially brittle inclusions like manganese sulfide stringers, near the center of the metal. When the steel is further processed by rolling or extrusion, these defects are flattened and spread out parallel to the forming direction. Large inclusions are most likely to be found near the center of plate or piping. These inclusions affect susceptibility to embrittlement and hydrogen blistering. They also affect toughness. This is why specifications for steel often require Charpy impact and tensile tests from midthickness locations [70].

Oxygen Steel is deoxidized while in the liquid state by the addition of silicon or aluminum that "kills" the steel. Calcium additions cause the resulting silicon oxide or aluminum oxide inclusions to be rounded and produce fewer stringers [61]. Semikilled steels, e.g. A53 pipe and A285 plate, have more oxide inclusions than fully killed A106 pipe and A516 plate [71–75]. These oxide inclusions lower notch toughness and other mechanical properties in the through-thickness (short transverse) direction.

Sulfur Sulfur affects the toughness of steels due to the presence of manganese sulfide stringers and other inclusions. The purity of steels has improved over recent decades. Sulfur levels in modern steels are often less than they were in the 1970s (less than 0.010% versus 0.025% in the 1970s). This is one reason why modern pipeline steels are considered to be less susceptible to brittle behavior of all types.

Cast Irons

Cast irons are alloys of iron and carbon having approximately ten times the carbon content of typical carbon steels. This carbon is in the form of graphite flakes or rounded nodules (Figure 4.48). The graphite is very weak, and this makes cast irons very brittle. They are seldom used on upstream oil and gas operations except for water service. Even in water service it is common to use cast steel for critical applications like fire water.

Any component available in cast iron can be made from cast steel, but usually with a cost penalty. Most upstream operators consider the increased reliability more important than the small cost penalties. If cast iron is specified, it should be clearly stated that nodular cast iron (also called ductile cast iron in the United States) is the intended material. The long flakes of graphite shown in Figure 4.48a cause gray cast iron (also called cast iron) to be too brittle for any corrosion or fluid-handling service.

(a)

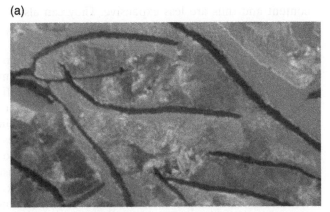

(b)

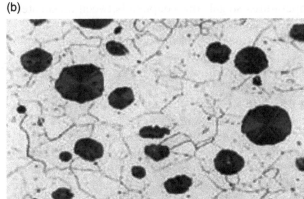

Figure 4.48   Cast iron microstructures: (a) gray cast iron and (b) nodular or ductile cast iron.

CORROSION-RESISTANT ALLOYS (CRAs)

NACE defines CRAs as alloys "…whose mass-loss rate in produced fluids is at least an order of magnitude less than that of carbon and low-alloy steel, thus providing an alternative method to using inhibition for corrosion control" [76]. There are a wide variety of CRAs and a number of different ways they can be classified. This section discusses the following alloy groups:

- Iron–nickel alloys
- Stainless steels
- Nickel-based alloys
- Cobalt-based alloys
- Titanium
- Copper
- Aluminum

CRAs contain alloying-addition metals that are more expensive than iron, and it is common for metal producers to attempt to deliver products with the minimums of any specified alloying additions. This means that newly purchased CRAs may have lower corrosion resistance than older alloys purchased to the same specification, because at one time most CRAs were delivered with close to the medium alloying content allowed by appropriate specifications/standards for the alloying additions in question. This phenomenon is the reason why, as discussed below, 317 stainless steel (UNS S31700) seems to be replacing the more common 316 stainless steel (UNS S31600). Both alloys have molybdenum additions, and the somewhat higher Mo content for 317 (3–4% Mo) guarantees that the Mo content will be at least 3%, which is also possible with 316, which has a Mo range from 2 to 3%, but recently ordered 316 is likely to have very close to the minimum of 2% Mo.

While the above paragraph uses 317 and 316 stainless steels as examples, the same phenomenon is true for most CRAs.

This degradation in quality (corrosion resistance) of CRAs is different than the situation for low-alloy carbon steels. Recent advances in steelmaking and quality controls normally lead to better carbon- and low-alloy steels. Keep in mind that carbon- and low-alloy steels are specified for their mechanical properties, and their resistance to weight-loss corrosion (as opposed to environmental cracking) is so low that they are not specified for corrosion applications without supplemental corrosion control methods, e.g. protective coatings, corrosion inhibitors, or cathodic protection (Chapter 6).

ISO 21457 and NACE 1F192 provide guidance on the use of CRAs in upstream oil and gas operations [64, 76].

Iron–Nickel Alloys

These alloys are more corrosion resistant than carbon steels, but their relative ductility at low temperatures is the more important reason for their use in oilfield applications.

Iron–nickel alloys are used for aboveground cryogenic storage tanks for LNG, because these alloys have sufficient ductility at LNG temperatures (−162 °C [−260 °F]). The addition of nickel to iron results in alloys where the transition of high-temperature austenite to α-ferrite is retarded. Alloys having 5–10% nickel will have a mixture of high-iron α-ferrite, which is subject to embrittlement at low temperatures, and austenite (γ-iron), which is ductile at the temperatures found in LNG storage and similar applications. The dual-phased structure still has ferrite, but the resulting alloy is ductile enough for static structures like storage tanks.

LNG storage tank walls are typically constructed from welded iron–nickel alloys (9% nickel alloys are most commonly used) [77–80], with piping and similar attachments made from austenitic stainless steel, which is more expensive but has better resistance to thermal fatigue. Welds on these LNG tanks use nickel-based alloys [81, 82]. Corrosion is not a problem at cryogenic temperatures, so galvanic coupling between nickel steel and stainless steel is not a problem.

The heat treatment of these 9-nickel alloys depends on section size and the specific chemistry of the non-nickel alloying additions, which are also being considered for H_2S service [83].

Austenitic stainless steel and aluminum alloys are also ductile at LNG temperatures, and they are sometimes used for building smaller storage tanks, but large containment vessels are usually welded from iron-9 nickel because of expense considerations. This technology has been in use since the 1940s and is well established worldwide.

Stainless Steels

Stainless steels are usually defined as alloys having a minimum of 11% chromium in addition to other alloying additions [84]. The names attached to the various classes of stainless steels usually derive from the predominant crystal structure that determines their mechanical properties. Stainless steels can have a wide variety of mechanical properties, but the reason for using them is for corrosion resistance, and this will be emphasized in the following discussions.

The commonly recognized classes of stainless steels are:

- Martensitic stainless steels
- Ferritic stainless steels
- Austenitic stainless steels
- Duplex stainless steels
- Precipitation-hardening stainless steels

Martensitic Stainless Steels Martensitic stainless steels are used in upstream production and pipelines more than any other class of CRAs [85–87]. There are two reasons for this: the martensitic structure produces strong, tough alloys, and, compared with other stainless steels, the martensitic alloys have the lowest alloying

content and thus are less expensive. They can also be heat treated to approximately the same strengths as carbon steel tubular goods.

API casing, tubing, and line pipe specifications identify two different chromium contents for iron–chromium alloys – 9 and 13% chrome. The 9% chrome has insufficient alloying to be a true stainless steel and is seldom specified. The use of 13% chrome alloys has increased in recent years due to increased production in more aggressive environments and concerns that the cost of corrosion monitoring, corrosion inhibitors, repairs, and, most important, lost production outweigh the added cost of CRAs. As one example, the materials cost of a carbon steel subsea pipeline, including both metal acquisition and welding, is approximately 25% of the total cost. Changing to CRAs doubles the materials cost, but it only increases the total initial cost of the pipeline by approximately 25%. Many operators consider this additional initial cost justified.

Martensitic stainless steel tubular goods are usually specified using API specifications, while other industrial specifications based on UNS or other international standards are common for other applications [88].

Table 4.8 shows some of the most commonly used martensitic stainless steels used in oil and gas production.

The limited alloying content of martensitic stainless steels means that, while they are more corrosion resistant than carbon steels, they cannot withstand aggressive environments. They are mostly used in applications where CO_2 corrosion is a problem and H_2S, if present, is at relatively low concentrations. ANSI/NACE RP0175/ISO 15156 places limits on the hardness that these alloys can have in H_2S service depending on alloy type and application [11].

The amount of chromium that can be added to martensitic stainless steels is limited, because austenite, which is necessary for heat treatment to produce martensite, does not form at chromium contents greater than 12–17% Cr (Figure 4.49). The upper limits are determined by other alloying additions, and the highest

TABLE 4.8   Nominal Composition of Selected Martensitic Stainless Steels Used in Oil and Gas Production

UNS Number	Name	C max	Fe	Cr	Ni	Mo	Other
S41000	410	0.15	Bal	12.5			
S41425	Super 13Cr	0.05	Bal	13.5	5.5	1.75	Cu 0.3
S41426	Super 13Cr	0.03	Bal	12.5	5.5	2.25	Ti 0.01, V 0.5
S41427	Super 13Cr	0.03	Bal	12.5	5.3	2	Ti 0.01, V 0.3
S42000	420	0.15	Bal	13			
K90941	9Cr 1Mo	0.15	Bal	9			
J91150	CA 15	0.15	Bal	12.75			
	API L80-9Cr	0.15	Bal	9	0.5 max	1	
	API L80-13Cr	0.15–0.22	Bal	13	0.5 max		

commercially available martensitic stainless steel routinely approved for H$_2$S service, UNS S42500, has 15% chrome [11]. Iron–chrome alloys with higher chrome contents have ferritic microstructures and are discussed in the section on ferritic stainless steels.

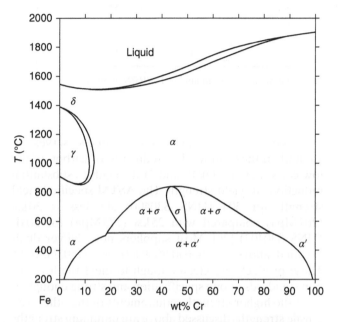

Figure 4.49 The iron–chromium phase diagram.

Figure 4.50 External corrosion of 13Cr L80Cr tubing stored outdoors for several years.

Storing API 13Cr tubulars is very important. Most 13Cr failures come from storage problems. Figure 4.50 shows 13Cr tubing that was stored outdoors for several years. Note how the exterior of the tubing has corroded, but the head has not. The reason the box end of the tubing has not corroded is because most 13Cr downhole tubing has couplings made of another alloy with a different hardness to avoid galling or seizing during makeup. Premium couplings are usually made from more CRAs.

The obvious corrosion shown in Figure 4.50 emphasizes that "stainless steel" is not immune to corrosion. Martensitic stainless steels have the lowest alloying contents of any CRAs, and they corrode, even in atmospheric corrosion. This corrosion susceptibility has led many organizations to develop storage guidelines for CRA tubular goods, which are often stored in field locations for long periods before they are used.

Downhole OCTGs are the most common oilfield applications for martensitic stainless steels. Wellhead equipment is often made from cast or forged versions of these alloys. Surface applications beyond the wellhead are limited because oxygen ingress can cause severe pitting in these alloys. The use of martensitic stainless steels for internal corrosion control of subsea pipelines is increasing, but problems with welding specially developed versions of these alloys with low-carbon contents has limited their widespread acceptance [44].

Ferritic Stainless Steels Ferritic stainless steels find limited use in oilfield applications. They have higher chromium concentrations than martensitic stainless steels, but their limited strength due to the lack of tempered martensite makes them less desirable for many applications. Table 4.9 shows some ferritic stainless steels listed in ANSI/NACE MR0175/ISO 15156 [11].

Austenitic Stainless Steels Austenitic stainless steels are the most commonly used stainless steels on a worldwide basis. Their use in oilfield applications is limited by their strength and their susceptibility to pitting and stress corrosion cracking in chloride and other halide containing environments. The maximum temperature limit for avoidance of stress corrosion cracking is widely assumed to be 60 °C (140 °F).

TABLE 4.9 Nominal Composition of Selected Ferritic Stainless Steels Approved for Use in H$_2$S Oilfield Service

UNS Number	Name	C Max	Fe	Cr	Mo	Other
S40500	405 SS	0.08	Bal	11.5–14.5		Al 0.10–0.30
S43000	430 SS	0.12	Bal	16.0–18.0		
S44635	26-1 Cb	0.10	Bal	25.0–27.0	0.75–2.50	Nb 0.05–0.20

TABLE 4.10   Nominal Composition of Representative Austenitic Stainless Steels

UNS Number	Name	C max	Cr	Ni	Mo	Other	Minimum PREN[a]
S30400	304 SS	0.08	19	9.25			18
S30303	304L	0.03	19	9.25			17
S31600	316 SS	0.08	17	12	2.5		23
S31603	316L SS	0.03	17	12	2.5		23
S31700	317 SS	0.08	19	13	3.5		28
S32100	321 SS	0.08	18	10.5		$Ti_{min} = 5 \times C$	17
S34700	347 SS	0.08	18	11		$Nb_{min} = 10 \times C$	17

[a] The compositions shown in this table are the averages between the minimum and maximum levels. PREN numbers are calculated based on minimum numbers.

The austenitic structure (FCC) is very ductile, and these stainless steels can be drawn into very thin tubing. Cold-worked austenitic stainless steels are usually annealed to relieve residual stresses before use in corrosive environments.

Table 4.10 shows several austenitic stainless steels. Type 304 (UNS S30400) stainless is the most common form of stainless steel and is the basis for all of these alloys, which are often called "18–8" stainless steels, because their nominal composition is based on the original German stainless steel compositions having 18% chrome and 8% nickel.

Unfortunately, 304 stainless steel is subject to pitting and crevice corrosion, so 316 stainless steel, which has molybdenum additions to limit this attack, was developed. Many organizations no longer use 304 stainless and consider 316 stainless to be their basic stainless steel for most process equipment applications.

Unfortunately, both of these alloys can become "sensitized" by improper welding procedures. Sensitization is caused by the formation of chromium carbides in grain boundaries. This reduces the chromium available to form protective passive films and results in localized grain boundary corrosion, primarily in the HAZs near welds. There are two alloying approaches to minimize this problem. One approach is to limit the carbon in the alloy. Types 304L and 316L stainless steel have lower maximum carbon contents, 0.03% C instead of 0.08% C, and this lowers the severity of sensitization. Another approach to the sensitization problem is to add "carbide getters" to the alloy. Types 321 and 347 stainless steels have titanium or niobium (which is also called columbium) added to the alloy. These alloys preferentially form titanium or niobium carbides instead of chromium carbides, and this limits sensitization. Market considerations lead North American steel producers to prefer to sell low-carbon stainless steels, while European producers have tended to use titanium or niobium additions. Recent consolidations in the steel industry may change this pattern in the future.

Carbon in austenitic stainless steels serves to strengthen these alloys. The reduction of carbon in the low-carbon grades (304L and 316L) leads to substantial reductions in yield strength. The ASTM specified yield strength for 304 (UNS S30400) stainless is 30 ksi (207 Mpa) compared with 25 ksi (173 Mpa) for 304L (UNS S30403) [86]. Most suppliers now deliver dual-certified alloys, e.g. S30403/S30400 or S31603/S31600. These products are clean enough to meet the low-carbon standards of one specification and strong enough to meet the higher-strength requirements of the other. The tensile strengths discussed above are minimum strengths for fully annealed alloys, and most austenitic stainless steels have much higher yield strengths depending on forming method and application.

Alloy UNS S31700 (317 stainless steel) and the low-carbon version UNS 31703 (317 SS) are less popular than the 316 and 316L stainless grades, but they are becoming more widely used. The alloy contents shown in Table 4.10 are simplifications of the actual standards and, with the exception of the carbon content maximums, are averages between the specified maximum and minimum contents. As steelmaking controls have improved, the delivered alloys are likely to be very close to the minimum specified content for expensive additions like chromium and nickel. Alloys 317 and 317L have higher specified chromium, nickel, and molybdenum contents and are therefore more corrosion resistant in most environments.

The alloys shown in Table 4.10 are only a representative sampling of the austenitic stainless steels that are available. Other austenitic alloys are available that are more suitable for machining, high-temperature pressure vessels, or welding. Most of them also have standardized UNS numbers.

Austenitic stainless steels are widely used in heat exchangers, pressure vessels, and process equipment. ANSI/NACE MR0175/ISO 15156 prescribes environmental limits for austenitic stainless steels for use in seal rings and gaskets, compressors, gas lift service, and for special components such as valve stems, pins and shafts, surface and downhole screens, control-line tubing hardware (e.g. set screws, etc.), injection tubing, and injection equipment.

Austenitic stainless steels are considered to be nonmagnetic alloys, but they can have some magnetism due to delta ferrites, which are often present in welds.

These alloys can also become embrittled due to the presence of sigma phase (see Figure 4.49), usually in welds [34].

It is not unusual to require welded connections in process equipment between carbon steel and stainless steels. When this becomes necessary, it is advisable to use stainless steel filler metals with higher alloying contents than the stainless steel being joined, e.g. types 308 (UNS S30800) or 309 (UNS S30900). This ensures that the weld bead, when diluted with the carbon steel base metal, will have a similar composition or be cathodic to both the carbon steel and the stainless steel components being joined [89].

At one time it was common for process equipment to be fabricated from carbon steel pressure vessel steel and then to be clad, using a variety of processes, with CRAs. This trend is changing, as the costs of cladding have escalated, and the trend is now to fabricate many of these vessels out of CRAs (Figure 4.51).

The discussion above concerned 300-series stainless steels based on Fe–Cr–Ni compositions, and these are the most commonly used stainless steels for process equipment. The high cost of nickel has led to the development of 200-series Fe–Cr–Mn stainless steels. These alloys are seldom used in oilfield applications because of inferior pitting corrosion resistance, especially in chloride-containing environments. The exception to this statement is UNS S20910 – normally referred to at Nitronic 50 alloy, which is included in Part 3 – Corrosion

Resistant Alloys of ANSI/ANSI/NACE MR0175/ISO 15156. Even though this alloy has a 200-series designation, because it does have Mn, it also has substantial Ni content. The nominal composition is Cr 22, Ni 12.5, Mn 5, Mo 2.25, Si 1, C 0.06, and Fe balance. This alloy has higher Cr and Ni contents than most 300-series austenitic stainless steels. It can also be used at higher strength/hardness levels than other austenitic stainless steels – HRC 35 instead of the HRC 22 limit for Fe–Cr–Ni austenitic stainless steel alloys [11].

Highly Alloyed Austenitic Stainless Steels These alloys are frequently called superaustenitic stainless steels, and some of them have such high alloying content that they no longer fit the strict definition of steel, which refers to iron-based (more than 50% iron) alloys. Table 4.11 shows some of these alloys.

Like all austenitic alloys, these metals can only be strengthened by alloying and, in thin sections, by cold working.

These alloys are used in similar applications to those for austenitic or duplex stainless steels where the environment is considered to be more aggressive. They can also be used in H₂S environments at strength/hardness levels above those for conventional austenitic stainless steels [11].

Grade 904L stainless steel is a nonstabilized austenitic stainless steel with low-carbon content. This high-alloy stainless steel has copper to improve its resistance to strong reducing acids, such as sulfuric acid. The steel is also resistant to stress corrosion cracking and crevice corrosion. Grade 904L is nonmagnetic and offers excellent formability, toughness, and weldability.

Grade 904L contains high amounts of expensive ingredients, such as molybdenum and nickel. Today, most of the applications that employed grade 904L have been replaced by low-cost duplex stainless steel 2205. The suitable alternatives to 904L stainless steels are shown in Table 4.12.

Figure 4.51 A 316 stainless steel separator to be used offshore in place of a clad vessel [90]. *Source*: Reproduced with kind permission of John R. Curry.

TABLE 4.11 Nominal Composition of Selected Highly Alloyed Austenitic Stainless Steels

USN	Name	C max	Cr	Ni	Mo	Cu	Minmum PREN[a]
S31254	254SMO	0.02	20	18	6.25	0.75	29.5
N08029	20 Cb3	0.07	20	35	2.5	3.5	29
N08367	AL6XN	0.03	21	24.5	6.5		20
N08904	904L	0.02	21	25.5	4.5	1.5	32

[a] The compositions shown in this table are the averages between the minimum and maximum levels. PREN numbers are calculated based on minimum numbers.

Duplex Stainless Steels Duplex stainless steels have a mixture of BCC ferrite crystals and FCC austenite crystals. The duplex phase structure is achieved by lowering the nickel content compared with austenitic stainless steels. The relative percentage of each phase depends on the alloy chemistry, but most of these alloys are intended to have approximately equal amounts of each phase in the alloy (Figure 4.52).

Duplex stainless steels may have similar pitting corrosion resistance to austenitic stainless steels, but their stress corrosion cracking resistance is often superior. Austenite is susceptible to chlorides, and ferrites may crack in H_2S environments. The combination of both phases in one alloy means that cracks that initiate in one phase are often blunted and stopped once they reach the other phase (Figure 4.53). Nonetheless, their use is sometimes limited to a maximum temperature of 65 °C (150 °F) in the presence of chlorides.

Table 4.13 shows the two most commonly used duplex stainless steels. ANSI/NACE SP0175/ISO 15156 requires that duplex stainless steels used in H_2S service be solution annealed (brought to thermodynamic equilibrium) condition [11]. This limits their strength but improves their corrosion resistance.

Some of the advantages of duplex stainless steels, when they are substituted for austenitic stainless steels, are:

- The lower-nickel content reduces costs. Nickel is substantially more expensive than chromium.
- Increased strength and hardness. ANSI/NACE MR0175/ISO 15156 allows higher hardness levels instead of the limitation of HRC 22 for many other alloys. The specific HRC limitations depend on the

TABLE 4.12 Possible Alternative Grades to Grade 904L Stainless Steels

Grade	Reasons for Choosing Grade 904L
316L	A lower cost alternative, but with much lower corrosion resistance
6Mo	A higher resistance to pitting and crevice corrosion is needed
2205	A very similar corrosion resistance, with the 2205 having higher mechanical strength, and at a lower cost to 904L (2205 not suitable for temperatures above 300 °C)
Super duplex	Higher corrosion resistance is needed, together with a higher strength than 904L

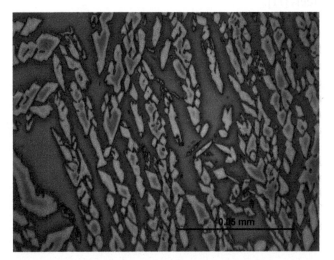

Figure 4.52 Microstructure of a duplex stainless steel. The dark phase is ferrite and the light phase is austenite. *Source*: Photo courtesy J. Ribble, Materials Evaluation & Technology Corporation, www.metco-ndt.com.

Figure 4.53 Stress-corrosion cracking in duplex stainless steel. *Source*: Chaung et al. [87]. Reproduced with permission of NACE International.

TABLE 4.13 Nominal Composition of Selected Duplex Stainless Steels

UNS Number	Common Name	Type	C Max	Cr	Ni	Mo	Fe	N	Minimum PREN[a]
S31803	2205	Duplex stainless steel	0.03	23	5.2	3	Bal	0.15	31
S32750	2507	Super duplex stainless steel	0.03	25	7	4	Bal	0.28	38

[a] The compositions shown in the table are the averages between the minimum and maximum levels. PREN numbers are based on minimum numbers.

alloy, and the latest versions and updates to the controlling documents should be checked.

- Increased resistance to stress corrosion cracking in some environments.

Limitations compared with austenitic stainless steels [77, 91, 92] are:

- Because of the BCC ferrite phase, these alloys become brittle at low temperatures and cannot be used in cryogenic service (e.g. for LNG). They can be used as an ambient-temperature arctic conditions.
- The austenite phase may be subject to environmental cracking in chloride-containing environments.
- The ferrite phase is subject to hydrogen embrittlement from excess cathodic protection and possibly other sources.

Recent revisions of ANSI/MR0175/ISO 15156 have altered the acceptable hardness levels for duplex stainless steels, and the acceptable levels seem to depend on both the alloy and the application. New alloys are being introduced, and the advantages and limitations of these alloys are poorly understood and not reflected in current industrial standards [85, 93]. Unfortunately, the newness of these materials causes difficulties in fabrication, because techniques for forming and welding these alloys are not well understood [91].

Duplex stainless steels are used for a wide variety of subsurface applications. Alternatives to the use of duplex stainless steels include superaustenitic stainless steels and, for lower cost and less corrosion resistance in many environments, austenitic stainless steels such as 317L.

Welding is important with duplex stainless steels to avoid the unwanted formation of brittle sigma phase, which also causes depletion of chromium from the surrounding austenite and can cause sensitization to intergranular corrosion. They can also form a chromium-rich α' (alpha prime) phase at temperatures from 300 to 600 °C (572–1112 °F) known as 475 °C embrittlement [94].

Precipitation-hardened Stainless Steels Precipitation-hardened stainless are chrome–nickel alloys having less nickel than austenitic stainless steels. They can be either austenitic or martensitic in their annealed condition. In most cases, the hardening process to produce high strength involves heat treatment to produce a high-martensite alloy. If they are to be used in H_2S service, they must then be tempered to remove the brittle martensite and convert it into tempered martensite, a fine-grained ferrite-plus-austenite structure.

Table 4.14 shows common grades of precipitation-hardening stainless steels. A chief advantage of precipitation-hardening stainless steels is that they can be machined to close dimensional tolerances and then heat treated for mechanical strength with minimal distortion. These alloys are normally available in forgings and similar products and are used for wellhead equipment, control devices, and similar applications. Their reduced alloying content when compared with many other CRAs limits their use in thin sections, and they are not normally supplied in tubular form. ANSI/NACE0175/ISO 15156 lists a number of typical applications and the temperature and environmental limits placed on these alloys in low partial pressure H_2S service. The same standard also prescribes different maximum hardness levels depending on alloy and application [12].

Precipitation-hardening stainless steels are considered to be intermediate in corrosion resistance between martensitic stainless steels and the austenitic and duplex stainless alloys. The most common of these alloys in oilfield use is UNS S17400 (17-4 PH).

Nickel-based Alloys

There are two different types of nickel-based alloys, solid-solution alloys and precipitation-hardened alloys. Pure nickel and the solid-solution alloys are FCC in structure and, similar to austenitic stainless steels, they are ductile but limited in the strength they can obtain. This is less of a limitation for upstream oilfield operations, which tend to be confined to temperatures of 450 °F (230 °C) and less, than in processing and refining where high-temperature processing is common and high-temperature strength is important.

Table 4.15 shows selected nickel-based alloys. These alloys tend to be more expensive than the iron-based alloys discussed above, and their use tends to be restricted to very corrosive environments where other alloys are unsuitable.

TABLE 4.14 Nominal Composition of Selected Precipitation-Hardening Stainless Steels

UNS	Name	C max	Fe	Cr	Ni	Mo	Other	Comments
S66286	A286	0.08	Bal	14.75	25.5	1.25	Ti = 2.12, B 0.001– 0.01, V 0.10–0.50	Austenitic
S17400	17-4 PH	0.07	Bal	16.25	4.0			Martensitic
S15500	15-5 PH	0.07	Bal	14.75	4.5		Cu = 3.5	Martensitic
S15700	PH 15-7 Mo	0.09	Bal	15	7	2.5		Martensitic

TABLE 4.15 Nominal Composition of Selected Nickel-based Alloys

USN	Name	Cr	Ni	Mo	Cu	Fe	Other	Comments
Selected Solid Solution Alloys Available as Tubular Products								
N04400	400		66		Bal			
N06022	C-22	21	Bal	13.5		4	W = 3	
N06255	SM2550	24.5	49.5	7	—	Bal	—	
N06625	625	21.5	Bal	9			Nb = 3.65	
N06985	G3	22.3	Bal	7	2.0	19.5	—	
N08028	28	27	31	3.5	1			
N08032	NIC 32	21.5	32	4.5	—	Bal	—	
N08042	NIC 42M	21.5	42	6	2.2	Bal	Ti = 0.9	
N08825	825	21.5	42	3	2.2	Bal	Ti = 0.9	
N10276	C276	15.5	Bal	16	—	5.5	W = 3.8	
Selected Precipitation-Hardening Alloys								
N05500	K-500		66		Bal		Ti = 0.6, Al = 2.7	High-strength version of N04400 used for bolts and other high-strength applications
N07718	718	19	52.5	3		Bal	Nb = 5.1, Ti = 0.9, Al = 0.5	API SPEC6A718 has additional requirements to avoid embrittling phases
N07725	725	20.7	57	8.3		Bal	Nb = 3.4, Ti = 1.4	
N07750	X-750	15.5	Bal	—	1	7	Nb = 3.3, Ti = 2.5, Al = 0.7	
N09925	925	21.5	42	3	2.3	Bal	Ti = 2.2, Al = 0.3	

Solid-solution Nickel-based Alloys   These alloys tend to be used for process equipment and similar topside applications where their relatively low strength, due to their single-phased FCC structure, is less important. They are also used for downhole tubular goods.

Precipitation-hardening Nickel-based Alloys
Precipitation-hardening nickel-based alloys can be fabricated into high-strength components such as bolts and other fasteners.

Unfortunately, these high-strength grades can be subject to hydrogen embrittlement, even the relatively low-levels of hydrogen produced on bare metal surfaces due to seawater cathodic protection systems [94–96]. Common applications for these higher-strength nickel-based alloys include wellhead and Christmas tree components, excluding bodies and bonnets, valve and choke components, springs, and bolts.

Both groups of nickel-based alloys find extensive use in seawater applications where their resistance to crevice corrosion and pitting allows their selection for seawater piping, pump shafts and impellers, and valves and valve inserts.

Cobalt-based Alloys

The relatively expensive alloys listed in Table 4.16 are used in very corrosive environments and situations where little or no corrosion can be tolerated. They can be used in H_2S service in the cold-worked and age-

TABLE 4.16 Nominal Composition of Selected Cobalt-based Alloys

UNS Name	Name	Cr	Ni	Co	Mo	Fe	Mn	W
R30003	Elgiloy	20	15.5	40	7	Bal	2	
R30004	Havar	20	13	42.5	2.4	Bal	1.6	2.8
R30035	MP-35N	20	35	Bal	9.7			

hardened condition at hardnesses up to HRC 55-60 for use in diaphragms, pressure-measuring devices, seals, and springs [11].

Cobalt-based alloys are finding increased use in wireline and downhole instrumentation packages. The wireline must be allowed to degas at surface temperatures for several days to allow dissolved hydrogen to be released. Carbon steels and other very-high-strength alloys might not be suitable in these environments, even for the limited times associated with these applications.

Titanium Alloys

Titanium applications in the oilfield are of two types. Most titanium is used because of the alloys' excellent resistance to corrosive environments. These applications tend to be topside and primarily in water-handling systems where their corrosion resistance is an advantage in seawater piping and heat exchangers. Sometimes titanium is used for downhole or subsea applications where the corrosion resistance, while still excellent, is less important and the strength-to-weight ratio of titanium, the highest of any commercially available alloys, becomes the reason for their choice. Table 4.17 shows typical titanium alloys used for both corrosion resistance and for mechanical properties reasons.

Corrosion-resistance applications tend to use commercially pure titanium (Ti grades 1–4) or, for heat exchangers and other complicated devices, titanium alloyed with palladium, which greatly improves the crevice corrosion resistance at elevated temperatures. Ruthenium is also added to titanium alloys for the same purposes and is considered to be a less expensive crevice corrosion addition.

The thermal conductivity of titanium is lower than for other commonly used heat exchanger tubing alloys. This means that titanium is frequently used with thinner gages than other alloys. The excellent corrosion resistance allows for this reduction, but heat exchanger tubing bundles frequently fail by fatigue caused by the turbulent fluid flow past their surfaces. The flexibility of thin titanium tubes must be accommodated by supplying more support baffles than would be required for some other tubing alloys.

Titanium plate-frame heat exchangers are often used offshore, because they offer significant weight savings over other materials [97]. The relatively high strength of titanium allows operation at pressures that cannot be achieved with aluminum heat exchangers, which offer similar weight savings.

While titanium alloys have generally good corrosion resistance, there are several cautions that should be observed with their use. Titanium will be cathodic to most other alloys. Depending on the relative size of the other metal, galvanic corrosion of the other alloy, typically carbon steel, will result. Unfortunately, hydrogen embrittlement of the cathodic titanium can also occur. Most organizations have decided to not mix metal systems, and it is common for seawater piping systems, heat exchangers, and ancillary equipment to be constructed from titanium to avoid any galvanic coupling problems.

Other environments that cause corrosion problems include hydrofluoric and uninhibited hydrochloric acidizing treatments. Methanol can cause stress corrosion cracking, although the presence of water will inhibit this corrosion.

Sometimes titanium is used for downhole or subsea applications where the corrosion resistance, while still excellent, is less important and the strength-to-weight ratio of titanium, the highest of any commercially available alloys, becomes the reason for their choice. Table 4.18 shows representative uses of titanium in offshore applications.

TABLE 4.17 Nominal Composition and Mechanical Properties of Selected Titanium Alloys

UNS Number	ASTM Grade	Alloy Composition	Minimum Tensile Strength		Minimum Yield Strength		Sour Service Approved?[a]
			ksi	MPa	ksi	MPa	
Alloys Primarily Chosen for Corrosion Resistance							
R50250	1	Unalloyed Ti	35	240	25	170	
R50400	2	Unalloyed Ti	50	345	40	275	Yes
R50550	3	Unalloyed Ti	65	450	55	380	
R50700	4	Unalloyed Ti	80	550	70	480	
R52400	7	Ti-0.15Pd	50	345	40	275	
R52250	11	Ti-0.15Pd	35	240	25	170	
R52402	16	Ti-0.05Pd	50	345	40	275	
R52252	17	Ti-0.05Pd	35	240	25	170	
R53400	12	Ti-0.3-Mo-0.8Ni	70	480	50	345	Yes
Alloys Primarily Chosen for Mechanical Properties							
R56400	5	Ti-6Al-4V	130	895	120	825	
R56401	23	Ti-6Al-4V ELI	115	790	110	755	
R56405	24	Ti-6Al-4V-0.05Pd	130	895	120	825	
R56403	25	Ti-6Al-4V-0.5Ni-0.05Pd	130	895	120	825	Yes
R56323	28	Ti-3Al-2.5V-0.1Ru	90	620	70	480	Yes
R56404	29	Ti-6Al-4V-0.1Ru	120	825	110	755	Yes
R56260	[b]	Ti-6Al-6Mo-2Sn-4Zr	170	1170	160	1100	Yes
R58640	[b]	Ti-3Al-8V-6Cr-4Mo-4Zr	170	1170	160	1100	Yes

[a] Listing as sour service approved means that the alloy is listed in table D.11 of NACE/ISO 15156, Part 3 for Corrosion-Resistant Alloys.
[b] No assigned ASTM titanium grade.

TABLE 4.18   Selected Uses of Titanium in Offshore Oil and Gas Production [98]

Application	Company	Project	Titanium Alloy Grade
Taper stress joints	Placid Oil	Green Canyon	23 (Ti-6Al-4V ELI)
Taper stress joints	Ensearch	Garden Banks	23
Taper stress joints	Oryx Energy	Neptune	23
Fire water systems	Norsk Hydro	Troll B (Oil) Brage, Visund	2 (commercially pure)
Fire water systems	Elf Petroleum	Froy TCP	2
Fire water systems	Statoil	Sleipnir West, Siri	2
Fire water systems	Statoil	Norne	2
Seawater lift pipes	Statoil	Sleipnir	2
		Veslefrikk	2
Ballast water systems	Mobil	Statfjord A/B Beryl	2
Ballast water systems		Hibernia	2
Penetration sleeves	Statoil	Sleipnir West	2
Penetration sleeves	Norsk Hydro	Oseberg	2
Penetration sleeves	Mobil	Statfjord	2
Freshwater pipework	Elf	Frigg	2
Seawater pipework	Esso	Jotun	2
Seawater pipework	Norsk Hydro	Njord, Visund	2 (110 tons)
Seawater systems, fire, ballast and produced water pipework	Statoil	Asgard B	2 (300 tons)
Gravity-based system	Statoil	Troll A (Gas)	2 (500 tons)
Drilling riser	Statoil (Conoco)	Heidrun	23
Booster lines	Statoil (Conoco)	Heidrun	9 (Ti-3Al-2.5V)
Anchor system pipework	Statoil (Conoco)	Heidrun	2
Penetrations and manholes	Statoil (Conoco)	Heidrun	2

Source: Reproduced with permission of Springer Nature.

Titanium alloys are difficult to weld and require special cleaning procedures to ensure that an oxide-free surface is available during welding and that the weld bead does not oxidize and prevent adequate fusion to the parent metal.

Copper Alloys

Pure copper is FCC and very ductile but has low yield and tensile strengths. For these reasons pure copper is only used for electrical conductivity and for electrical conduit. Piping and other structural applications are made from one of several alloy groups. Copper alloys have better thermal conductivity than other alloys, and this makes them the alloy system of choice for many heat-transfer applications, although their relative weight and erosion–corrosion susceptibility sometimes justifies the use of titanium, aluminum, or stainless steel for these applications. Copper alloys also have natural biofouling resistance, and this means that copper alloys are often used for piping systems where fouling cannot be tolerated, e.g. stagnant seawater firewater piping.

Copper-based alloys have three major environmental limitations, and extensive research has been devoted to

alloy development and design methods to minimize these limitations. Dealloying, the loss of one constituent of an alloy leaving an altered residual structure, is a potential problem for most copper-based alloys. This has been minimized by alloying addition controls and by the development of more resistant alloy systems, e.g. cupronickels, which, while not immune to this problem, are much less likely to have the problem than the earlier copper–zinc brasses that they have largely replaced for many condenser, heat exchanger, and piping applications [99, 100]. Copper alloys are also subject to stress-corrosion cracking in ammonia-containing environments. The third limitation is erosion corrosion. Maximum allowable fluid velocities are much lower for copper alloys than for carbon steel and most other, usually stronger and harder, alloys.

ANSI/NACE RP0175/ISO 15156 places no restrictions on the use of copper-based alloys in H_2S environments. This document, which addresses cracking problems associated with H_2S, does comment that, while many copper-based alloys have been used successfully in downhole environments, they can suffer other forms of corrosion in sour oilfield environments, particularly if oxygen is present [11]. The reference to oxygen is very

TABLE 4.19 Nominal Compositions of Selected Copper Alloys

UNS Number	Name	Cu	Al	Sn	Zn	Ni	Other
Wrought Copper Alloys							
C17200	Copper–beryllium	Bal					Be=2, Cu+Ni≥0.2, Co+Ni+Fe≤0.6
C44300	Admiralty brass	71		1	28		
C46400	Naval brass	60		0.8	39.2		
C2600	Cartridge brass	70			30		
C26130	Arsenical 70/30 brass	70			30		As=0.05
C28000	Muntz metal, 60%	60			40		
C61300	Aluminum bronze	92.65	7	0.35			
C61400	Aluminum bronze	91	7				Fe=2
C63000	Nickel aluminum bronze	82	10			5	Fe=3
C68800	Aluminum brass	73.5	3.4		22.7		Co=0.4
C70600	Cupronickel, 90–10	90				10	
C71500	Cupronickel, 70–30	69.5				30	Fe=0.5
C72200	Cupronickel plus Cr	82.2				16.5	Cr=0.5, Fe=0.8
Cast Copper Alloys							
C86300	Manganese bronze	63	6		25		Fe=3, Mn=3
C90300	Tin bronze	88		8	4		
C90500	Tin bronze	88		10	2		
C90700	Tin bronze	89		11			
C95400	Aluminum bronze	85	11				Fe=4
C95500	Nickel aluminum bronze	80	11			4.3	Fe=4
C95800	Nickel aluminum bronze	81	9			5	Fe=4, Mn=1
C95900	Aluminum bronze	Bal	12.8				Fe=4
C96200	Cast copper–nickel	88.6				10	Fe=1.4

important. Unlike most other oilfield alloys, copper is resistant to acids, because the equilibrium potential for copper is noble to the hydrogen reduction reaction at all pHs, and oxygen is the most likely source of a reducible chemical to balance corrosion-related oxidation of the metal. Unfortunately, copper-based alloys have alloying additions that depend on passive films to limit corrosion, and these alloys can corrode in some acids, although the corrosion rate is often very slow.

Some representative copper-based alloys are listed in Table 4.19.

Wrought Copper-based Alloys Most wrought copper-based alloys are used for piping systems and heat transfer, e.g. water-cooled condensers. At one time most wrought copper-based alloys were brasses made from alloying copper with zinc. Zinc strengthened the alloy, but it lowered the overall corrosion resistance and made the alloy subject to dezincification, a form of dealloying. In recent years the trend has been to use copper–nickel alloys for most condensers and other copper-based piping systems, especially in seawater service [101].

All copper alloys are considered to have excellent antifouling characteristics, and this makes them desirable for firewater systems, where fouling in stagnant seawater could lead to debris plugging nozzles and other

Figure 4.54 Underdeposit corrosion of 90-10 cupronickel tubing in offshore Gulf of Mexico firewater system.

tight restrictions during emergencies. Unfortunately most metals, to include copper alloys, are subject to underdeposit attack, a form of crevice corrosion. This is shown in Figure 4.54 for 90-10 cupronickel (UNS C70600) tubing removed from an offshore platform firewater system. The tubing corroded at the six o'clock position due to debris collecting at this location.

Problems like this have led some organizations to specify fiberglass and similar materials for firewater systems in recent years. The smoothness of the plastic piping interior retards biofouling attachment, and polymers are not susceptible to underdeposit attack.

Aluminum brass, which is a copper–zinc alloy with aluminum additions, was at one time used onshore in freshwater for erosion–corrosion resistance near condenser tube inlets, but this is no longer the practice for large condensers [102]. Velocity limits to prevent erosion in copper-based piping systems vary from $4\,ft\,s^{-1}$ $(1.2\,m\,s^{-1})$ for pure copper up to $15\,ft\,s^{-1}$ $(4.5\,m\,s^{-1})$ for 70-30 cupronickel (UNS C71500) [101].

Copper beryllium alloys (e.g. UNS C17200) are unusual for copper-based alloys, because they are very strong. Yield strengths for UNS C17200 can range from 32 to 100 ksi (220–700 MPa) depending on component size and heat treatment. Oilfield applications are primarily for nonsparking tools. The tools are not as strong as steel tools and wear out faster, but their increased cost is justified for safety reasons. Copper–beryllium alloys are also used for nonmagnetic measurement-while-drilling tools for directional drilling, both downhole and in pipelines. A partial list of other oilfield applications includes nonmagnetic drill string components, subsea valve gates, springs, and fasteners.

Cast Copper-based Alloys   At one time most cast copper-based alloys were copper–tin alloys called bronzes. Today the term bronze is applied to a wide variety of copper-based alloys having little or no tin. Bronze most often refers to copper-based alloy systems where the major alloying element is neither zinc (brasses) nor nickel (cupronickels). The relative scarcity of tin and the associated high costs of copper–tin alloys have led to the development of aluminum bronzes (copper-based alloys with aluminum additions) and nickel–aluminum bronzes.

While bronzes are available as wrought products, they are most commonly used for castings. The phase diagrams of all bronzes are similar to the copper–tin diagram shown in Figure 4.55. The melting temperatures of bronzes drop much quicker for these alloys than for brasses and cupronickels. This makes them easier to melt and pour. Equally important, the low-temperature microstructure of these alloys is two phased, which means that, compared with brasses and cupronickels, they are harder and more erosion resistant. Thus they are better suited for the complicated geometries and fluid flow patterns found in pumps, valves, and similar equipment. The relative thickness of castings compared to fluid transfer piping and heat-transfer tubing means that any reduced corrosion resistance of these alloys can usually be tolerated.

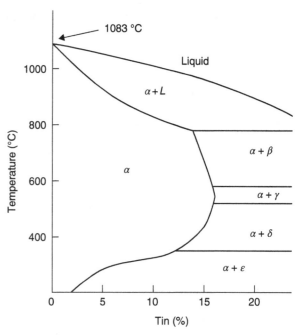

Figure 4.55   The copper–tin phase diagram.

The most common alloys for large seawater pumps are nickel–aluminum bronzes (e.g. UNS C95500 or C95800). Foundry practice for these alloys is critical, because they can form unwanted microstructures that are subject to selective phase attack, a form of dealloying [103–105].

Copper-based Alloys in Hydrocarbon Service   Copper alloys are acceptable in accordance with the ANSI/NACE MR0175/ISO 15156 guidelines for environmental cracking in oilfield H_2S service. The latest 2016 version of the standard says that copper alloys have been used without restrictions on the key environmental parameters of H_2S concentration, chloride concentration, and pH. However, the same standard cautions that environmental degradation, such as weight-loss corrosion, may be a problem, particularly if oxygen is present. Furthermore, the standard cautions that some copper-based alloys may be sensitive to galvanically induced hydrogen stress cracking (GHSC).

The lack of published literature on copper alloys in aggressive wellbore environments led to a laboratory study of several copper alloys under consideration for gate valve and wellhead applications. The overall corrosion rates for the copper-based alloys tested, aluminum bronze (UNS C63000 and C62730), beryllium–copper alloy (UNS C17200), and copper–nickel alloy (UNS C96900) were substantially higher (by orders of magnitude) than for the baseline alloys, 17-4PH stainless steel (UNS S17400), type 410 stainless steel (UNS S41000), type 660 stainless steel (UNS

S66286), nickel-based alloy 718 (UNS N07718), and nickel-based alloy 625 (UNS N06625), tested in the same production-simulation environments. This laboratory exposure test at 350 °F (177 °C) suggests that copper-based alloys should not be used in aggressive high-temperature H_2S environments. The use of the same alloys at lower temperatures and pressures may be acceptable, but testing in accordance with a testing protocol such as recommended by NACE or other standards is advisable [106].

Aluminum Alloys

Aluminum alloys have had limited uses in upstream oilfield operations, but their use is increasing. Aluminum is a very reactive metal, and the natural oxide film that forms on aluminum surfaces is usually much thicker than on the other CRAs discussed in this chapter. This makes aluminum very resistant to atmospheric corrosion, even in marine environments.

Aluminum has several other very desirable characteristics. Many aluminum alloys have high strength-to-weight ratios. This is a major advantage for offshore platform topside structures such as helicopter decks (helidecks) and modular crew quarters and offices. It is ductile at low, even cryogenic temperatures. This makes aluminum piping and process equipment an excellent choice for LNG processing and similar low-temperature applications.

Unfortunately, aluminum is also an amphoteric metal, which means it corrodes much faster in both acidic (low pH) and caustic (high pH) environments. High-pH environments are relatively less important, because most minerals have limited solubility at high pHs, and the metal tends to stay dry underneath a protective mineral scale, but many production fluids have low pHs making aluminum unsuitable for some applications. Aluminum alloys have no restrictions on cracking susceptibility in H_2S environments [11]. This is somewhat surprising, because aluminum is known to suffer hydrogen embrittlement in some environments [107, 108]. Aluminum is also subject to liquid metal embrittlement, and mercury must be removed from natural gas streams before they enter aluminum piping and heat exchangers [109].

Aluminum is often used for jacketing thermally insulated piping. Moisture ingress can cause corrosion at the six o'clock position where liquid collects and is exposed to moist air containing CO_2. (Therefore components should be designed to allow moisture to escape, usually through drainage holes placed at regular intervals along the bottom of the jacketing.) A similar situation is shown in Figure 4.56, which shows corrosion at the bottom of an aluminum moist air ventilation tube.

Figure 4.56 Aluminum corrosion at the six o'clock position where acidic atmospheric condensate water collected in a low-temperature ventilation system.

Aluminum Alloy Designations   The alloy designation system for aluminum alloys requires a chemistry designation and a temper designation. Most industrial users use the system developed by the Aluminum Association in North America. An abbreviated version of these classifications is shown in Table 4.20. The table only shows the alloy designations for wrought alloys, but the system for cast alloys is very similar. In both cases, a complete specification would describe the chemistry and temper in a single number, e.g. aluminum 6061-T6, which is the most commonly specified heat treatable alloy of aluminum in the usually specified strongest temper.

The most commonly used wrought aluminum alloys are listed in Table 4.21. Note that each alloy has both a chemistry designation and one or more typical temper designations. The UNS numbers for these alloys are also listed, although they are seldom used except in the writing of international standards. The UNS system does not allow for temper designations, although some users will simply add the temper designation to the alloys, e.g. Aluminum Association alloy 5052-H32 would be UNS A95052H32.

Note that the aluminum alloys fall into two groupings – those that can be thermally treated for strength and those that can only be strain hardened (work hardened). Sheet and plate aluminum alloy components are typically the work hardened 3xxx and 5xxx alloys, while extrusions for structural members and tubular products are typically produced from the thermally treated (age-hardened) 6xxx alloys.

Typical Applications for Various Aluminum Alloy Groups   Commercially pure aluminum alloys, the 1xxx group, are the most CRAs. They are so soft that they are

TABLE 4.20   Wrought Aluminum Alloy Designation Systems

Numbering System for Wrought Aluminum Alloys Used in Oil and Gas Production

1xxx	Super- or commercial-purity aluminum	Not heat treatable, used as corrosion-resistant cladding on stronger aluminum alloys
2xxx	Copper is the major alloying addition	Heat treatable
3xxx	Manganese is the major alloying addition	Work hardening
4xxx	Silicon is the major alloying addition	Work hardening, but Al–Si alloys are usually used as castings (4xxx alloys are typically the cladding for braze clad products)
5xxx	Magnesium in the major alloying addition	Work hardening
6xxx	Magnesium and silicon are the major alloying additions, usually at approximately 2:1 magnesium to silicon	Heat treatable

Temper Designations for Wrought Aluminum Alloys

F	As fabricated, with no specified control over hardening processes
O	Annealed to soft state
H	Strain hardened
H1	Strain hardened only
H3	Strain hardened and stabilized
H32/H34	Strain hardened and stabilized to ¼ hard (H32) or ½ hard (H34) condition
H116	For 5xxx alloys in marine service, this temper is for alloys that have been strain hardened as the final operation in manufacture and meet specified levels of exfoliation and intergranular corrosion
H321	For 5xxx alloys in marine service, this temper is for alloys that have been thermally stabilized as the final operation in manufacture and meet specified levels of exfoliation and intergranular corrosion
T	Thermally treated for mechanical properties
T5	Cooled from hot working and thermally aged
T6	Solution treated and thermally aged

TABLE 4.21   Typical Wrought Aluminum Alloys Used in Oil and Gas Production

UNS Number	Aluminum Association Number	Typical Use
A92024	2024-T6	Drill pipe
A92014	2014-T6	Drill pipe
A93105	3105-H14	Thermal jacketing for insulated piping
A93003	3003-H14	Housing/office modules, thermal jacketing for insulated piping, and brazed aluminum heat exchanger components
A95052	5052-H32 or -H34	Housing/office modules and brazed aluminum heat exchanger components
A95083	5083-H116 or -H321	Plate for helidecks, hulls for workboats, and brazed aluminum heat exchanger components
A95086	5086-H116 or -H321	Plate for helidecks, hulls for workboats, and brazed aluminum heat exchanger components
A95454	5454-H32 or -H34	Plate for helidecks, hulls for workboats, and brazed aluminum heat exchanger components
A95456	5456-H116 or -H321	Plate for helidecks, hulls for workboats
A96061	6061-T5 or -T6	Extrusions for railings and ladders and brazed aluminum heat exchanger components
A96063	6063-T5 or -T6	Extrusions for railings and ladders and brazed aluminum heat exchanger components
A96005	6005-T5 or -T6	Extrusions for railings and ladders
A96105	6105-T5 or -T6	Extrusions for railings and ladders
A96082	6082-T5 or -T6	Extrusions for railings and ladders

not used in oilfield applications except as corrosion-resistant surfaces on clad structural-alloy plate.

Most users consider the 5xxx and 6xxx alloys to be the most suitable for marine applications, and their use in nonmarine applications is also widespread. A typical workboat or helideck would have plate components of 5xxx alloys and the supporting structure/frame made from 6xxx alloys. Aluminum fasters could also be made from 6xxx alloys, but aluminum has serious galling problems, and it is common to use specially designed connections when mechanically joining aluminum structural components.

Clad aluminum alloys are available. They were originally developed with commercially pure aluminum surfaces for corrosion control over stronger structural aluminum alloy plate. Many complex aluminum heat exchangers, to include those used in oilfield heat exchangers, have low-melting temperature exteriors with structural alloys in the center. These clad products are used to manufacture brazed aluminum structures such as heat exchangers.

Brazed aluminum heat exchangers are frequently used offshore for gas processing. The parting sheets, which separate the fluids, are usually made from 3xxx alloys, and they are brazed together using a low-melting point aluminum–silicon alloy from the 4xxx series [110]. The aluminum–silicon alloys have very low-melting temperatures in comparison with other aluminum alloys, and they are usually used as castings. They are seldom used for structural purposes in oilfield applications, and this is the reason they are not listed in Tables 4.20 and 4.21.

Welding of aluminum is possible, but it is more difficult than for carbon steel and other iron-based alloys. Complex machinery and process vessels are factory welded, but it is common to use mechanical connections in oilfield construction and assembly, e.g. of helidecks, while similar structures from carbon steel would be welded.

There have been limited attempts to develop and market aluminum drill pipe. Drill pipe is usually made from 2xxx alloys containing copper as the main alloying addition. The 2xxx alloys have some of the highest fatigue strengths of any aluminum alloys, and this property, combined with their low weight, makes them acceptable for many applications in other industries. They were the first high-strength aluminum alloys developed. Their corrosion resistance, while adequate for many aerospace applications, is not sufficient for oilfield operations except in circumstances like drill pipe, where they are only exposed to downhole fluids for limited periods of time and can be inspected for corrosion and other damage, typically wear and fatigue cracking, between uses [111, 112].

Additional Considerations with CRAs

CRAs are used whenever the increased capital cost is justified by reduced maintenance and inspection costs or when increased reliability is necessary. Many high-volume gas wells are so corrosive that carbon steels are not considered, and the question becomes which CRAs should be used. While some gas fields have high-H_2S levels, most gas well corrosion is due to CO_2.

Oil fields are generally not corrosive until the water cut increases and/or the system "sours" due to the increased production of H_2S. For this reason, it is common to specify that all equipment be constructed from materials considered resistant to H_2S-related cracking in accordance with the appropriate guidelines in ANSI/NACE MR0176/ISO 15156 [9–11]. These precautions do not prevent other forms of environmental degradation, and corrosion can be the unfortunate result.

The increased alloying content of CRAs means that they cost more than carbon steels. A way of reducing costs is to use CRA cladding or lining on carbon steel components. This long-standing practice for wellhead and process equipment has been extended to downhole equipment and tubular goods. The CRA can be applied by a number of processes. If CRAs are mechanically bonded using either explosive bonding or thermal shrinking, the composite structures are referred to as a lined pipe instead of clad pipe [76].

ANSI/NACE MR0175/ISO 15156-3 places restrictions on the use of cladding, linings, and weld overlays [11]:

> Unless the user can demonstrate and document the likely long-term in-service integrity of the cladding or overlay as a protective layer, the base material, after application of the cladding or overlay shall comply with ANSI/NACE MR0175/ISO 15156-2 or this part of ANSI/NACE MR0175/ISO 15156, as applicable.
> This may involve the application of heat or stress-relief treatments that can affect the cladding, lining, or overlay properties.
> Factors that can affect the long-term in-service integrity of a cladding, lining, or overlay include environmental cracking under the intended service, the effects of other corrosion mechanisms, and mechanical damage.

Similar restrictions are placed on wear-resistant coatings or hard facings.

Pitting Resistance Equivalent Numbers (PRENs)
Many organizations use the PREN formula in NACE MR0176/ISO 15156 as a basis for determining the relative corrosion resistance of oilfield alloys. The PREN formula in this standard is [11]:

$$PREN = w_{Cr} + 3.3(w_{Mo} + 0.5w_W) + 16w_N$$

where

w_{Cr} is the weight percentage of chromium in the alloy.

w_{Mo} is the weight percentage of molybdenum in the alloy.

w_W is the weight percentage of tungsten in the alloy.

w_N is the weight percentage of nitrogen in the alloy.

The NORSOK version of PREN does not include tungsten in the calculation [25].

Larger values of PREN are considered to indicate greater resistance to pitting corrosion. The contribution of other alloying additions, e.g. nickel and copper, is not considered in this formula, and most authorities recommend using PREN as a general guideline. It is important to remember that the PREN numbers shown in NACE MR0175/ISO 15156 were included to aid in classifying alloys into different categories. Many experts are of the opinion that they are unreliable indicators of corrosion resistance except in a very general sense, i.e. if two alloys have widely separate PREN numbers, then it is logical to conclude that the higher PREN number will indicate greater corrosion resistance, but for numbers that are close, e.g. within 5 PREN numbers, the performance may be affected by alloying parameters not considered by the PREN formula, e.g. the nickel content and the microstructure.

Other PREN number formulas have been developed for other applications, e.g. seawater, but their use is not widespread in oilfield applications.

It is common to rank alloys by PREN and then to consider their relative resistance to cracking in environments of interest. The logic behind this ranking procedure is that alloys will first develop pits that serve as stress raisers for the initiation of subsequent cracking [113].

Temperature Criteria Pitting and crevice corrosion are temperature dependent, and a number of tests have been developed to determine the temperature at which these forms of corrosion are likely to occur. Most of these tests involve exposing metal samples with artificial crevices to increasing temperatures until a predetermined amount of crevice or pitting corrosion is observed. The idea is to limit the use of an alloy to temperatures where these forms of corrosion are unlikely. Unfortunately, the published data for critical pitting temperatures (CPTs) and critical crevice corrosion temperatures (CCTs) produce widely conflicting data (up to ±20 °C [36 °F]), so the use of published data, even if obtained by following an ASTM or other standard, is questionable. Service environments may produce pitting or crevice corrosion at temperatures significantly lower than those determined in controlled laboratory tests. The best use of these tests is for their intended purpose of ranking alloys insofar as their pitting or crevice corrosion tendencies [114–116].

Alloy Selection The cheapest and fastest method of selecting alloys for corrosive environments is to rely on published information. The drawbacks to this approach are that changes in environments, even within the same field, can cause significant changes in corrosion resistance [117]. Several NACE and ISO standards provide guidance on alloy selection, and these guidelines are often followed [9–11, 25, 64, 76]. Unfortunately, the most commonly followed guidelines, ANSI/NACE MR0175/ISO 15156 Parts 1–3 only cover resistance to hydrogen-related cracking, and many organizations are surprised when materials chosen in accordance with "NACE guidelines" suffer other forms of environmental degradation, e.g. pitting or chloride-related stress corrosion cracking. It is possible that organizations that start using ISO 21457 will have more comprehensive guidance on alloy selection, as this standard includes references to the ANSI/NACE/ISO standard, but it also gives guidance on methods of avoiding weight-loss corrosion [64].

For aggressive conditions and/or new fields, it is common to select candidate alloys according to the following pattern:

- Selection for general corrosion resistance.
- Selection for localized attack resistance.
- Selection of CRAs by environmental specification, e.g. in accordance with MR0175/ISO 15156 [1].

This preliminary selection allows an initial screening of candidate alloys that are then often evaluated in a series of controlled laboratory exposures, in environments as close as possible to the anticipated field conditions, before final alloy selection [1, 85]. Important environmental considerations include the aging of a field for oil wells and the presence of organic acids in gas wells [1].

Note that some alloys, e.g. 3xx stainless steels, cannot be used above certain temperatures because of a concern for chloride stress corrosion cracking. These and other alloys cannot be used for downhole tubulars because of strength considerations, and this limits most downhole alloy selections to considerations between carbon steel, martensitic stainless steel, and nickel-based alloys.

The most commonly specified CRAs are martensitic stainless steels, commonly referred to as 13Cr or super-13Cr alloys. They are widely used in downhole applications and for subsea pipelines. They are seldom used for topside surface applications, because they are subject to pitting corrosion.

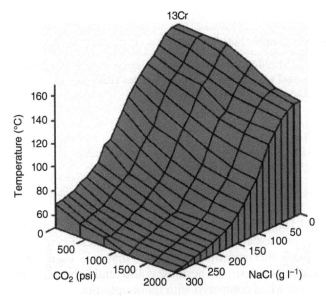

Figure 4.57 Corrosion resistance of 13Cr (UNS S42000) in the absence of oxygen and H_2S [86]. *Source:* Reproduced with permission of John Wiley & Sons.

A common progression of alloy selection, useful for high-temperature gas wells and other corrosive environments, is:

Carbon steels → martensitic stainless steels (13Cr and "Super 13Cr") → austenitic stainless steels (often limited because of temperature considerations and relatively low strength) → duplex stainless steels → high-nickel austenitic alloys.

This is reflected in Figures 4.57–4.59, which show the results of laboratory tests in simulated oilfield environments. The figures show environments where corrosion rates will be less than or equal to $0.05\,mm\,yr^{-1}$ (2 mpy) with no sulfide corrosion cracking or stress corrosion cracking. The limitations shown for these and other nickel-containing alloys are probably conservative [86].

Other alloy systems are used for specialized applications, both downhole and topside, but these are the most commonly specified alloys for downhole OCTG and other applications.

Topside applications have fewer strength-related and temperature-related limitations, and a general summary of the kinds of alloys often used for topside applications is shown in Table 4.22.

POLYMERS, ELASTOMERS, AND COMPOSITES

Polymers are organic materials. Elastomers and composites used in oilfield applications are materials based on polymers.

Elastomers are rubbery materials that are capable of recovering their original shape after being deformed. They are used in seals and similar applications.

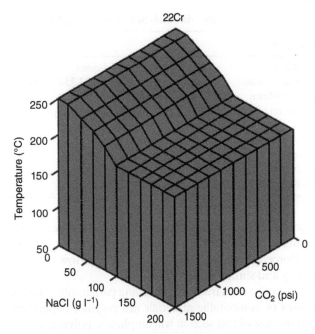

Figure 4.58 Corrosion resistance of duplex stainless steel 2205 in the absence of oxygen and H_2S [86]. *Source:* Reproduced with permission of John Wiley & Sons.

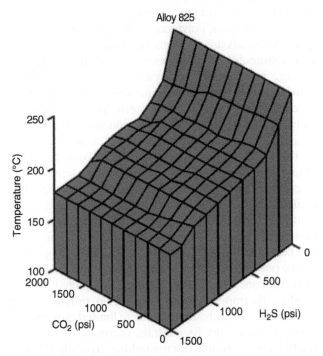

Figure 4.59 Corrosion resistance of nickel based alloy 825 (UNS N08825) in the absence of oxygen and H_2S [86]. *Source:* Reproduced with permission of John Wiley & Sons.

Composites are materials made from two or more constituent materials with significantly different properties. When combined together in a composite, the constituent materials maintain their separate identities but contribute to the overall performance of the

TABLE 4.22 Topside Materials Selection Examples

Equipment		Typical Material Selection		
Flowlines	Super duplex	UNS S31254 254 SMO	Carbon steel with inhibitors	
Produced water	UNS S31254 254 SMO	Carbon steel with inhibitors	Glass reinforced plastic	
Flare systems	UNS31603 SS316L	Low temperature – carbon steel		
Water injection	Carbon steel (<5 ppb oxygen)			
Heat exchanger	Titanium	Super duplex		
Chemical injection	Titanium	PVDF-PE	Carbon steel	
Firewater	Cupronickel	Glass reinforced plastic		
Gas-cooling heat exchangers	Aluminum			

composite. Most oilfield composites have fiber reinforcements and polymer matrixes. Fiberglass and other fiber-reinforced polymers (FRPs) are often used in low-pressure piping and similar applications.

As can be seen from the preceding paragraphs, all three classes of nonmetallic materials are polymer dependent, and the rest of this section will emphasize polymers.

Polymers are organic materials consisting of large molecules formed from small precursors called monomers. The term polymer is a combination of the terms poly, signifying many, and mer, which signifies repeat units or molecules of the original monomer(s). Some polymers are homopolymers, polymers made from one monomer type, and others are copolymers, which are formed from two or more different monomers that are joined together in large molecules of copolymeric materials. Chemical names for polymers are combinations of the name of the precursor monomer(s) and the prefix "poly," e.g. polyethylene is formed from ethylene monomer. The atomic weight of polymers is typically in the thousands [4].

There are a number of ways of classifying polymers. One of the most common is based on their temperature responses. Thermoplastic polymers soften and melt when heated and stiffen and solidify at lower temperatures. By contrast, thermosetting polymers undergo chemical reactions upon heating during manufacture or fabrication and become solids with no melting points. Thermoplastics can be remelted and recycled, whereas thermosets cannot be melted once the setting reaction is completed.

Most thermoplastics are long-chain polymers that may become branched but will have limited crosslinking between the chains. By contrast, thermosets have relatively higher amounts of crosslinking, typically 10–50% of the chain repeat units, which leads to stiffening and relatively brittle behavior [4].

Common oilfield thermoplastic materials include high-density polyethylene (HDPE) and polyvinyl chloride (PVC). Both of these materials are used for oilfield piping, and HDPE is also used for storage tanks up to several thousand gallons in capacity. Fluoropolymers, commonly referred to by trade names such as Teflon®

and Halar®, are also thermoplastics, even though they can be used at relatively high temperatures. Epoxies are typical thermosetting polymers that are used when hardness is desired. Thermosets also tend to be relatively brittle when compared with thermoplastics.

Elastomers are rubbery flexible polymers used for o-rings, gaskets, and seals of many types. They usually are made from thermosetting polymers with limited crosslinking. The limited crosslinking allows these materials to deform without breaking and, if the load is released, they recover their original shape.

Plastics are polymers that can be molded or shaped. They are stiffer than elastomers, usually due to higher molecular weights. While all plastics are polymers, many polymers are not plastics, even though the terms are often used interchangeably.

The long chains in polymers are normally based on covalent carbon–carbon bonds. The most notable exception is silicones, a class of polymers with silicon–oxygen bonds as the repeat units in the polymer chain. This is shown in Figure 4.60, which shows the structures of several common polymers.

Additives to polymers are used as stabilizers against oxidation and ultraviolet (UV) degradation, flame retardants, fillers, plasticizers in flexible piping, and reinforcements. The presence of these additives means that materials from different suppliers may have different degradation mechanisms and environmental vulnerabilities. This is why NACE and other organizations have developed standards for testing polymers, composites, and elastomers for suitability to various oilfield environments [117–125].

Unlike metals and ceramics, the mechanical properties of polymers are strongly dependent on loading rates. There are standards for the loading rates to be used in determining hardness and strength, and these differ in some ways from the methods used in determining similar properties in metals. Virtually all polymers are subject to creep at ambient temperatures, in marked contrast to most metals. This is one reason for the widespread use of reinforcement additives in many polymers.

The increasing use of polymers is primarily due to their relatively good corrosion resistance and, in the case of composites, their excellent strength-to-weight ratios. The smooth surfaces of polymeric piping are also advantageous in minimizing biofouling. This is a major reason why they are increasingly used for firewater systems, where debris from biofouling can plug valves and similar small openings [126, 127].

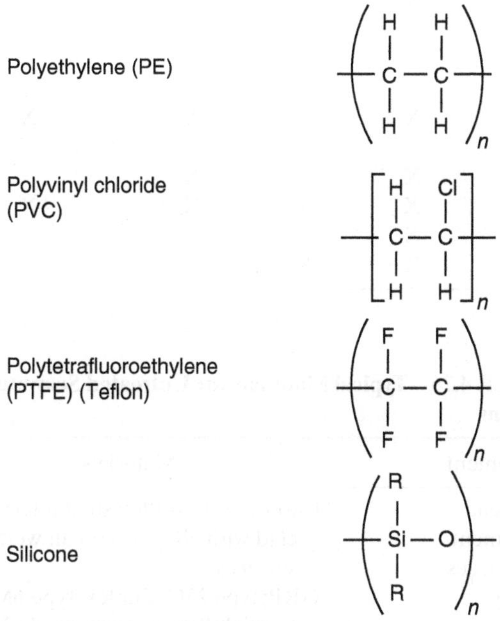

Polymeric piping is more subject to mechanical damage than metals, and this is the reason why polymeric piping is not used for many aboveground hydrocarbon piping systems. Even composites are more subject to mechanical damage, and, unlike metals, composites will tend to fracture instead of bending when overloaded. This has been a major problem with the installation and use of relatively large composite storage tanks. They have been known to crack due to shipping and construction-site handling.

All polymers are permeable to gases. This limits their use for storing liquids if air contamination is a concern, e.g. produced water intended for reinjection. It also means that polymers will absorb gases, and to a lesser extent water and small hydrocarbon liquid molecules. Rapid decompression of gaskets, o-rings, and liner materials on metal piping is a concern with their use, and standards for testing and rating these polymers have been developed [117, 123, 125]. Swelling due to absorption of gases and liquids can also occur.

Polymers have relatively low upper temperature limits, and they also become brittle when cold. They should only be used in their prescribed temperature ranges.

Polymer degradation includes UV degradation and chemical attack, which can be swelling, softening due to

Figure 4.60 Selected polymer repeat units. The R in the silicone structure stands for a radical. It is usually CH_3, but other versions of silicone are also available.

TABLE 4.23 Elastomer Deterioration Modes

Failure Mode	Description
Fracture/rapid tearing	Exceeded strength properties. Consider extremes of operational environment (pressure, elevated temperature, load, etc.) remembering strength may reduce due to aging and fluid absorption
Rapid gas or explosive decompression	Gas dissolved at high pressure comes out of solution and forms bubbles in the material when pressure is lost. Bubbles may cause fracture of the material or of an interface
Stress relaxation	Loss of force (strength) over time under contact deformation resulting in loss of sealing. Physical and chemical aging effects – usually the latter in long-term performance. This effect may be countered by swelling due to thermal expansion and/or absorption of fluids
Creep	Increased deformation over time due to both physical and chemical aging effects. Chemical aging usually governs long-term performance. Can lead to extrusion failures in seals
Swelling	Absorption of fluids resulting in excessive stress if constrained (e.g. seals) or deformation and weaking if unrestrained. Enhanced by thermal expansion effects. Governed by compatibility of the material and the fluid
Thermal contraction	Caused by reduction in temperature that may also produce hardening and increased stress relaxation
Chemical degradation (aging)	Chemical changes due to reaction with oxygen or other oxidizers or ongoing vulcanization (anaerobic aging). Resultant changes may include stiffness or excessive fatigue forces in flexible joints
UV and ozone cracking	Surfaces exposed to UV or ozone prior to installation or during service must be resistant, e.g. hose covers, piping, and polymer-reinforced composite structures like walkways
Fatigue crack growth	Fatigue resistance may be reduced by elevated temperatures, aging, and swelling by fluids
Abrasion erosion	Loss of material by rubbing or fluid flow with abrasive medium
Bond failure	Hose and fittings and metal plates are boded to elastomer layers. Inadequate bonding may be due to manufacturing conditions or degradation caused by fluid ingress and corrosion

Source: From Campion et al. [128].

TABLE 4.24 Deterioration Possibilities for Various Elastomeric Components

Failure Mode	Static Seal	Packers and Plugs	Repair Clamps	Dynamic Seal	Hoses	Flexible Joints	Valve Sleeves	Pulsation Bladder and Bellows
Fracture/rapid tearing	X	X	X	X	X	X	X	X
Explosive decompression	X		X		X	X		
Stress relaxation	X		X	X	X			
Creep/extrusion	X	X	X					
Swelling	X	X	X	X	X			
Thermal contraction	X	X						
Chemical degradation (aging)	X	X	X	X	X	X	X	X
UV and ozone cracking					X		X	
Fatigue crack growth				X	X	X	X	X
Abrasion/erosion				X	X			
Bond failure					X	X		

Source: From Campion et al. [128].

TABLE 4.25 Design Information for Materials Selection

Information to be Provided
Project design basis, Annex A
Corrosion-prediction model
Future changes in reservoir H_2S-content
Methodology or model for pH calculation of produced water
Formation water analysis
Content of mercury in production fluids or gas
The oxygen content in deaerated seawater for injection
Erosion-prediction model
Temperature limitations for use of stainless steels in marine atmosphere
Limitations in mechanical properties and use of materials
Temperature limitations for nonmetallic materials
Environmental requirements regarding use of corrosion inhibitors
Model for inhibitor evacuation, corrosion inhibition test methods, and acceptance criteria
Use of external coatings to increase maximum temperature for stainless steel (SS)
Applicable standard for cathodic protection (CP) design to be defined
Strength and hardness limitation of fasteners in marine atmosphere

Source: Adapted from ISO 21457: 2010 [64].

leaching of additives, or oxidation. The attack can be of the polymer matrix or of additives and reinforcements. Low temperatures can cause brittleness in otherwise flexible polymers and elastomers.

Unfortunately nondestructive testing methods to monitor polymer degradation are not available beyond routine inspection for swelling or discoloration.

Table 4.23 lists common elastomer deterioration modes. Other polymers will have similar problems, but

TABLE 4.26 Typical Materials for Untreated Seawater Systems

Equipment	Materials
Wellhead equipment/ Xmas trees	Carbon or low-alloy steel internally clad with alloy 625 on all wetted surfaces
Piping	GRP; type 25Cr duplex; type 6Mo; CuNi 90/10[a]; titanium grade 2
Vessels	Carbon or low-alloy steel[b] with internal organic coating or lining in combination with cathodic protection; GRP; type 6Mo; type 25Cr duplex
Pumps	Type 25Cr duplex
Valve body/bonnet	Carbon or low-alloy steel clad with alloy 625; type 25Cr duplex
Valve internals	Type 25Cr duplex or alloys with equivalent or better corrosion resistance

[a] CuNi 90/10 is not compatible with CRAs or more noble materials with respect to galvanic corrosion.
[b] Carbon steel clad with CRA may be used as alternative to solid CRA.
Source: Adapted from ISO 21457: 2010 [64].

the use of elastomers in high-pressure sealing applications means that these are the most important examples for upstream oil and gas operations. The components most likely to fail are listed in Table 4.24.

Polymer degradation can arise from:

- Incorrect material selection
- Inadequate design
- Inadequate specification
- Incorrect installation
- Operation outside design limits
- Careless handling

In summary polymeric materials, while relied upon much less heavily than metals, are subject to failure modes not found in metals. Because most engineers, even materials engineers, have limited understanding of polymers, it is especially important to carefully test and monitor supplies to be sure they have the desired properties and will not deteriorate in unexpected ways.

Materials Selection Guidelines

The previous sections of this chapter have placed emphasis on compliance with ANSI/NACE MR0195/ISO 15156, but other problems including weight-loss corrosion and fatigue are also important. ISO 21457 was introduced in 2010 with the objective of covering the noncontroversial issues related to materials selection in oil and gas production systems. The developers of this standard intended it to cover approximately 80% of materials selection decisions for both onshore and offshore gas and oil production and to supplement the information contained in NORSOK M-001, which is directed at offshore oil and gas production [25, 115, 116]. This would allow the design organization to focus on complex and project-specific materials selections. Tables 4.25 and 4.26 show some examples of information contained in ISO 21457. Additional tables in this standard show typical materials for many common components of offshore oil and gas production systems and also suggest temperature limits for specific alloys in various environments.

REFERENCES

1 Kane, R. (2006). Corrosion in petroleum production operations. In: *Metals Handbook, Volume 13C – Corrosion: Environments and Industries* (ed. S.D. Cramer and B.S. Covino Jr.), 922–966. Materials Park, OH: ASM International.

2 Craig, B.D. (2004). *Oilfield Metallurgy and Corrosion*, 3. Denver: MetCorr.

3 Shoesmith, D.W. (1987). Effects of metallurgical variables on aqueous corrosion. In: *Metals Handbook, Volume 13 – Corrosion* (ed. S.D. Cramer and B.S. Covino Jr.), 45–49. Materials Park, OH: ASM International.

4 Callister, W.D. and Renthwisch, D.G. (2008). *Fundamentals of Materials Science and Engineering: An Integrated Approach*. Hoboken, NJ: Wiley.

5 Davis, J.R. (2000). *Corrosion: Understanding the Basics*, 33. Materials Park, OH: ASM International.

6 *Incomplete Weld in a Carbon Steel Gas Main Pipe*. Corrosion Testing Laboratories, Inc. http://www.corrosionlab.com/Failure-Analysis-Studies/29074.incomplete-weld.carbon-steel.htm (accessed 21 June 2016).

7 He, M.L. (2007). Metallographic interpretation of steel forging defects. *Microscopy and Microanalysis* 13 (Supplement S02): 1050–1051. doi: 10.1017/S143192760707105X (About DOI), Published online: 05 August 2007.

8 ASTM A962. *Common Requirements for Bolting Intended for Use at Any Temperature Range from Cryogenic to the Creep Range*. West Conshohocken, PA: ASTM International.

9 ANSI/NACE MR0175/ISO 15156-1. *Petroleum and Natural Gas Industries – Materials for Use in H_2S-Containing Environments in Oil and Gas Production – Part 1: General Principles for Selection of Cracking-Resistant Materials*. Houston, TX: NACE International.

10 ANSI/NACE MR0175/ISO 15156-2. *Petroleum and Natural Gas Industries – Materials for Use in H_2S-Containing Environments in Oil and Gas Production – Part 2: Cracking-Resistant Carbon and Low-Alloy Steels, and the Use of Cast Irons*. Houston, TX: NACE International.

11 ANSI/NACE MR0175/ISO 15156-3. *Petroleum and Natural Gas Industries – Materials for Use in H_2S-Containing Environments in Oil and Gas Production – Part 3: Cracking-Resistant CRAs (Corrosion-Resistant Alloys) and Other Alloys*. Houston, TX: NACE International.

12 Hertzberg, R.W. (1996). *Deformation and Fracture Mechanics of Engineering Materials*, 4. New York: Wiley.

13 API 5CT/ISO 11960:2004. *Specification for Tubing and Casing*. Washington, DC: API.

14 ASTM A370. *Test Methods and Definitions for Mechanical Testing of Steel Products*. West Conshohocken, PA: ASTM International.

15 ASTM E8/E8M. *Test Methods for Tension Testing of Metallic Materials*. West Conshohocken, PA: ASTM International.

16 ISO 15579. *Metallic Materials – Tensile Testing at Low Temperature*. Geneva: ISO.

17 Bannister, A.C. (19 February 1999). Sub-task 2.3 report: yield stress/tensile stress ratio: results of experimental programme. In: *Structural Integrity Assessment Procedures for European Industry*. Rotherham: British Steel plc.

18 API 5L. *Specification for Line Pipe*. Washington, DC: API.

19 ASTM E18. *Test Methods for Rockwell Hardness of Metallic Materials*. West Conshohocken, PA: ASTM International.

20 Milliams, D.E. and Tuttle, R.N.. *ISO 15156/NACE MR0175 – A New International Standard for Metallic Materials for Use in Oil and Gas Production in Sour Environments*, NACE 03090. Houston, TX: NACE International.

21 ISO 6507. *Vickers Hardness Test – Part 1: Test Method*. Geneva: ISO.

22 ISO 6508-1. *Rockwell Hardness Test – Test Method (Scales A, B, C, D, E, F, G, H, K, N, T)*. Geneva: ISO.

23 ISO 18265. *Tables for Comparison of Hardness Scales*. Geneva: ISO.

24 ASTM E140. *Hardness Conversion Tables for Metals – Relationship Among Brinell Hardness, Vickers Hardness, Rockwell Hardness, Superficial Hardness,*

Knoop Hardness and Scleroscope Hardness. West Conshohocken, PA: ASTM International.

25 Pavlina, E.J. and Van Tyne, C.J. (2008). Correlation of yield strength and tensile strength with hardness of steels. *Journal of Materials Engineering and Performance* 17 (6): 888–893.

26 Eliassen, S. and Smith, L. (2009). *Guidelines on Materials Requirements for Carbon and Low Alloy Steels for H₂S-Containing Environments in Oil and Gas Production, EFC,* 3. European Federation of Corrosion.

27 ASTM E23. *Standard Test Methods for Notched Bar Impact Testing of Metallic Materials.* West Conshohocken, PA: ASTM International.

28 API Specification 5D. *Specification for Drill Pipe.* Washington, DC: API.

29 NORSOK M-001. *Materials Selection.* Lysaker, Norway: Standards Norway.

30 Subramanian, K., Penso, J., and Pordal, H. Computational methods to support brittle fracture assessment of PSVs involving autorefrigeration. In: *ASME 2015 Pressure Vessels and Piping Conference,* 19–23. Boston, MA: Paper No. PVP-2015-45201, PP. V007TO07AOO8, 5 pages.

31 Bai, Y. and Bai, Q. (2005). *Subsea Pipelines and Risers,* 270. Amsterdam: Elsevier.

32 Bauman, J. *Materials Selection and Corrosion Issues on an Offshore Producing Platform,* NACE 04126. Houston, TX: NACE International.

33 Brooks, C.R. and Choudhury, A. (2002). *Failure Analysis of Engineering Materials.* New York: McGraw-Hill.

34 API 571. *Damage Mechanisms Affecting Fixed Equipment in the Refining Industry.* Washington, DC: API.

35 API 579/ASME FFS-1. *Fitness for Service.* New York: ASME.

36 ASME B31.3. *Process Piping Design.* New York: ASME.

37 ASME. *BPV Code: Section VIII, Division 1: Design and Fabrication of Pressure Vessels.* New York: ASME.

38 Ellenberge, J.P., Chuse, R., and Carson, B.E. (2004). *Pressure Vessels: The ASE Code Simplified.* New York: McGraw-Hill.

39 API 650. *Welded Steel Tanks for Oil Storage.* Washington, DC: API.

40 API 602. *Design and Construction of Large, Welded, Low-Pressure Storage Tanks.* Washington, DC: API.

41 Phaal, R. and Wiesner, C.S. (1993). *Toughness Requirements for Steels.* Cambridge, UK: Woodhead Publishing, Ltd.

42 Escoe, A.K. (2006). *Pipeline and Pipeline Assessment Guide.* Houston, TX: Gulf Publishing.

43 Tipper, C.F. (1962). *The Brittle Fracture Story.* Cambridge, UK: University Press.

44 Palmer, A. and King, R. (2006). *Subsea Pipeline Engineering.* Tulsa, OK: PennWell.

45 DNVGL-RP-C203. (April 2016). *Fatigue Design of Offshore Steel Structures.* Olso, Norway: DNV GL.

46 ANSI/ASME B31G. *Manual for Determining the Remaining Strength of Corroded Pipelines.* New York: ASME.

47 Tomar, M.S. and Fingerhut, M. *Reliable Application of RSTRENG Criteria for In-the-Ditch Assesment of Corrosion Defects in Transmission Pipelines,* NACE 06176. Houston, TX: NACE International.

48 Tada, H., Paris, P.C., and Irwin, G.R. (2000). *The Stress Analysis of Cracks Handbook.* New York: ASME.

49 NACE SP0502. *Pipeline External Corrosion Direct Assessment Methodology.* Houston, TX: NACE International.

50 Smart, J.S. (1980). Corrosion failure of offshore steel platforms. *Materials Performance* 19 (5): 41–48.

51 Barsom, J.M. and Rolfe, S.T. (1999). *Fracture and Fatigue Control in Structures: Applications of Fracture Mechanics,* 3. West Conshohocken, PA: ASTM.

52 Kundu, T. (2008). *Fundamentals of Fracture Mechanics.* Boca Raton, FL: CRC Press.

53 Anderson, T.L. (2005). *Fracture Mechanics,* 3. Boca Raton, FL: CRC Press.

54 ASTM A333. *Seamless and Welded Steel Pipe for Low-Temperature Service and Other Applications with Required Notch Toughness.* West Conshohocken, PA: ASTM International.

55 Borenstein, S.W. (1994). *Microbiologically Influenced Corrosion Handbook.* Cambridge, UK: Woodhead Publishing, Ltd.

56 NORSOK M-601 (April 2008). *Welding and Inspection of Piping,* 5. Lysaker, Norway: Standards Norway.

57 Belanger, A. and Barker, T. Multiple data inspection of hard spots and cracking. In: *ASME 2014, 10th International Pipeline Conference,* vol. 2, IPC2014-33060. Calgary, Alberta, Canada: Pipeline Integrity Management.

58 Byars, H.G. (1999). *Corrosion Control in Petroleum Production,* 2. Houston, TX: NACE International.

59 Smith, L.M. and Celant, M. (1998). *CASTI Handbook of Cladding Technology,* 2. Edmonton, Alberta, Canada: CASTI Publishing.

60 Gutierrez, I. (2016), 3D printing and NACE MR0175/ISO 15156 (5 May 2016). http://oilandgascorrosion.com/3d-printing-nace-mr0175-iso-15156 (accessed 22 March 2018).

61 ASTM E527. *Numbering Metals and Alloys in the Universal Numbering System.* West Conshohocken, PA: ASTM International.

62 SAE HS-1086/ASTM DS-561. *Metals & Alloys in the Unified Numbering System.* Troy, MI: SAE International.

63 ASTM E527/SAE J1086. *Recommended Practice for Numbering Metals and Alloys.* West Conshohocken, PA: ASTM International.

64 ISO 21457:2010. *Materials Selection and Corrosion Control for Oil and Gas Production Systems.* Geneva: ISO.

65 Skar, J. and Olsen, S. (2016). *Development of the NORSOK M-001 and ISO 21457 Standards – Basis for Defining Materials Application Limits,* NACE-2016-7433. Houston, TX: NACE.

66 Maddox, R.N. and Morgan, J. (1998). *Gas Conditioning and Processing, Volume 4: Gas Treating and Sulfur Recovery,* Campbell Petroleum Series, 4. Norman, OK: John M. Campbell.

67 Fruehan, R.J. (1998). Overview of steelmaking processes and their development. In: *The Making, Shaping, and*

Treating of Steel, 1–12. Pittsburgh, PA: AISE Steel Foundation.

68 BS 4515-1:2009. *Specification for Welding of Steel Pipelines on Land and Offshore. Carbon and Carbon Manganese Steel Pipelines*. London: British Standards Institute.

69 Bailey, N. (1994). *Weldability of Ferritic Steels, Welding Institute Edition*, 147. Cambridge, UK: Woodhead Publishing, Ltd.

70 Kor, G. and Glaws, P. (1998). Ladle refining and vacuum degassing. In: *The Making, Shaping, and Treating of Steel*. Pittsburgh, PA: AISE Steel Foundation.

71 ASTM A53. *Pipe, Steel, Black and Hot-Dipped, Zinc-Coated, Welded and Seamless*. West Conshohocken, PA: ASTM International.

72 ASTM A285/A285M. *Pressure Vessel Plates, Carbon Steel, Low- and Intermediate-Tensile Strength*. West Conshohocken, PA: ASTM International.

73 ASTM A106/A106M. *Seamless Carbon Steel Pipe for High-Temperature Service*. West Conshohocken, PA: ASTM International.

74 ASTM A516/A516M. *Pressure Vessel Plates, Carbon Steel, for Moderate- and Lower-Temperature Service*. West Conshohocken, PA: ASTM International.

75 ASTM A333/A333M. *Seamless and Welded Steel Pipe for Low-Temperature Service*. West Conshohocken, PA: ASTM International.

76 NACE 1F192(2000 Revision). *Use of Corrosion-Resistant Alloys in Oilfield Environments*. Houston, TX: NACE International.

77 ASTM A353/A353M. *Pressure Vessel Plates, 9 Percent Nickel, Double-Normalized and Tempered*. West Conshohocken, PA: ASTM International.

78 ASTM A553/A553M. *Pressure Vessel Plates, Alloy Steel, Quenched and Tempered 8 and 9 Percent Nickel*. West Conshohocken, PA: ASTM International.

79 ASTM A844/A844M. *Steel Plates, 9% Nickel Alloy for Pressure Vessels, Produced by the Direct-Quenching Process*. West Conshohocken, PA: ASTM International.

80 BS 10028-4. *Flat Products Made of Steels for Pressure Purposes. Nickel Alloy Steels with Specific Low Temperature Properties*. London: British Standards Institute.

81 Gioielli, P.C. and Zettlemoyer, N. (2007). S-N fatigue tests of 9% nickel steel weldments. *Proceedings of the Seventeenth International Offshore and Polar Engineering Conference*, Lisbon, Portugal (1–6 July 2007).

82 Kuppan, T. (2000). *Heat Exchanger Design Handbook*, 839–1082. Marcel Dekker.

83 Zeemann, A. and Emygdio, G. *9% Ni Alloy Steel for H_2S Service*, NACE C2014-4361. Houston, TX: NACE International.

84 Sedriks, J. (2003). Corrosion resistance of stainless steels and nickel alloys. In: *Metals Handbook, Volume 13A – Corrosion: Fundamentals, Testing, and Protection* (ed. S.D. Cramer and B.S. Covino), 697–702. Materials Park, OH: ASM International.

85 Craig, B.D. (1995). *Selection Guidelines for Corrosion Resistant Alloys in the Oil and Gas Industry*, Technical Publication No. 10073. Toronto: The Nickel Development Institute.

86 Kelly, J. (2004). Stainless steels. In: *Handbook of Materials Selection* (ed. M. Kutz), 77. Hoboken, NJ: Wiley.

87 Chaung, H.E., Watkins, M., and Vaughn, G.A. (1996). Stress-corrosion cracking resistance of stainless alloys in sour environments. In: *Corrosion Resistant Alloys in Oil and Gas Production* (ed. J. Kolts and S.W. Ciraldi), 779–785. Houston, TX: NACE International.

88 Lyons, W.C. and Plisga, G.J. (2005). *Standard Handbook of Petroleum & Natural Gas Engineering*, 6-371–6-405. Houston, TX: Gulf Professional Publishing.

89 Carbon Steel Handbook, 101467. (2007). Palo Alto, CA: EPRI.

90 Curry, J. (2015). Improvements in the steels used in oil and gas processing equipment over the last half century (7 December 2015). https://www.petroskills.com/blog/entry/improvements-in-the-steels-used-in-oil-and-gas-processing-equipment (accessed 19 April 2018).

91 Mahajanam, S., Rincon, H., and McIntyre, D. (2010). Metallurgical examination of defects in duplex stainless steel pipe fittings. *Materials Performance* 49 (4): 56–61.

92 Bhattacharya, A. and Singh, P.M. (2008). Role of microstructure on the corrosion susceptibility of UNS S32101 duplex stainless steel. *Corrosion* 64: 532–540.

93 API TR 938C (2011). *Use If Duplex Stainless Steels in the Oil Refining Industry*, 2. Washington, DC: API.

94 Srhidhar, N. and Cragnolino, G. (1992). Chapter 5: stress-corrosion cracking of nickel-base alloys. In: *Stress-Corrosion Cracking* (ed. R. Jones), 131–180. Materials, OH: ASM International.

95 Efird, K.D. (1985). Failure of monel Ni-Cu-Al alloy K-500 bolts in seawater. *Materials Performance* 24: 37.

96 Wolfe, L.H. and Joosten, M.W. (1988). Failures of nickel/copper bolts in subsea applications. *SPE Production Engineering* 3 (3): 382–386.

97 Manning, F.S. and Thompson, R.E. Chapter 12: heating and cooling. In: *Oilfield Processing of Petroleum: Volume One – Natural Gas*, 265–343. Pennwell.

98 Schutz, R.W., Baxter, C.F., Boster, P.L., and Fores, F.H. (2001). Applying titanium alloys in drilling and offshore production systems. *Journal of the Minerals, Metals, and Materials Society* 53 (4): 33–35.

99 Lenard, D.R., Martin, J., and Heidersbach, R. (1999). Dealloying of cupro-nickels in stagnant seawater. Paper No. 314, Corrosion/99, Orando (April 1999).

100 Verink, E. and Heidersbach, R. (1972). Evaluation of the tendency for dealloying in metal systems. In: *Localized Corrosion-Cause of Metal Failure*, ASTM STP 516 (ed. M. Henthorne), 303–322. West Conshohocken, PA: ASTM.

101 Warren, N. (2006). *Metal Corrosion in Boats*, 3, 81. Dobbs Ferry, NY: Sheridan House.

102 Gaffoglio, C.J. (1982). A new copper alloy for utility condenser tubes. *Power Engineering* 86 (8): 60–62.

103 Lenard, D.R., Bayley, C.J., and Noren, B.A. (2008). Electrochemical monitoring of selective phase corrosion of nickel aluminum bronze in seawater. *Corrosion* 64: 764–772.

104 Han, Z., He, Y., and Lin, C. (2000). Dealloying characterizations of Cu-Al alloy in marine environment. *Journal of Materials Science Letters* 19: 393–395.

105 Culpan, E. and Foley, A. (1982). The detection of selective phase corrosion of nickel aluminum bronze by acoustic emission techniques. *Journal of Materials Science* 17: 953–964.

106 Haeberle, T. and Maldonado, J. *Alloys in Simulated Oil and Gas Production Environments*, NACE04106. Houston, TX: NACE International.

107 Kamoutsi, H., Haidemenopoulos, G.N., Bontozoglou, V., and Pantelakis, S. (2006). Corrosion-induced hydrogen embrittlement in aluminum alloy 2024. *Corrosion Science* 48 (5): 1209–1224.

108 Speidel, M.O. (1984). Hydrogen embrittlement and stress corrosion Cracknig of aluminum alloys. In: *Hydrogen Embrittlement and Stress Corrosion Cracking* (ed. A.R. Troiano, R. Gibala and R.F. Hehemann), 271–296. American Society for Metals.

109 Wilhelm, S.M. (2008). Risk analysis for operation of aluminum heat exchangers contaminated by mercury. *Proceedings, 4th Global Congress on Process Safety, AIChE Spring Meeting*, New Orleans.

110 Campbell, J.M. (2004). *Gas Conditioning and Processing, Volume 2: The Equipment Modules*, Campbell Petroleum Series, 8 (ed. R.A. Hubbard), 92–93. Norman, OK: John M. Campbell.

111 ISO 15546:2007. *Aluminum Alloy Drill Pipe*. Geneva: ISO.

112 Saakiyan, L.S., Efremov, A.P., Klyarovskii, V.M., and Orfanova, M.N. (1985). Corrosion behavior of aluminum drill pipe under interoperation storage conditions. *Materials Science* 21 (1): 83–84.

113 Hibner, E.L. and Tassen, C.S. *Corrosion Resistant OCTGs and Matching Age-Hardenable Bar Products for a Range of Sour Gas Service Options*, NACE 01102. Houston, TX: NACE International.

114 ASTM G48. *Test Methods for Pitting and Crevice Corrosion Resistance of Stainless Steels and Related Alloys by Use of Ferric Chloride Solution*. West Conshohocken, PA: ASTM International.

115 Steinsmo, U., Rogne, T., Drugli, J.M., and Gartland, P.O. (1997). Critical crevice temperature for highly-alloyed stainless steels in chlorinated seawater applications. *Corrosion* 53 (1): 26–32.

116 Sridhar, N., Dunn, D.S., Brossia, C.S., and Cragnolino, G.A. (2005). Chapter 19: crevice corrosion. In: *Corrosion Tests and Standards: Application and Interpretation*, 2, ASTM Manual 20e (ed. R. Baboian), 221–232. West Conshohocken, PA: ASTM International.

117 NORSOK M-710. *Qualification of Non-metallic Materials and Manufacturers*. Lysaker, Norway: Standards Norway.

118 NACE TM0187. *Evaluating Elastomeric Materials in Sour Gas Environments*. Houston, TX: NACE International.

119 NACE TM0192. *Evaluating Elastomeric Materials in Carbon Dioxide Decompression Environments*. Houston, TX: NACE International.

120 NACE TM0296. *Evaluating Elastomeric Materials in Sour Liquid Environments*. Houston, TX: NACE International.

121 ISO 23936-1. *Non-metallic Materials in Contact with Media Related to Oil and Gas Production – Part 1: Thermoplastics*. Geneva: ISO.

122 ISO 23936-2. *Non-metallic Materials in Contact with Media Related to Oil and Gas Production – Part 2: Elastomers*. Geneva: ISO.

123 NACE TM0297. *Effects of High-Temperature, High-Pressure Carbon Dioxide Decompression on Elastomeric Materials*. Houston, TX: NACE International.

124 NACE TM0298. *Evaluating the Compatibility of FRP Pipe and Tubulars with Oilfield Environments*. Houston, TX: NACE International.

125 NACE RP0491. *Worksheet for the Selection of Oilfield Nonmetallic Seal Systems*. Houston, TX: NACE International.

126 Dexter, S. (2003). Microbiologically influenced corrosion. In: *Metals Handbook, Volume 13A – Corrosion: Fundamentals, Testing, and Protection, Environments and Industries*, 399–416. Materials Park, OH: ASM International.

127 API RP 14G. *Recommended Practice for Fire Prevention and Control on Fixed Open-Type Offshore Production Platforms*. Washington, DC: API.

128 Campion, R.P., Thomson, B., and Harris, J.A. (2005). Elastomers for Fluid Containment in Offshore Oil and Gas Production: Guidelines and Review, HSE Research Report 320.

5

FORMS OF CORROSION

INTRODUCTION

There are many systems for classifying corrosion, but the most common categories follow a pattern popularized by M. Fontana in a series of articles he wrote for *Chemical and Engineering News* in the 1940s and later in the textbooks he published while a professor at Ohio State University [1].

The forms listed in Dr. Fontana's textbook are:

- Uniform attack
- Galvanic or two-metal corrosion
- Pitting
- Crevice corrosion
- Intergranular corrosion
- Selective leaching
- Erosion corrosion
- Stress corrosion
- Hydrogen damage

An advantage of this form of corrosion classification is that it can usually be confirmed by visual inspection, which allows identification of possible remedial measures without laboratory analysis [1]. The reason that the terminology for forms of corrosion described in the above list has gained widespread use is that they are tied to appropriate methods of corrosion control. The above forms of attack are common in the chemical process industry, where Fontana started his career. They also require water in some form as a part of the environment. In addition to the above forms of attack, his textbook discusses high-temperature corrosion,

corrosion of metals in elevated-temperature gaseous environments.

All of the above forms are commonly encountered in oilfield applications and are discussed in some manner in other corrosion textbooks and reference materials [2–20]. Some of the terminology has changed in recent years, and this chapter will attempt to use the terminology most likely to be discussed in oilfield literature.

The most important departure from Dr. Fontana's classification system is that this book discusses environmentally assisted cracking instead of stress corrosion cracking (SCC). The reason for this departure from Fontana's system is because most authorities on SCC consider the mechanism, at least for carbon and low-alloy steels, to be hydrogen embrittlement (HE) caused by reactions with the environment. This SCC-related HE is sometimes termed environmental hydrogen embrittlement (EHE) to distinguish it from HE caused by manufacturing processes, often termed internal hydrogen embrittlement (IHE).

Other authors use slightly different classifications of corrosion. Figure 5.1, from a general corrosion textbook, emphasizes the idea that some of Fontana's forms of corrosion are easily identified, whereas others may require laboratory microscopic examination to determine what has happened [16]. Fontana suggested that careful examination of the corroded surfaces or equipment can often identify the problem and how it can be corrected [1]. Note how Figure 5.1 emphasizes that most corrosion is localized. This means that weight-loss corrosion tests and other means of characterizing corrosion to produce average corrosion rates have limited usefulness in

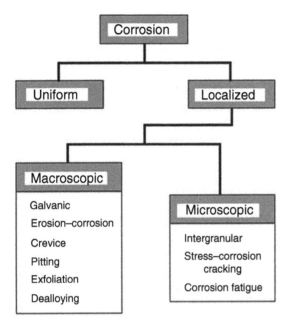

Figure 5.1 Forms of corrosion. *Source*: Davis [16]. Reprinted with permission of ASM international.

real-life industrial settings. The localized nature of corrosion is especially important in oil and gas production and in flowline and pipeline operations.

It is unfortunate that many life prediction and corrosion monitoring procedures assume uniform general corrosion and ignore the idea that localized corrosion of various types is the most likely source of corrosion problems on large-scale equipment. The oil and gas industry concentrates on pitting corrosion, the most likely source of piping and vessel leaks, and environmentally assisted cracking (SCC), which is a potential source of sudden fluid releases or structural failures.

The end of this chapter discusses terms and forms of corrosion unique to the oilfield environment. It would be possible to discuss most of these oilfield-related terms in discussions of the more universal terminology used in other industries, but it is important to understand the unique circumstances associated with these other forms, or terms, for corrosion.

It is also important to remember that any metal, or metallic system, can have, and usually does have more than one form of corrosion occurring simultaneously. Successful corrosion control requires that all forms of corrosion likely to occur in a system are addressed.

GENERAL CORROSION

The term general corrosion is often used instead of the earlier term uniform attack. It is intended to describe situations where the overall surface of a

Figure 5.2 Localized general corrosion caused by fog condensation dripping from overhead equipment onto process piping.

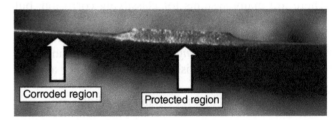

Figure 5.3 Cross section of a carbon steel tray in an amine sweetening unit. *Source*: Byars [9]. Reproduced with permission of NACE International.

metal undergoes attack. The metal gradually becomes thinner until the structure fails. This attack is seldom uniform in nature, and that is why the earlier term, uniform corrosion, has fallen into disfavor. This is shown in Figure 5.2, which shows localized general corrosion caused by water dripping onto process piping in a gas processing plant.

Most general corrosion is localized in nature, but on occasion uniform attack does occur. Uniform corrosion of carbon steel occurs when the surface is not protected by passive films (formed by reactions of the metal with the environment), scales (deposits of minerals from the environment), or protective coatings. This happens in many acidic environments and can be explained by examination of the potential–pH diagram for iron (Figure 2.11). Figure 5.3 shows uniform corrosion on a tray removed from an amine sweetening unit. The top center of the photo was protected by a mounting clip and did not corrode. The top side to the left and right was corroded by acidic liquids flowing across the tray, and the bottom of the tray was exposed to acidic vapor condensation and also corroded [9].

Figure 5.4 General corrosion along the bottom of a gas well flowline where acidic condensate thinned the bottom of this horizontal piping. It is obvious that the acidic water flowed at two different levels during the lifetime of this equipment. *Source*: Byars [9]. Reproduced with permission of NACE International.

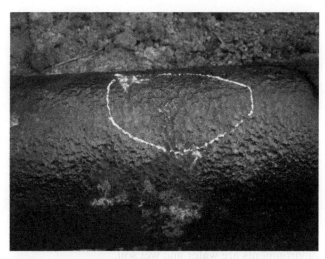

Figure 5.5 General attack of pipeline exterior beneath disbonded pipeline coating.

Uniform attack corrosion rates are controlled by the transport of reducible species to the metal surface. This transport results in linear corrosion kinetics, i.e. doubling the time of exposure will produce twice the amount of corrosion. Whenever this happens it is relatively easy to predict corrosion rates based on inspection reports, exposure tests prior to construction, weight-loss sampling with corrosion coupons, electrochemical monitoring, etc.

Unfortunately, most general attack is not uniform, and this can lead to areas of greater and lesser metal loss. This is shown in Figure 5.4, which shows general corrosion along the bottom of a gas well flowline. Similar corrosion patterns are found in condensate return lines of steam systems where air leaks cause the liquid in the lines to become acidic. This corrosion pattern is sometimes termed condensate or CO_2 channeling.

The rippled surface of the pipeline exterior shown in Figure 5.5 is a common appearance for general corrosion. In this case the corrosion was caused by a disbonded coating that led to corrosion by groundwater seeping underneath the debonded coating. Debonded coatings leading to relatively wide areas of general corrosion are a major problem with the pipeline industry. Modern pipeline coatings are much less likely to produce this corrosion pattern.

General corrosion is the most common form of corrosion and accounts for most of the corrosion experienced worldwide, to include in upstream oil and gas operations. It is relatively unlikely to be of major technical concern, because it can be monitored and replacements can be planned. It is important to understand when corrosion degradation will be general,

or "uniform," in nature and to understand that most corrosion-associated equipment failures will result from the other forms of corrosion, which tend to be localized in nature.

General corrosion is more likely to occur in locations where acidic water collects. This can be at the bottom of horizontal piping, as shown in Figures 5.2 and 5.4, but it can also occur in locations where condensation occurs due to thermal gradients. It is important to monitor corrosion rates in the appropriate location including the bottom of horizontal flowlines and the top of lines where uninhibited condensate can collect. Figure 5.2 shows localized corrosion in a very large gas processing plant in a coastal location. The plant operators had installed atmospheric corrosion rate-monitoring panels at various locations around the plant, but this atmospheric monitoring was not useful in determining where condensation from overhead equipment was likely to produce corrosion.

General corrosion control is normally controlled by design, e.g. adding thicker metal corrosion allowances (Figure 1.4), by including appropriate drainage and inspection capabilities, by the use of protective coatings or corrosion inhibitors, or by the selection of corrosion-resistant alloys (CRAs). Cathodic protection and modification of the environment, e.g. dissolved gas removal, can also be effective.

NORSOK M-001, ISO21457, and other design codes suggest corrosion allowances for design purposes, but, as shown in Figures 5.2–5.4, most corrosion, even general corrosion, is localized, and design codes cannot account for localized variations in corrosion rates [18, 21].

The general unsightliness of general atmospheric corrosion often leads to remedial action before the loss of metal becomes an engineering concern. Unfortunately, this is not always the case. Inspection for corrosion in hard-to-reach locations is a major problem in most industries.

GALVANIC CORROSION

Galvanic corrosion can result from electrical contact between two different metals. It can also be caused by any situation that produces changes in electrochemical potential, e.g. differences in temperature, chemicals in the environment, etc. In order for galvanic corrosion to occur, the anode and cathode must be in electrical contact and exposed to a continuous electrolytic environments. The most common electrolytic environments are water and wet soil.

Galvanic Coupling of Two or More Metals

When two corroding metals are electrically connected in the same electrolyte, the more active metal will tend to have more oxidation and corrode at a faster rate, while the less active, or noble metal, will have diminished oxidation and corrosion. This is the principle of cathodic protection and of galvanic coatings, e.g. zinc coatings or galvanizing, on metal surfaces. Unfortunately, most galvanic couples are between carbon steel, the most common structural metal, and more CRAs, which tend to be cathodic and to increase the corrosion of the carbon steel. Table 5.1 is based on work by the International Nickel Company at their former seawater laboratory in North Carolina. It shows the relative galvanic relationships between metals and alloys in quiescent seawater at their North Carolina harbor facility. This table is widely cited to indicate the relative potentials of metals in seawater worldwide [1].

Note that no voltage numbers are indicated on this table. This is because of slight fluctuations in the potential depending on salinity, dissolved oxygen levels, and other seawater variables. The brackets in the table indicate metals that are considered to be galvanically compatible. Many operators have adopted the policy of having fluid-handling systems based on stainless steel, copper alloys, titanium alloys, and so forth and try to not mix the alloys in a given process stream whenever possible.

A typical example of galvanic corrosion is shown in Figure 5.6, which shows a brass valve connected to galvanized steel piping. The galvanic corrosion on the pipe exterior is obvious.

Control of the water chemistry on the inside is necessary to prevent general corrosion of the galvanized steel, which is seldom galvanized on the pipe interior. The lack of corrosion on the pipe interior emphasizes that all forms of corrosion, to include galvanic corrosion, cannot

TABLE 5.1 Galvanic Corrosion

↑	Platinum
	Gold
Noble or	Graphite
cathodic	Titanium
	Silver
	Hastelloy C (62Ni, 17Cr, 15Mo)
	18-8 Mo stainless steel (passive)
	18-8 Stainless steel (passive)
	Chromium stainless steel 11–30%Cr (passive)
	Inconel (passive) (80Ni, 13Cr, 7Fe)
	Nickel (passive)
	Silver solder
	Monel (70Ni, 30Cu)
	Cupronickels (60-90Cu, 40-10Ni)
	Bronzes (Cu–Si)
	Copper
	Brasses (Cu–Zn)
	Nickel (active)
	Tin
	Lead
	Lead–tin solders
	18–8 Mo stainless steel (active)
	18–8 Stainless steel (active)
	Ni-resist (high-Ni cast iron)
	Chromium stainless steel, 13%Cr (active)
	Cast iron
	Steel or iron
	2024 Aluminum (4.5Cu, 1.5Mg, 0.6Mn)
Active or	Cadmium
anodic	Commercially pure aluminum (1100)
↓	Zinc
	Magnesium and magnesium alloys

Source: From Fontana [1].

occur without a chemically reducible species in the environment to consume the electrons liberated by oxidation of the anodic reaction. The lack of reducible species is the main reason why galvanic corrosion is not common in downhole oil well equipment, where there is little oxygen and other reducible species are unlikely.

Figure 5.6 Galvanic corrosion of galvanized piping in connection with bronze valve.

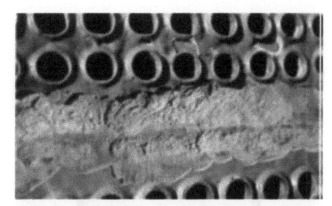

Figure 5.7 Painted carbon steel tube sheet in a heat exchanger with copper-alloy tubes [19]. *Source*: Reproduced with permission of Springer.

Area Ratio

Whenever two metals are joined in a galvanic couple, the total oxidation reaction, and the consequent corrosion rate, of the anode is increased, while the corrosion rate on the cathode is reduced. If the anode is small, then the corrosion will be significantly increased, and this is the reason why many authorities caution that if galvanic couples cannot be avoided, then the anode must always be the larger component in the galvanic couple. This is the normal situation, because most structures are made from carbon steel, and many connections, e.g. instrumentation, tend to be of CRAs, if only to avoid galling effects on threaded connections.

Figure 5.7 shows corrosion on a heat exchanger tube sheet at coating defects of a carbon steel tube sheet with copper-alloy tubes. The unfavorable area ratio between the exposed carbon steel anodes and the copper tube cathodes makes the corrosion deeper than it would be if the tube sheet had not been painted. Two alternatives to this situation are possible [19]:

- Painting the inside of the tubing near the inlet galvanic couple locations for approximately 10 times the tubing diameter or cladding the tube sheet with an alloy galvanically compatible with the tubing alloy.
- Sizing the tube sheet, baffles, and supports thick enough so that the galvanic corrosion does not affect performance. This second approach is common in many industries and is shown in Figure 5.8.

Figure 5.9 shows the results of galvanic corrosion on a scuba tank. The exterior of the tank is covered with a protective coating and is only wetted for short periods

Figure 5.8 Titanium tubes in a cooling water heat exchanger with carbon steel baffles and water box. Deposits on the titanium tubes are the salts (scale) formed from the cooling water, and not corrosion [19]. *Source*: Reproduced with permission of Springer.

of time. The brass valve does not cause serious corrosion on the carbon steel exterior for two reasons. The time of immersion of a scuba tank is usually very short and is limited by the tank capacity and the energy of the diver. Scuba tanks spend most of their service lives exposed to atmospheric corrosion and relatively little time immersed in seawater or other waters. Even if the tank were immersed for longer periods, the surface area of the valve is relatively small compared with the large tank.

The corrosion of this tank occurred on the interior, where the galvanic couple between the brass valve and the carbon steel tank body involved similar surface areas of wetted metal. Moisture inside scuba tanks collects near the bottom of the tank, and this tank, like

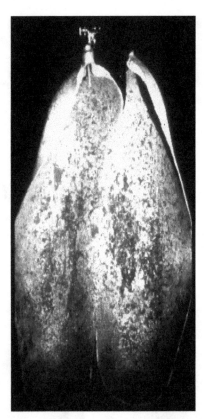

Figure 5.9 Galvanic corrosion of a scuba tank. *Source*: Photo courtesy NACE International.

Figure 5.10 Corrosion of carbon steel structure connected to stainless steel fasteners.

many recreational tanks, was not stored in the vertical (valve up) position that would have kept the liquid condensate at the bottom of the tank. Bimetallic galvanic corrosion led to enough wall thinning to produce rupture due to internal corrosion.

The area ratio concept must be used with caution and a careful consideration of the structures involved. It is important to always avoid accelerated corrosion on the critical component, the one that, if it corrodes, will lead to system failure.

Weld Filler Metals   Filler metals for welding should always be cathodic to the base metals being joined. Welds are used for immersion and atmospheric exposure, and in either situation the most critical location in the galvanic couple is the weld and adjacent area. The filler metal should always be cathodic to the base metal being joined.

Fasteners   Unlike welded connections, threaded connections should seldom be used in immersion environments. They are usually exposed to atmospheric corrosion. It is not unusual for engineers to avoid specifying corrosion-resistant fasteners because the most common corrosion-resistant fasteners have anodic

coatings, usually zinc or cadmium, and the concern is that the small fastener will be connected to a large carbon steel structure. This is a misguided caution. The fastener is the critical component in these applications, and failure of the fastener is to be avoided if at all possible. In addition, these very thin metallic coatings are only appropriate for atmospheric exposure, and the wetted area around a fastener is usually very small, limiting the effect of the "large anode."

Figures 5.10 and 5.11 show two different structures with threaded connections, leading to corrosion. In Figure 5.10 a galvanized carbon steel pipe support is connected to a concrete pad with stainless steel fasteners. The galvanizing is almost gone from the carbon steel, and the corrosion of the carbon steel at the carbon steel–stainless steel interface is obvious. Even though the carbon steel has corroded, the structure is in no danger of failing. The situation is different on the aluminum hinge shown in Figure 5.11. In this case, the formerly galvanized carbon steel fastener is corroding, and the adjacent aluminum-to-stainless steel connection shows minimal signs of deterioration. The inherent corrosion resistance of aluminum in a marine atmosphere means that the aluminum is protected by a passive film and the effects of joining it to stainless steel are minimal. The supposedly more compatible aluminum–zinc galvanic couple on the galvanized carbon steel fastener was ineffective. The thin zinc coating did not last very long in a marine atmosphere, and the carbon steel fastener is corroding and will eventually fail.

Galvanic corrosion with threaded fasteners can be avoided. Figure 5.12 shows a connection between a carbon steel piping and CRA piping. CRA fasteners are used at the carbon steel flanges. No increased corrosion of the carbon steel pipe due to the connection has

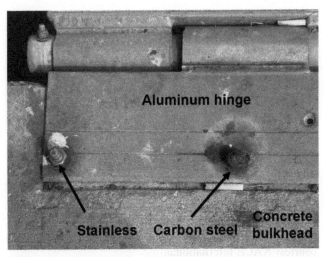

Figure 5.11 Aluminum hinge connected to concrete bulkhead with stainless and carbon steel fasteners.

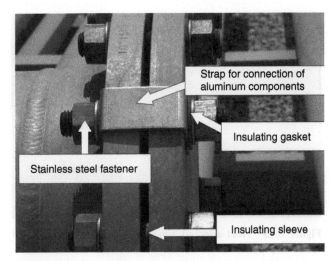

Figure 5.13 Electrical isolation of aluminum piping from stainless steel fasteners.

Figure 5.12 CRA fasteners on flange connecting carbon steel piping to CRA piping.

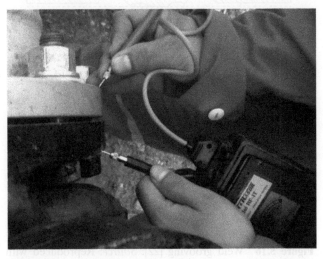

Figure 5.14 Continuity testing to be sure that electrical isolation has been achieved.

occurred. All of the apparent corrosion is due to coating defects. The reason that no galvanic corrosion has occurred is because the two metals are electrically insulated. CRA-threaded connections are routinely insulated from carbon steel pipes using electrical insulation kits like those shown in Figure 5.13. These kits include reinforced gaskets to separate the fasteners from the metal being connected and sleeves that fit around the shank of the fastener from the flanges. Electrical isolation supplies are routinely available from many piping and corrosion control suppliers.

Once an insulation kit has been installed, it is important to check against electrical continuity using a simple ohmmeter. This is shown in Figure 5.14. It is important to use reinforced gaskets in these insulation installations, because nonreinforced insulators may creep and

cause electrical shorting. It is also important to periodically check these insulated joints because they can become shorted due to motion between piping components.

Metallurgically Induced Galvanic Corrosion

There are a number of metallurgically induced corrosion cells possible. This discussion mentions a few that have been reported in oilfield environments.

Heat-affected Zone (HAZ) Corrosion Welding produces changes in metallurgical structure. The idea that filler metals should always been cathodic to the base metal being joined has already been discussed in the discussion

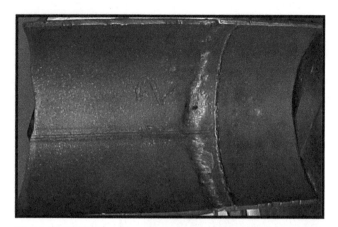

Figure 5.15 Heat-affected zone corrosion on carbon steel crude oil pipeline.

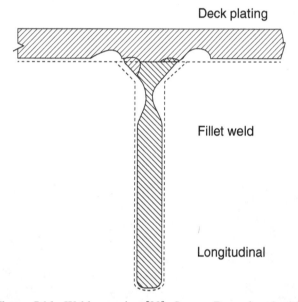

Figure 5.16 Weld grooving [22]. *Source*: Reproduced with permission of DNV GL AS.

of the area effect. Unfortunately, many welding procedures can produce situations where the heat-affected zones (HAZs) become anodic to the surrounding metal. This is shown in Figure 5.15. Note how the corrosion in Figure 5.15 is located parallel to the field-installed weld and that no corrosion is associated with the longitudinal seam weld that was made under controlled conditions in the pipe. This form of corrosion is sometimes called grooving corrosion when it occurs in structural members (Figure 5.16) or preferential weld corrosion on pipelines [19–23].

Problems with pipeline girth welds are common, and this is why pipeline inspections, e.g. internal pig inspections, concentrate on these areas as potential corrosion sites. Many organizations also decided not to

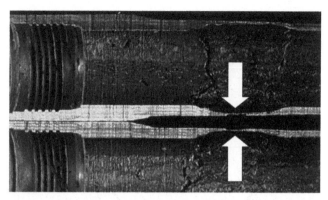

Figure 5.17 Ringworm corrosion in oilfield production tubing. Note the internal corrosion in the heat-affected zone, a short distance from the welded connection. *Source*: Photo courtesy NACE International.

use electrical resistance welded (ERW) pipelines because of corrosion problems along the longitudinal seam welds. This produced problems in pipeline steels called selective seam corrosion. This was a problem prior to the 1970s, when ERW pipe manufacturing processes used low frequencies (60 Hz). The problem seems to have been minimized due to the development of high-frequency ERW procedures and improved quality control, and high-frequency ERW piping is now the standard for most large-diameter pipelines [23–28].

Ringworm Corrosion   Figure 5.17 shows a phenomenon called "ringworm corrosion" that was a major corrosion concern in the Permian Basin of West Texas in the 1940s and early 1950s. The metal in the HAZs of oil country tubular goods (OCTGs – tubing, casing, and drill pipes) sometimes corrodes near the upset head or welded tool joint. In upset tubing the metal needs to be heated to the austenite stable region (above approximately 750 °C [1382 °F]) so that it can be deformed. Welded connections have similar HAZs that can lead to the same problem [9, 28]. The problem was solved in the 1950s by introducing the practice of full-length normalizing – heat treating the entire joint to a suitable (austenizing) temperature and then air cooling (normal cooling procedure) so that the entire joint has the same microstructure and corrosion resistance. This problem has reappeared worldwide in recent years, because engineers and purchasing organizations have not learned the lessons of decades past.

Lüders Band Corrosion   Lüders bands (also called stretcher marks, Hartmann lines, or Piobert lines) are localized bands of plastic deformation that can occur on carbon steels and other materials in regions of localized plastic deformation [29]. They form in carbon

Figure 5.18 Lüders band corrosion on oilfield production tubing (OCTG). *Source*: Photo courtesy NACE International.

steel and other materials when the initial resistance to deformation is overcome and localized yielding (plastic deformation) occurs. This localized deformation is usually at approximately 45° to the primary stress axis and may form ripples, Lüders bands, when the deformation reaches the metal surface. These deformations are an indication that part of the metal has been stressed more than other regions, which are in a lower energy state and less susceptible to corrosion. If the differences in stress are not removed, they can lead to a corrosion pattern termed "Lüders band" corrosion. This is shown in Figures 5.18 and 5.19. Once again, full-length normalizing is the recommended solution to this problem. While Figures 5.18 and 5.19 show downhole tubing, this is also a potential problem on pipeline steel.

While the corrosion patterns shown in Figures 5.18 and 5.19 are due to plastic deformation during pipe mill processing, this problem can also occur as a result of deformation in the field. It is important that piping, especially for large-diameter pipelines, be handled and bent very carefully to avoid localized regions of high stress and the formation of Lüders bands [30–32].

Environmentally Induced Galvanic Corrosion

Changes in environments produce galvanic corrosion cells. Typical examples are the differences between the potentials in deep water, which is usually colder and has

Figure 5.19 Close-up view of Lüders band corrosion on OCTG.

less oxygen, and surface water, which is warmer and high in oxygen. Galvanic differences also occur offshore from major rivers, where the surface water may be fresh and have low salinity for several meters before the lower, denser salt water becomes prevalent. Temperature gradients can also cause changes in potential. The reasons for these potential differences are easy to explain based on the principles of the Nernst equation (Chapter 2). Most oilfield environments, e.g. wet soil or brackish water, are too complex to model using the Nernst equation, and field measurements are necessary to confirm the presence of these potential gradients, which are often termed concentration cells [2].

The following examples are important in onshore pipelines and similar situations:

Pipeline Under Road Crossing This is an obvious situation where the moisture, access of air, and soil compaction levels can combine to produce differences in corrosive environments underneath roads or similar moisture and permeation barriers and the land on either side of the road. This is shown in Figure 5.20, which indicates that the relatively high oxygen rates adjacent to the roadway lead to anodic corrosion beneath the pavement. While this is a common situation, it is also possible in certain climates that pipelines beneath roadways and other moisture permeation areas may be less corroded than the wetter areas nearby. Because it is hard to predict the complex effects of differential environments at road crossings, it is considered good practice to monitor pipeline potentials or to use other means of monitoring corrosion at road crossings.

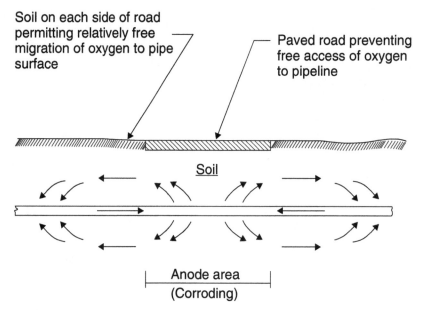

Soil on each side of road permitting relatively free migration of oxygen to pipe surface

Paved road preventing free access of oxygen to pipeline

Soil

Anode area (Corroding)

Figure 5.20 Differential aeration cell on a pipeline beneath a paved road. *Source*: Beavers [33]. Reproduced with permission of NACE International.

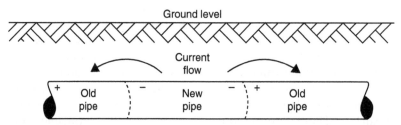

Ground level

Current flow

Old pipe

New pipe

Old pipe

Figure 5.21 New pipe connected to old pipe producing a galvanic corrosion cell. *Source*: Byars [9]. Reproduced with permission of NACE International.

Old Pipe Connected to New Pipe New sections of pipe in an old pipeline are often anodic to the older pipe, either because the old pipe is covered with heavy layers of rust that are cathodic to the new pipe or because the new pipe is placed in a more aggressive environment. One of the reasons for placing a new pipe, as shown in Figure 5.21, is as a repair of corroded pipe. This is an indication that the environment where the repair is located is more corrosive than nearby areas. Even if that were not the case, e.g. when the new pipe is the result of a modification, i.e. placing a new connection to an existing line, the new pipe is likely to be located in less compacted soil that is more permeable to water and air. Some organizations have adopted the policy that whenever a new pipe installation occurs, they will install galvanic anodes in the location. The cost of the anode is minimal compared with the cost of the construction, and it may help and cannot hurt.

Water-depth Corrosion Cells Offshore structures, pipelines emerging from buried or immersed locations,

and similar structures can have oxygen concentration cells, explained by the Nernst equation (Chapter 2). Figure 5.22 shows two oxygen concentration cells that occur on offshore platform legs – at the air/water interface and at the mud line [3].

Many offshore platform legs also have macroscopic thermogalvanic cells between relatively warm surface waters and colder deep water. Depending on location, surface waters may be as warm as 25–30 °C (77–86 °F), while deep ocean waters (below the thermocline) approach 4 °C (39 °F). The deep water is cold, but it is also under high pressure and away from air, so, depending on the location, the deep water may be depleted in oxygen or even saturated with oxygen. The important thing to remember in this discussion is that while corrosion rates are usually higher near the surface where the environment is exposed to air, sometimes deep waters can also be corrosive. Pipelines and other ocean-bottom or river-bottom structures can have a variety of galvanic cells due to the differences in temperature and exposure to different dissolved oxygen levels.

Polarity Reversal

Zinc-coated steel (galvanized steel) is used for corrosion control in atmospheric exposures and in fresh water. It was reported in 1939 that zinc sometimes becomes cathodic to carbon steel in fresh water at elevated temperatures. This caused concern in a number of circles and changed engineering practice, e.g. the construction of domestic hot water heaters. The reversal occurs in some fresh waters at temperatures above 60 °C (140 °F). Research in the 1950s indicated that this polarity reversal, where galvanized steel suffers accelerated attack, occurs in waters high in carbonates and nitrates but is unlikely to occur in waters with chlorides and sulfates, such as seawater or formation water [2, 34, 35]. The only other commercially important polarity reversal is with tin plating. Tin cans have an anodic coating (the tin) in deaerated organic acids (food containers), but this relationship reverses upon exposure to air, and tin acts like a noble coating.

Many authorities advise designers to check the polarity of metals in electrolytes of interest, but, as a practical matter, virtually all corrosion-related designers assume that the potentials shown in Table 5.1 are valid and testing to determine relative galvanic relationships, while possible, is rarely done.

Conductivity of the Electrolyte

Figure 5.23 shows how galvanic corrosion is concentrated near the two-metal interface in tap water while it extends for a long distance in seawater. At one time a "rule of thumb" used in seawater stated that galvanic corrosion effects would only extend approximately 10 pipe diameters into seawater heat exchangers [34]. Later research at the Ocean City Research Laboratory showed that the effects of changing the header material on seawater-cooled heat exchangers extended much farther than previously thought.

Control of Galvanic Corrosion

The obvious way to control galvanic corrosion is to avoid the use of different alloy systems in the same electrolyte. Oil companies follow this guideline by not mixing alloy families in process streams, e.g. using only aluminum, copper, stainless steel, or carbon steel whenever possible.

Dielectric (insulating) connections, like those shown in Figures 5.13 and 5.14, are effective means of preventing galvanic corrosion on atmospherically exposed flanged connections. This is especially important when connecting aluminum piping, as shown in Figure 5.13, because nongalling aluminum fasteners are not available. As stated above, it is generally inadvisable to use bolted connections on submerged or buried piping systems. The surrounding soil or water may conduct electricity across the isolating fittings. This can also happen in atmospherically exposed flanges if the fluid inside the piping is conductive. Dielectric unions along piping systems are often overcome by electrical grounding, and this is a further reason to not mix alloys whenever possible.

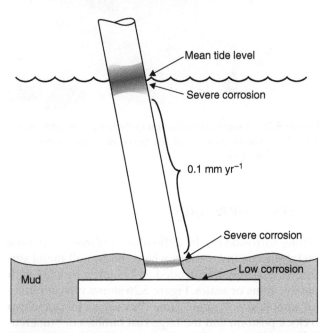

Figure 5.22 Oxygen concentration cells on an offshore platform leg. *Source*: Bradford [3]. Reproduced with permission of CASTI Publishing, Inc.

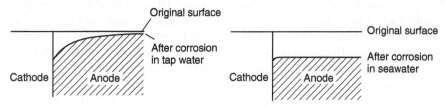

Figure 5.23 Effect of electrolyte conductivity on the distribution of galvanic corrosion [34]. *Source*: Reproduced with permission of John Wiley & Sons.

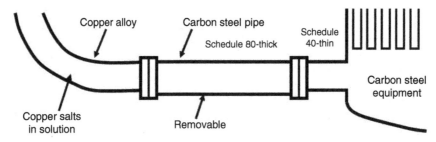

Figure 5.24   Removable component between copper-alloy piping and carbon steel equipment.

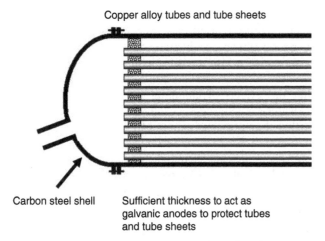

Figure 5.25   Galvanic corrosion of thick-shelled carbon steel heat exchanger to lower corrosion rates of copper-alloy tubes and tube sheets, figure based on ideas presented in figure A.39 of Ref. [37]. *Source*: Adapted from Iranian Petroleum Standard E-TP-706 [37].

Figure 5.26   Pitting corrosion on carbon steel potable water pipe. Note the deepest pit at the two o'clock position on this 100 mm (4 in.) pipe.

When connections of two incompatible metals are unavoidable, it is important that the smallest (or thinnest in the case of heat exchanger tubing) metal must be cathodic to the surrounding metals. Cathodic protection of unavoidable two-metal contacts is also possible [36].

One seldom discussed option is the use of "sacrificial" nipples in piping systems where galvanic corrosion will occur, e.g. between CRA valves and carbon steel or galvanized piping. A short, easily removed component of relatively thick carbon steel adjacent to the cathodic corrosion-resistant metals removes the need for welding when replacing the corroded carbon steel nearest to the CRA. It is important that the replaceable component, usually a pipe nipple, be connected mechanically so that no welding will be required when replacing the corroded component [37]. This idea is illustrated in Figure 5.24. Similar images are discussed in many fundamental corrosion courses, to include the NACE Basic Corrosion course.

Galvanic corrosion can sometimes prove useful by protecting thin-walled heat exchanger tubes using carbon steel heat exchanger shells and headers as anodes (Figure 5.25) [37].

PITTING CORROSION

Pitting corrosion can be defined as localized attack on a metal surface in locations where the overall metal surface is relatively uncorroded and is often covered with passive films or scales. Figure 5.26 shows typical pitting on a potable water pipe. Note the deepest pit in the two o'clock position and the large rust bubbles or "tubercules" that have formed over the pits. It is common that the rust tubercules can impede water flow and may be more significant than the relatively shallow pits, when compared with the size of the tubercules that form over them. These tubercules are porous corrosion products with scale deposits and can prevent access of corrosion inhibitors or biocides to the metal substrates. The most common way of removing deposits like shown in this picture is by mechanical removal using pipeline pigs or similar devices.

In 2000 a major gas transmission line near Carlsbad, New Mexico, exploded due to internal pitting corrosion. Figure 1.1 at the beginning of this book shows some of the damages from this rupture. Figure 5.27 shows pitting corrosion and comes from the Pipeline Accident Report for this incident, which concluded that condensed water with dissolved salt and bacteria caused wall thinning to the extent that the pipeline bursts [38]. The results of this fatal incident led to the development of NACE SP0106, Control of Internal Corrosion in Steel Pipelines and Piping Systems [39].

Pitting corrosion is perhaps the most common form of oilfield corrosion after general attack. Unlike general attack, which can be monitored and predicted, pitting corrosion may start and propagate quickly in relatively short time periods, leading to significant damage. Monitoring for pitting corrosion requires frequent inspection or sampling, because no corrosion may occur for a long time followed by relatively aggressive pit initiation and growth.

Occluded Cell Corrosion

Pitting is only one of several forms of corrosion that has similar mechanisms. In 1970 B.F. Brown suggested the term "occluded cell corrosion" to encompass the mechanisms of pitting, crevice, stress corrosion, intergranular corrosion, filiform corrosion, and exfoliation [40]. He also suggested that corrosion fatigue, which shares several characteristics with the other forms of corrosion, had enough unique characteristics that it should be considered separately. Since this suggestion in 1970, many researchers have confirmed Brown's suggestions that all of these forms of corrosion shared several characteristics including acidification near the corroding anode and concentration of halides, usually chloride ions, in the corroding location. Figure 5.28 summarizes the results of research in many worldwide laboratories. Figure 5.29, from a corrosion textbook, compares the macroscopic features of these three forms of occluded cell corrosion. As suggested by Brown, the chemistry changes inside the occluded cell are similar for all three forms of corrosion.

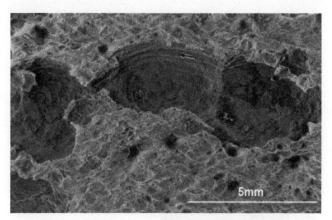

Figure 5.27 Pits on the pipeline steel involved in the 2000 Carlsbad, New Mexico, natural gas pipeline rupture [38].

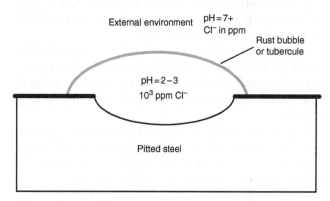

Figure 5.28 Pit with pH and chloride concentration changes indicated. *Source*: From Martin [41].

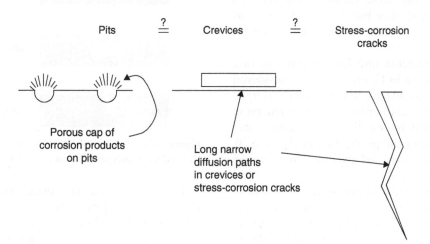

Figure 5.29 Schematic illustration showing similarities between pitting corrosion, crevice corrosion, and stress corrosion cracking [42]. *Source*: Reproduced with permission of Springer.

Note that the pH of the bulk environment in Figure 5.28, tap water, is indicated to be slightly basic (pH > 7). This is common for potable water supplies worldwide, because it was learned in the 1920s that high-pH water would produce calcium carbonate scales that retard corrosion in fresh water [2]. The pH inside the rust bubble or tubercule is shown to be between 2 and 3. This is a commonly reported pH for most occluded cell corrosion, although pHs below 1 have also been reported [40]. The low pH is a result of oxidation. All oxidation reactions lower the pH of the environment, just as all reduction reactions increase the pH. The high concentration of chlorides is due to the rapid migration of negatively charged anions, of which chlorides are the most common, to balance the electrical charge in the local low-pH environment inside the rust bubble or tubercule. Similar changes in pH and chloride levels have been reported in virtually all examples of occluded cell corrosion as well as in fatigue cracks, which are now considered to be occluded cells by many researchers. The combination of low pH and increased chloride levels inside occluded cells means that once this form of corrosion starts, it is likely to proceed at an accelerated rate.

Removal of occlusions, e.g. mechanical removal of the rust bubble or tubercule, can slow or stop pitting corrosion, and this is a reason for pipeline pigging and other mechanical means of cleaning piping systems.

Figures 5.26 and 5.27 show pitting corrosion on carbon steel. Other alloys may not form rust bubbles or tubercules, but they all have low pHs and concentrated halides (chloride, bromide, etc.) inside the pits.

Pitting Corrosion Geometry and Stress Concentration

Many corrosion pits, like those shown in Figures 5.26 and 5.27, are relatively shallow, but they can still serve as stress concentrators that initiate SCC or corrosion fatigue.

The stress concentration and loss of cross section caused by the pit shown in Figure 5.30 was considered severe enough that it could have led to corrosion fatigue of the North Sea platform where it was found. The pit in question was repaired by welding using underwater divers [43]. Shortly after the publication of Figure 5.30, another North Sea platform collapsed due to corrosion fatigue with many fatalities.

Other pits can have different geometries as shown in Figures 5.31–5.33. Film-protected alloys are more likely to form pits with relatively small surfaces and widespread corrosion beneath the pit entrance, but environmental factors, e.g. scale deposits or biofilms, can also produce similar pitting patterns in carbon steel.

Figure 5.30 Corrosion pit on a weld of an offshore platform. The overlying marine growth and tubercules were mechanically removed to allow visual inspection using a remotely operated vehicle (ROV) with a television camera. *Source*: Smart [43]. Reproduced with permission of NACE International.

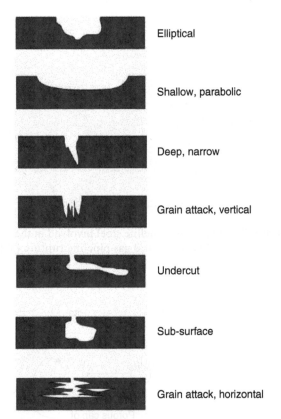

Figure 5.31 Pit morphology. *Source*: NACE TM0106 [44]. Reproduced with permission of NACE International.

Pitting corrosion often occurs in clusters or colonies. If the individual pits are too close to other pits, the total effect is similar to having one larger defect on the metal surface. The pitting colony shown in Figure 5.34 is an example of a large number of pits that must be evaluated to determine their overall effect on the safe operating

Figure 5.32 Internal corrosion pits formed on carbon steel pipe on an offshore platform.

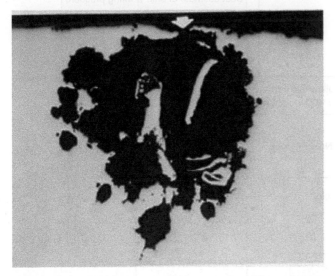

Figure 5.33 Pitting on stainless steel tubing. *Source*: McDanels [45]. Reproduced with permission of ASM International.

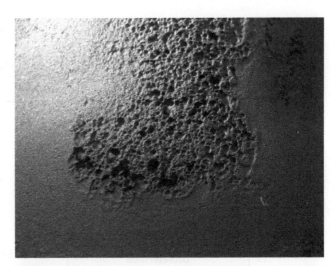

Figure 5.34 Colony of corrosion pits on stainless steel piping in a gas processing plant at a seaside location.

tom of the image, but other deep pits are also evident and must be measured. Pit gages similar to those shown in Figure 5.36 are used for this purpose.

Pitting Initiation

Pits form at defects on metal surfaces. These are often microscopic in nature and cannot be detected by field-level inspection devices. Examples of pit initiation sites include impurities or grain boundaries on the metal surface and mechanical damage to surface films, either passive films or scales [4].

Pitting Resistance Equivalent Numbers (PRENs)

There are a number of published pitting resistance equivalent numbers (PRENs). The most commonly used PREN is from ANSI/NACE RP0176/ISO 15156 [50]:

$$\text{PREN} = w_{\text{Cr}} + 3.3\left(w_{\text{Mo}} + 0.5w_{\text{W}}\right) + 16w_{\text{N}}$$

where

w_{Cr} is the weight percentage of chromium in the alloy.
w_{Mo} is the weight percentage of molybdenum in the alloy.
w_{W} is the weight percentage of tungsten in the alloy.
w_{N} is the weight percentage of nitrogen in the alloy.

Larger values of PREN are considered to indicate greater resistance to pitting corrosion. A more complete discussion of PRENs is available in Chapter 4.

While PRENs can be used to compare the supposed resistance of different alloy groups with pitting, they offer only general guidelines on pitting and crevice

pressure of the contained fluid. Guidance is available on how to determine if pits are clustered so close that they are considered to act like larger defects (stress risers) in several different codes and standards [46–49]. Figure 5.35 shows the decision process recommended by DNV for determining the safe working pressure on piping and pipelines. Note the necessity to determine if corrosion and other defects are located close enough so that they must be considered as single larger defects. This procedure can be used for clusters of corrosion pits, but it does not work for cracks from stress corrosion or from welding defects [49]. These procedures require the determination of the deepest pit in each colony. Most of the deeper pits in Figure 5.34 are located near the bot-

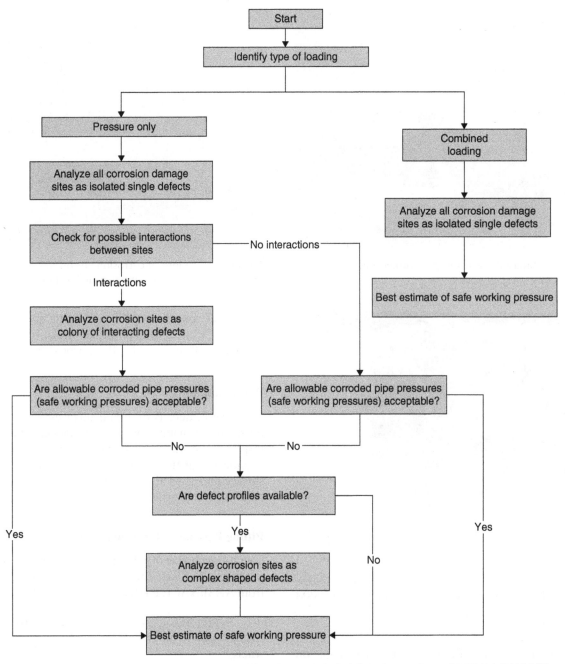

Figure 5.35   Flowchart of the assessment procedure for corroded piping. *Source*: From DNV-RP-F101 [49].

corrosion resistance and do not consider all alloying constituents, e.g. nickel, that contribute to localized corrosion resistance.

Figure 5.37 shows deep pitting within months of a plate composed of Alloy 825 (UNS N08825), an alloy with a PREN in the low 30s. It would normally be considered a very corrosion-resistant alloy. Even the most CRAs are subject to pitting and crevice corrosion in elevated-temperature brines like seawater or formation waters.

Pitting Statistics

Figure 5.38 shows the distribution of pits on a pipeline where a shrink sleeve coating of a girth weld was ineffective. The markings on the pipe indicate that the deepest pits vary from 0.100 in. (2.5 mm) to 0.210 in. (5.3 mm) in depth. This is a more than 100% variation in pit depth, and most of the exposed surface in this location is relatively uncorroded. The wide variability in pit depths was the subject of an early study of pipeline pitting by Gordon

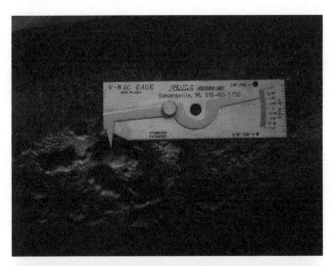

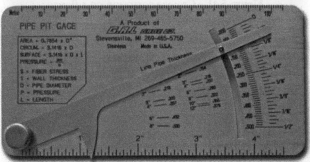

Figure 5.36 Pit gages used for determining corrosion pit depths on the exterior of corroded pipelines and similar equipment.

Figure 5.37 Pitting corrosion of an Alloy 825 (UNS N08825) heat exchanger baffle exposed to seawater [51]. *Source*: Courtesy of NACE International.

Scott, an API fellow and later NACE president working at the National Bureau of Standards [52]. The conclusions from this study, which involved excavation of miles of pipeline in several states, can be summarized as follows:

- The larger the number of pits, the deeper the maximum pit.
- Larger areas of inspection will produce deeper maximum pit depths.

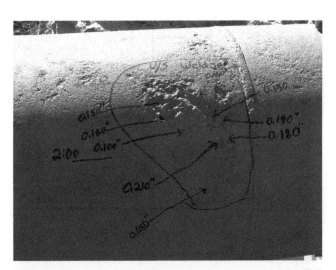

Figure 5.38 Pitting on carbon steel pipeline at a location where shrink sleeve protection of a girth weld was ineffective. *Source*: Photo courtesy R. Norsworthy, Polyguard Products, Inc.

These results, from some of the earliest scientific research into pipeline corrosion, indicate that small samples of any type, whether coupons used to monitor corrosion control effectives or limited inspections of pipeline exteriors using external corrosion direct assessment (ECDA) methods, are unlikely to identify the deepest pits in any real system [52, 53].

Prevention of Pitting Corrosion

Pitting corrosion on carbon steels is minimized by the use of cathodic protection, protective coatings, and corrosion inhibitors. The same approach is applied for martensitic stainless steels (13Cr alloys) used as OCTGs. Other CRAs have varying pitting and crevice corrosion resistances. Stainless steels and similar alloys benefit from the addition of molybdenum (e.g. UNS S30400 is more susceptible than UNS 31600, which has 2½ Mo added for pitting and crevice corrosion resistance). Titanium alloys are generally considered to be immune to pitting corrosion, but palladium or molybdenum additions are found to be helpful in adding resistance at elevated temperatures.

CREVICE CORROSION

The mechanisms of crevice corrosion are essentially the same as for pitting corrosion; the only important difference is that the crevice, which serves as the corrosion initiation site, is readily visible to the unaided eye.

Figure 5.39 shows locations of crevice corrosion susceptibility on a bolted connection. Problems with

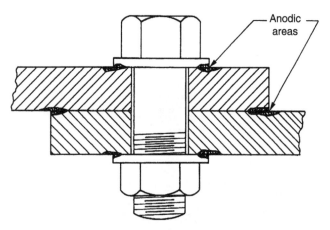

Figure 5.39 Crevice corrosion locations on a bolted connection [54]. *Source*: Reproduced with permission of John Wiley & Sons.

Figure 5.41 Crevice corrosion on a flange.

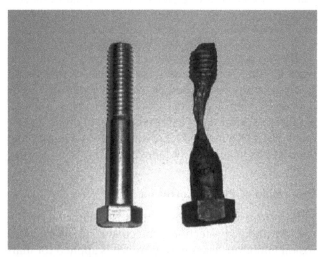

Figure 5.40 Crevice corrosion of a 10-year-old bolt on a water-control valve [55]. *Source*: Reproduced with permission of corrosionsource.

Figure 5.42 Crevice corrosion on a heat exchanger header plate.

crevice corrosion are a major reason why bolted connections are seldom used in submerged applications, although cathodic protection to minimize crevice corrosion is possible [36, 54]. Bolts also suffer crevice corrosion as shown in Figure 5.40. The 10-year-old carbon steel bolt on the right in Figure 5.40 was installed on a valve in a water supply system.

Figure 5.41 shows crevice corrosion on a flange after the bolts and attached piping has been removed.

Heat exchangers are another type of equipment with significant crevice corrosion problems. The corrosion of a header plate is shown in Figure 5.42. The corrosion shown in Figure 5.42 is not a major operational concern; it can be repaired using a CRA weld overlay that is similar in chemistry to the header plate. Of more concern is

corrosion of heat exchanger tubing, which is much thinner than header plates and the shells. Figure 5.43 shows corrosion at the header plate–tubing interface. Note that the corrosion is concentrated in the relatively thick header and is not a threat to the much thinner tubing. It is common to make the headers out of metals that are slightly anodic to the tubing, so that any corrosion that may occur will be on the relatively thick header plates and not on the thinner tubing.

Crevice corrosion is also called under-deposit attack and has a variety of other names as well. The corrosion in the six o'clock position on the cupronickel firewater line shown in Figure 4.54 is an example of under-deposit crevice corrosion in a Gulf of Mexico firewater system. Many microbially influenced corrosion (MIC) situations could also be described as under-deposit attack, although they are most often termed MIC-related pitting.

Figure 5.43 Crevice corrosion on a stainless steel heat exchanger [54]. *Source*: Reproduced with permission of John Wiley & Sons.

Figure 5.44 Corrosion underneath a pipe support.

Figure 5.45 Crevice corrosion underneath the restraining strap on a pipeline. *Source*: Photo courtesy J. Byrd, Byrd Coating Consultants, Wellington, Florida, reproduced with permission.

Figure 5.46 CUPS due to contoured fiberglass pads producing crevices if improperly sealed. *Source*: Reproduced with permission from Deepwater Corrosion Services Inc. [56].

Corrosion Under Pipe Supports (CUPS)

CUPS is a form of crevice corrosion. In recent years many industries, to include the upstream oil and gas industry, plus their inspection service contractors, have become aware of corrosion associated with pipe supports. Pipe supports must allow movement due to thermal expansion and contraction and due to fluid flow-induced vibrations. Pipes are often placed on steel supports, but it is common to place a soft gasket between the pipe and support. The supported piping or pipelines are exposed to atmospheric corrosion, and moisture trapped near the supports or motion-constraining devices causes corrosion that is hard to inspect. Figures 5.44–5.48 show CUPS. The problem in Figure 5.44 is apparent to

visual inspection, but the corrosion in Figure 5.45 was only revealed when the restraining strap was moved and the underlying pipe was exposed. While visual inspection showed the CUPS in Figures 5.44 and 5.45, nondestructive inspection techniques are also available that can find CUPS when visual inspection cannot [57, 58].

CUPS is influenced by:

- Moisture accumulation – this can be rain, condensation, or dripping from overhead equipment.
- Pipe operating temperatures.
- Thermal expansion and contraction – this can damage the coating by fretting or crack the coating.

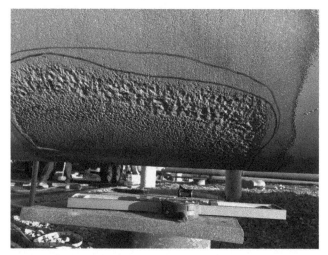

Figure 5.47 Pitting corrosion at a pipe support. Note how the pits are deeper away from the bottom (six o'clock) position.

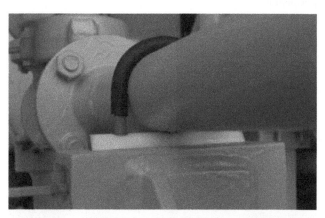

Figure 5.48 Commercially available pipe restraint intended to minimize CUPS. *Source*: Reproduced with permission from Deepwater Corrosion Services Inc. [56].

Products like the pipe restraint device shown in Figure 5.48 are commercially available to limit CUPS. The contact surfaces above and below the pipe are softer than the pipe or its coating and also provide slippery nonabsorbent surfaces that prevent coatings damage.

Pack Rust

The accumulation of rust in crevices between metal surfaces can create stresses sufficient to distort the metal. Broken welds and popped rivets can also occur. Figures 5.49–5.52 show examples of pack rust. Note the broken metal caused by pack rust in Figures 5.49b and 5.51. The distortion of metal caused by pack rust is evident in Figures 5.50 and 5.52.

Figure 5.50 Distortion of structural steel caused by pack rust. *Source*: Reproduced with permission from Sabnis [59].

(a)

(b)

Figure 5.49 Pack rust on a flange causing the flange protector to distort (a) and eventually break (b). *Source*: Reproduced with permission from Deepwater Corrosion Services Inc. [56].

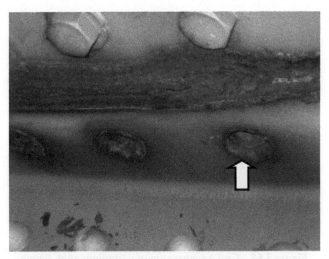

Figure 5.51 Popped rivets caused by pack rust. *Source*: Reproduced with permission from Sabnis [59].

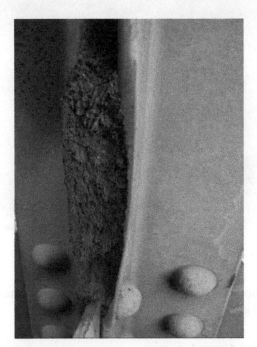

Figure 5.52 Close-up image of the metal distortion caused by pack rust. *Source*: Reproduced with permission from Sabnis [59].

Crevice Corrosion Mechanisms

Many authorities describe crevice corrosion in terms of electrochemical concentration cell corrosion [2, 4]. The electrochemical potential differences between inside a crevice and outside a crevice can be described as oxygen concentration cells and metal ion concentration cells. The relatively high oxygen concentrations available for reduction outside a crevice are used to explain why oxygen concentration cells cause crevice corrosion inside a crevice. The Nernst equation can also be used to

explain how high metallic ion concentrations inside a crevice would lead to metal ion concentration corrosion immediately adjacent to the edge of the crevice. While this is an interesting possibility, metal ion concentration cell crevice corrosion has never been reported outside the laboratory.

Crevice corrosion is usually, but not always, a topside problem [60]. Leaking gaskets, mechanical motion leading to openings in bolted connections, and a variety of other sources of moisture ingress produce corrosive conditions inside crevices. The chemistry of the electrolyte inside a crevice is typical of all occluded cells, and increasing metal ion concentrations, reduced pHs, and the migration of chloride and other negatively charged anions all contribute to the corrosivity inside the crevice. Relatively high oxygen concentrations outside the crevice lead to accelerated attack, because oxidation of the metal must be balanced by nearby reduction reactions, and oxygen reduction is the most common source of reducible chemicals. This is the reason why crevice corrosion is usually minimal in downhole environments. Complicated geometries, e.g. on pumps, are common downhole, but the relative lack of a reducible chemical species in many oil wells limits corrosion. Unfortunately, gas wells, which are acidic and corrosive, can have crevice corrosion, and the use of CRAs frequently becomes necessary on wellhead equipment and other complicated surfaces exposed to corrosive condensates [61].

Under-deposit corrosion is another term for some types of crevice corrosion. Debris of any type will often accumulate at low spots and cause corrosion underneath deposit. The stagnant environment beneath the deposits is different than the environment beyond the deposits, and corrosion results. Figure 4.54 shows under-deposit corrosion of 90–10 cupronickel tubing in offshore Gulf of Mexico firewater system.

Even aluminum, which normally has excellent marine atmospheric corrosion resistance, can suffer crevice corrosion if atmospheric moisture condensation can enter crevices (Figure 5.53).

Alloy Selection

Most of the above discussion and Figures 5.41, 5.42, and 5.44–5.52 concerned crevice corrosion on carbon steel. Crevice corrosion can be a problem on a wide range of CRAs also. Molybdenum (Mo) additions are often used to increase the pitting and crevice corrosion resistance of stainless steel tubing. Type 316 (UNS S31600) stainless steel (2½ Mo) is a standard grade often specified instead of the similarly available Type 304 (UNS S30400) (no Mo requirements) for most marine atmospheric applications, and higher-alloy grades are also available at increased costs. The increased Mo in Type 317 stainless steel (UNS S31700) has 3½ Mo, which guarantees

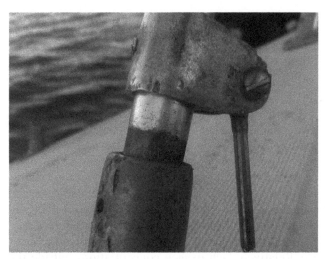

Figure 5.53 Crevice corrosion of aluminum shaft beneath a fitting.

Figure 5.54 Crevice corrosion of a titanium alloy flange.

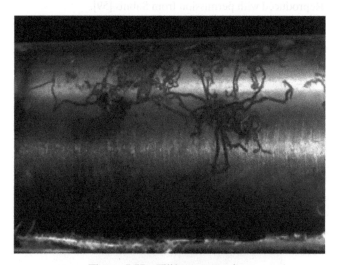

Figure 5.55 Filiform corrosion.

the Mo content will be at least 3% (bottom of the acceptable Mo content range). This alloy is replacing 316 in many services, because of its increased Mo content.

The 6 Mo austenitic stainless steel alloys with PREN numbers of 40 and higher and a nominal Mo mass fraction of 6–8% are also more resistant to both pitting and crevice corrosions. Examples include UNS S31254, UNS N08367, and UNS N08926 [21].

Duplex stainless steels (2507, UNS 32750) have been shown to be even better than austenitic stainless steels [62].

Mounting stainless steel and similar alloy tubing on aluminum racks is also a means of limiting the corrosion of stainless steels and similar alloys. The aluminum, which is anodic to stainless steels, will corrode slowly in marine atmospheric service, and this will protect the tubing being supported. Periodic replacement of the corroded aluminum supports is easier than tubing replacement [62].

It is important to always specify the correct alloy to control crevice corrosion. Titanium, which is considered immune to crevice corrosion in ambient-temperature seawater, may corrode at elevated temperatures. Titanium alloys with palladium, molybdenum, or ruthenium additions are more resistant to crevice corrosion. Figure 5.54 shows crevice corrosion of a titanium flange. The engineers specified titanium, but did not specify the alloy, so the organization installed a commercially available Ti-6Al-4V (UNS R56400) part in seawater service. This particular application, which was not at elevated temperatures, could have used commercially pure titanium (ASTM Grades 1–4) that would have worked quite well. The same major oil company now uses palladium additions (ASTM Grades 7 or 11) on all applications to avoid any possibility of crevice corrosion.

Filiform Corrosion

Filiform corrosion (filamentary corrosion underneath protective coatings on metal surfaces) is a special type of crevice corrosion. The long, thin filaments of corrosion products extend through defects in the protective coating and provide galvanic contact between the oxygen-deficient anode near the metal-coating interface and the relatively high oxygen environment outside the coating. The progress of this form of corrosion is a simple case of differential aeration cells that can be explained by the Nernst equation discussed in Chapter 2. Figures 5.55 and 5.56 show filiform corrosion underneath protective coatings. The corrosion starts at coating defects, typically along edges, and proceeds underneath

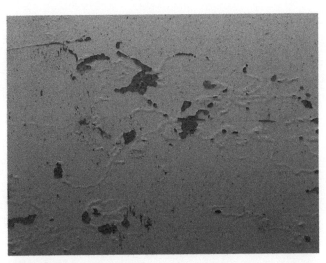

Figure 5.56 Filiform corrosion thread underneath a protective coating. *Source*: Photo Courtesy of NACE International.

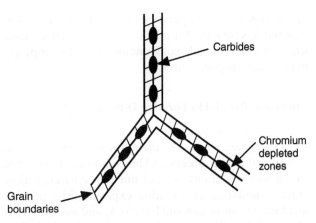

Figure 5.57 Sensitized regions in austenitic stainless steel.

the coating. It may eventually produce coating disbonding and blistering, leading to penetration of the metal substrate (Figure 5.56).

Filiform corrosion is caused by nonadherent coatings on poorly prepared metal surfaces. Appropriate surface preparation, to include removal of organic contaminants on the surface, is the primary means of control [63].

INTERGRANULAR CORROSION

Most grain boundaries are more reactive than the surrounding base metal due to the presence of impurities and other defects. The increased reactivity of grain boundaries over crystal matrices is usually negligible and of minimal consequence. Intergranular corrosion occurs when significant differences in alloy chemistry near grain boundaries cause significant attack and the alloy disintegrates. Intergranular corrosion can be a significant problem in stainless steels and in some aluminum alloys. Other alloy systems, e.g. highly alloyed austenitic iron–chromium–nickel alloys, can also have this problem.

Stainless Steels

Most upstream intergranular corrosion of stainless steels is due to improper welding practices. The temperatures involved in welding create HAZs where carbon in the stainless steel can react with chromium to form chromium carbides. This is most likely to happen in grain boundaries, where carbon is most likely to be concentrated. The result is chromium

depletion from the surrounding grain boundary regions and the creation of three composition ranges in the metal:

- Chromium carbides concentrated in the grain boundaries.
- Chromium-depleted zones in the grain boundary regions.
- Bulk crystals with no segregation and the overall composition of the alloy.

This is shown in Figure 5.57, which is common to virtually all corrosion engineering textbooks.

The continuous chromium-depleted regions are anodic to the larger unaffected grains, and this unfavorable area ratio causes increased corrosion in the grain boundaries. The HAZs are shown in Figure 5.58. Note that the newly recrystallized grains (labeled weld decay zone in Figure 5.58) are located approximately halfway through the HAZ.

Figure 5.59 shows intergranular corrosion in welded 316L (UNS S31603) stainless steel. Note how the corrosion is in the HAZ and not immediately adjacent to the weld bead (Figure 5.59a). The intergranular nature of the attack is apparent under the microscope (Figure 5.59b). Note how some of the attack in Figure 5.59b is producing sharp cracks in the grain boundaries. While 316L has reduced carbon content (0.03% max vs. 0.08% max for 316), it is obvious that this carbon content was still enough to produce sensitization in the HAZ grain boundaries.

Manufacturing operations can undo sensitization by post-weld heat treatment, but this is very difficult in field operations. Once again, the use of carefully approved welding procedures is very important.

Refineries and other high-temperature operations can sometimes operate at temperatures above 500 °C (950 °F) where sensitization can occur. Most downstream operations are at temperatures too low for this to happen, but it is possible for manufacturers to deliver sensitized stainless steel components due to improper manufacturing procedures.

Corrosion Parallel to Forming Directions

Many metal objects are more prone to corrosion on surfaces perpendicular to the metal forming direction (the rolling or drawing direction). This is shown in Figure 5.60, which shows a stainless steel nut that corroded along grain boundaries in a marine exposure. The nut was machined from hexagonal bar stock, and sulfide stringers parallel to the original bar stock longitudinal direction allowed intergranular crevice corrosion on the facing side (to the right in the picture) of the nut. Most machining-grade metals have deliberate additions of a soft second phase to expedite machining processes.

Sheet, plate, and tubing are also more corrosion susceptible in the through-thickness direction. This is why corrosion rates are faster on cut surfaces, e.g. where tapping is necessary for instrumentation connections. No inappropriate heating was involved in causing the corrosion shown in Figure 5.60. While the metal shown in Figure 5.60 was stainless steel, this directionally oriented corrosion is common in many other alloy systems.

Aluminum

Aluminum alloys get their strength from alloying additions. This makes the grain boundaries of aluminum susceptible to corrosion, and the formation of aluminum oxide pushes the metal apart in a direction

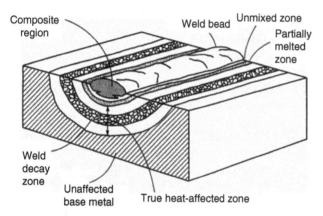

Figure 5.58 Heat-affected zones in welded stainless steel [64]. *Source*: Reprinted with permission of ASM International®. All rights reserved. www.asminternational.org.

Figure 5.60 Intergranular corrosion along the forming direction on a stainless steel nut.

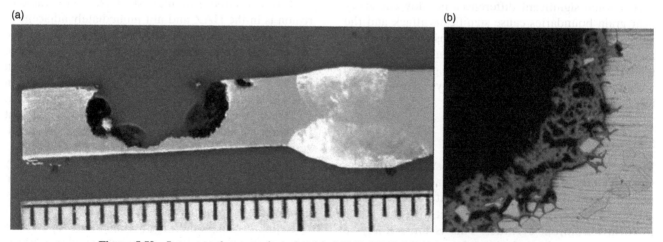

Figure 5.59 Intergranular corrosion of welded 316L (UNS S31603) stainless steel (a). Note the sharp intergranular indications in the magnified view (b).

(a)

(b)

Figure 5.61 Exfoliation of aluminum at coastal locations. (a) Protective passive layer removed by thermal expansion and contraction causing rubbing against steel anchor bolt. (b) Galvanic corrosion with stainless steel fastener.

perpendicular to the rolling direction. This is shown in Figure 5.61, which shows exfoliation (the loss of leaves) and intergranular corrosion of an aluminum guard rail in a coastal marine environment. Note how the galvanized bolt in Figure 5.61a has rubbed against the bolt-hole on the aluminum rail and caused corrosion where the bare metal was exposed after the protective aluminum oxide passive film was removed. This picture is also an example of the problems associated with improper design for thermal expansion and contraction, which caused the relative motion between the expanding and contracting aluminum and the fixed bolt location. This problem of expansion and contraction can be expected to occur on helidecks and other sunlight-exposed structures. Aluminum is not welded in most field applications, so exposure of through-thickness grain boundaries at bolted connections is common.

Other Alloys

Nickel-based alloys, e.g. C-276 (UNS N10276) and the casting alloys CW-12MW (UNS N30002), CW-6M (UNS N30107), and CW6MC (UNS N26625), can have intergranular corrosion problems if they are not properly solution annealed.

Carbon steel and other alloys can also suffer grain boundary attack and exhibit exfoliation, but welded stainless steels and aluminum exposed to wet atmospheres are the most likely alloys to have this problem in the oilfield. All metals are especially prone to this form of corrosion on surfaces perpendicular to the forming direction, where the grain boundaries are closer.

DEALLOYING

Dealloying is a corrosion process where one constituent of an alloy is removed, leaving an altered residual structure. It was first reported in 1886 on copper–zinc alloys (brasses) and has since been reported on virtually all copper alloys as well as on cast irons and many other alloy systems [1, 2]. Alternate terms for dealloying include parting, selective leaching, and selective attack. Terms such as dezincification, dealuminification, denickelification, and so forth indicate the loss of one constituent of the alloy, but the general term dealloying has gained wider use in recent years.

Figure 5.62 shows a typical example of dealloying. The chrome-plated brass valve corroded at breaks in the coating. The dark regions on the brass are regions where virtually all of the zinc has been removed, leaving a porous copper structure with virtually no mechanical strength and no change in surface profile.

Dealloying is also a problem with cast irons. While both diffusion and noble metal deposition are discussed as mechanisms associated with dealloying in copper-based alloys, the mechanism of dealloying in cast irons involves the dissolution of the iron-rich phases, leaving a porous matrix of graphite and iron corrosion products. Figure 5.63 shows porous graphite plugs in a cast iron water pipe.

Mechanism

Dealloying has been shown to occur by at least two different mechanisms. Sometimes the entire alloy dissolves, and one constituent redeposits on the corroded metal surface. In other circumstances diffusion removes only the more corrosion-susceptible constituent, leaving an altered porous matrix. Both mechanisms have been shown to occur simultaneously on the same metal surface [65, 66].

Selective Phase Attack

Selective phase attack is a form of dealloying where some phases of an alloy are more corrosion susceptible than the overall alloy. It is an oilfield problem with large bronze castings. The cross section of a nickel–aluminum bronze pump component is shown in Figure 5.64. The problem is caused by improper foundry procedures that produce susceptible phases in otherwise corrosion- and erosion-resistant alloys used for large seawater pumps. Quality control checks on bronze castings need to include chemical analysis to ensure that the composition is within specifications [67, 68].

Susceptible Alloys

Virtually all copper-based alloys are susceptible to dealloying. Stagnant seawater is more corrosive and can produce dealloying even in cupronickels, which are generally quite corrosion resistant (Figure 5.65) [65].

Cast irons are also susceptible. Ductile cast iron is less susceptible, but it is not immune.

Residual cementite in corrosion product layers on carbon steel has also been reported [69, 70]. This corrosion process is similar to the dealloying of cast iron (Figure 5.63) and could also be considered a form of selective phase attack.

Control

Dealloying is normally controlled by alloy selection. Cupronickels, while not immune, are probably the best wrought alloys for seawater service. Nickel–aluminum bronzes are used as castings. This means that even if

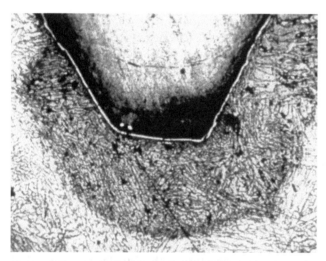

Figure 5.62 Dezincification of a chrome-plated scuba tank valve. *Source*: Heidersbach [65]. Reproduced with permission of NACE International.

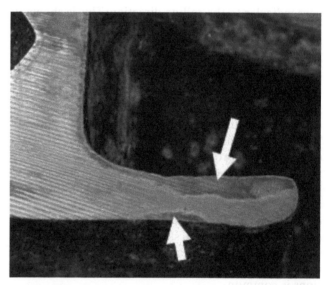

Figure 5.64 Selective phase attack of nickel–aluminum bronze. *Source*: Lenard [67]. Reproduced with permission of NACE International.

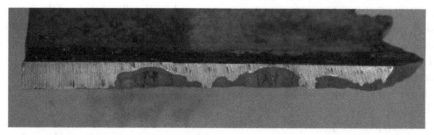

Figure 5.63 Dark, graphitic corrosion on the exterior of a cast iron water main. *Source*: Photo courtesy Testlabs International, Ltd., Winnipeg, Canada, www.testlabs.ca.

they suffer selective phase attack, bronze parts are generally so thick that the problem does not lead to catastrophic failure. All copper alloys are subject to dealloying in stagnant seawater, perhaps due to H_2S-generating biofilms.

EROSION CORROSION

Erosion is the mechanical removal of solid surfaces due to impact with harder materials. The removal of surface films, usually mineral scale or passive films, leaves the underlying metal exposed to corrosion. Virtually all erosion in oilfield applications thus becomes a combination of mechanical damage (erosion) and chemical reactions of the substrate with the environment (corrosion). The division between purely erosion and erosion-corrosion damage is hard to determine in the field, and will not be further discussed. Both erosion (mechanical attack) and corrosion (chemical attack) lead to degradation of oilfield equipment, and the relative contributions of one or the other degradation mechanism are usually irrelevant to control of the degradation of the associated equipment.

Erosion corrosion is the result of a combination of an aggressive chemical environment and high fluid surface velocities. This can be the result of fast fluid flow past a stationary object, or it can result from the quick motion of an object in a stationary fluid, such as happens when a ship propeller churns the ocean. Other terms include flow-enhanced or flow-accelerated corrosion, which also include mechanisms not related to erosion corrosion [4]. In erosion corrosion mechanical effects predominate [69]. Wellhead components like the one shown in Figure 5.66 sometimes fail within weeks due to sand production or erosion from small liquid droplets from a gas stream.

Surfaces that have undergone erosion corrosion are generally fairly clean, unlike the surfaces from many other forms of degradation (Figures 5.67 and 5.68).

Mechanism

Erosion corrosion is often the result of the wearing away of a protective scale or coating on the metal surface. Many people assume that erosion corrosion is associated with turbulent flow. This is true, because all practical piping systems require turbulent flow. The fluid

Figure 5.66 Eroded wellhead component. *Source*: Byars [9]. Reproduced with permission of NACE International.

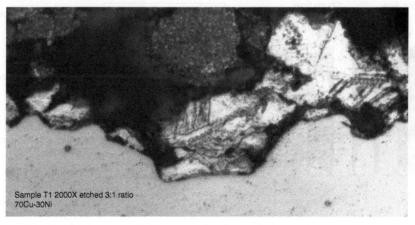

Sample T1 2000X etched 3:1 ratio
70Cu-30Ni

Figure 5.65 Dealloying on 70–30 cupronickel condenser tube in stagnant seawater. *Source*: Heidersbach [65]. Reproduced with permission of NACE International.

would not flow fast enough if lamellar (nonturbulent) flow were maintained.

Most, if not all, erosion corrosion is caused by multiphase fluid flow [71]. The flow regime maps shown in Figure 5.69 indicate the distribution of liquid (dark areas) and vapor (light areas) in vertical and horizontal flow. Slug flow has serious velocity-related problems, but none of these patterns produce erosion corrosion in straight piping in the absence of entrained solids. Where a flow pattern changes, e.g. at a rough pipe connection, a wellhead, and so forth, liquid droplets or impinging gas bubbles, which can collapse and produce shock waves that spall the protective surface film, or solid particles can cause accelerated attack by removing the protective film, either a passive film, mineral scale, or corrosion inhibitor film. These flow regime maps do not indicate the effects of entrained solids, e.g. sand, corrosion products, or scale, all of which are known to accelerate erosion corrosion.

Velocity Effects and ANSI/API RP14E

Most metals have a critical velocity, which is the highest fluid velocity that can be tolerated before erosion corrosion will occur. For topside equipment piping this is defined by a formula in ANSI/API RP 14E [73]:

$$V_{max} = \frac{C \times A}{\rho^{0.5}} \tag{5.1}$$

where

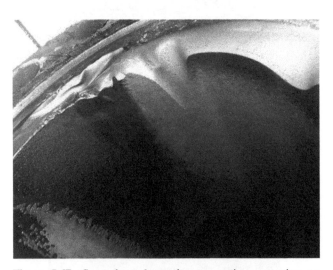

Figure 5.67 Smooth surfaces due to erosion corrosion on carbon steel piping downstream from a defective valve.

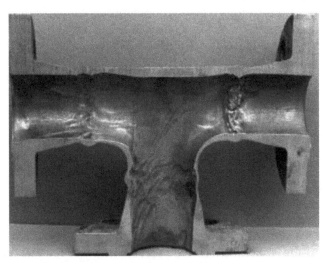

Figure 5.68 Wavy surface typical of erosion corrosion. *Source*: Bradford [3]. Reproduced with permission of CASTI Publishing, Inc.

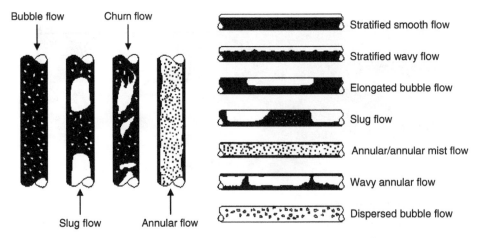

Figure 5.69 Multiphase fluid flow regimes in straight runs of vertical or horizontal piping [72]. *Source*: Reproduced with permission of Elsevier.

TABLE 5.2   Erosion–corrosion Variables for Choosing *C*-factors in ANSI/API RP 14E

Systems	Material Choice
Seawater	Carbon steel without inhibition
Single-phase production (all liquids)	Carbon steel with inhibition
Multiphase production (oil wells or gas wells)	13Cr and modified 13Cr
Dry gas injection (no corrosion, no liquids)	Duplex stainless steel
Methanol (no corrosion)	Super duplex stainless steel Ni-based CRAs

	SI	FPS
V_{max} = critical (maximum) velocity	$m\,s^{-1}$	$ft\,s^{-1}$
ρ = density	$kg\,m^{-3}$	$lbm\,ft^{-3}$
A = conversion constant	1.23	1

The ANSI/API recommended values for the *C*-factor are:

$C = 100$ for solids-free continuous service.
$C = 125$ for solids-free intermittent service.
$C = 150$–200 for solids-free, noncorrosive continuous service.
$C = 250$ for solids-free, noncorrosive intermittent service.

The same recommended practice suggests a minimum velocity in two-phase flow of approximately $10\,ft\,s^{-1}$

$(3\,m\,s^{-1})$ to minimize slugging in separation equipment. This is more important if elevation changes are involved.

The practice does not consider fluid properties such as viscosity, effects of solid particles, substrate materials properties such as hardness, and geometric properties such as elbows and flow constrictions. All of these properties are known to affect erosion-corrosion resistance [73]. Particulate erosion by sand is most likely to cause problems in production systems, and this is why sand production monitors are often added to topside piping systems [74].

Some companies have developed proprietary in-house guidelines on how to calculate maximum allowable velocities. Variables included in some of these guidelines are shown in Table 5.2. *C*-factors range from 200 for continuous use of carbon steel with no inhibitor in multiphase oil and gas wells (higher than the RP14E guidelines) to 450 for nickel-based alloys in all of the environments and other CRAs in dry gas injection. Other models are used for sand erosion, which is not considered in the RP 14E process. The total procedure is modeled in Figure 5.70.

The question of appropriate maximum velocities for downhole applications is a subject of continuing controversy and ongoing research. The subject is complicated, with over 30 parameters reported in the literature, and no consensus on this subject is likely [74–91].

Most of the previous discussion has concerned topside piping and downhole OCTG applications, but other oilfield applications have erosion–corrosion

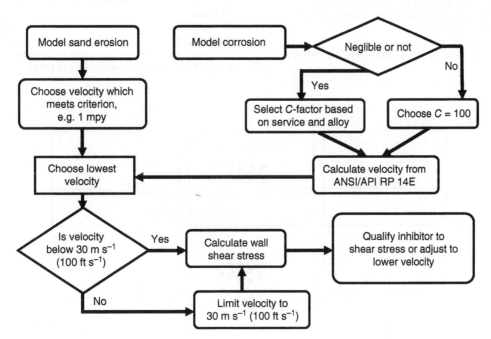

Figure 5.70   Erosion-corrosion decision process for downhole tubing.

problems as well. Condenser tubes have erosion problems at the tube inlets, and Tables 5.3 and 5.4 show recommended water velocity limits for condenser tubes and heat exchangers in seawater. Note how the recommendations in the much newer DNV recommended practice (Table 5.4) are lower (more conservative) than the numbers shown in Table 5.3.

TABLE 5.3 Suggested Velocity Limits for Condenser Tube Alloys in Seawater

Alloy	Design Velocity That Should Not Be Exceeded	
	(ft s⁻¹)	(m s⁻¹)
Copper	3[a]	0.9[a]
Silicon bronze	3[a]	0.9[a]
Admiralty brass	5[a]	1.5[a]
Aluminum brass	8[a]	2.4[a]
90–10 copper–nickel	10[a]	3.0[a]
70–30 copper–nickel	12[a]	3.7[a]
Ni–Cu alloy 400	No maximum velocity limit[b]	
Type 316 stainless steel	No maximum velocity limit[b]	
Ni–Cr–Fe–Mo alloys 825 and 20Cb3	No maximum velocity limit[b]	
Ni–Cr–Mo alloys 625 and C-276	No velocity limits	
Titanium	No velocity limits	

[a] In deaerated brines encountered in the heat recovery heat exchangers in desalination plants, the critical velocities can be increased from 1 to 2 ft s⁻¹ (0.3–0.6 m s⁻¹).
[b] Minimum velocity 5 ft s⁻¹ (1.5 m s⁻¹).
Source: LaQue [92]. Reproduced with permission of John Wiley & Sons.

Materials

Erosion corrosion of carbon steel is due to erosion of scale on the surface. Duplex and austenitic stainless steels have passive films that quickly reform, and these materials are erosion-corrosion resistant. Martensitic stainless steels, 13Cr alloys, are intermediate between the other two and show the effects of both erosion and corrosion [74].

Cavitation

The erosion corrosion that has been discussed so far has been due to moving fluids or solids impacting against a stationary metal surface. Cavitation is somewhat different, because it usually causes damage due to rapid movement of a metal surface in such a manner that a liquid, e.g. in a pump, undergoes a rapid loss of pressure that causes the liquid to form vapor bubbles. This release of vapor bubbles is not harmful, but if the same bubbles collapse against a metal surface, as shown in Figure 5.71, damage of the surface film(s) results in fresh metal exposures, which then corrode. Cavitating pump impellers and housings can undergo rapid attack. Designing pumping systems to avoid the occurrence of cavitation, normally by maintaining a positive head on the liquid, is one means of avoiding this problem. Another is to use hard-facing alloys on pump components. Figure 5.72 shows a

TABLE 5.4 DNV Recommended Maximum Flow Velocities in Heat Exchangers [93]

Tube Material	Maximum Flow Velocity (m s⁻¹)
Al brass	2.1
Cu–Ni 90/10	2.4
Cu–Ni 70/30	3.0

Source: Reproduced with permission of DNV.

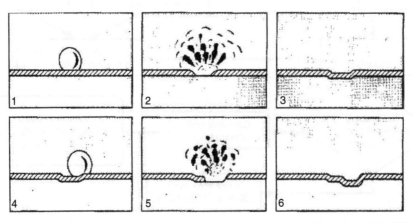

Figure 5.71 Cavitation bubble collapse and subsequent corrosion. *Source*: Courtesy of NACE International.

Figure 5.72 Erosion corrosion due to cavitation on a stainless steel pump impeller. *Source*: Bogaerts [82]. Reproduced with permission of NACE International.

pump impeller that were damaged due to cavitation. Note how the damage is on the downstream (low-pressure) locations.

Areas of Concern

Erosion corrosion is a possibility whenever changes in fluid flow patterns occur, especially when they are accompanied by concurrent changes in pressure or temperature. This can be downstream of flow restrictions, where additional turbulence and phase changes have been introduced, as well as at locations of local flow disruption.

Wellheads Erosion corrosion is normally handled in wellhead and Christmas tree equipment by making the equipment from erosion-resistant materials. The additional thickness of the castings common in wellhead equipment allows for some metal loss. It is common to use hard-facing liners on carbon steel components in these locations. This reduces the cost compared with making the entire assembly of an erosion–corrosion-resistant alloy.

Pumps Erosion corrosion in pumps is treated in a number of ways. Like wellhead equipment, many pump components are made from castings and are fairly thick. This provides erosion tolerances that are sometimes sufficient, provided that inspection during downtimes is possible.

Cavitation damage can be minimized by placing pumps in locations where sufficient positive head is available to prevent cavitation.

Damage to large pumps is often repaired using hard-facing alloys, typically nickel–cobalt alloys, that are applied by welding or flame spray processes. The use of

erosion-resistant alloys is also important. Many large seawater pumps made from bronzes, e.g. nickel–aluminum bronzes, may have erosion-corrosion problems caused by improper foundry techniques that produce unwanted phases in the alloy that then undergo selective phase attack (a form of dealloying), leaving soft surfaces that can then be eroded.

Downhole Applications While the ANSI/API recommended practice is written for topside service piping systems, it has also been used for downhole production tubing and for injection wells. If the recommended maximum velocities are too conservative, they can cause major losses of production. Setting the limits too high means erosion, possible equipment failure, and potential loss of production. Most companies consider the guidelines too conservative and operate with *C*-factors of 400 or greater and injection water (not multiphased fluid) velocities of up to 50 ft s^{-1} (15 m s^{-1}) for CRAs (e.g. 13Cr) [76].

Downhole tubing can have erosion-corrosion problems caused by localized turbulence near joints. This is shown in Figure 5.73, which shows erosion corrosion of downhole tubing from an offshore production platform in the North Sea. This platform received major attention when downhole erosion was reported shortly after production started [86, 87]. Downhole multiphase fluid flow regimes are seldom as simple as shown in Figure 5.69. Deviations from vertical flow can often exceed 45° and can sometimes approach horizontal. This means that inspection tools must check in the most likely locations for damage and asymmetrical damage of downhole tubing has been reported [86].

Note how most of the tubing surfaces shown in Figure 5.73 are not corroded. This is due to a combination of protective iron carbonate scales from the production fluid and the action of corrosion inhibitors where the scales have been breached. In situations where the scale is eroded, like suggested in Figure 5.70, the corrosion inhibitor dosages may be inadequate to cover the exposed metal surfaces, and the fluid velocities may be too fast and erode the inhibitor films from the exposed metal [86–89].

Condenser Inlets Condenser inlets are another area of erosion–corrosion concern. The first few centimeters of condenser tubes are prone to erosion corrosion, and a common method of minimizing this problem is to use plastic inserts that expand upon wetting and line the tube near the inlet. The loss of heat transfer is minimal, because the thermally insulating polymers extend only a short distance into tubing that is usually several meters long. This is more of a problem with soft condenser tube materials, e.g. cupronickel and other copper alloys. The

Figure 5.73   Eroded downhole tubing from an offshore production platform.

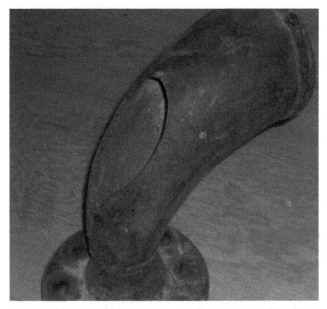

Figure 5.74   Erosion corrosion of an elbow in natural gas piping.

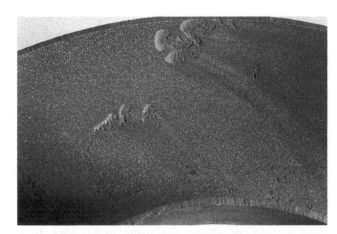

Figure 5.75   Erosion corrosion in a carbon steel steam pipe. *Source*: Bogaerts [82]. Reproduced with permission of NACE International.

trend to the use of titanium for seawater piping offshore has minimized this problem, because titanium is much more erosion resistant.

Erosion in Elbows and Bends in Piping   Figures 5.74–5.76 show erosion corrosion piping near elbows and bends. The additional turbulence at sharp bends in piping causes accelerated erosion, especially when solids are entrained in the system. Liquid droplets can also impinge at piping bends and produce similar erosion patterns. Notice the localized erosion damage. It is very important to inspect in the proper locations to monitor if erosion is occurring. Placing an ultrasonic probe only a few centimeters away from the damage would miss it entirely. This problem has caused many utility systems to develop erosion modeling software to allow plant inspectors to determine where their periodic inspections should occur. The miles of piping in a typical power plant are too extensive to allow 100% inspection.

Steam injection systems in oilfield operations are even more complicated than power plant piping, and software for predicting where inspections should occur is not available. Any potential inspections are complicated, because most erosion-subject steam injection piping is covered with insulation and the quality of steam (presence or absence of water droplets) is likely to be lower in injection systems.

One potential remedy to minimize erosion corrosion in steam piping is to increase the radius of any bends in the piping. This, of course, means increased installation costs and space limitations, especially offshore, will often prevent this approach.

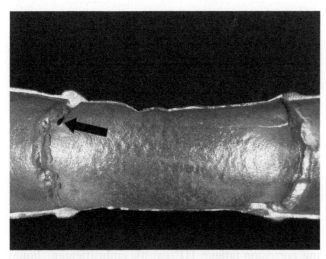

Figure 5.76 Interior view of erosion corrosion on welded Monel piping at 90° elbow. The arrow indicates the leakage location. *Source*: Reproduced with kind permission of A. Zeemann, *Materials Life: The Materials Image Data Base.*

Figure 5.77 Erosion corrosion of a stainless steel seal in a steam line. Time to failure was hours.

Seals and Control Surfaces Seals are locations where erosion corrosion can be very rapid. Improperly placed seals, or the relative motion of components at sealed joints, can create small openings that can be rapidly eroded, often within hours, once high-pressure steam or other fluids start leaking. This is shown in Figures 5.77 and 5.78. In both cases the eroded material was a relatively soft austenitic stainless steel, and the use of a harder erosion-resistant material (e.g. chrome cobalt alloys) was appropriate.

Additional Areas of Concern It has been estimated that up to 15% of failures in oil and gas production are due to erosion corrosion in gravel packs, nozzles, and

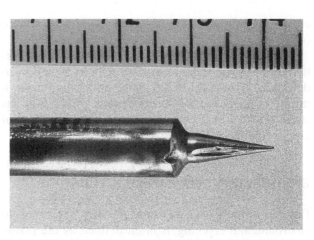

Figure 5.78 Erosion corrosion through a pin (spindle) in a pressure-reducing valve for a high-pressure boiler. Time to failure a few days. *Source*: Bogaerts [82]. Reproduced with permission of NACE International.

TABLE 5.5 Equipment Features Susceptible to Erosion Corrosion [89]

Chokes	Most vulnerable
Sudden constrictions	
Partially closed valves	
Standard radius elbows	
Weld intrusions	
Pipe bore mismatches at flanges	
Reducers	
Long radius elbows and miter elbows	
Blind tees	
Straight pipes	Least vulnerable

Source: Feyerl [89]. Reproduced with permission of NACE International.

Christmas trees before they reach a first separator [89]. The equipment features most likely to experience erosion corrosion are shown in Table 5.5. Erosion corrosion is most likely to happen with particulates, e.g. sand production, although liquid droplets and cavitation can also produce problems.

Erosion and Erosion-corrosion Control

Erosion corrosion can be controlled by the use of harder alloys (including flame-sprayed or welded hard facings) or by using more CRAs. Alterations in fluid velocity and changes in flow patterns can also reduce the effects of erosion corrosion. Chemical treatment with corrosion inhibitors may require much higher dosage levels than are required in the absence of erosion corrosion. This is because erosion removal of protective films may expose much higher bare metal surface areas. If solid-particle erosion is involved, most corrosion inhibitors will adhere (chemisorb) to the particles as well as to bare metal.

Prediction of erosion-corrosion locations and severity is limited, and there is no clear consensus on how to determine erosion thresholds [90]. For this reason, monitoring in likely erosion locations once production has started is the primary means of controlling the effects of erosion corrosion [91].

ENVIRONMENTALLY ASSISTED CRACKING

Environmentally assisted cracking, often shortened to environmental cracking, is often defined as the brittle failure of an otherwise ductile material due to the presence of tensile stresses and a specific environment. The stresses involved in environmental cracking can be from applied loads or from residual stresses caused by manufacturing and construction processes [2]. The overall macroscopic stresses associated with this cracking are generally much below the yield stresses of the alloys in question. This cracking can result in the sudden rupture of structures, especially pressure vessels such as pipelines, and this sudden failure can lead to significant safety concerns.

Because of safety concerns and the widespread occurrence of environmental cracking in many industries and environments, it is the form of corrosion most studied in research laboratories worldwide. Unfortunately, despite decades of research, no consensus on the mechanism(s) of environmental cracking is available, and the classification of environmental cracking remains controversial.

At one time SCC was considered an anodic phenomenon, and HE, due to cathodic hydrogen charging, was considered to be another form of environmental cracking [4, 50, 93]. This idea is controversial, and no clear consensus on the mechanisms is likely [94]. While the mechanisms of some forms of environmental cracking remain unclear, approaches to minimize the problem are available and will be emphasized in this discussion. In the absence of a consensus on this subject, this discussion will treat environmental cracking in accordance with the guidance available in ANSI/NACE MR0175/ISO RP010156, the most widely used oilfield standards on the subject [50, 95, 96]. Because ANSI/NACE MR0175/ISO RP010156 is only concerned with H_2S-related cracking, the terminology used in ISO 21457 is used in discussing liquid metal embrittlement, chloride SCC, and other cracking phenomena not H_2S related [21].

There are a wide variety of terms used to describe environmental cracking to include SCC, season cracking, corrosion fatigue, HE, caustic embrittlement, liquid metal embrittlement, etc. Many oilfield personnel will use the term SCC as a synonym for all of the above terms.

One clear consensus is that most environmental cracking, with the exception of corrosion fatigue, occurs due to a combination of tensile stresses and specific corrosive environments. Corrosion fatigue can occur in any corrosive environment. A few oilfield environments are listed in Table 5.6. It is apparent from Table 5.6 that hydrogen sulfide (H_2S) and chlorides are common to environmental cracking in most oilfield environments. Despite all of the research on environmental cracking, no screening tests have been developed that identify new environments, and all of the alloy–environment combinations have been identified due to field failures.

At one time environmental cracking was considered to be a problem of certain alloys in certain environments. Unfortunately, every structural alloy system has some environment where cracking is known to occur [97]. Note that in the following discussion, there are some mechanisms where the stresses come from chemical

TABLE 5.6 Metals and Environmental Cracking Environments [72]

Metal	Environment	Factors That Increase Risk of SCC
Carbon steels	Hydrogen sulfide	Increasing H_2S, moderate temperatures, more acidic, higher strength/hardness, higher stress levels
Carbon steels	Carbonates	Higher strength
Carbon steels	Chlorides	Higher strength, higher stress levels, more acidic
Copper alloys	Ammonia	Higher strength, higher stress levels
Martensitic stainless steels	Hydrogen sulfide	Increasing H_2S, moderate temperatures, more acidic, higher strength/hardness, higher stress levels
Austenitic stainless steels	Chlorides	Higher strength, higher chloride levels, higher stress levels, more acidic, higher temperatures, presence of H_2S
Duplex stainless steels	Chlorides	Higher strength, higher chloride levels, higher stress levels, more acidic, higher temperatures, presence of H_2S
Titanium	Alcohol	Higher stress levels, lower water content

Source: Reproduced with permission of Elsevier.

reactions either on the surface or within the metal. This concept of environmental cracking is a major departure from the idea of tensile stresses (applied or residual) found in the general (as opposed to oilfield) literature.

Stress Corrosion Cracking (SCC)

SCC is defined in two different oilfield standards as "cracking of metal involving anodic processes of localized corrosion and tensile stress (residual and/or applied)" [21, 95]. The idea of anodic processes is in dispute by many authorities, especially for SCC of high-strength ferritic steels. Many authorities attribute SCC to cathodic HE processes and sometimes use the term "environmental hydrogen embrittlement" (EHE) instead of SCC [94]. The exact mechanism is irrelevant to field applications.

The term SCC is used for most aqueous environmental cracking that is not clearly associated with hydrogen or hydrogen sulfide. The most common characteristic of environments that cause this form of corrosion is the presence of chlorides, although copper-based alloy stress corrodes in ammonia and other nitrogen-containing environments. Many environments that cause SCC are only mildly corrosive insofar as other forms of corrosion are concerned. It is not unusual to see SCC on surfaces that otherwise seem to be uncorroded, or only mildly corroded, at least to the unaided eye.

Authorities agree that three conditions must be met for SCC [1–4, 94, 98–103]:

- A specific corrosive environment.
- Tensile stress, which can be due to applied loads or from manufacturing/assembly processes.
- A material susceptible to SCC in the specific environment.

These ideas are shown schematically in Figure 5.79. Figure 5.80 shows piping locations where SCC is most likely to occur due to residual stresses associated with welding and other fabrication processes.

SCC initiation sites include pits, metallurgical defects, surface discontinuities, intergranular corrosion, and other stress raisers (Figure 5.81).

SCC is often unaccompanied by visible corrosion products and may appear to be the result of strictly mechanical causes. This is shown in Figures 5.82 and 5.83. Note the "river branching pattern" or "spider web" of cracks in Figure 5.83. This branched-crack appearance is direct evidence of SCC, and no other metallurgical failure mode produces this pattern of cracking. The absence of this branching should not be taken as evidence of a lack of SCC, because many metals, to include high-strength steels, may reach a critical flaw size and

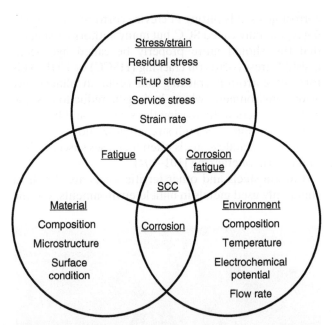

Figure 5.79 Venn diagram showing how SCC depends on stress, material, and specific environments.

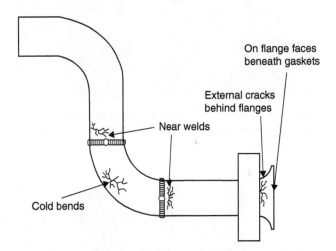

Figure 5.80 Piping locations where SCC is most likely to occur.

fail by SCC before the crack has extended long enough to start branching. Cracks as short as a fraction of an inch (several mm) have been known to produce SCC failures in high-strength steel and other materials.

The growth of stress corrosion cracks is discontinuous and is thought to usually involve initiation and first-stage propagation, secondary steady-state propagation, and final rapid failure. HE has been proposed as the mechanism whereby these stages occur, although this is disputed by some authorities [94]. The strongest evidence for hydrogen-related mechanisms is in high-strength ferritic steels [94]. In the presence of active corrosion, usually pitting or crevice corrosion, the cracking is generally called SCC, but many authorities suggest that this should more properly be called hydrogen-assisted stress corrosion cracking (HSCC) or EHE. This form of corrosion cracking can occur in almost any acidic environment where hydrogen reduction is the cathodic reaction. Increasing the strength and hardness levels of carbon steel, high-strength low-alloy steels, and martensitic stainless steels often increases susceptibility to this form of cracking [101–104].

Carbon steels and other ferritic steels are the most commonly used metals in oilfield environments, and the

possibility of hydrogen involvement in the SCC mechanism(s) is an indication that corrosion control methods based on cathodic protection must be used with caution. This is one reason why pipelines, where corrosion control is normally by a combination of protective coatings and cathodic protection, are seldom constructed out of high-strength (yield stresses greater than 80 ksi [550 MPa]) steel. Most cathodic reduction reactions on carbon steel pipelines are due to oxygen

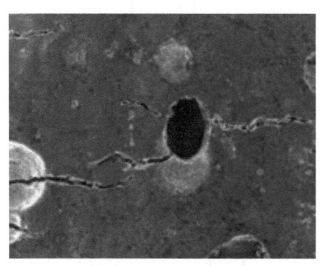

Figure 5.81 Multiple SCC cracks originating from corrosion pit in a steam turbine disc [99]. *Source*: Reproduced with permission of Elsevier.

Figure 5.82 Stress corrosion cracking underneath insulation on a stainless steel condensate line. *Source*: Courtesy of NACE International.

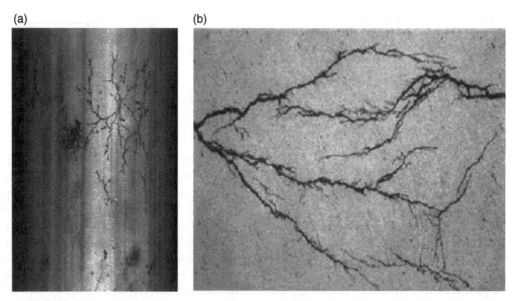

Figure 5.83 Multiple branching cracks typical of many stress corrosion cracking failures. Part (a) shows a macroscopic exterior view. Part (b) shows a metallographic cross section of the same metal. *Source*: Bogaerts [82]. Reproduced with permission of NACE International.

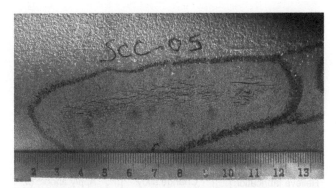

Figure 5.84 Clustered "colonies" of stress corrosion cracks on the outside of a carbon steel pipeline. *Source*: Merle [105]. Reproduced with permission of NACE International.

reduction, but the possibility of hydrogen reduction, especially beneath disbonded protective coatings, is a potentially serious concern. The possibility of HE due to cathodic protection on high-strength fasteners at flanges on immersed blowout preventers and similar equipment is also a concern.

Unlike the controversy associated with mechanisms of SCC in steels, there is no dispute that SCC in titanium alloys can involve hydrogen, and hydrides of titanium have been detected in titanium after SCC events.

Stress corrosion cracks often appear in groups or colonies on otherwise uncorroded surfaces. This is shown in Figure 5.84, which shows intersecting cracks on the outside of a buried pipeline during external excavation and examination. Cracks, either internal or external, will often grow together until they reach a critical flaw size that may lead to final rupture. This is shown in Figure 5.85, where small cracks have joined together and a circumferential crack has grown around a pipeline.

Cracks often branch as they progress into the metal. There are a number of reasons for this branching, which is shown in Figure 5.86 for the same pipeline that is shown in Figure 5.85. Note how the crack splits at approximately the mid-wall location in the pipeline and how the outside surface is generally roughened, even though only one SCC crack seems to have started at the surface shown in this picture.

SCC cracks can progress in an intergranular (between the crystals) or a transgranular (across the crystals) manner. This is one means that failure analysts have used to determine the causes of cracking, e.g. in pipeline steels where the pH of the surface moisture may influence whether cracking is intergranular or transgranular [105]. Figure 5.87 shows the transgranular nature of cracks in the pipeline steel shown in Figures 5.85 and 5.86. While this analysis has proven useful in analyzing the causes of SCC in buried pipelines, many cracks will change mode during the crack progression [4].

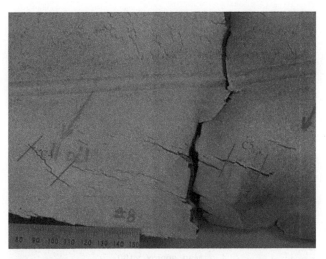

Figure 5.85 Small cracks joining together and intersecting circumferential cracks on the exterior of a pipeline. *Source*: Sutherby [106]. Reproduced with permission of NACE International.

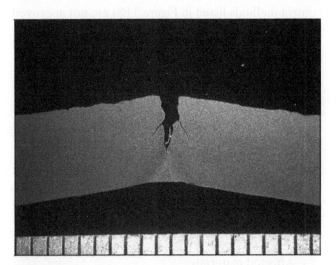

Figure 5.86 Typical cross section of secondary cracks showing mid-wall bifurcation. *Source*: Sutherby [106]. Reproduced with permission of NACE International.

SCC cracking is controlled by avoiding metal–environment combinations that cause this problem. Additional methods of SCC control include various methods of stress relief, e.g. post-weld heat treatment, protective coatings, corrosion inhibitors, and cathodic protection. Much of this control, e.g. in pipelines, is devoted to avoiding the formation of stress risers, small defects on metal surfaces that serve as initiation sites for SCC cracks, which, once started, may be difficult to control [92]. Figure 5.88 is a summary of SCC control methods as applied in many industries [16].

Plated metallic coatings or conversion coatings on carbon and low-alloy steels (one of the environmental

Mag. approx. ×150

Figure 5.87 Transgranular SCC crack in pipeline steel. *Source*: Merle [105]. Reproduced with permission of NACE International.

control methods listed in Figure 5.88) are not approved for use in H₂S environments [50]. These thin coatings are likely to have defects (coating holidays) that would allow hydrogen entry into the metal. The standard for CRAs (Part 3 of the same standard) does provide for the use of CRA cladding or weld overlays in similar environments [50]. Cladding and overlays are much thicker than plating or conversion coatings.

Chloride Stress Cracking Chloride stress corrosion cracking (CLSCC) is a concern for austenitic stainless steels (300 series) and other alloys [15, 104]. This is just one example of a specific corrosive environment that can cause SCC of susceptible materials (Figures 5.79 and 5.82). The following generalizations pertain to CLSCC [107]:

- Affected materials:
  ○ All 300-series stainless steels.
  ○ Duplex stainless steels are more resistant.
  ○ Nickel-based alloys are not immune, but are generally resistant.
- Critical factors:
  ○ Increasing temperatures increase susceptibility. Ambient-temperature SCC is possible but most problems occur above 60 °C (140 °F).
  ○ Initiation from pitting or crevice corrosion. More severe conditions are necessary to initiate CLSCC than are needed to sustain it.
- Insulated piping:
  ○ CLSCC (Figure 5.82) is often due to leaching of chloride ions from the insulation [108].
- Appearance:
  ○ The metal usually shows no visible signs of corrosion (Figures 5.82, 5.84 and 5.85).

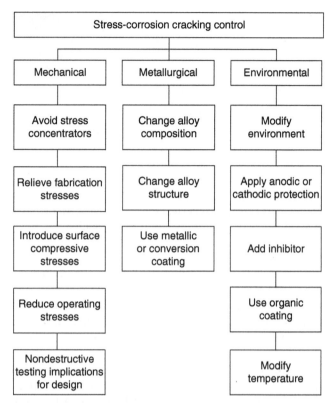

Figure 5.88 SCC control methods. *Source*: Davis [16]. Reproduced with permission of ASM International.

- Fracture surfaces usually have a brittle appearance.
- Prevention/Mitigation:
  ○ Use resistant materials.
  ○ Low-chloride water and proper drying out after hydrotesting.
  ○ Proper coatings under insulation.
  ○ Avoid stagnant water/condensation locations.
- Inspection and monitoring:
  ○ Visual and surface inspection methods.

Caustic Embrittlement Caustic embrittlement is one of the first forms of environmental cracking or SCC to receive widespread industrial attention. Boiler water is treated to a pH known to reduce corrosion rates (typically about pH = 8–9) by treatment with sodium hydroxide. If water evaporation produces higher-pH fluids, e.g. in crevices or at points where tubes meet restraints such as headers or baffles, the pH can increase to levels as high as 14, and this can cause intergranular cracking of carbon steels. Caustic cracking is controlled by the use of buffers, which prevent the buildup of high pHs, and by the substitution of ammonium hydroxide for sodium hydroxide. At high temperatures the ammonia evaporates, which deprives the hydroxide (OH⁻) ions of the necessary cation to balance electrical charge and prevents the buildup of excessive pHs [109]. The principals of boiler water treatment were developed in the 1920s by the US Bureau of Mines, and recent advances have come from a variety of companies that specialize in boiler water treatment chemicals. Most oilfield operators rely on guidance from these specialized water treatment suppliers.

Hydrogen Embrittlement and H₂S-related Cracking

The small size of hydrogen atoms means that hydrogen can readily dissolve in most metals. The dissolution of hydrogen into metals comes from two common sources: the reduction of hydrogen ions at cathodes in electrochemical cells due to corrosion or electroplating and the entry of hydrogen into metals from environments having hydrogen-entry promoters such as H_2S and cyanides. These environments produce small amounts of monatomic hydrogen that usually combine to form molecular (diatomic) hydrogen molecules, which are too large to dissolve into metals. In the short time that monatomic hydrogen atoms exist on surfaces, small amounts of monatomic (nascent) hydrogen dissolve into the metal substrate and follow diffusion paths from locations of high concentrations (the source surface) to regions of lower concentration (the metal interior and usually the opposite surface). This dissolution is usually at interstitial sites (between the atomic locations of the metal) or along grain boundaries. Several forms of hydrogen degradation are associated with the recombination of internal hydrogen atoms to form hydrogen molecules, which are too large for interstitial diffusion through the metal lattice. Other forms of HE are due to mechanisms that are not presently understood and the subject of research controversies.

HE and other hydrogen-related problems can occur in any H_2S-containing environment and in electroplated metals even in the absence of environmental hydrogen. Another source of monatomic hydrogen is welding, and improper welding procedures can introduce monatomic hydrogen into metals.

There are a wide variety of hydrogen and H_2S-related cracking phenomena encountered in upstream environments. Many of the classifications discussed below follow terminology in ANSI/NACE MR0175/ISO 15156, which emphasizes the choice of materials for use in H_2S environments [50, 95, 96].

Hydrogen Embrittlement (HE) Small quantities of hydrogen inside certain metals make them susceptible to subcritical crack growth under stress. Metals can also have major decreases of yield strength and undergo brittle failures in hydrogen-containing environments. Both processes are commonly called HE. Oilfield metals with HE problems include high-strength steels,

aluminum, and titanium, although most problems occur in high-strength steels. The exact mechanisms of HE in steels have not been established, and no iron hydrides have ever been reported, but brittle intermetallic hydrides have been found in titanium and other hydrogen-embrittled metals. Ferritic steels, e.g. carbon steels and low-alloy steels, are considered to be more susceptible than austenitic alloys [96].

The initial 1975 version of NACE RP0175 limited H_2S exposures to metals having hardnesses of HRC22 or less, which, depending on the size and shape of the metal, correlates to yield strengths of approximately 80 ksi (550 MPa). The API carbon and low-alloy bolt standard, API 20E, limits hardness to HRC 34, unless another international standard, such as ANSI/NACE MR0175/ISO 15156, calls for a lower level of hardness [110].

The sudden propagation of brittle fractures may be time delayed and occurs months, even years, after exposure to hydrogen. This is a characteristic failure mode in plated metal components. It is thermodynamically impossible to electroplate metals such as zinc or chromium onto steel without also generating hydrogen gas, some of which invariably dissolves into the steel. The standard way of compensating for this inevitable introduction of hydrogen into the metal substrate is to use an elevated-temperature bakeout procedure of several hours, depending on metal thickness, at temperatures around 175–205 °C (350–400 °F) [94, 111].

Higher-strength steels are considered to be more susceptible than lower-strength alloys, and work hardening seems to be preferable to heat treatment, e.g. in high-strength cables [2]. Concerns for HE are the reasons that very high-strength wire, used in downhole wireline applications, must be allowed to outgas for days between downhole trips. Many authorities consider the highly cold-worked metals used in these wires to be less susceptible to hydrogen effects than thicker metals, usually heat treated for strength, at the same strength (or hardness) levels. The presence of multiple defects, primarily dislocations, is thought to serve to minimize the accumulation of hydrogen in any one location and to minimize formation of subsurface molecular (diatomic) hydrogen considered to be associated with HE.

Hydrogen charging, the introduction of monatomic hydrogen into metals, can come from the breakdown of water at elevated temperatures in welding processes. This is the reason for protective, water-impermeable coatings on most welding rod [112].

Charging can also occur on cathodes at defects exposing steel underneath anodic protective coatings, e.g. zinc plating. This is one reason that galvanizing (zinc coatings) were not allowed on high-strength fasteners [113].

Pickling processes (immersion of the steel in acids to remove mineral scales and rust) also cause hydrogen charging and are the other concern with using zinc or other metallic coatings on these high-strength steels [94, 113, 114].

HE is considered to be a relatively low-temperature phenomenon, and most failures seem to occur at temperatures below 100 °C (212 °F). Prolonged exposure to high temperatures can have counteracting effects. More hydrogen may be generated on metal surfaces due to accelerated corrosion or other chemical reactions, but atomic diffusion and subsequent outgassing is also enhanced.

Hydrogen Stress Cracking (HSC)   HSC is a term used in ANSI/NACE RP0176/ISO 15156 to describe cracking in a metal due to the presence of hydrogen in a metal along with residual or applied tensile stresses [95]. It is used to describe cracking in metals that are not sensitive to sulfide stress cracking (SSC) but that are embrittled when galvanically coupled as cathodes to corroding anodes. The term galvanically induced HSC is applied to this mechanism of cracking. The discussion above on HE of galvanized bolts would be considered to involve HSC by many oilfield authorities, although this is not the terminology used outside the oilfield industry [94].

Hydrogen-induced   Cracking   (HIC)   Hydrogen-induced cracking (HIC), also known as stepwise cracking (SWC), in carbon and low-alloy steels is caused by atomic hydrogen diffusing into the steel and forming hydrogen molecules internally at trap sites, such as vacancies in the metal, grain boundaries, dislocations, and second-phase particle boundaries, to include inclusions [113–117]. HIC is a form of hydrogen-related cracking that does not require tensile stresses, either applied or residual, to produce cracking. This lack of applied stresses is the main differentiation between HIC and SCC. HIC is a major concern in the H_2S environments covered by ANSI/NACE SP0176/ISO 15156 and in other environments, e.g. strong mineral acids (H_2S is a weak mineral acid), known to produce HE [99].

Figure 5.89 shows HIC in pipeline steel. Note how the cracks form parallel to the rolling direction of the steel and how they tend to be planar – relatively long and flat. Modern pipeline steels are treated with calcium that reduces the total volume of inclusions and tends to

form hard, rounded inclusions that should present fewer locations for HIC cracks to form [118].

Figure 5.90 shows environmental conditions where HIC is considered important for pipeline steels. The combinations of carbon dioxide and H_2S in the environment shown in this figure are approximate, but they are used for determining environments where HIC-resistant steels are necessary. Many operators assume that virtually all oilfields will eventually sour and require H_2S-resistant materials, but some gas fields are depleted so quickly that this may not be necessary.

HIC susceptibility has been shown to be greatest in high-sulfur steels and to be less likely in modern, low-sulfur steels. However, in low-sulfur steels other considerations, such as the presence of ferrite–pearlite banding, also promote HIC [96]. Condensate waters in gas systems lack the mineral buffering available from formation waters that accompany crude oil. For this reason, gas systems are likely to have acidic environments and are usually more corrosive than crude oil systems. If the gas contains H_2S, then the gas systems will require H_2S-resistant materials of construction [119].

Sulfide Stress Cracking (SSC)   This common form of environmental cracking (SSC) is considered to be a form of HIC. It requires a residual or applied tensile stress and the combined presence of water and H_2S, which indicates that some organizations would consider this to be a form of SCC, although this is not the interpretation in ANSI/NACE RP0175/ISO 15156 [89].

Susceptible alloys, especially steels, react with hydrogen sulfide, forming metallic sulfides and monatomic nascent hydrogen. This monatomic hydrogen forms as a reduction reaction product and diffuses into the metal matrix causing internal cracking. High-nickel contents, which limit surface corrosion and also cause the microstructure to have austenite, a more resistant phase than the ferrite common in carbon and low-alloy steels, greatly improve the resistance to SSC. SSC is worst around 80 ± 20 °C (176 ± 36 °F). Above this temperature hydrogen is more mobile and likely to diffuse from the metal before forming internal defects, leading to cracking. The term SSC is applied to cracking in liquid water environments.

Soft zone cracking (SZC) is a form of SSC that occurs in steels having localized "soft zones" of low-yield strengths in otherwise harder/stronger materials [95]. These zones are normally associated with HAZs in welded steels. Localized plastic deformation yielding can increase susceptibility to hydrogen-related cracking in these steels.

Stress-oriented   Hydrogen-induced   Cracking (SOHIC)   SOHIC is the result of small cracks formed approximately perpendicular to the principal stress

Figure 5.89   Hydrogen-induced cracking (HIC) in pipeline steel. *Source*: Photo courtesy NACE International.

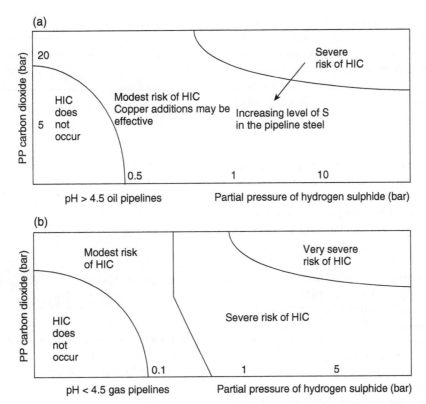

Figure 5.90 Environmental conditions where HIC is important in (a) crude oil and (b) natural gas pipelines. *Source*: Palmer [118]. Reproduced with permission of Pennwell Publishing.

(applied or residual) in steel, resulting in "ladder-like" cracking linking sometimes small preexisting HIC cracks [95]. The result is relatively long cracks in the through-thickness direction of a structure, e.g. a pipeline. This is shown in Figure 5.91, where small HIC–SWC cracks have been connected by perpendicular connecting cracks, resulting in considerably less resistance to applied stresses. Note that SWC is apparent near the center of the steel, whereas SOHIC, due to joining of smaller cracks, appears near the top of the sample where the stresses were higher.

SOHIC has been reported in the parent material of longitudinally welded pipe and in the HAZs of welds in pressure vessels. It is a relatively unusual phenomenon usually associated with low-strength ferritic pipe and pressure vessel steels.

Susceptibility to SOHIC can be complicated, as low-sulfur modern steels, thought to be resistant to this form of damage, have been found to have additional problems associated with the metallurgical treatments of newer steels used in repair and new construction [119–122].

Hydrogen Blistering The small cracks discussed above in sections on HIC, SSC, and SOHIC are sometimes referred to as small blisters. Another use of the term hydrogen blistering is for hydrogen gas blisters that

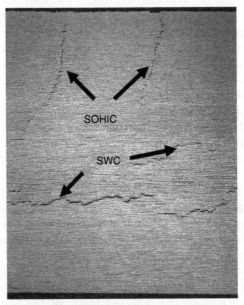

Figure 5.91 SOHIC cracking near the surface and SWC near the center of hydrogen-cracked line pipe steel.

form inside structures that become so large that their deformations can be seen macroscopically as surface deformations. These larger blisters form in the same way as the smaller cracks. Hydrogen diffuses into the steel and then gets trapped at locations where migration is hindered, and the recombination of thermodynamically unstable monatomic hydrogen to thermodynamically stable diatomic hydrogen molecules is most likely to happen. The most likely sites for this to happen are near large ceramic inclusions (impurities) within the steel. Figure 5.92 shows schematically how these inclusions are likely to be arranged in steel plate and other thick sections. As a result of the hot rolling process, ceramic inclusions near plate surfaces are more likely to be broken up and become smaller. Near the center of plate steel, there has been less plastic deformation, and both the steel crystals and the inclusion particles tend to be larger. As hydrogen migrates through the steel, the chances for monatomic atoms to meet and become trapped are greater near the larger inclusions near the center of the plate. This is the reason why plate steel, especially the relatively low-quality, inexpensive plate steel used for storage tanks and similar structures, is more likely to produce hydrogen blistering. Many of these blisters form near the center of plate steel and can become several inches (centimeters) in horizontal dimension.

Common locations for hydrogen blisters include the bottoms of aboveground storage tanks where acidic waters can concentrate and the walls of tanks and process equipment. Figure 5.93 shows blisters in the wall of a CO_2 scrubber, and Figure 5.94 shows blisters in a crude oil pipeline that collected sour crude for many years.

The quality of steel determines the resistance to hydrogen cracks and blisters of all types. Pressure vessel steels are not immune to this problem, but it is much less common in the highly processed and refined low-alloy steels used for most process pressure vessels.

Typical treatments for hydrogen blistering depend on the application. For storage tank bottoms, which are loaded in compression, it is fairly common to locate the boundaries of the blisters with ultrasonic inspection. Then, at a predetermined distance from the boundaries, a plate is welded over the blister, and the blister is left in place. Other organizations carefully pierce the blister using nonsparking drills to relieve the pressure before welding a patch over the surface. Similar procedures are sometimes used on tank walls, and monitoring the growth of blisters has been reported [82].

In pressure vessels, it is common to grind the blisters out of the metal. This grinding to a predetermined distance beyond the detected limits of the blister is intended to remove the overlying "blister" steel, which serves no structural purpose, and also to eliminate any microcracking that has not been detected by the inspection process. The remaining wall thickness is then determined. While welding repairs are sometimes performed to replace the missing metal, it is more common to determine the maximum allowable operating

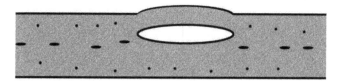

Figure 5.92 Schematic representation of ceramic inclusions and hydrogen blister formation in plate steel.

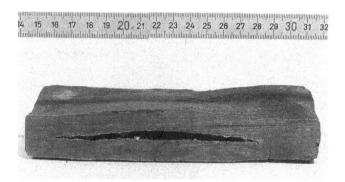

Figure 5.93 Hydrogen blister formed on wall plate of a CO_2 scrubbing tower. *Source*: Bogaerts [82]. Reproduced with permission of NACE International.

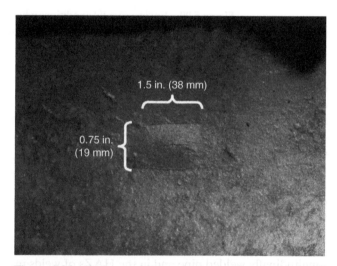

Figure 5.94 Hydrogen blister on the inside of a crude oil pipeline.

pressure of the equipment and to downgrade the service if necessary. If reductions in operating pressure are inadvisable, then external repairs (sleeves, clamps, etc.) are used to increase the effective wall thickness in the area of concern. The same downgrading or external reinforcement procedures are used for pipelines, but it is seldom feasible to grind out the blisters, especially if they are internal.

The best way to avoid hydrogen blistering is by the use of higher-quality, lower-inclusion steel, and this is the reason that most hydrogen blisters are encountered in aboveground storage tanks and other large structures where the economics of construction from higher-grade steel are not justified and inspection and repair of occasional blisters is an accepted alternative.

Hydrogen Attack High-temperature hydrogen environments can cause hydrogen to diffuse into the metal and react with the carbon in the iron carbides (cementite) in steel. The resulting formation of gaseous methane can cause blistering similar to HIC or hydrogen blistering. This form of attack, while a concern in refining, is seldom a concern in upstream operations [1, 15].

Internal Hydrogen Embrittlement (IHE) HE can be caused during manufacturing before exposure to corrosive environments. This embrittlement is often termed internal hydrogen embrittlement (IHE) to differentiate it from HE caused by corrosive environments, EHE.

Environmental H_2S is one of several hydrogen-entry promoters known to increase the amount of hydrogen that is absorbed into steel. Cyanides, common in electroplating baths, are the other most commonly encountered hydrogen-entry promoters, and the problems with electroplated parts are the reason why ASTM and other standards have been developed for removing of hydrogen that enters steel parts during electroplating. Hydrogen can also enter steel during pickling processes, where steels are immersed in acid baths to remove scale and other surface contaminants. The high temperatures associated with welding can dissociate atmospheric water into elemental hydrogen and cause weld-related IHE. Other sources of hydrogen during manufacturing and construction include heat-treating atmospheres, the breakdown of organic lubricants, and grinding [94].

Control of IHE includes the use of higher-quality steels that have fewer inclusions that can serve as hydrogen trap sites, the use of copper-coated welding rods to keep moisture from entering rod during storage prior to use, shielding gases to keep moisture away from liquid weld beads, and hydrogen bakeout procedures to remove any hydrogen that entered the steel before it is placed in service [94, 111, 113, 114].

Liquid Metal Embrittlement (LME)

Liquid metals can attack solid metals and produce fractures similar to those found in other forms of environmental attack. Both intergranular and transgranular attacks are known to occur. This form of attack can happen during welding and other processes, but careful removal of low-melting coatings, e.g. zinc coatings, prior to welding normally solves this problem.

Most concerns with LME in upstream operations are offshore and relate to safety in fires and to cracking of brazed aluminum heat exchangers, which are used in cooling natural gas to reduce the volume before injecting it into a pipeline for transmission to shore.

Many operators have banned the use of aluminum or galvanizing for offshore platform structures, e.g. handrails, ladders, etc., because of a concern that these liquid metals could cause embrittlement of carbon steels and stainless steel if the lower-melting zinc or aluminum were to melt in a fire. The counterargument to this thinking is that by the time temperatures have reached the melting point of these metals, it is too late for the equipment concerned. No firm guidelines on this question exist, and many operators are now using aluminum, e.g. for decks, ladders, handrails, and helidecks. The thin layers of zinc on galvanized structures are not considered adequate corrosion protection for most offshore structural members. Galvanizing is commonly used for corrosion resistance in bolts and similar fasteners, often with an overlay of an organic coating for ease of disconnect. In some locations, cadmium, which is banned for toxicity reasons by some authorities, is the preferred anodic coating material for fasteners, although it is seldom applied to larger components.

Brazed aluminum heat exchangers are used for offshore cooling of natural gas, because of their weight savings over other metals. To avoid cracking by liquid mercury, which is eventually found in virtually all natural gas formations, it is standard practice to place mercury removal processes in the gas treatment stream prior to the brazed aluminum heat exchangers used for cryogenic cooling [123].

Corrosion Fatigue

Corrosive environments lower the fatigue life of many oilfield components and structures [124, 125]. This is shown schematically in Figure 5.95, which shows the elimination of the fatigue or endurance limit due to the presence of a corrosive environment. It is possible to cyclically stress a component so quickly that the effects of corrosion are minimal, and it is also possible to introduce components so corrosive that they corrode to failure before the effects of cyclic stresses become significant. It is very hard to predict the cumulative effects of corrosion and fatigue. The simplistic ideas shown in Figures 4.21 and 5.94 assume that the cyclic stresses are the same on each repetition. This is approximately true for many rotating shafts, sucker rods, etc., but it is far from true for wave loading on offshore structures and for many other oilfield applications [124, 125].

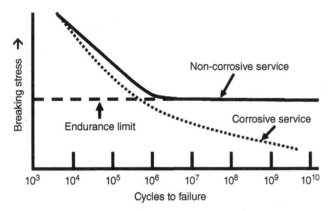

Figure 5.95 The reduction in fatigue resistance due to a corrosive environment.

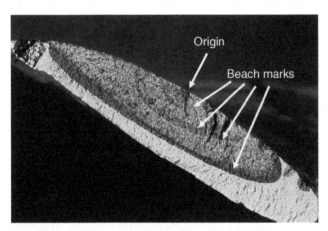

Figure 5.96 Beach marks on the surface of a corrosion fatigue surface from a marine propeller. *Source*: Photo courtesy NACE International.

Fatigue cracks, like many other forms of environmental cracks, often start at stress raisers such as corrosion pits, machined notches, or surface scratches. Once corrosion starts, localized corrosion fatigue cracks develop occluded cell concentration differences compared with the external environment, and the interaction between chemistry and mechanical loading becomes very complicated and virtually impossible to predict or model. It thus becomes important to monitor likely fatigue sites for crack initiation and propagation in attempts to detect damage and correct the situation before fracture occurs.

Figure 5.96 shows a typical corrosion fatigue fracture surface. The origin of the fracture surface is at the upper center of the picture, and the crack grew radially from this origin, leaving indications of the crack progression, called beach marks or clamshell markings, which are indications that the equipment tarnished to different levels, possibly due to intermittent operation of the equipment or changes in the cyclic loading level. Once the crack reached a critical flaw size, the part broke,

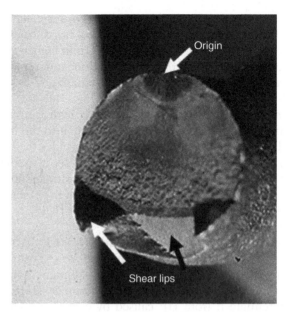

Figure 5.97 Corrosion fatigue and shear lip formation on a sucker rod fracture surface. *Source*: Photo courtesy NACE International.

leaving a shiny fracture surface at the bottom of the picture. It is apparent that other surface defects started additional cracks in the lower right of this picture. The appearance of beach marks is a characteristic feature of many corrosion fatigue failures. Note how the fatigue fracture spreads over most of the metal surface before the final tensile overload indicated by the shiny surfaces at the bottom of the picture.

Another corrosion fatigue fracture surface is shown in Figure 5.97. This oilfield sucker rod has a clearly defined origin, and beach marks indicate how the fracture spread from this origin due to cyclic loading. After the crack progressed across approximately one-half of the rod, the surface became rough, due to the fact that the crack remained open during compression strokes and did not polish the surface as much as near the origin. Once the crack reached a critical flaw size, the rod broke, forming shear lips at approximately 45° to the fracture surface. The shadows formed by the shear lips on the lower left and right of the picture appear as dark triangles in Figure 5.97. The bottom center of the picture shows where the metal pulled apart by shear, but the shear lip is down and away from the viewer. Most overload failures, to include fatigue failures, will form shear lips on the final fracture surface. A side view of this phenomenon is shown in Figure 5.98.

Most fatigue failures progress for significant distances before final overload failures, and inspection to detect and monitor fatigue cracking is an important part of corrosion fatigue control. Methods of preventing

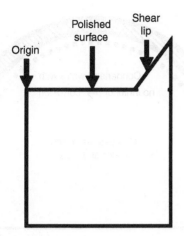

Figure 5.98 Schematic showing the side view of a typical corrosion fatigue rod failure.

corrosion, such as the use of more CRAs, corrosion inhibitors, and cathodic protection, are also helpful.

OTHER FORMS OF CORROSION IMPORTANT TO OILFIELD OPERATIONS

There is no universally accepted terminology for corrosion, and the oil and gas industry has adopted a number of terms for corrosion that frequently overlap with the terms discussed in previous sections of this chapter. This section discusses some forms of corrosion commonly discussed in the oilfield literature and some other forms of corrosion likely to be encountered in oil and gas production.

Oxygen Attack

Oxygen attack normally refers to pitting corrosion due to the presence of dissolved oxygen in production fluids. This seldom occurs in downhole environments, and when it does it is usually due to inadequate treatment of injection waters that were exposed to oxygen-containing air in topside processing and storage. Once the topside injection water has been treated down to 5 ppb or less, oxygen attack, if it occurs, is due to leaking seals or gaskets. Once these are repaired, the corrosion rate becomes minimal. The most common treatments for oxygen attack are mechanical deaeration and treatment of the water with oxygen scavengers.

Sweet Corrosion

This term, which is becoming outdated, refers to corrosion in environments where the corrosion is due to the presence of dissolved CO_2. Most production fluid corrosion, especially in natural gas production, is due to CO_2. The

Figure 5.99 Mesa corrosion of a 9-chrome sucker rod.

most common treatments for sweet, or CO_2, corrosion are the use of corrosion inhibitors. These inhibitors become ineffective as the downhole temperatures and pressures increase [60].

Sour Corrosion

The term sour corrosion refers to corrosion that occurs in production fluids due to the presence of H_2S. H_2S is a weak mineral acid, and most corrosion reactions are relatively minor in these environments.

Figure 3.3 compares the effects of oxygen, CO_2, and H_2S on corrosion. It is obvious that H_2S is the least corrosive of the three gases. Most problems with H_2S are associated with hydrogen-related cracking, and, while the environments associated with H_2S cracking are considered to be "sour," the cracking is seldom referred to as sour corrosion.

Mesa Corrosion

Mesa corrosion refers to corrosion in CO_2 environments that produce relatively flat surfaces where the metal is protected by carbonate films, usually siderite ($FeCO_3$). This corrosion is characterized as fairly deep pits in the form of sharp-edged holes that are considered to look like the flat-topped "mesa" (table topped) mountains found in the Southwestern United States [126]. Mesa corrosion is shown in Figures 5.99–5.101. All three pictures show relatively flat surfaces with localized pitting where the partially protective carbonate films break down. Most control of CO_2 (sweet) corrosion is by the use of corrosion inhibitors.

Top-of-line (TOL) Corrosion

The increased use of multiphase offshore pipelines and gathering lines has led to interest in top-of-line (TOL) corrosion, which can also occur onshore [127–135]. This form of corrosion is shown schematically in Figure 5.102. Condensate, containing water high in CO_2 and possibly acetic acid or other organic acids, aggressively attacks the top of horizontal pipelines where no corrosion inhibitor is present.

Figure 5.100 Mesa corrosion on carbon steel plate. *Source*: Bogaerts [82]. Reproduced with permission of NACE International.

Figure 5.101 Mesa corrosion at a threaded connection on downhole tubing. *Source*: Bogaerts [82]. Reproduced with permission of NACE International.

This condensation happens as the pipeline cools with distance from the compressor station. While corrosion inhibitor can be maintained in the water phase at the bottom of the line, it is difficult to apply this inhibitor to the top of the line. Much effort concentrates on developments in modeling of where TOL corrosion is likely to occur and how delivery of corrosion inhibitors can be maintained. Research efforts include modeling

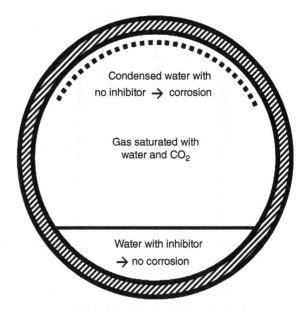

Figure 5.102 Top-of-line corrosion in a multiphase pipeline.

the temperature profiles of subsea pipelines. This is shown in Figure 5.103, which shows the water dropout profile for a subsea pipeline.

Efforts to apply corrosion inhibitor to the inside top of multiphase pipelines include periodic pigging with slugs of corrosion inhibitor. TOL corrosion is an area of continuing research. Additional modeling includes pH profiles and other parameters that affect pipeline corrosion [126–135].

Channeling Corrosion

Channeling corrosion is also known as 6 o'clock corrosion and bottom-of-line corrosion. It normally occurs in the lower interior of horizontal piping and is often attributed to relatively low flow velocities in multiphase fluids (Figure 5.104) or in water injection lines (Figure 5.105). Other examples of this 6 o'clock corrosion are shown in Figures 3.30, 4.54, and 5.4.

The corrosion in Figure 5.104 was caused by low flow velocities in an aging oilfield. The line was sized for higher production rates, and as production slowed the flow velocities of oil and entrained water decreased until the water separated from the water-in-oil emulsion and collected on the bottom on the line. This is a common problem as oilfields age [137].

The water injection pipeline in Figure 5.105 developed deposits on the bottom, which served to promote biofilms and MIC. The MIC-produced channeling could also have been classified as under-deposit (crevice) corrosion. Like most oilfield corrosion, several factors and explanations of what happened are possible.

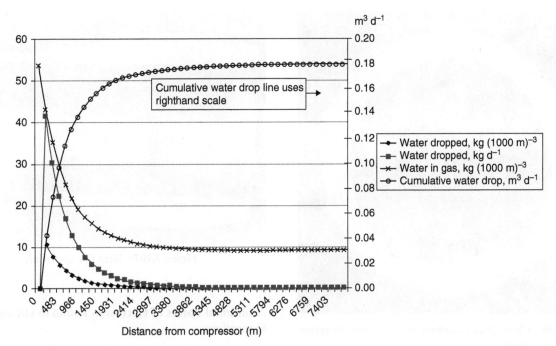

Figure 5.103 Water dropout versus distance from pipeline compressor [127]. *Source*: Reproduced with permission of NACE International.

Figure 5.104 Channeling or six o'clock corrosion at the bottom of a gathering line. Arrows point to leak (bottom arrow) and profile. Note that the rust scale hides the channel for most of the distance between the two arrows.

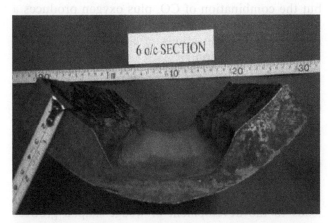

Figure 5.105 Severe channeling of a water injection pipeline. *Source*: Heidersbach and van Roodselaar [136]. Reproduced with permission of NACE International.

Figure 5.4 showed the effects of condensate corrosion in a gas well flowline. It is very similar in cross section to Figure 5.106 from a condensate return line in a steam-generating power plant [138]. Both of these images show the effect of acid corrosion. In Figure 5.4 the natural gas condensate contains organic acids and is naturally corrosive. In Figure 5.106 the line contains steam and condensed water. It is important to limit the ingress of oxygen

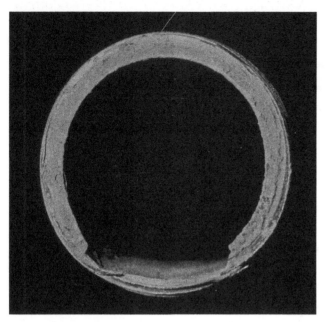

Figure 5.106 Channeling corrosion of a condensate return line at a power plant.

Figure 5.107 Wire line corrosion.

into CO_2-containing flowlines of any type. CO_2 is a weak acid and only slightly corrosive (Figure 3.3), but the combination of CO_2 plus oxygen produces a low-pH environment (due to the CO_2) that, with relatively high concentrations of air (oxygen), causes accelerated corrosion.

Grooving Corrosion: Selective Seam Corrosion

Grooving corrosion or selective seam corrosion are terms used to discuss weld-line corrosion grooves in vintage low-frequency electric resistance welding (LR-ERW) or electric flash welding (EFW) processes. It is localized corrosion attack along the weld bondline that leads to the development of wedge-shaped grooves that may become filled with corrosion products. Improvements in welding procedures seem to have eliminated this problem in pipe manufactured after approximately 1970 [139–144].

Wireline Corrosion

Downhole inspection devices are often suspended using high-strength wires. These wires can cut into protective coatings and wear away corrosion inhibitor and protective scale films. An example of wireline corrosion is shown in Figure 5.107.

Additional Forms of Corrosion Found in Oil and Gas Operations

The forms of corrosion discussed in this section are found in other industries, but they also occur in oilfield operations.

Acid (Hydrogen) Grooving   Highly concentrated oxidizing mineral acids such as sulfuric acid are stored in carbon steel tanks. The concentrated acid forms a semi-protective film, and general uniform corrosion is normally acceptable for most sulfuric acid service. The gradual thinning should be monitored so that replacements can be planned at appropriate intervals.

Hydrogen grooving results when condensate, which is acidic but more corrosive than the highly concentrated liquid being stored, drips from locations where it forms. This is typically the top of lines and below locations where acid is added to the tanks or piping systems. The corrosive condensate dissolves the passive film and creates grooves as shown in Figure 5.108.

Standard advice for control of hydrogen grooving is to use thick-walled piping, avoid stagnant situations where hydrogen gas bubbles can evolve in the same location over long periods of time and erode the passive film, and inspect locations where wall thinning is likely [82, 145, 146]. Alloys, generally high-nickel alloys, are used for piping systems where the effects of fluid motion or of dilution, which makes the acid more ionic and therefore more corrosive, must be used in this type of service.

Alkali Attack   Steam injection is a major means of secondary recovery in relatively viscous oilfields. Feedwater for steam production is normally treated to minimize the presence of minerals that would cause boiler deposits and limit heat transfer. If the boiler water

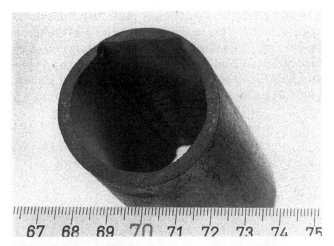

Figure 5.108 Hydrogen grooving in sulfuric acid piping system. *Source*: Bogaerts [82]. Reproduced with permission of NACE International.

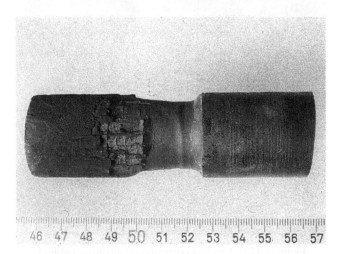

Figure 5.109 Alkali attack at crevice in boiler water system. *Source*: Bogaerts [82]. Reproduced with permission of NACE International.

treatment is inadequate or in dead legs and crevices, the pH of the water can increase to such levels that corrosive hot mineral deposits form in crevices (Figure 5.109). These molten salts are corrosive. This is shown in Figure 5.109. If mineral deposits due to precipitation of scale are not controlled, overheating of boiler tubes can occur, leading to tube swelling (creep) and rupture. These problems can be avoided by appropriate boiler feedwater treatment.

Contact Corrosion   Contact corrosion is the result of small particles of suitable materials embedded in stainless steel. Carbon steel is a very common contaminant. It often comes from grinding operations that leave small particles of carbon steel, which are hard, embedded in

Figure 5.110 Contact corrosion on stainless steel piping caused by grinding debris from nearby carbon steel. *Source*: Bogaerts [82]. Reproduced with permission of NACE International.

the relatively soft stainless steel. The carbon steel particles form galvanic cells and quickly corrode, leaving pits on the surface that can then promote corrosion of the stainless steel.

Figures 5.110–5.112 show contact corrosion. The thin 316 stainless piping in Figure 5.110 corroded in a warm humid atmosphere after only 3 years. Eventual failure of the system would result without remedial action, most likely replacement of the piping in question. The valve shown in Figure 5.111 was discovered during post-construction inspection. While the discoloration on the valve material and on the stainless steel cabinet in Figure 5.112 is unsightly, no degradation in performance is likely.

Contact corrosion can be prevented by carefully choosing grinding media that do not contain metallic iron, by shielding stainless steel from contamination from nearby grinding operations, by avoiding the storage of stainless steel equipment on carbon steel racks, or by dissolving the embedded particles in aggressive pickling acids and then repassivating the stainless steel with nitric acid. As stated above, contact corrosion is not an engineering concern on thick-walled castings like the one shown in Figure 5.111, but it can lead to perforation in thinner piping like shown in Figure 5.110.

It should be noted that some organizations, especially in Europe, use the term contact corrosion for galvanic corrosion, i.e. galvanic contact corrosion.

End Grain Attack   Figures 4.28 and 4.29 explained how the crystallographic structure is different in the three principal directions of plate metals. End grain attack has been shown in Figures 5.60 and 5.61, which showed intergranular corrosion, and in Figure 5.101, which shows mesa corrosion. The point to remember is that the ends of crystals have closer spacing and are more likely to corrode than the flatter surfaces parallel to the rolling or forming direction.

Many organizations use exposure samples of various types to monitor or measure corrosion rates. These small samples have different ratios of end grain vs. parallel-to-forming direction crystal exposure, and this invariably produces different corrosion rates.

Fretting Corrosion   Fretting corrosion happens when small oscillations in metal-to-metal contact abrade the protective films on metal surfaces and produce accelerated corrosion. It is similar to, and often considered to be, a form of erosion corrosion. The concept of fretting corrosion is shown in Figure 5.113. This is a form of corrosion that is often found in standby equipment, where

repetitive vibrations wear away protective films in the same location. This is shown in Figure 5.114 where a bearing race on a pump has suffered fretting corrosion. One means of limiting this form of corrosion is to operate the standby equipment for short periods at planned intervals, e.g. weekly. The bearings are unlikely to stop in

Figure 5.112   Contact corrosion on a control cabinet.

Figure 5.111   Contact corrosion on stainless steel valve components.

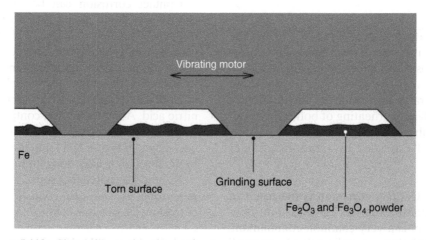

Figure 5.113   Sketch illustrating the mechanism of fretting corrosion. *Source*: Ahmad [13]. Reproduced with permission of Elsevier.

the same position after each operation, and localized accelerated attack can be avoided.

Fretting corrosion can be found on many systems that are subject to vibrations. Figure 5.115 shows fretting corrosion from a shackle pin on an FPSO mooring chain. Routine inspection for fretting corrosion should detect this form of corrosion before it leads to system failure.

Fretting corrosion is a major concern in heat exchangers, where vibrations cause damage at intermediate support baffles (Figure 5.116). Ultrasonic inspection is used to monitor wall loss at the baffle locations.

Downhole wear of moving parts, like the pump component shown in Figure 5.117, is more likely in deviated wells. Many offshore wells are more horizontal than vertical, and this can become an increasing problem.

Stray Current Corrosion   Stray current corrosion can be caused by improperly grounded welding equipment, reversed polarity on impressed current cathodic protection systems, telluric currents, and a number of other

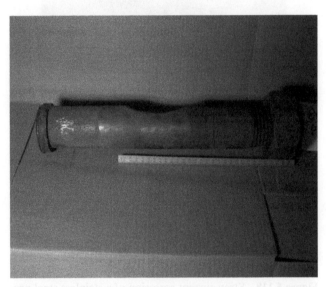

Figure 5.114 Fretting corrosion on roller bearing race. *Source*: Bogaerts [82]. Reproduced with permission of NACE International.

Figure 5.115 Fretting corrosion of a shackle pin from an FPSO mooring chain. *Source*: Photo courtesy Ammonite Corrosion Engineering, Calgary, AB, Canada.

(a)

(b)

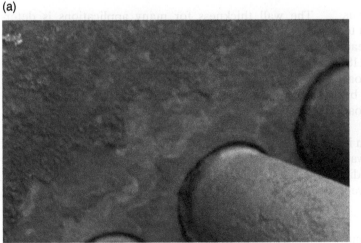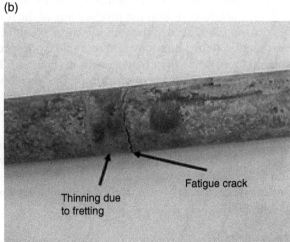

Figure 5.116 Fretting corrosion location at intermediate support baffle on heat exchanger. (a) Loose fit tubing. (b) Fretting and fatigue crack of heat exchanger tube.

Figure 5.117   Uneven wear on downhole pump component.

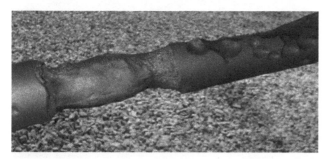

Figure 5.118   Stray current corrosion of a stainless steel propeller shaft. *Source*: Photo courtesy of Dudley Gibbs, Dudley's Marine Electric, Humacao, Puerto Rico, USA.

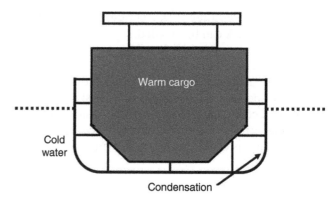

Figure 5.119   Cross section drawing of double-hulled tanker showing annular space where condensation leads to accelerated "thermos bottle corrosion" for tankers carrying warm cargo in cold waters.

sources. Many organizations rely on electrical isolation systems like those shown in Figures 5.13 and 5.14.

Stray current corrosion can produce any of the forms of corrosion discussed above, but it is more common to be diagnosed in situations where rapid widespread general corrosion, like that shown in Figure 5.118, occurs. The stray current corrosion of the small boat propeller shaft shown in Figure 5.118 was caused by improper grounding of the electrical system on the boat in question.

Oilfield applications where stray current corrosion is likely to occur include pipelines sharing rights of way and other situations where impressed current cathodic protection causes stray current corrosion on nearby well casings [2, 147, 148].

Thermos Bottle Corrosion   The introduction of double-hulled tankers has created new corrosion concerns related to the operation of these tankers in cold water, where cold water on the exterior hull creates a "thermos bottle effect" and increased condensation, leading to corrosion in the annular spaces between the external hull and the interior cargo tanks. This is shown schematically in Figure 5.119.

Many floating production, storage, and offloading vessels (FPSOs) are converted tankers. When operating in cold waters, the thermos bottle effect can be expected to lead to accelerated corrosion. Most corrosion control in the annular spaces on FPSOs is by protective coatings, and it is important that the annular spaces be designed so that inspection and maintenance can be accomplished. In addition to corrosion in these annular spaces, fatigue is an important concern in these vessels [149–155].

ADDITIONAL COMMENTS

The wall thickness for many applications is determined by calculating the wall thickness that meets mechanical requirements, such as pressure and weight of equipment, and adding an extra thickness called the corrosion allowance to account for the metal loss expected during the equipment design life [49]. Penetration rates may vary, so, in the absence of design codes that specify another procedure, it is common to assign a safety factor of 2, e.g. the corrosion allowance is often twice the anticipated general corrosion penetration. While this very conservative corrosion allowance is generally acceptable in relatively small chemical process plants, this allowance (twice the anticipated necessary wall thickness) would be prohibitively expensive for most oil and gas production applications, and corrosion allowances are generally much smaller for oil and gas production.

TABLE 5.7 Metal Failure Frequency for Various Forms of Corrosion

Forms of Corrosion Failure	Occurrence (%)	
General	31	
Stress corrosion cracking	24	0.65
Pitting	10	
Intergranular corrosion	6	
Erosion corrosion	7	24%
Weld corrosion	5	
Temperature (cold wall, high temperature, and hot wall)	4	
Corrosion fatigue	2	
Hydrogen (embrittlement, grooving, blistering, and attack)	2	
Crevices	2	
Galvanic	2	
Dealloying	1	11%
End grain attack	1	
Fretting	1	
Total	100	100%

Source: Landrum [142]. Reproduced with permission of NACE International.

The problem with corrosion allowances is that most corrosion is localized, and localized corrosion rates, e.g. for pitting or stress corrosion, are many times deeper than the rates associated with general corrosion. Approximately 70% of corrosion in a typical process plant will be due to forms of corrosion not addressed by corrosion allowances. Table 5.7 shows the forms of corrosion reported by corrosion engineers from a major chemical company over a period of several years. The general trends probably relate to oil and gas production. The major differences are probably the increase in frequency of HE and other H_2S-related forms of cracking. Most concerns of engineers and management are related to pitting and environmental cracking (termed stress corrosion cracking in Table 5.7).

General corrosion can be detected and repaired on a routine maintenance basis. Other forms of corrosion, e.g. SCC or HE, can result in sudden equipment failure and shutdown. Inspection and compensation for these events is a major concern that must be considered in design, inspection, and maintenance procedures.

Corrosion allowance was never meant to provide enough steel to let a corrosion problem go unattended, but a corrosion allowance does provide a little time so that one can detect corrosion problems and devise a remedy.

REFERENCES

1 Fontana, M.G. (1986). *Corrosion Engineering*, 3. New York: McGraw-Hill.
2 Revie, R.W. and Uhlig, H.H. (2008). *Corrosion and Corrosion Control*, 4. Hoboken, NJ: Wiley-Interscience.
3 Bradford, S. (2001). *Corrosion Control*, 2. Edmonton, Alberta, Canada: CASTI Publishing, Inc.
4 Trethewey, K.R. and Chamberlain, J. (1995). *Corrosion for Science and Engineering*, 2. Essex, England: Longman Scientific & Technical.
5 Roberge, P.R. (2006). *Corrosion Basics: An Introduction*, 2. Houston, TX: NACE International.
6 Roberge, P.R. (2007). *Corrosion Inspection and Monitoring*. Hoboken, NJ: Wiley-Interscience.
7 Cramer, S.D. (2003). *ASM Handbook, Volume 13A – Corrosion: Fundamentals, Testing, and Protection* (ed. B.S. Covino) (eds.). Materials Park, OH: ASM International.
8 Revie, R.W. (ed.) (2000). *Uhlig's Corrosion Handbook*, 2. New York: Wiley.
9 Byars, H.G. (1999). *Corrosion Control in Petroleum Production, TPC Publication 5*, 2. Houston, TX: NACE International.
10 Craig, B.D. (2004). *Oilfield Metallurgy and Corrosion*, 3. Denver: Metcorr.
11 Scully, J.C. (1990). *The Fundamentals of Corrosion*, 3. New York: Pergamon Press.
12 Jones, D.A. (1996). *Principles and Prevention of Corrosion*, 2. Prentice-Hall, Inc.
13 American Petrolium Institute (API) (1990). *Corrosion of Oil and Gas-Well Equipment, Book 2 of Vocational Training Series*, 2. Washington, DC: API.
14 Ahmad, Z. (2006). *Principles of Corrosion Engineering and Corrosion Control*. Oxford, England: Butterworth-Heinenmann.
15 API RP571. *Damage Mechanisms Affecting Fixed Equipment in the Refining Industry*. Washington, DC: American Petroleum Institute.
16 Davis, J.R. (2000). *Corrosion: Understanding the Basics*. Materials Park, OH: ASM International.
17 Chilingar, G., Mourhatch, R., and Al-Qahtani, G. (2008). *The Fundamentals of Corrosion and Scaling for Petroleum and Environmental Engineers*. Houston, TX: Gulf Publishing.
18 NORSOK M-001. *Materials Selection*. Lysaker, Norway: Standards Norway.
19 Groysman, A. (2010). *Corrosion for Everybody*. New York: Springer.
20 NACE TPC Publication No. 5. (1979). *Corrosion Control in Petroleum Production*. Houston, TX: NACE International.
21 ISO 21457-2010. *Materials Selection and Corrosion Control for Oil and Gas Production Systems*. Switzerland, Geneva: ISO.
22 DNV-RP-C101. *Allowable Thickness Diminution for Hull Structure of Offshore Ships*. Oslo, Norway: Det Norske Veritas.
23 McIntyre, D., Case, R., Mahajanam, S. et al. *Impact of Corrosion Inhibition on the Mitigation of Preferential Weld Attack in Sea Water Injection Pipelines*, NACE 2014-3847. Houston, TX: NACE International.

24   Kiefner, J.F. (1992). Installed pipe, especially pre-1970, plagued by problems. *Oil and Gas Journal* 90 (32): 45–51.

25   Kiefner, J.F. (1992). Pressure management key to problematic ERW pipe. *Oil and Gas Journal* 90 (33): 80–81.

26   Bond, C.M. and Woolin, P. (2005). *Preferential Weld Corrosion: Effects of Weldment Microstructure and Composition*, NACE 05277. Houston, TX: NACE International.

27   Kiefner, J.F. and Kolovich, K.M. (2012). *ERW and Flash Weld Seam Failures*, Battelle Final Report No. 12-139. Worthington, OH: Kiefner & Associates, Inc.

28   Manuel, R.W. (1947). Effect of carbide structure on the corrosion resistance of steel. *Corrosion* 3 (9): 197.

29   Hertzberg, R.W. (1995). *Deformation and Fracture Mechanics of Engineering Materials*, 4, 29–30. New York: Wiley.

30   Aguirre, F., Kyriakides, S., and Yun, H.D. (2004). Bending of steel tubes with Lüders bands. *International Journal of Plasticity* 20 (7): 1199–1225.

31   Carr, M., Cosham, A., Bruton, D., and MacRae, I. (2009). A structural reliability assessment of the strains developed during lateral buckling. *Proceedings of the Nineteenth International Offshore and Polar Engineering Conference*, Osaka, Japan (21–26 June 2009).

32   Schlipf, J. (1986). Phenomenological theory of Luders bands. *Journal of Materials Science* 21 (11): 4111–4116.

33   Beavers, J.A. (2001). Chapter 16: Fundamentals of corrosion. In: *Peabody's Control of Pipeline Corrosion*, 2 (ed. R.L. Bianchetti), 297–317. Houston, TX: NACE International.

34   Zhang, X.G. (2000). Chapter 8: Galvanic corrosion. In: *Uhlig's Corrosion Handbook*, 2 (ed. R.W. Revie), 137–164. New York: Wiley.

35   Porter, F.C. (1991). *Zinc Handbook: Properties, Processing, and Use in Design, International Lead Zinc Research Organization*, 112. New York: Marcel Dekker Inc.

36   Allen, B., Heidersbach, R., and Mealy, S. (1980). Cathodic protection of stainless steels against crevice corrosion in seawater, Paper No. 3856. *Proceedings, Offshore Technology Conference*, Houston, Texas (May 1980).

37   Iranian Petroleum Standard E-TP-706 (2005). *Corrosion Consideration in Design*. Iranian Petroleum Standard.

38   National Transportation Safety Board (2003). *Pipeline Accident Report, Natural Gas Pipeline Rupture and Fire Near Carlsbad, New Mexico, August 19, 2000*, NTSB/PAR-03/01 (11 February 2003). Washington, DC: National Transportation Safety Board.

39   NACE (2006). *NACE Standard SP0106-2006 Control of Internal Corrosion in Steel Pipelines and Piping Systems*. Houston, TX: NACE International.

40   Brown, B.F. (1970). Concept of the occluded corrosion cell. *Corrosion* 26 (8): 249–250.

41   Martin, J., Heidersbach, R., and MacDowell, L. (2009). Pitting corrosion. https://corrosion.ksc.nasa.gov/pittcor.htm (accessed 26 May 2017).

42   McCafferty, E. (2009). *Introduction to Corrosion Science*, 302. New York: Springer.

43   Smart, J. (1980). Corrosion failure of offshore steel platforms. *Materials Performance* 19: 41–48.

44   NACE TM0106 (2016). *Detection, Testing and Evaluation of Microbially Influenced Corrosion (MIC) on External Surfaces of Buried Pipelines*. Houston, TX: NACE International.

45   McDanels, S.J. (1998). Failure analysis of launch pad tubing from the Kennedy Space Center. In: *Microstructural Science*, vol. 25 (ed. S.D. Cramer and B.S. Covino), 125–129. Materials Park, OH: ASM International.

46   ASME B31.4. *Pipeline Transportation Systems for Liquid Hydrocarbons and Other Liquids*. New York, NY: ASME.

47   ASME B31.8. *Gas Transmission and Distribution Piping Systems*. New York, NY: ASME.

48   ISO/DIS 13623. *Pipeline Transportation Systems*. Switzerland: ISO.

49   DNV-RP-F101. *Corroded Pipelines*. Oslo, Norway: DNVGL.

50   ANSI/NACE MR0175/ISO 15156-3. *Petroleum and Natural Gas Industries – Materials for Use in H_2S-Containing Environments in Oil and Gas Production – Part 3: Cracking-Resistant CRAs (Corrosion-Resistant Alloys) and Other Alloys*. Houston, TX: NACE International.

51   Bauman, J. (2004). *Materials Selection and Corrosion Issues on an Offshore Producing Platform*, NACE 04126. Houston, TX: NACE International.

52   Scott, G.N. (1933). Exhibit "A", Report of A.P.I. Research Associate to Committee of Corrosion of Pipe Lines, Part I – Adjustment of soil-corrosion pit-depth measurements for size of sample, and Part II – A preliminary study of the rate of pitting of iron pipe in soils. *American Petroleum Industry Bulletin* 204–220.

53   NACE RP0502. *Pipeline External Corrosion Direct Assessment Methodology*. Houston, TX: NACE International.

54   Verink, E.D. (2000). Chapter 5: Designing to Prevent Corrosion. In: *Uhlig's Corrosion Handbook*, 2 (ed. R.W. Revie), 95–109. New York: Wiley.

55   Raymond, D. (July 2002). Cathodic protection in cold climate, valve case studies, water division. *Corrosioneering – The On-Line Corrosion Journal*. http://www.corrosionsource.com/corrosioneering/journal/Jul02_Raymond/Jul02_Raymond_1.htm (accessed 10 September 2009).

56   Britton, J. (1998). New paint preservation technologies for offshore & marine equipment. http://stoprust.com/technical-papers/30-marine-paint-preservation (accessed 27 May 2017).

57   Garcia, V., Boyero, C., and Jimenez-Garrido, J. (2016). Corrosion detection under pipe supports using EMAT medium range guided waves. *19th World Conference on Non-Destructive Testing*, Munich, Germany (2016). http://www.ndt.net/article/wcndt2016/papers/th2d2.pdf (accessed 27 May 2017).

58   LeBer, L., Benoist, G., and Dainelli, P. (2016). Corrosion detection and measurement using advanced ultrasonic tools. *19th World Conference on Non-Destructive Testing*, Munich, Germany (2016). http://www.ndt.net/article/wcndt2016/papers/th3d1.pdf (accessed 27 May 2017).

59   Sabnis, G. Crevice corrosion & pack rust: a serous structural problem. https://www.slideshare.net/harishsharma46/delhi-sabnis-corrosion-termarust-on-crevice-corrosion-1 (accessed 15 April 2018).

60 Kane, R. (2006). Corrosion in petroleum production operations. In: *Metals Handbook, Volume 13C – Corrosion: Corrosion in Specific Industries*, 922–966. Materials Park, OH: ASM International.

61 Rhodes, P.R., Skogsberg, L.A., and Tuttle, R.N. (2007). Pushing the limits of metals in corrosive oil and gas well environments. *Corrosion* 63 (1): 63–100.

62 Dam, A., Okeremi, A., and Speed, C. (2013). Pitting and crevice corrosion of offshore stainless steel tubing. *Offshore Magazine* 73 (5): 122–126.

63 DIN EN ISO 4628-10. *Paints and Varnishes – Evaluation of Degradation of Coatings – Designation of Quantity and Size of Defects, and of Intensity of Uniform Changes in Appearance – Part 10: Assessment of Degree of Filiform Corrosion (ISO 4628-10:2003)*, German version EN ISO 4628-10:2003. Brussels: European Committee for Standardization.

64 Davis, J.R. (2006). *Corrosion of Weldments*. Materials Park, OH: ASM International.

65 Heidersbach, R. (1982). Chapter 7: Dealloying corrosion. In: *NACE Handbook, Volume 1, Forms of Corrosion– Recognition and Prevention* (ed. C.P. Dillon), 99–104. Houston, TX: NACE International.

66 Lenard, D.R., Martin, J., and Heidersbach, R. (1999). Dealloying of cupro-nickels in stagnant seawater. *Corrosion/99*, Paper No. 314, Orlando (April 1999).

67 Lenard, D.R., Bayley, C.J., and Noren, B.A. (2008). Electrochemical monitoring of selective phase corrosion of nickel aluminum bronze in seawater. *Corrosion* 64: 764–772.

68 Michels, H.T. and Kain, R.M. (2003). *Effect of Composition and Microstructure on the Seawater Corrosion Resistance of Nickel Aluminum Bronze*, NACE 03262. Houston, TX: NACE International.

69 Crolet, J.L., Olsen, S., and Wilhelmsen, W. (1994). Influence of a layer of undissolved cementite on the rate of the CO_2 corrosion of carbon steel. *CORROSION/94*, Paper No. 4. Houston, TX: NACE International.

70 Crolet, J., Thevenot, N., and Nesic, S. (1998). Role of conductive corrosion product in the protectiveness of corrosion layers. *Corrosion* 54: 194–203.

71 Chexal, B., Horowitz, J., Jones, R. et al. (1996). *Flow-Accelerated Corrosion in Power Plants*, EPRI-EDF TR-106611. Pleasant Hill, CA: Electric Power Research Institute.

72 Lyons, W.C. and Plisga, G.J. (2005). *Standard Handbook of Petroleum & Natural Gas, Engineering*. Burlington, MA: Gulf Publishing.

73 ANSI/API RP 14E. *Design and Installation of Offshore Products Platform Piping Systems*. Washington, DC: API.

74 Martin, J.W., Sun, Y., Alvarez, J. et al. Design and Operations Guidelines to Avoid Erosion Problems in Oil and Gas Production Systems. Houston, TX: NACE International.

75 Barton, N. (2003). Erosion in hydrocarbon production systems: review document, *Health & Safety Executive (UK) Research Report 115*. http://www.hse.gov.uk/research/rrpdf/rr115.pdf (accessed 18 September 2009).

76 Salama, M.M. (1993). Saltwater injection systems can tolerate higher velocities. *Oil and Gas Journal* 91 (28): 102–103.

77 Ferng, Y.M., Ma, Y.P., and Chung, N.M. Application of local flow models in predicting distributions of erosion-corrosion locations. *Corrosion* 56 (2): 116–126.

78 McLaury, B.S., Shirazi, S.A., Mazumder, Q.H., and Viswanathan, V. (2006). *Effect of Upstream Pipe Orientation on Erosion in Bends for Annular Flow*, NACE 06572. Houston, TX: NACE International.

79 Smart, J.S. (1990). *A Review of Erosion-corrosion in Oil and Gas Production*, NACE90010. Houston, TX: NACE International.

80 Salama, M.M. and Venkatesh, E.S. (1983). Evaluation of API RP 14E erosional velocity limitations for offshore gas well. *Proceedings Offshore Technology Conference*, Houston, TX (1983), OTC 4485.

81 Bardal, E. (2004). *Corrosion and Protection*. London: Springer Verlag Limited.

82 Bogaerts, W. and Agema, K.S. (1991). *Active Library on Corrosion*. Houston, TX: NACE-Elsevier, NACE International.

83 NORSOK L-002. *Piping System Layout, Design and Structural Analysis*. Lysaker, Norway: Standards Norway.

84 NORSOK P-100. *Process Systems*. Lysaker, Norway: Standards Norway.

85 DNVGL-RP-0501. *Managing Sand Production and Erosion*. Oslo, Norway: DNVGL.

86 Houghton, C.J. and Westermark, R.V. (1983). North Sea downhole corrosion: identifying the problem; implementing the solutions. *Journal of Petroleum Technology* 35 (1): 239–246.

87 Houghton, C.J. and Westermark, R.V. (1983). Downhole corrosion mitigation in Ekofisk (North Sea) field. *Materials Performance* 22 (1): 16–23.

88 Neville, A., Wang, C., Ramachandran, S., and Jovancicevic, V. (2003). *Erosion-corrosion Mitigation Using Chemicals*, NACE 03319. Houston, TX: NACE International.

89 Feyerl, J., Mori, G., Holzleitner, S. et al. (2008). Erosion-corrosion of carbon steels in a laboratory: three-phase flow. *Corrosion* 62 (2): 175–186.

90 Shirazi, S.A., Mclaury, B.S., Shadley, J.R. et al. (2015). Erosion–corrosion in oil and gas pipelines. In: *Oil and Gas Pipelines: Integrity and Safety Handbook* (ed. R.W. Revie). Hoboken, NJ: Wiley. doi: 10.1002/9781119019213.ch28.

91 Kolts, J., Joosten, M., and Singh, P. (2006). *An Engineering Approach to Corrosion/Erosion Prediction*, NACE 06560. Houston, TX: NACE International.

92 LaQue, F.L. (1975). *Marine Corrosion: Causes and Prevention*. New York: Wiley.

93 DNV Recommended Practice. *Corrosion Protection of Ships*. Oslo, Norway: DNVGL.

94 Chung, Y. (2016). Distinguishing hydrogen embrittlement from stress corrosion crakcing. *Materials Performance* 55 (3): 64–67.

95 ANSI/NACE MR0175/ISO 15156-1. *Petroleum and Natural Gas Industries – Materials for Use in H_2S-containing Environments in Oil and Gas Production – Part 1: General Principles for Selection of Cracking-resistant Materials*. Houston, TX: NACE International.

96   ANSI/NACE MR0175/ISO 15156-2. *Petroleum and Natural Gas Industries – Materials for Use in H₂S-containing Environments in Oil and Gas Production – Part 2: Cracking-Resistant Carbon and Low-alloy Steels, and the Use of Cast Irons*. Houston, TX: NACE International.

97   Brown, B.F. (June 1977). *Stress Corrosion Cracking Control Measures*, NBS Monograph 156. Washington, DC: National Bureau of Standards.

98   Iannuzzi, M. (2011). Environmentally assisted cracking (EAC) in oil and gas production. In: *Stress Corrosion Cracking: Theory and Practice* (ed. V.S. Raja and T. Shoji), 570–607. Woodhead Publishing.

99   Turnbull, A. (June 2014). Corrosion pitting and environmentally assisted small crack growth. *Proceedings of the Royal Society A* 470 (2169). http://rspa.royalsocietypublishing.org/content/royprsa/470/2169/20140254.full.pdf (accessed 15 April 2018).

100 Jones, R.H. (2017). Chapter 1: Mechanisms of stress-corrosion cracking. In: *Stress-Corrosion Cracking: Materials Performance and Evaluation* (ed. R.H. Jones), 1–42. Materials Park, OH: ASM International.

101 Schweitzer, P.A. (1989). *Corrosion and Corrosion Protection Handbook*, 2, 295. New York: Marcel Dekker.

102 Cheng, Y.F. (2013). *Stress Corrosion Cracking of Pipeline Steels*. Hoboken, NJ: Wiley.

103 Parkins, R.N. (2000). A review of stress corrosion cracking of high pressure gas pipelines. *NACE Corrosion 2000*, Paper 363. Houston, TX: NACE International.

104 Cottis, R. (2000). *Stress Corrosion Cracking – Guides to Good Practice in Corrosion*, 1–16. Middlesex, UK: National Physical Laboratory.

105 Merle, A. and Ehlers, P. (2006). A statistical predictive model to prioritize site selection for stress corrosion cracking direct assessment. *NACE Northern Area Western Conference*, Calgary, Alberta, Canada (6–9 February 2006), Figure 1.

106 Sutherby, R., Hamre, T., and Purcell, J. (2006). Circumferential stress corrosion cracking rupture case study. *NACE Northern Area Western Conference*, Calgary, Alberta, Canada (6–9 February 2006).

107 Parrott, R. and Pitts, H. (2011). Chloride stress corrosion cracking in austenitic stainless steel. *HSE Research Report 902*.

108 Ahluwalia, H. (2006). CUI: an in-depth analysis. In: *ASM Metals Handbook, Volume 13C*, 654–658. Materials Park, OH: ASM International.

109 GE Handbook of Industrial Water Treatment. http://www.gewater.com/handbook/index.jsp (accessed 29 May 2017).

110 API 20E. *Alloy and Carbon Steel Bolting for Use in the Petroleum and Natural Gas Industries*. Washington, DC: API.

111 ASTM B850. *Post-Coating Treatments of Steel for Reducing the Risk of Hydrogen Embrittlement*. West Conshohocken, PA: ASTM International.

112 Maroef, I., Olson, D.L., Eberhart, M., and Edwards, G.R. (2002). Hydrogen trapping in ferritic steel weld metal. *International Materials Reviews* 47 (4): 191–223.

113 Townsend, H. (1975). Effects of zinc coatings on the stress corrosion cracking and hydrogen embrittlement of low-alloy steel. *Metallurgical Transactions A* 6A: 877–883.

114 Brahimi, S. (2014). *Fundamentals of Hydrogen Embrittlement in Steel Fasteners*. Montreal, Canada: IBECA Technologies Corp.

115 West, A.J. and Louthan, M.R. (1979). Dislocation transport and hydrogen embrittlement. *Metallurgical and Materials Transactions A* 10 (11): 1675–1682.

116 Kane, R.D. (2003). Corrosion in petroleum refining and petrochemical operations. In: *Metals Handbook, Volume 13C – Environments and Industries*, 967–1014. Materials Park, OH: ASM International.

117 NACE TM0103. *Laboratory Test Procedures for Evaluation of SOHIC Resistance of Plate Steels Used in Wet H₂S Service*. Houston, TX: NACE International.

118 Palmer, A.C. and King, R. (2006). *Subsea Pipeline Engineering*, 206. Tulsa, OK: PennWell.

119 Pargeter, R. (2007). *Susceptibility to SOHIC for Linepipe and Pressure Vessel Steels – Review of Current Knowledge*, NACE 07115. Houston, TX: NACE International.

120 Louthan, M.R. (2008). Hydrogen embrittlement of metals: a primer for the failure analyst. *Journal of Failure Analysis and Prevention* 8 (3): 289–307.

121 NACE TM0177. *Laboratory Testing of Metals for Resistance to Sulfide Stress Cracking and Stress Corrosion Cracking in H₂S Environments*. Houston, TX: NACE International.

122 Iannuzzi, M. (2015). High strength low alloy steels and hydrogen. http://www.aboutcorrosion.com/2015/11/08/high-strength-low-alloy-steels-and-hydrogen-presentation (accessed 27–28 October 2015).

123 Campbell, J.M., Lilly, L.L., and Maddox, R.N. (2004). *Gas Conditioning and Processing, Volume 2: The Equipment Modules*, 8 (ed. R. Hubbard), 93. Norman, Oklahoma: John M. Campbell Co.

124 DNV-RP-C203. *Fatigue Design of Offshore Steel Structures*. Oslo, Norway: DNVGL.

125 ABS (February 2014). *Fatigue Assessment of Offshore Structures*. Houston, TX: American Bureau of Shipping.

126 Nyborg, R. (2003). *Understanding and Prediction of Mesa Corrosion Attack*, NACE 03642. Houston, TX: NACE International.

127 Edwards, M.A. and Cramer, B. (2000). *Top of Line Corrosion – Diagnosis, Root Cause Analysis, and Treatment*, NACE 00072. Houston, TX: NACE International.

128 Singh, P., Frenier, H.N., Goldmann, E.R. et al. (2013). *Corrosion Mechanism in a Three-phase North Slope Pipeline*, NACE 2824-2013. Houston, TX: NACE International.

129 Nyborg, R. and Dugstad, A. (2009). *Top of Line Corrosion and Water Condensation Rates in Wet Gas Pipelines*, NACE 07555. Houston, TX: NACE International.

130 Gunaltun, Y.M., Supriyatman, D., and Achmad, J. (1999). *Top of Line Corrosion in Multiphase Gas Lines: A Case History*, NACE 99036. Houston, TX: NACE International.

131 Vitse, F., Nesic, S., Gunaltun, Y. et al. (2003). *Mechanistic Model for the Prediction of Top-of-the-line Corrosion Risk*, NACE 03633. Houston, TX: NACE International.

132 Gunaltun, Y.M. and Belghazi, A. (2001). *Control of Top of Line Corrosion by Chemical Treatment*, NACE 01033. Houston, TX: NACE International.

133 Mendez, C., Singer, M., Camacho, A. et al. (2005). *Effect of Acetic Acid pH and MEG on the CO$_2$ Top of the Line Corrosion*, NACE 05278. Houston, TX: NACE International.

134 Cai, J., Pugh, D., Ibrahim, F. et al. (2009). *Top-of-line Corrosion Mechanisms for Sour Wet Gas Pipelines*. NACE C2009-09285. Houston, TX: NACE International.

135 Frenier, W. and Wint, D. (2014). *Multi-faceted Approaches for Controlling Top of line Corrosion (TLC) in PIpelines*, SPE-169630-MS. Scotland: Society of Petroleum.

136 Heidersbach, K. and van Roodselaar, A. (2012). *Understanding, Preventing, and Identification of Microbial Induced Erosion-corrosion (Channeling) in Water Injection Pipelines*, NACE C2012-0001221. Houston, TX: NACE International.

137 Alyeska Pipeline Service Company (2011). *Low Flow Impact Study*. Final Report (15 June 2011).

138 Buecker, B. (2000). *Fundamentals of Steam Generation Chemistry*, 142. Tulsa, OK: PennWell.

139 US Department of Transportation PHMSA Fact Sheet (2011). Selective seam corrosion. https://primis.phmsa.dot.gov/comm/FactSheets/FSSelectiveSeamCorrosion.htm (accessed 29 May 2017).

140 Baker, M. (2004). Low frequency ERW and lap welded longitudinal seam evaluation. TTO Number 5. Final Report, Revision 3 (April 2004).

141 Baker, M. and Fessler, R. (2008). *Pipeline Corrosion*. Final Report (November 2008).

142 Beavers, J.A., Brossia, C., and Denzine, R. (2014). Development of selective seam weld corrosion test method. *10th International Pipeline Conference, Volume 2: Pipeline Intergrity Management*, Calgary, Alberta, Canada (29 September to 3 October 2014), Paper No. IPC2014-33562, pp. V002T06A057: 9 pages.

143 Mead, R. (2006). Non-destructive evaluation of low-frequency electric resistance welded (ERW) pipe utilizing ultrasonic in-line inspection technology. *ECNDT 2006*, Berlin.

144 NTSB/PAR-11/01, PB2011-916501 (2011). *Pacific Gas and Electric Company, Natural Gas Transmission Pipeline Rupture and Fire, San Bruno, California, Septembeer 9, 2010*. Report adopted 30 August 2011. Washington, DC: National Transportation Safety Board.

145 NACE SP0284 (2006). *Design, Fabrication, and Inspection of Storage Tank Systems for Concentrated Fresh and Process Sulfuric Acid and Oleum at Ambient Temperatures*. Houston, TX: NACE International.

146 Landrum, R.J. (1989). *Designing for Corrosion Control*. Houston, TX: NACE International.

147 Hanson, H. (2003). Stray current corrosion. In: *Metals Handbook, Volume 13A – Corrosion: Fundamentals, Testing, and Protection*, 214–215. Materials Park, OH: ASM International.

148 ANSI/NACE RP0502. *Pipeline External Corrosion Direct Assessment Methodology*. New York: American National Standards Institute Houston, TX: NACE International.

149 Wang, G., Spence, J.S., Olson, D.L. et al. (2005). Chapter 25: Tanker corrosion. In: *Handbook of Environmental Degradation of Materials* (ed. M. Kutz), 523–546. Norwich, NY: William Andrew, Inc.

150 Rauta, D., Gunner, T., Eliasson, J. et al. (2004). Double hull tankers and corrosion protection. *Transactions, Society of Naval Architects and Marine Engineers (SNAME)* 112: 170–190.

151 MacMillan, A., Fischer, K.P., Carlsen, H., and Goksoyr, O. (2004). Newbuild FPSO corrosion protection – a design and operation planning guide. *Proceedings, Offshore Technology Conference*, Houston, TX (3–6 May 2004), Paper 16048. https://doi.org/10.4043/16048-MS.

152 Srinivasan, S. and Singh, S. (2008). *Design of a Non-Ship-Shaped FPSO for Sakhalin-V Deepwater (Russia)*, SPE-114882-MS. Society of Petroleum Engineers.

153 RIGZONE. (2009). StatoilHydro taps sevan for cylindrical FPSO feasibility study. http://www.rigzone.com/news/article.asp?a_id=81208 (accessed 27 August 2016).

154 Cammaert, G. (2008). Floating production advances: managing FPSOs in ice. http://www.epmag.com/Magazine/2008/9/item8295.php (accessed 26 May 2017).

155 European Martime Safety Agency (2005). *Double Hull Tankers: High Level Panel of Experts Report* (3 June 2005).

6

CORROSION CONTROL

Most corrosion control in oil and gas production, as well as in other industries, is by the use of protective coatings, water treatment and corrosion inhibitors, and cathodic protection. These services are often provided by contractors that specialize in one or more of these corrosion control methods. It is important that employees of operating companies understand the principles involved in these corrosion control techniques, and many advances in these methods are developed by oil and gas companies in cooperation with specialty contractors.

Coatings and inhibitors have much proprietary technology that changes with time and, equally important, are dependent on the quality of local suppliers. Cathodic protection does not vary as much worldwide, and this is why it is dealt with in greater detail.

This chapter is intended to cover the principles of the three corrosion control methods in the discussion. Additional information relating to specific types of oilfield equipment is discussed in Chapter 8.

PROTECTIVE COATINGS

Protective coatings are the most commonly used means of corrosion control. They are the standard means of controlling external corrosion on everything from offshore structures to pipelines and process vessels. They may also be used on storage tank, pipeline, and storage vessel interiors. The reasons for their widespread use include the ease and low cost of application. Most protective coatings are applied by liquid paint systems, but metallic coatings and wraps are also used. Ceramic coatings are used in some industries, but their brittle nature limits their use in oilfield applications.

Liquid coatings can be lacquers, varnishes, or paint. The first two are usually single-phase liquids, but paints, which are more complex, are generally used because of their greater protective qualities. Most paints are based on organic chemistry, but inorganic coatings, e.g. inorganic zincs and thermally sprayed aluminum, are also available and widely used.

Linings are protective coatings that are applied in thicker films, usually 5 mm (0.2 in.) or more. They are usually applied as flexible solid films and find use on the interiors of storage and ballast tanks, process vessels, and large-diameter piping. Their use is relatively limited in oilfield applications.

The associated costs of applying a protective coating system to an existing structure are typically:

- Surface preparation   50%+
- Permits and scaffolding   30–35%
- Materials   ~10%
- Inspection and other costs   ~10%

More expensive coating materials may have longer service lives, and this means that the total costs of protective coatings over the service life of a structure may be lower than if less expensive coatings were to be applied.

Paint Components

The components of paint coatings are pigments, binders, volatile vehicles, and additives.

Pigments are usually inorganic minerals or metal particles. They provide opacity and color, but they also provide corrosion protection. Their low permeability to water and oxygen migration provides the corrosion protection. Maximum

Metallurgy and Corrosion Control in Oil and Gas Production, Second Edition. Robert Heidersbach.
© 2018 John Wiley & Sons, Inc. Published 2018 by John Wiley & Sons, Inc.

protection is provided by paints with high volumes of pigment in the cured paint film. Primer coatings are sometimes named after their pigments, e.g. zinc-rich primers.

Metallic zinc pigments can provide cathodic protection to steel substrates at coating holidays. Aluminum is less likely to provide this protection. Aluminum-pigmented paints have advantages over zinc at higher temperatures, e.g. in flares and on the exteriors of hot piping and process vessels.

Binders are necessary to hold pigment particles together and to provide adhesion to the underlying substrate, either protected metal, in the case of primers, or underlying paint films. Paint coating types are often classified by the binder, e.g. polyurethanes, epoxies, vinyls, and so forth.

The binder/pigment ratio is an important parameter in determining the effectiveness of a paint film. Too much binder produces high gloss but may produce chalking after environmental exposure. Too little binder means that the pigment will not be adequately wetted, leading to paint film porosity and loss of corrosion resistance. The best corrosion protection is obtained with paints that provide high pigment volumes but still ensure adequate wetting of the pigments.

Volatile vehicles, either water or organic solvents or dispersants, dissolve or disperse the binder and allow the coating to spread. Modern coatings have fewer volatile components due to environmental and health concerns with volatile organic compounds (VOCs).

Some paints cure by evaporation of the vehicle, while others, e.g. epoxies, cure by chemical reactions that form thermosetting polymer binders. The reaction-based coatings tend to have fewer volatile components.

Other constituents added to paints include plasticizers, which lower the brittleness of the cured film, and anti-skimming and anti-settling agents necessary to keep the paint useable after transport before final use.

Coating Systems

It is common for coating systems to have several layers that are usually characterized as primers, intermediate or midcoats, and topcoats. It is important that all layers of a coating system are compatible so that interlayer adhesion and unwanted chemical reactions between the layers are avoided. This is normally accomplished by using materials from the same manufacturer for all layers of the coating system [1].

Primers, the first coating to be applied, provide adhesion of the paint film to the substrate. They also provide most of the corrosion protection and, if necessary, are designed so that they can "key" or bond to the outer coats. Some primers will also contain corrosion inhibitors or metallic pigments, usually zinc flakes, which provide some cathodic protection to the underlying steel substrate at coating holidays [1].

Intermediate or midcoats provide a barrier to water passage. They may also smooth out the surface prior to the application of the topcoat. They also serve as bonding interfaces allowing adhesion to both the primer and the topcoat.

Topcoats provide the desired color to the coating system. Unlike the lower coatings, which need to bind to subsequent coatings and are usually rough on a microscopic scale, most topcoats also provide a smooth surface, which promotes water runoff.

Environmental regulations associated with VOCs have caused many coating manufacturers to alter the chemistries of their products [1].

Corrosion Protection by Paint Films

All paint films are permeable to moisture and oxygen to some extent, but their effect on lowering corrosion rates is primarily due to the low permeability of the coatings compared with that of the uncoated environment [2–7]. The barrier concept is shown in Figure 6.1, which also shows the important properties of the primer, intermediate coat, and topcoat. Inorganic pigments in the primer provide most of the moisture and oxygen ingress barrier effect. Intermolecular spacings in polymers are much larger than in inorganic pigments, and most of the moisture permeates through the organic binders.

No coating system is perfect, and coating holidays, places where the coating is missing or has been removed, are locations where most corrosion occurs. One way of slowing corrosion at coating holidays is to have corrosion inhibitors in the pigment. This idea is shown in Figure 6.2. Several pigments that have been used for this purpose are listed below:

- Zinc chromate.
- Zinc phosphate – the only pigment on this list not banned for environmental reasons.
- Red lead.
- Calcium plumbate – contains lead.
- Coal tar.

Concerns with environmental damage have limited the use of corrosion inhibitors in pigments, and the use of chromates, the most effective of these pigments, has been largely replaced due to concerns with heavy metal pollution. Slow-release corrosion inhibitors are intended to release oxidizing agents, which passivate the surface at holidays, but the nonavailability of chromates for this purpose has greatly reduced their effectiveness.

Metallic pigments, either zinc or aluminum flakes, are added as pigments to many primers. They are virtually impermeable to moisture and oxygen migration. Zinc pigments also provide a measure of cathodic protection once the coating is breached. This cathodic protection is greatly reduced if the primer is overcoated, but this is

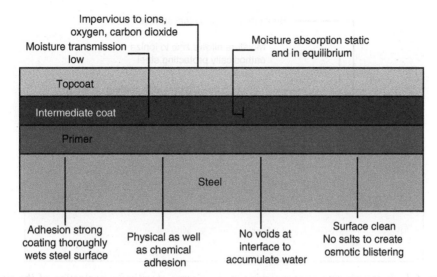

Figure 6.1 Protective coating system serving as a moisture and oxygen permeation barrier. *Source*: Photo courtesy of NACE International Protective Coatings and Liners course. Reproduced with permission.

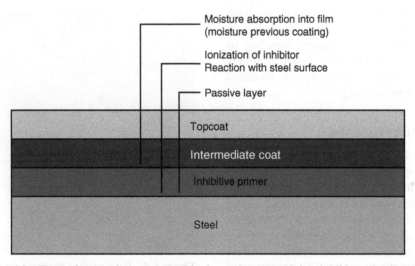

Figure 6.2 Protective coating system with slow-release corrosion inhibitors in the primer coat. *Source*: Photo courtesy of NACE International Protective Coatings and Liners course. Reproduced with permission.

often necessary for color coding or wear resistance reasons. Inorganic zinc primers are not subject to ultraviolet (UV) damage, so there is no need to overcoat them except for the reasons above. Organic zinc primers are also widely used, and they do benefit from overcoats [7]. Figure 6.3 illustrates the idea of using zinc-rich primers for cathodic protection at coating holidays.

Inorganic zinc (IOZ) and metalized coating primers have greater tolerance than many organic coatings for salt residues on the surfaces being coated. IOZ primers have porous surfaces that require a misty tack coating (usually of organic zinc) prior to applying full topcoats (usually organic zinc coatings) at holidays and other locations where IOZ coating primers must be repaired.

Desirable Properties of Protective Coating Systems

In addition to providing corrosion protection, coating systems should also have:

- Strong, durable bonding to the substrate.
- Flexibility, because organic coatings have different coefficients of thermal expansion than metals, and coating flexure is inevitable.
- Toughness or the ability to withstand mechanical shock and loading.

The choice of coating systems often requires compromises between these characteristics. For example,

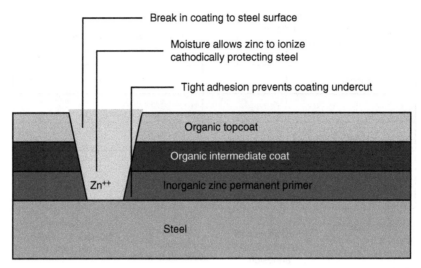

Break in coating to steel surface

Moisture allows zinc to ionize
cathodically protecting steel

Tight adhesion prevents coating undercut

Organic topcoat

Organic intermediate coat

Zn⁺⁺   Inorganic zinc permanent primer

Steel

Figure 6.3 Inorganic zinc primer serving as the source of cathodic protection of the steel substrate at a coating holiday. *Source*: Photo courtesy of NACE International Protective Coatings and Liners course. Reproduced with permission.

very hard coatings such as those often used on pipeline exteriors are often difficult to repair if damaged in shipping and construction. The necessary bonding to undamaged coatings is hard to achieve. This problem is also apparent at field welds, where organic coatings must be removed prior to welding.

Developments in Coatings Technology

Modern coating systems are generally longer lasting than those that were available in the past. Worker safety and environmental concerns have led to the development of new coatings having higher solids and lower VOC contents. The higher solids content in modern coatings frequently leads to the need for better surface preparation, as many of these coating systems are less tolerant of surface contamination than the systems they are replacing. Electrostatic spraying is an application that was once confined to manufacturing of relatively small items, but recent developments allow for the use of electrostatic spraying in major new construction and rehabilitation. This technique is especially useful on complicated geometries where it is difficult to apply even coatings with other techniques.

Surface Preparation

Proper surface preparation, the most expensive part of any coatings project, is necessary for coatings to bond properly to metallic and other substrates [8, 9]. Most premature coatings failures can be attributed to improper surface preparation or to allowing of the properly prepared surface to degrade before coatings are applied. Properly coated offshore platform deck

Figure 6.4 Poor surface preparation led to a lack of coating adhesion and corrosion on this offshore platform deck in only three years.

coating systems sometimes last 30 years before recoating becomes necessary, but Figure 6.4 shows corrosion that resulted from poor surface preparation on an offshore platform deck after only 3 years. This failure could have been easily avoided.

The principal surface condition factors that are known to influence this performance are the presence of rust and mill scale; surface contaminants including salts, dust, oils and greases; and surface profile, which must have enough roughness to allow mechanical adhesion between the primer and the bare metal surface but low enough so that paint covers the high spots with adequate cover.

The choice of surface preparation techniques and the necessary levels of cleanliness for new construction depend on the generic type of primer (chemical nature of the binder), the severity of the environment, and the desired coating service life. Coatings should be applied in accordance with manufacturer's recommendations, and these recommendations should include surface preparation guidelines listing minimally acceptable surface preparation conditions. General guidelines for selected generic coatings are shown in Table 6.1. NACE/SSPC surface preparation standards are listed in Table 6.2. Note that while white metal blast cleaning, NACE No. 1/SSPC-SP 5, is the cleanest possible substrate for any of the coating systems, it is only required for inorganic zinc primers. Substantial cost savings can be achieved if the surface preparation requirements for these other coatings are relaxed, but these savings may be achieved with reductions in long-term performance of the coating.

Abrasive Blasting Dry abrasive blasting is the most commonly specified method of preparing steel surfaces for protective coatings. Tables 6.3 and 6.4 provide comparisons on the relative costs of dry abrasive blasting to the cleanest condition, white metal, and other surface

preparation methods. While the best surface preparation produces the longest-lasting coatings, the reduction in costs associated with less-thorough surface preparation techniques can be significant.

Dry abrasive blasting is usually the preferred means of surface preparation, but wet abrasive blasting and high-pressure water jetting are becoming more common due to environmental and waste disposal concerns. The term "abrasive blast cleaning" in the past was usually assumed to imply dry abrasive blasting using sand or similar abrasive particles applied by high-pressure air sources. In recent years the increased use of wet abrasive blasting has necessitated the introduction of the terms "dry abrasive blasting" and "wet abrasive blasting." NACE and other organizations now address both techniques in recent standards [10–12].

TABLE 6.1 Minimally Acceptable Surface Preparation Levels for Selected Generic Coatings

Generic Coating Type	Recommended Minimum Surface Preparation
Alkyds and oil-based coatings	NACE No. 3/SSPC-SP 6 Commercial Blast Cleaning
Waterborne acrylics	NACE No. 3/SSPC-SP 6 Commercial Blast Cleaning
Epoxy	NACE No. 3/SSPC-SP 6 Commercial Blast Cleaning
Zinc-rich epoxy	NACE No. 3/SSPC-SP 6 Commercial Blast Cleaning
Inorganic zinc	NACE No. 1/SSPC-SP 5 White Metal Blast Cleaning

TABLE 6.2 NACE/SSPC Joint Surface Preparation Standards

NACE No. 1/SSPC-SP 5	White Metal Blast Cleaning
NACE No. 2/SSPC-SP 10	Near-White Metal Blast Cleaning
NACE No. 3/SSPC-SP 6	Commercial Blast Cleaning
NACE No. 4/SSPC-SP 7	Brush-Off Blast Cleaning
NACE No. 5/SSPC-SP 12	Surface Preparation and Cleaning of Metals by Waterjetting Prior to Recoating
NACE No. 6/SSPC-SP 13	Surface Preparation of Concrete
NACE No. 8/SSPC-SP 14	Industrial Blast Cleaning
NACE No. 10/SSPC-PA 6	Fiberglass-Reinforced Plastic (FRP) Linings Applied to Bottoms of Carbon Steel Storage Tanks
NACE VIS 7/SSPC-VIS 4	Guide and Visual Reference Photographs for Steel Cleaned by Waterjetting
NACE VIS 9/SSPC-VIS 5	Guide and Reference Photographs for Steel Surfaces Prepared by Wet Abrasive Blast Cleaning

TABLE 6.3 Comparative Costs, Relative Performance, and Other Considerations for Various Surface Preparation Techniques

Surface Preparation	Cost (%)	Performance (%)	Dust (%)	Debris (%)
White metal	100	100	100	100
Near white	80	90–95	100	80
LP WC	10	70–80	0	3–5
UHP WJ	25	90	0	5–10
Wet abrasive to white metal	120–150	95	0	110–125

LP WC, Low Pressure Water Cleaning, is cleaning performed at pressures less than 5000 psi (34 MPa). Minimum pressure is 3500 psi; use of a rotating tip is mandatory. Hand scraping of blisters and other defects may be required. Chemical decontamination is mandatory.
UHP WJ, Ultrahigh Pressure Water Jetting, is cleaning performed above 25000 psi (170 MPa). Use of a rotating tip is mandatory. Chemical decontamination is mandatory.
Wet abrasive to white metal: Cleaning to SSPC SP 5, White Metal Blast Cleaning, use of water ring for dust control is mandatory. Chemical decontamination is mandatory.

TABLE 6.4 Comparison of Abrasive Blasting Costs for General Construction

Blast Method	Relative Cost (%)
NACE No. 1 White Metal Blast	100
NACE No. 2 Near White Metal Blast	70
Commercial Blast, SSPC No. 6/NACE No. 3	40
Brush Blast – loose rough previous coat	20

The primary functions of blast cleaning are:

- Remove material from the surface that can cause early coating system failure.
- Provide a suitable surface profile (roughness) to enhance adhesion of the coating system.

The hierarchy of blasting standards is as follows:

- White metal blast cleaning
- Near-white metal blast cleaning
- Commercial blast cleaning
- Industrial blast cleaning
- Brush-off blast cleaning

White metal blast cleaning is considered the best surface preparation, but some authorities suggest that similar coatings system performance can be achieved with near-white or commercial blasting at substantial cost savings [13].

Abrasive blasting leaves a rough anchor pattern on the metal surface, while water jetting usually does not. This has limited water jetting to cleaning metal surfaces for recoating, but modern developments with high-pressure water jetting are overcoming this lack of anchor pattern, and it is likely that the use of water jetting will supplant abrasive blasting for many projects. Dry abrasive blasting to provide an anchor pattern is sometimes used on bare metal surfaces after they have been cleaned by water jetting. This blasting also removes any "flash rusting" that may have formed on wet metal surfaces.

The term "rust back" is sometimes applied to rusting that forms when dry blast-cleaned steel is exposed to moisture, contamination, or corrosive atmospheres. Both flash rusting and "rust back" must be evaluated and corrected if the amount of newly formed rust is excessive. Corrosion inhibitors can be added to the water used in wet abrasive blasting or by spraying corrosion inhibitor onto the surface immediately after cleaning [14, 15].

Neither blasting nor water jetting can remove grease and other organic contaminants from the surface, and solvents or other cleaning agents must be used prior to either blasting or water jetting.

Figure 6.5 The surface of a pipeline ready for recoating in the field.

All of the NACE/SSPC blast-cleaning standards contain the following sections:

- Procedure before blast cleaning
- Blast leaning methods and operations
- Procedure following blast cleaning
- Inspection

Inspections of blast-cleaned surfaces include measurements of the surface profile. If a surface is too smooth, the primer will not develop adequate bonding to the substrate. If it is too rough, the coating may be too thin at high spots and pinpoint corrosion may occur. Figure 6.5 shows the rough surface and anchor pattern of a blast-cleaned pipeline prepared for recoating.

Visual standards are used by coatings inspectors to determine the condition of metal surfaces before and after surface preparation. The condition of the metal surface prior to cleaning/preparation alters the appearance of the cleaned surface. For this reason SSPC has developed visual standards with photographs showing metal surfaces, with varying degrees of corrosion and pitting, before and after cleaning. Figure 6.6 shows examples from SSPC-VIS 1, which applies to dry abrasive blast cleaning, the most commonly used cleaning method on oilfield structures [16, 17]. Similar visual standards are available for other means of surface preparation.

Water Jetting   Water jetting is currently used primarily for preparing surfaces for recoating. This limitation is because most water jetting systems cannot produce an adequate surface profile or anchor pattern. This limitation has been largely overcome, but water jetting is still used much less than dry abrasive blasting for coatings surface

(a)   (b)   (c)

Figure 6.6 Standard images from left-to-right of deeply pitted steel surface. (a) Before dry abrasive blasting, (b) cleaned to near-white, and (c) cleaned to the white metal condition [16,17].

Figure 6.7 Water jetting surface preparation prior to recoating. *Source*: Photo courtesy of Hammelmann Corp., Dayton, Ohio, reproduced with permission.

TABLE 6.5 Comparison of Water Jetting and Abrasive Blast Cleaning Standards and Surface Conditions

Surface Finish Grade	ISO 8501-1	SSPC	NACE
White metal	Sa 3	SP 5	No. 1
Near-white metal	Sa 2½	SP 10	No. 2
Brush-off	Sa 1	SP 7	No. 4
Solvent cleaning		SP 1	
Power tool cleaning	St 2 or 3	SP 3[34]	
Power tool cleaning to bare metal		SP 11	
HPWJ and UHPWJ		SP 12	No. 5
WJ-1 clean to bare substrate			
WJ-2 very thorough or substantial cleaning			
WJ-3 thorough cleaning			
WJ-4 light cleaning			
Wet abrasive blasting		TR 2	6G198

preparation. The NACE/SSPC surface preparation standards listed in Table 6.4 describe water jetting only for recoating purposes. This is shown in Figure 6.7, where water jetting is being used for surface preparation on the outside of a large marine vessel prior to recoating.

The advantages of water jetting, which uses high-pressure water, are that it removes most contaminants, has no sparking or dust hazards, and removes soluble salts. Corrosion inhibitors are sometimes added to the water to prevent flash rusting, although opinions differ on how deleterious these thin rust areas are to primer-to-metal adhesion.

Table 6.5 provides approximate comparisons between water jetting and abrasive blast cleaning standards.

Surface Cleaning   Neither abrasive blasting nor water jetting can remove grease and other organic surface contaminants. These problems must be removed with various commercial products suited to this purpose [18].

Soluble salts may also be present on the metal surface. They are not removed by dry abrasive blasting and can lead to osmotic blistering of newly applied coatings. Soluble salts are normally removed by water washing or water jetting, often with proprietary chemical additions sold for this purpose added to the water. It is common to check for the presence of soluble salts by various methods, and most coatings contracts will specify a maximum level of acceptable chloride or salt contamination on the surface. The amount of salts detected will vary depending on the detection method, and the method should be specified in the coatings contract.

ISO, NACE, and other standards discuss the effects of salt deposits on coatings performance and offer guidelines on how they can be removed [19]. Once a salt removal process has been applied, a variety of tests are available to determine the presence, or absence, of residual salts [20, 21]. Figure 6.8 shows two different plastic devices widely used to wet metal surfaces and extract soluble salts. A controlled volume of distilled water is placed in the containers, which have a defined

(a)   (b)

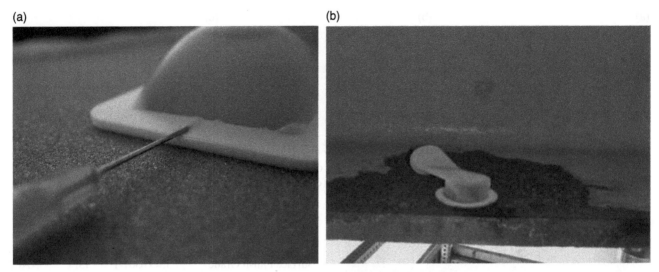

Figure 6.8 Two different devices for collecting soluble salts from steel surfaces. (a) Adhesive "Bresle" patch. (b) Flexible sleeve.

surface area, and, after the soluble salts dissolve in the water, a water sample is extracted and the salt content determined. Unfortunately the various test methods yield conflicting results, and this means that coatings specifications/project documents must clearly state how salt levels should be measured [22, 23]. This also means that, while it is widely understood that salt contamination of steel surfaces can degrade coating performance, no consensus is likely on how much salt contamination can be tolerated for various coating system/metal substrate combinations.

The water collected from salt-contaminated metal surfaces can be analyzed for specific ions or for total conductivity [20, 24]. While there are advocates and standards for both methods, the consensus seems to be that conductivity is the most reliable approach, even though it cannot identify which ions are on the metal surface.

While chlorides are not the only soluble salt contamination found on metal surfaces, washing that removes chloride salts will normally remove any other salty contaminant that could cause coating degradation and lack of adhesion.

Purposes of Various Coatings

Protective coating systems usually consist of multiple layers of paint having different purposes, but some systems are intended to be single-layer coatings or multiple layers of the same type of coating. These choices are shown in Figure 6.9.

The various layers of a coating system must be compatible with each other, and this is normally done by using the same coating manufacturer for all the layers, typically primer, intermediate, and topcoat.

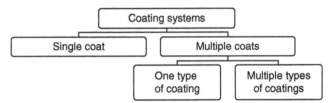

Figure 6.9 Single- and multicoat protective coating systems. *Source*: Courtesy of NACE International, reproduced with permission.

A single-coat system is usually a relatively thin system and usually provides minimal protection appropriate for temporary protection during shipping and manufacturing. These single-film coatings are often referred to as "shop coats" and should be replaced before service in any aggressive environment. Some single-film systems are much thicker, such as the trowel-applied coating shown in Figure 6.10, which is used for protection of pipeline joints during construction. Other single-film coatings are sometimes applied as maintenance coatings in the splash zone where surface preparation is difficult, thick films are necessary, and interfilm adhesion is difficult to achieve. Problems with single-coat systems include solvent entrapment in thick systems, difficulty in maintaining desired coating thickness, and potential holidays and misses.

Multiple-coat systems can be layers of the same product or a variety of products that combine the desired properties of each layer, e.g. primer, barrier intermediate coat, and top coat.

Coatings are classified in two different ways. Some classifications specify coatings by binder chemistry, e.g. epoxy, polyurethane, etc. The chemistry of these binders is understood by coatings manufacturers and many

Figure 6.10 Trowel application of thick-film single coating of coal-tar urethane for field application at pipeline joints. *Source*: Courtesy of NACE International, reproduced with permission.

field-oriented coatings professionals. Most field-oriented coatings professionals do not understand the chemistry of coatings; they just know what works and what does not work. Coatings manufacturers list their products by generic binder types or by appropriate applications.

Generic Binder Classifications

Alkyd Paints These paints provide minimal protection and are seldom used in oilfield applications. Note that many manufacturers will call these systems paints to distinguish them from the more protective chemistries listed as "coatings" – implying *protective* coatings.

Bituminous Coatings These asphalt or coal-tar-based coatings are seldom used in oilfield applications on metal substrates with the exception of field-applied repairs and girth-weld coatings on buried pipelines. At one time coal-tar coatings were standard coatings for pipelines and other oilfield structures, but environmental and occupational health concerns have limited their use in recent years [25].

Vinyl Coatings Vinyl coatings are widely used protective coatings used for corrosion resistance in chemical service. They are easy to apply and form tight homogeneous films over the substrate. Intercoat adhesion is also excellent. They do soften slightly when covered with some crude oils. Their flexibility allows them to accommodate the motion of the steel substrate, e.g. during ship or platform launching [26, 27].

They are physically drying one-component paints and usually cannot withstand temperatures greater than 75–80 °C (165–175 °F). They are relatively easy to apply and quick drying, but they have low solids content and have relatively high VOC contents.

Film thickness is typically only 50 µm (2 mils), and this, combined with their relative softness, means they cannot withstand mechanical abuse. It also limits their use on rough previously coated surfaces.

Chlorinated Rubber Coatings Chlorinated rubbers were once widely used for marine and other oilfield applications [27], but worker safety and environmental concerns have limited their use in recent years.

Epoxy Coatings These coatings are the workhorses of the marine coatings and oilfield industries. Most epoxies require near-white blast coating surface preparation, although some versions can be applied underwater as repair coatings. With proper surface preparation, they have excellent substrate adhesion combined with good impact and abrasion resistance and can achieve high film thicknesses.

Epoxies chemically cure. They are available in systems where the reactants are mixed during application as well as single-component products, which require more careful shipping and storage control and careful consideration of application and curing temperatures. These chemical curing-related considerations mean that the use of epoxy coatings requires highly trained and experienced coatings contractors.

A wide variety of end-product properties can be achieved by varying the components, and epoxies are sold as pure epoxies, phenolic epoxies, coal-tar epoxies, solvent-free epoxies, waterborne epoxies, and so forth.

All epoxies are subject to UV degradation. This means they must be topcoated for most atmospheric exposures.

Epoxy Mastic Coatings These coatings have been marketed in recent years as easy-to-apply, inexpensive thick-film coatings. They have relatively short service performance compared with other coatings. Because of this and the fact that coating materials are minor costs (approximately 10%) of coatings applications projects, their reduced cost is not justified for many applications. They are not generally recommended for atmospheric or marine exposures, but they may be appropriate for buried service, e.g. as repair coatings on pipelines.

Acrylic and Polyurethane Coatings These coatings are usually used as colored topcoats over epoxy-based primers and intermediate coatings. They have excellent hardness and very good UV and chemical resistance.

They are two-component or moisture-curing systems that require temperature and other application controls. The choice between various topcoats is often determined by pot life, application conditions, and curing time before they can be used.

Polyester Coatings Glass flake-reinforced polyester high-build coatings are used for high wear and abrasion applications such as walkways and tidal and splash zones of steel structures.

Vinyl Ester Coatings These coatings find their best uses as tank linings. They are often applied as thick coatings (2 coats of 750 μm – 30 mils, total 1500 μm – 60 mils). Many linings are applied as sheet materials, but these linings can be applied by airless spray with two-component liquids that quickly cure. Glass flake reinforcement improves abrasion resistance. They should only be applied over blast-cleaned steel (NACE No. 2/SSPC-SP 10).

Inorganic Zinc Coatings Inorganic zinc (IOZ) coatings have metallic zinc pigments. They are widely used as primers for long-lasting atmospheric exposure protection. They are not generally recommended for buried or submerged service.

The zinc pigments make mechanical contact with each other and with the steel substrate. This allows them to provide galvanic protection of exposed steel at holidays, and they usually have at least 75% by weight metallic zinc in the dry film [6]. The precise minimum zinc content for effectiveness depends on which of several commercially available inorganic binders is used.

IOZ primers will become dull gray upon weathering and sometimes require topcoats for visibility or color coding purposes. These topcoats reduce the effectiveness of the cathodic protection provided by the underlying inorganic zinc primer. Unlike organic primers, inorganic zincs are immune to UV degradation, so sunlight exposure is never a problem, and many IOZ primers with no topcoats have been used for decades with minimal degradation.

Surface preparation for inorganic zinc primers requires at least a commercial blast cleaning (NACE No. 3/SSPC-SP5), but white metal blasting (NACE No. 1/SSPC-SP 5) is preferred and results in better performance. Topcoating of inorganic zinc primers requires cleaning to remove any reaction products that may have formed.

Repairs to IOZ primers require the use of organic zinc coatings (also called zinc epoxies). IOZs do not adhere well to prior IOZ coatings, but organic zincs are the opposite – they will adhere to IOZs and to many other materials. Repainting IOZ requires that zinc salts be removed from the surface to prevent blistering. This

TABLE 6.6 Comparison of Inorganic Zinc and Galvanized Coatings

Inorganic Zinc	Galvanizing
Not a metallic coating	Metallic coating
Excellent corrosion resistance	Excellent corrosion resistant
Chemically bonded to steel substrate	Chemically bonded to steel substrate
Individual particles	Continuous zinc
Medium abrasion resistance	Limited abrasion resistance
Slower reaction with acids	Faster reaction with acids
Long life span	Shorter life span

is a special problem for offshore platforms, where the availability of clean freshwater is at a premium.

Table 6.6 compares the performance of inorganic zinc and galvanizing coatings. For all but the most benign environments, e.g. onshore atmospheric exposures, galvanizing is not practical, at least in part because of the thin coatings applied by standard commercial galvanizing operations.

Zinc is an amphoteric metal, and it corrodes at unacceptable rates in both acids and bases. This means that no zinc coating pigments can used for direct exposure to acids, e.g. drilling muds and many other completion and workover fluids, or to bases. Topcoats are necessary in these environments.

Organic Zinc Primers Organic zinc primers are used for many of the same applications as inorganic zinc primers. Their organic binders mean that, unlike inorganic zinc coatings, organic zinc primers must be topcoated to protect them from UV degradation. The organic binder also means that the cathodic protection provided by the metallic zinc pigments is less effective because the intermetallic contacts between the pigment particles and between the pigments and the steel substrate is are less effective due to the resistivity of the organic binder.

Organic zinc primers are fast drying and their overcoating interval is relatively short. They are compatible with all of the topcoat systems used in oilfield applications. Problems occur with alkyds, but alkyds are seldom used in oilfield applications due to their limited environmental resistance.

The surface preparation for organic zinc primers requires at least a near-white blast cleaning (NACE No. 2/SSPC-SP 10), and they are less tolerant of surface salts than inorganic zinc primers.

Organic zinc coatings are used to recoat and spot paint IOZ and galvanized surfaces.

Polyurea Coatings Polyurea coatings show much promise and will become more prevalent. The advantages of polyureas include their quick setting, insensitivity to ambient temperature and weather changes during application, and good mechanical properties. Unfortunately, their quick-setting properties limit their adhesion to metal substrates, and they are not currently used as protective coatings for large metal structures. It is likely that these limitations will be overcome, and polyureas will become standard protective coatings for many oilfield applications when that happens [28].

Coatings Suitable for Various Service Environments or Applications

Several organizations suggest appropriate coatings systems for various environments and applications. Tables 6.7 and 6.8 show the SSPC classification of environments and suggested coatings systems for these environments. NORSOK, the Norwegian standards organization, recommends the coating systems for offshore platform applications shown in Table 6.9. The referenced publications also contain suggestions on surface preparation

methods and the necessary surface conditions prior to coating application [8, 29].

Many coatings manufacturers and suppliers list coatings by application instead of by generic type. These listings generally follow terminology similar to that shown in Tables 6.6–6.8, and they often specify the recommended operating temperatures (usually as maximum acceptable sustained temperatures) that the coating systems can withstand. Coating systems from various manufacturers will perform differently, and it is common to test new coatings systems for performance in accordance with recommended accelerated exposure testing procedures [30].

Coatings Inspection

Surface preparation and other types of coatings inspection are normally performed by third-party inspection organizations. Surface preparation is the most important part of any coatings application project, and inspection by NACE or Norwegian FROSIO (the Norwegian Professional Council for Education and Certification of Inspectors of Surface Treatment)-certified inspectors is often required. These third-party inspectors are trained to ensure that surface preparation and coatings applications are conducted in accordance with established international standards as specified by the coatings inspection and application contract. Inspectors with this training are also involved with evaluation of existing coating systems to determine the extent of coating damage and recommend remedial measures [31, 32].

Documentation of coatings inspection is very important, because the third-party inspector must convince all interested parties that the surface preparation, coatings application, and inspections have all been conducted in accordance with contract requirements and industry standards. Owners want perfect coatings and contractors want payment; these conflicting interests must be resolved in accordance with pre-agreed procedures.

TABLE 6.7 SSPC Environmental Zones [29]

Zone	Description
0	Dry interiors
1A	Normally dry interiors
1B	Normally dry exteriors
2A	Frequent wetting by freshwater
2B	Frequent wetting by salt water
2C	Immersion in freshwater
2D	Immersion in salt water
3A	Acidic atmospheric exposure, pH <5
3B	Neutral chemical atmospheric exposure, pH 5–10
3C	Alkaline chemical exposure, pH 10 and higher
3D	Atmospheric exposure with solvents and hydrocarbons
3E	Severe chemical atmospheric exposure

Source: Reproduced with permission of SSPC.

TABLE 6.8 Selected Environmental Zones for Which SSPC Systems Are Recommended [29]

Painting System		Environmental Zone											
SSPC No.	Generic Type	0	1A	1B	2A	2B	2C	2D	3A	3B	3C	3D	3E
PS 11	Coal tar epoxy				X	X	X	X	X	X	X		
PS 12	Zinc-rich (no topcoat)		X	X	X	X	X	X		X		X	
PS 12	Zinc-rich (with topcoat)		X	X	X	X	X		X	X	X	X	
PS 19	Ship bottoms												
PS 21	Ship topsides		X	X	X	X							
CS 23	Metallic thermal spray		X	X	X	T	X	T	T	T	T		

T, recommended only with proper sealing or topcoating.
For zone 3E use specific exposure data to select a coating.
Source: Reproduced with permission of SSPC.

TABLE 6.9   Recommended Coating Systems for Various Locations on Offshore Platforms [9]

Application		Coating System	NDFT (μm)
Carbon Steel		1 coat zinc rich epoxy	60
	Operating temperature <120°C	1 coat two-component expoxy	200
	Structural steel	1 coat top coat	*75*
	Exteriors of equipment, vessels, piping, and valves (not insulated)	MDFT (μm)	335
	All carbon steel surfaces in noncorrosive areas (e.g. living quarters)		
	Deck areas		
Carbon Steel		Thermally sprayed aluminum or alloys of aluminum	200 min
	Operating temperature >120°C		
	All insulated surfaces of tanks, vessels, and piping		
	Flare booms	Sealer	
	Underside of bottom deck, jacket above splash zone, crane booms, and life boat stations are optional stations		
Walkways, escape routes, and other deck areas as specified		Nonskid epoxy screed	3000
Under epoxy-based primer		1 coat epoxy primer	50
		or	
		1 coat zinc rich epoxy	60
		1x epoxy tie coat	*25*
		MDFT (μm)	85
Under cement-based fire protection		1 coat zinc rich epoxy	60
		1 coat two-component epoxy	*200*
		MDFT (μm)	260
Application (if not specified in others)		1 coat epoxy primer	50
	Uninsulated stainless steel when painting is required	1 coat two-component epoxy	100
		1 coat top coat	75
	Aluminum when painting is required	MDFT (μm)	225
	Galvanized steel		
	Insulated stainless steel piping and vessels at temperatures <120°C	2 coats immersion grade expoxy phenolic	*2 × 150*
		MDFT (μm)	300
Submerged carbon steel and carbon steel in the splash zone		1 coat epoxy primer[a]	225
		1 coat two component epoxy	*225*
Submerged stainless steel and stainless steel in the splash zone		MDFT (μm)	450
Internal seawater filled compartments, e.g. ballast tanks			

[a] NORSOK states "two component epoxy for both layers."
Source: Reproduced with permission of NACE International.

Written standards, e.g. NACE, SSPC, ISO, and so forth, take preference, but only if they are specified in the contract. Specifications and/or approved contractor submittals always take precedence over standards [32].

Common inspection hold points or check points include:

- Presurface inspection
- Postsurface preparation
- Prepainting
- During and after application

The initial surface inspection determines the condition of the steel surfaces before operations begin. ISO standards for degree of rusting serve to indicate the relative initial surface condition before surface preparation and coatings begin [33]:

- A – Steel surface largely covered with adhering mill scale but little, if any rust.
- B – Steel surface has begun to rust and from which the mill scale has begun to flake.

• C – Steel surface on which mill scale has rusted away or from which it can be scraped, but with slight pitting visible under normal vision.
• D – Steel surface on which mill scale has rusted away and on which general pitting is visible under normal vision.

Other organizations have provided similar standards for grading the initial surface of metals prior to surface preparation [16, 17, 34].

Each type of paint coating will have a specific surface preparation requirement that must be followed.

Proper paint adhesion depends on the removal of all greases and other organic contaminants prior to surface preparation. Abrasive or water jetting cannot be relied upon to do this, so solvent cleaning is usually necessary to remove them. This contamination is normally detected visually, but UV lighting (300–399 nm) is sometimes used to aid in organic contaminant detection [35].

The presence of soluble salts on metal surfaces has been recognized for many years as leading to premature coatings failures. These salts are normally chlorides and sulfates, but other species may also be present. They are usually removed by power washing. Proprietary commercial products claim to aid in this salt removal. Chloride contamination receives most of the soluble salt attention, and any washing process that removes chloride salts is likely to also remove other salts.

No internationally recognized standards on the acceptable levels of salt contamination on a metal surface are available, although efforts are underway to develop them. Table 6.10 shows the limits suggested by NACE for offshore structures. These levels are to be measured using methods suggested by ISO and other standards [8, 20, 36–40].

TABLE 6.10 Recommended Maximum Allowable Total Soluble Chloride Ion Contents for Offshore Structures [26]

Coating Category	New Construction (mg m^{-2})	Maintenance (mg m^{-2})
Splash zone, exterior submerged zone, and ballast water tank	20[a]	20[a]
Atmospheric zone	20	50
Stainless steels	20	20

[a] The level of residual salt contamination on the surface has a very significant effect on the service life of the immersion-type coating systems. A clean surface should be obtained prior to coating. However, if the soluble chloride ion content is high and obtaining a clean surface is too costly, the allowable soluble chloride ion content shall be agreed to by the coating applicator and facility owner.
Source: Reproduced with permission of NACE International.

One of the problems with soluble salt detection and measurement is the wide variability associated with measuring the salt levels on rough surfaces. Replication is difficult. One promising method of diminishing this variability is the development of conductometric testing. Instead of using colorimetric titrations or similar methods of determining the salt content on a metal surface, the surface can be exposed to a source of deionized water and the resulting conductivity can be measured as a direct indication of the soluble salts that are present. This method has the added benefit of measuring all salts, not just chloride ions [37–40].

Steel surfaces must have a profile in order for the paint coatings to develop adequate surface contact and adhesion. This is one of the main results of abrasive blasting and is considered to be a limitation to many water jetting operations. If the surface profile is too deep, paint films will not adequately cover the highest points, and pinpoint rusting will occur. This idea is shown in Figure 6.11. Pinpoint rusting is unsightly, but it can also lead to eventual undercutting and wider corrosion problems. The surface profile necessary to ensure adequate bonding and to avoid pinpoint rusting will depend on the type of coating involved.

Anchor patterns and surface profile determinations are usually evaluated visually by placing laboratory-prepared optical comparators on the surface and comparing the adjacent abrasive-blasted surfaces to the comparator. Figure 6.12 shows one version of commercially available comparators used for this purpose. Comparisons are made with a specially constructed magnifying glass that allows comparison of the surface in question with four different grades of surface preparation. Other methods of surface profile determination include the use of replica tape for making a negative image of the surface in question. A soft plastic tape with a hard backing is applied to the area in question. After the tape is removed, the profile of the tape replica is measured using a specially calibrated microscope [29, 41–46].

Coating inspectors also use visual photographs available from NACE and SSPC to document the surface conditions to be expected after abrasive blasting or water jet cleaning [16, 47–50].

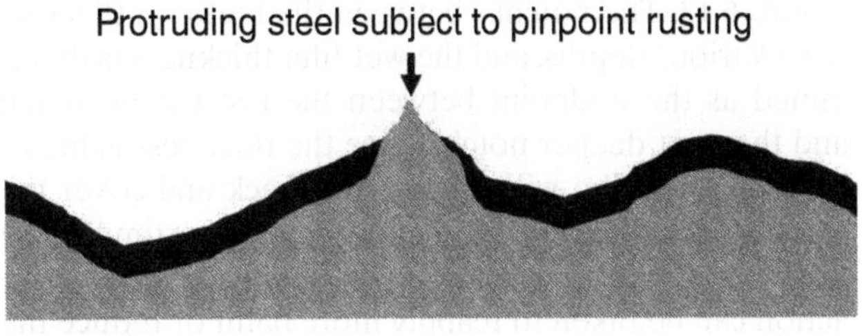

Protruding steel subject to pinpoint rusting

Figure 6.11 Protruding steel subject to pinpoint rusting.

Figure 6.12 Keane-Tator surface profile comparator disc. *Source*: Photo courtesy of KTA-Tator, Inc., reproduced with permission.

The appearance of grit-blasted surfaces will vary depending on the specific grit used and the steel substrate in question. It is often advisable to prepare a field sample for referee purposes prior to starting a surface preparation project. This field sample, representative of the agreed-upon surface cleanliness for the project may be a plate of metal or a part of the structure. These prepared areas can then be used as for reference by the contractor and the inspector. It is important that the surface be preserved in its "as-agreed" condition until the project is completed.

Inspection Before and During Coating Application
Application equipment and methods need to be inspected prior to and during painting operations. This includes making sure that pigments are stirred and suspended in the vehicle, proper mixes of two-component paints are maintained, etc. New low-VOCs paints make coating application more difficult and harder to apply.

Wet film thickness is normally determined manually by inserting a special tool into the wet film as shown in Figure 6.13. The notches between the teeth on this tool have various depths, and the wet film thickness is determined as the midpoint between the last coated notch and the next deeper notch. Once the thickness is measured, the wet film will usually flow back and cover the tooth marks. The dry film thickness can be estimated by knowing the percent solids in the paint, and corrective action can be taken to reapply more paint or reduce the application rate as needed. Thin films will not provide

Figure 6.13 Wet-film step or notch gage applied to a newly painted surface. *Source*: Courtesy of NACE International, reproduced with permission.

adequate coverage and many thick films will not cure correctly. If a more precise measurement is necessary, eccentric roller gages, lens gages, or needle micrometers can be used [51–53].

Inspection After Application The most common tests applied to dry films are thickness and adhesion measurements.

Magnetic thickness gages are usually used for dry film thickness determinations on steel substrates. They need to be calibrated, and they are less accurate on curved surfaces or thin (sheet vs. plate) carbon steels. They usually cannot be used on most stainless steels with the exception of the low-alloy 13-chrome martensitic stainless steels used for pipelines and similar applications, where they must be specially calibrated to compensate for the reduction in magnetic alpha ferrite in the metal [52, 53].

Nondestructive magnetic measurements can be confirmed by destructive tests, which are also used in cases of disputes. The most common destructive test uses a Tooke gage, which measures the exposed cut coating after it has been cut with a special scribe [54]. The exposed cut surfaces, shown in Figure 6.14, are viewed under a special magnifying glass, and the thickness of the cut edges are calibrated on the instrument.

Adhesion tests are destructive tests often performed on dry films in accordance with a variety of international standards [55–57].

Coating holidays are defects (usually holes, which may be microscopic) in the coating that expose the underlying metal to the corrosive environment. They are normally detected using electrical holiday detector. An electrode is passed over the surface being inspected, and, as it passes over a defect, the system passes current between the electrode and the metal being inspected [58, 59]. The operator can then mark the holiday for repair. This is shown in Figure 6.15. While Figure 6.15 shows the inspection of an internal coating in a large diameter pipe, the same procedure is performed on structural metal, pipeline exteriors, and any steel structure coated with a nonconductive coating system. Holiday detectors must be selected for the appropriate application, as high-voltage detectors used on thick-film exterior pipeline coatings would damage thinner coatings, and the lower-voltage instruments used for thin films would not penetrate thicker, more resistant coatings [58–60].

Any defects noted during surface preparation, during application, or after film drying should be noted and remediated before the next step in the coatings or construction process.

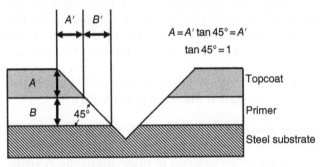

Figure 6.14 Cross section of a coating cut for Tooke gage measurement of dry film thicknesses.

Weather Conditions   Weather conditions must be measured before coatings operations can begin. Table 6.11 shows the conditions where condensation will form on a metal surface depending on the metal temperature, air temperature, and relative humidity. NORSOK and DNV specify that no final blast cleaning or coating applications can be conducted if the relative humidity is greater than 85%, the steel temperature is less than 3 °C (5 °F) above the dew point, and coating applications and curing temperatures must be above freezing for both air and metal temperatures [8, 60]. These Norwegian

Figure 6.15 Electrical holiday testing in the interior of a large-diameter water pipe. *Source*: J. Brodar [25]. Reproduced with permission of NACE International.

TABLE 6.11   Relative Humidity and Surface Condensation Temperatures on Uninsulated Metal Surfaces

Metal Surface Temperature (°F)	Surrounding Air Temperature (°F)																
	40	45	30	55	60	65	70	75	80	85	90	95	100	105	110	115	120
35	60	33	11														
40		69	39	20	8												
45			69	45	27	14											
50				71	49	32	20	11									
55					73	53	38	26	17	9							
60						75	56	41	30	21	14	9					
65							78	59	45	34	25	18	13				
70								79	61	48	37	23	22	16	13		
75									80	64	50	40	32	25	20	15	
80										81	66	53	43	35	29	22	16
85											81	68	55	46	37	30	25
90												82	69	58	49	40	32
95		% of Relative Humidity											83	70	58	50	40
100													84	70	61	50	
105														85	71	61	
110															85	72	
115																86	

Source: Baboian [61]. Reproduced with permission of NACE International.

standards are primarily intended for the cold and humid conditions found in the North Sea and nearby construction yards. NACE uses the same relative humidity recommendations and further recommends air temperatures above 5 °C (40 °F). Some paints can be applied at lower temperatures, and moisture-cured urethane coating systems may be used at higher humidities [26]. Coatings must always be applied in accordance with manufacturer's recommendations, and some coatings, e.g. epoxies, may require higher temperatures for proper curing.

Areas of Concern and Inspection Concentration

This section is intended to illustrate a number of practices necessary to ensure successful coatings applications, inspection, and repair. In any structure there are areas where coatings surface preparation, application, and repair are more difficult. There are also regions on a structure where corrosion damage, if encountered, will be more significant than the unsightly damage elsewhere on the same structure. These are areas that can sometimes be remediated prior to coating. If they cannot, then special attention must be paid during all phases of a coating project, as well as during operations when the difficult/critical areas should require additional emphasis.

Key Features That Should Be Remediated Before Coating   There are certain features, perhaps structural defects, on structures that should be removed before surface preparation and coating application. Several of these are shown in Figures 6.16–6.19. These features should be removed or minimized before surface preparation begins; otherwise they will lead to premature coatings failures [62].

Skip welds are often applied to hold equipment in place during construction and usually supply adequate strength for structural purposes. Unfortunately, if they are not sealed using continuous sealing welds, they allow crevices to form. This is a major problem for any structures intended for immersion service, because crevices allow moisture to collect and create corrosion problems even in atmospheric exposure. The sealing welds do not need to be strong, because this could cause metal distortion, as shown in Figure 4.35, but they should be watertight and inspected to ensure that no cracks or crevices remain.

Welds should be ground smooth and weld spatter should be ground flat so that it does not cause thin layers or bare spots in the coating (Figure 6.17) [62].

Voids due to bridged-over gouges or coating shrinkage will collect moisture, especially in immersion service, and will eventually lead to coating failure (Figure 6.18). The sharpness of the gouges or internal corners should be ground smooth or, in the case of internal corners like those shown in Figure 6.19, a rounded weld bead can increase the radius and provide more surface area for adhesion.

Riveted construction should be avoided and bolted connections limited whenever possible [62].

Stripe Coating   Certain key locations will remain difficult to clean, coat, and inspect. It is common practice to use stripe coatings in these locations. A stripe coat is a supplemental coating applied to ensure there is adequate protection of critical areas like flanges, edges, welds, fasteners, and other irregular

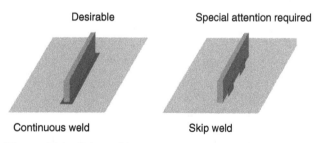

Figure 6.16 Skip welds versus continuous welds. *Source:* Reproduced with permission of NACE International.

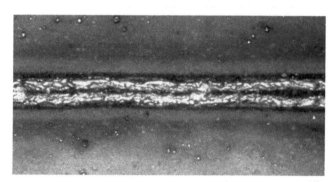

Figure 6.17 Weld spatter. *Source:* Reproduced with permission of NACE International.

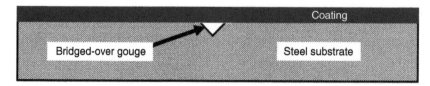

Figure 6.18 Gouge with bridged-over coating subject to moisture accumulation and corrosion.

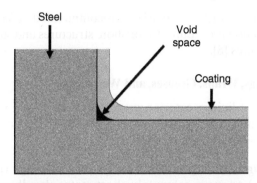

Figure 6.19 Void space caused by coating shrinkage.

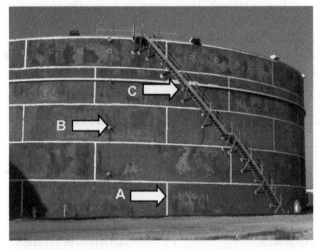

Figure 6.20 Stripe coating prior to painting on an aboveground storage tank. A – welds, B – through-wall fitting, C – ladder attachment.

Figure 6.21 Flange on the exterior of an aboveground storage tank. All corrosion was ground away and the edges rounded before abrasive blasting and stripe coating.

Figure 6.22 Construction aid on an offshore structure tank wall that was left in place and stripe coated prior to over coating. *Source*: Reproduced with permission of NACE International.

areas (Figure 6.20) [8, 63]. They should be allowed to set-to-touch before over coating.

Stripe coats are usually applied by brush or roller to ensure adequate thickness. Their color should contrast with the substrate and with the overcoat to allow for easy inspection to ensure adequate cover. It is considered good practice to grind all edges and to extend the striping at least 1 in. (2½ cm) from edges. This is especially important for maintenance coating, because exposed edges are more susceptible to corrosion and are likely to exhibit intergranular exfoliation (Figure 6.21).

Construction aids, such as shown in Figure 6.22, must also be stripe coated. While they are designed to be removed, they are often left in place.

Figure 6.23 shows equipment at a tropical pipeline receiving station with corrosion on fasteners, flanges, welds, and edges. All of these areas need to be ground smooth and stripe coated after surface preparation is completed.

Final Comments on Inspection Most North American organizations use NACE, ASTM, and SSPC standards for coatings inspection, while European and other organizations tend to use ISO standards, which are likely to be derived from standards originally developed in European countries. This practice is changing, and the trend toward increased use of ISO standards is likely to continue [32].

While standard industrial practice is to develop and maintain checklists of items to be inspected, these checklists do not substitute for inquisitive and intelligent inspection teams. Unanticipated situations are likely to occur on many projects, and it is the responsibility of the inspection organization to bring them to the attention of the appropriate parties.

Figure 6.23 Corrosion on a complicated geometry structure needing grinding and smoothing plus stripe coating after surface preparation.

Table 6.12 summarizes recommended coatings inspection procedures for offshore structures and similar structures [8].

Linings, Wraps, Greases, and Waxes

Internal linings, exterior wraps, and greases and waxes find limited but important uses for controlled oilfield corrosion.

Linings   Linings are relatively thick coatings (paint layers) or, more commonly, sheet materials adhered to or in intimate contact with the interior surface of a pipe or other container. Their purpose is usually to protect the metal surface from corrosion, although some tubular product liners are also used to minimize flow resistance and increase production rates – usually on downhole tubing where inside diameters are restricted and boundary layer effects are significant.

TABLE 6.12   Recommended Inspection and Testing for Protective Coatings on Recently Constructed Offshore Facilities and Similar Equipment

Test Type	Method	Frequency	Acceptance Criteria	Consequence
Environmental condition	Ambient and steel temperature Relative humidity Dew point	Before start of each shift + Twice per shift	In accordance with specified requirements	No blasting or coating
Visual examination	Visual for sharp edges, weld spatter, slivers, rustgrade, etc.	100% of all surfaces	No defects Refer to specified requirements	Defects to be repaired
Cleanliness	ISO 8501-1 Rust grades ISO 8502-3	100% of all surfaces	IAW with specifications Maximum quantity and size rating	Defects to be repaired Reblasting
	Dust on surfaces (pressure tape method)			Recleaning and retesting until acceptable
Salt Test	ISO 8502-6 or equivalent Flexible salt sleeve	Spot checks	Maximum conductivity corresponding to 20 mg m^{-2} NaCl	Recleaning and retesting until acceptable
Roughness	ISO 8503	Each component or once per 10 m^2	As specified	Reblasting
	Comparator or stylus			
Visual examination of coating	Visual to determine: – Curing – Contamination – Solvent retention – Pinhole popping – Sagging – Surface defects	100% of surface after each coat	As specified	Repair of defects
Holiday detection	NACE SP0188	IAW system specifications	No holidays	Repair and testing
Film thickness	SSPC-PA 2 calibration on smooth surface	SSPC-PA 2	SSPC-PA 2 and coating system data sheet	Repair, additional coats or recoating as appropriate
Adhesion	ISO 4624	Spot checks	Depends on coating system	Coating to be rejected

Source: Adopted from table 11.1 in NORSOK M-CR-501.

Linings are often used on tank bottoms and the lower interior side walls of large storage tanks as well as on cargo and product holds on ships, floating production storage and offloading vessels (FPSOs), etc. They are also used for lining pipelines, injection wells, production tubing, and other equipment tubular products [64–66]. Figure 6.24 shows a lining inside a flanged pipeline segment.

High-density polyethylene (HDPE) is the most commonly used liner material, but other thermoplastic materials are also used. HDPE liners are used for water injection pipelines and injection well tubing, multiphase oil and gas gathering lines, sour multiphase crude product pipelines, and oil transmission lines. Medium-density polyethylene (MDPE) is used for water disposal and injection lines, while polyamide 11 (PA-11), also known as nylon, finds use for elevated temperature sour gas and multiphase sour hydrocarbon gathering lines [65].

Many liners are used to prolong the life of existing structures that have already deteriorated significantly. This makes surface preparation difficult, but the surfaces to be lined should be as clean and obstruction free

as possible. A sizing plate is often run through the pipeline to confirm the minimum ID within the host pipe. Solid liners are then often inserted by pulling them through the system using an arrangement similar to that shown in Figure 6.25. Once the liner has been inserted, it is expanded against the liner wall, usually by fluid pressure. Most liners are thermoplastics, and they will set after a certain time conforming to the restraints of the structure surrounding them.

It is important that liners fit tightly, because gases, water, and other fluids will permeate all polymers and can accumulate at the liner–metal interface. Disbonding of the liner can occur if the fluid pressure in the system is suddenly released and the accumulated gas pressure between the liner and the wall does note permeate through the liner. Venting systems have been developed to prevent high-pressure buildup in the annulus between the liner and the structure. If the proper venting does not occur, disbonding such as shown in Figures 6.26 and 6.27 may result. Undetected disbonding may allow corrosive fluid accumulation behind the liner. Disbonded liners can also create significant flow restrictions.

Monitoring systems for checking the annular pressure behind linings have been noted. Through-wall X-rays can be used to identify precise locations where disbonding has occurred.

Wraps and Tapes   At one time, pipeline wraps were a standard means of coating the exterior of buried pipelines.

Figure 6.24 The flange end of a liner segment installed in an oilfield pipeline.

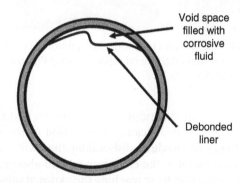

Figure 6.26 Disbonded liner caused by rapid pressure release in a fluid piping system.

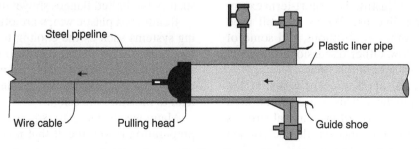

Figure 6.25 System for drawing unexpanded liner into a pipeline. *Source*: Reproduced with permission of NACE International.

Figure 6.27 In-service liner breach failure on a high-pressure sour gas pipeline [66]. *Source*: Reproduced with permission of NACE International.

Figure 6.29 Reinforced wraps used to coat flanges on aboveground piping systems.

Figure 6.28 Petrolatum tape with glass-reinforced outer wrap used at the air–soil interface on piping. *Source*: Photo courtesy Denso North America Inc.

Figure 6.30 Petrolatum tape wrap on cooling tower piping. *Source*: Photo courtesy Denso North America Inc.

Many years of field experience have shown that this is no longer advisable, because any motion of the pipeline is likely to produce disbonding that will expose unprotected metal to the environment and also prevent cathodic protection from reaching the exposed substrate underneath the disbonded wrap. The application of new long-distance pipeline wraps is now limited to occasional rehabilitation projects where the economics of wrapping is considered justified for short-term extensions on projected pipeline use. Wraps are still used extensively for many other applications, and some of them are discussed below. Other discussions appear in the section on pipelines later in this book.

Figure 6.28 shows the application of wraps at the soil–air interface on pipelines. This is a common area for corrosion problems because of mechanical stresses that can degrade pipeline coatings and the lack of soil

consolidation which prevents effective cathodic protection. The two-layer wrapping system shown in Figure 6.28 includes a mastic tape intended to ensure watertight bonding to the primary coating system, a fusion-bonded epoxy. The lighter-colored outer wrapping is glass reinforced for mechanical damage protection.

Wraps are also used to protect hard-to-coat areas such as the bolted flanges shown in Figure 6.29.

Reinforced plastic wraps are often used on rigid piping systems such as the cooling tower piping shown in Figure 6.30. These relatively brittle plastics should not be used on equipment subject to mechanical vibrations or large thermal excursions, because they can crack and lead to moisture intrusion.

Many wet surfaces are so complex and their surface preparation is so difficult that no other coating systems are appropriate. Figure 6.31 shows valves in an outdoor sump associated with a tropical-climate crude oil pipeline. Wax-filled wraps like the ones shown in Figure 6.32 are the preferred means of controlling exterior corrosion in these situations. The reinforced wraps are intended to provide a diffusion barrier against moisture intrusion, and the hydrophobic nature of the wax filler further reduces water intrusion.

Another advantage of wraps is they can be applied to wet surfaces with minimal surface preparation. Figure 6.33 shows a swimmer applying a wrap to the underwater and tidal portions of a marine piling.

Most wraps are intended to be moisture permeation barriers, but if they are disbonded or mechanically damaged, they often act as electrical insulators preventing cathodic protection currents from reaching the exposed metal surface. This happened on the pipeline shown in Figure 6.34, and extensive corrosion occurred underneath the wrap.

It is important that wraps have sufficient mechanical strength and substrate adhesion. If they do not, they can unravel from the structure as shown in Figure 6.35.

Figure 6.33 Repair wrap being installed on the tidal and splash zone portions of a marine piling. *Source*: Photo Courtesy Denso, Inc.

Figure 6.31 Wet valves in a sump on a crude oil pipeline.

Figure 6.32 Waxed tape on a valve in a manhole.

Figure 6.34 Pitting of underground pipeline where wrap was ineffective in preventing corrosion.

Figure 6.35 Low-build nonadhesive wrap coming loose from aboveground pipeline. *Source*: Photo courtesy Sandra Munoz, Corrosion y Proteccion, Cuernevaca, Mexico.

Greases and Waxes   These materials are hydrophobic and are often used for temporary coatings to limit corrosion. Waxes are often incorporated into wraps to enhance their ability to limit moisture migration.

Coatings Failures

All coatings will eventually fail. The reason for the following discussions is to provide some insight into why these failures occur. If they are the inevitable result of aging, then the coating system selection, surface preparation, and application were suitable for the intended service. If premature problems develop, understanding why these problems occurred will help coatings professionals decide how to prevent them. This understanding may also provide useful information justifying more expensive surface preparation or inspection procedures. Surface preparation is usually the most expensive and most important part of any coating application, and corrosion professionals must frequently justify more expensive procedures to management.

Reasons for coating failures, in their approximate order of importance, are:

- Poor surface preparation and cleanliness
- Poor coating application
- Poor or inadequate inspection
- Poor specifications (both construction and coating)
- Poor component design
- Murphy's law

Remember, most coating systems will work for their intended environment and most coatings failures are due to inappropriate surface preparation or application procedures or conditions.

The sections that follow provide a brief introduction to some of the types of coating failures likely to be

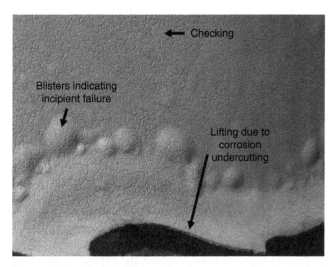

Figure 6.36 Marine piling with aging coating showing typical aging features and patterns.

encountered in oil and gas production. Complete books and handbooks on coatings failure identification and remedies are available [67, 68]. While failure analysis may be useful in determining how to avoid similar problems in the future, some coatings experts suggest that the repair methods for failed coatings do not depend on the failure mode, but instead will require the following [69, 70]:

- Understand of the equipment and substrate's operating parameters.
- Identify the optimal coating solution.
- Write a site-specific coating specification.
- Ensure the specification is followed via third-party inspection.

Normal Aging Failures   Normal protective coating aging failures may show the following phenomena:

- Checking, crazing, alligatoring, or cracking
- Blistering
- Lifting or undercutting of the paint film
- Edge effects
- Reverse impact damage
- Chalking and discoloration

Figure 6.36 shows a marine piling with typical aging of the protective coating. Several degradation processes are apparent.

Coatings inspectors are trained to evaluate coatings degradation in accordance with established international standards. Figure 6.37 shows one of many figures from ASTM D 610, one of the standards used to rate corrosion underneath protective coatings [34].

Checking, alligatoring, and cracking are terms that refer to similar phenomena; the only difference is the

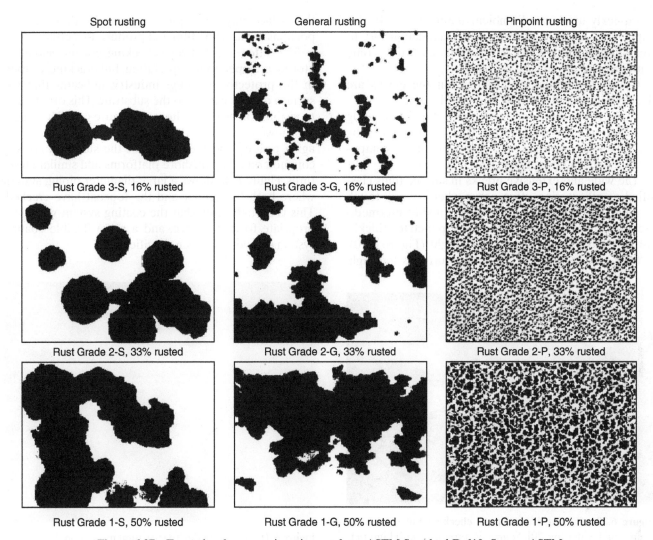

Spot rusting   General rusting   Pinpoint rusting

Rust Grade 3-S, 16% rusted   Rust Grade 3-G, 16% rusted   Rust Grade 3-P, 16% rusted

Rust Grade 2-S, 33% rusted   Rust Grade 2-G, 33% rusted   Rust Grade 2-P, 33% rusted

Rust Grade 1-S, 50% rusted   Rust Grade 1-G, 50% rusted   Rust Grade 1-P, 50% rusted

Figure 6.37 Example of spot rusting pictures from ASTM Standard D 610. *Source*: ASTM D610 [34]. Reproduced with permission of ASTM International.

depth of penetration of the coating defects. They can be defined as follows:

- Checking – slight breaks in the surface film.
- Crazing – similar to checking, but the cracks are generally wider and progress deeper into the film.
- Alligatoring – wide and extensive breaks in the surface. This is most common in bituminous pavements and other thick films, but it is also seen on oilfield equipment. The name is intended to indicate that the surface looks like an alligator's hide.
- Cracking – breaks in the film extend to the substrate, which often lead to corrosion.

Checking, the least extensive of these phenomena, often occurs as a coating dries or continues to react (Figure 6.38). It can be an indication of a coating setting

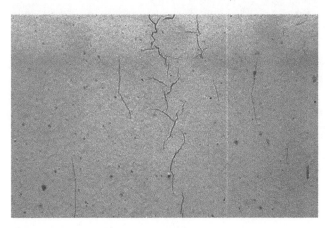

Figure 6.38 Checking on the surface of a protective coating [69, 70]. *Source*: Reproduced with kind permission of Brendon Fitzsimons, FITZ-COATINGS Ltd.

too quickly at elevated ambient temperatures. It also happens as coatings degrade, and degradation due to weathering, including UV degradation, starts to become significant on the coating surface.

Crazing is similar to checking, but the cracks are deeper and wider (Figure 6.39) [69, 70].

A coarse checking pattern on the surface of a coating is sometimes called alligatoring. It is typically caused by aging, sunlight exposure, and/or loss of volatile components as the coating sets and ages.

The wide and extensive breaks in surface films classified as alligatoring often occur when a hard, tough coating is applied over a softer, more pliable intermediate layer or primer. Some coatings, if applied too thickly, can alligator when exposed to sunlight (Figure 6.40). If the ambient temperature during curing is too high,

the surface may cure rapidly compared with the deeper portions of the same film and produce this effect.

The differences between checking, crazing, and alligatoring are open to interpretation, but cracking, as used in the protective coatings industry, indicates that the surface defects extend to the substrate. This often leads to visible corrosion, as shown in Figure 6.41. The cracking shown in Figure 6.41 is due to the combination of an aging coating and the flexure of the substrate. This is common on many offshore platforms and similar structures. Figure 6.42 shows a large-diameter piping system where both the primer and the topcoat have cracked. This is an indication that the coating system was inappropriate for the service, and a more flexible coating system should have been specified.

Figure 6.39 Crazing is similar to checking but cracks are deeper and wider [69, 70]. *Source*: Reproduced with kind permission of Brendan Fitzsimons, FITZ-COATINGS Ltd.

Figure 6.41 Cracking due to structural motion on the exterior wall of a vessel. Note the rust staining at locations where motion has occurred and at welds.

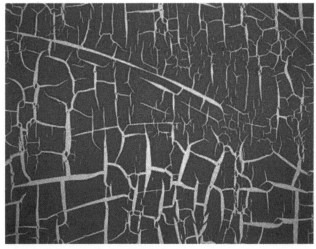

Figure 6.40 Alligatoring of a protective coating [69, 70]. *Source*: Reproduced with kind permission of Brendon Fitzsimons, FITZ-COATINGS Ltd.

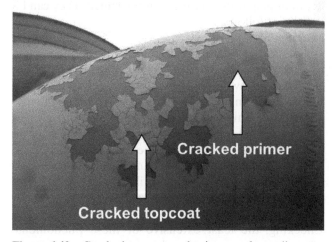

Figure 6.42 Cracked topcoat and primer on large diameter piping.

Blisters may be a normal result of aging coating systems or they may early indicators of improper coating application.

All paint films are permeable to moisture to some extent. Blistering occurs when the moisture at the film-to-substrate interface builds up to the extent that pressure is exerted and the coating–substrate interface disbonds. Water will always migrate through the paint film driven by osmosis, and when the osmotic pressure within the blister balances the coating adhesion around the blister circumference, the blister will cease to grow. Some blisters are associated with areas of corrosion, which may start at coating holidays, whether these are from the original coating process or due to mechanical damage after the coating has set and aged. Most blisters will show minimal corrosion beneath them due to the high pH of the water that collects in the resulting cavity. The oxygen levels in the blister water will also become very low, and thus further corrosion will not happen [71]. Blisters should not be broken unnecessarily, because this will remove the protective paint film and allow more aggressive fluids to attack the underlying substrate. When blisters occur late in the lifetime of a coating, they are indications of imminent coating failure. The isolated nature of some blisters is also an indication that blisters are located where the surface preparation (salt, solvents, and other organic contaminants) was less effective than at other locations that do not blister. The extensive blistering shown in Figure 6.43 is an indication of poor surface preparation [70, 71].

The extensive blisters shown in Figure 6.43 can be compared with the localized blistering shown in Figures 6.44 and 6.45. Figure 6.44 shows blisters on the belowground section of a pipeline vent riser. One of the blisters has been broken to show the uncorroded

steel underneath. If the soil had remained undisturbed by a construction project, the coating would have continued to provide corrosion protection for many more years. Figure 6.45 shows blisters forming near a scratch on the exterior of a buried pipeline coating. Most of the coating is undamaged, but blisters are forming parallel to the scratch, and the coating would eventually disbond and fail. The black spots in the blistered area are where the blisters have been broken to reveal discolored but otherwise uncorroded steel.

Coatings inspectors evaluate blistered coating surfaces by comparing their appearance to published photographs in international standards [71]. They then rate the frequency and size of blisters [72]. Figure 6.46 shows one photograph used for this purpose [72].

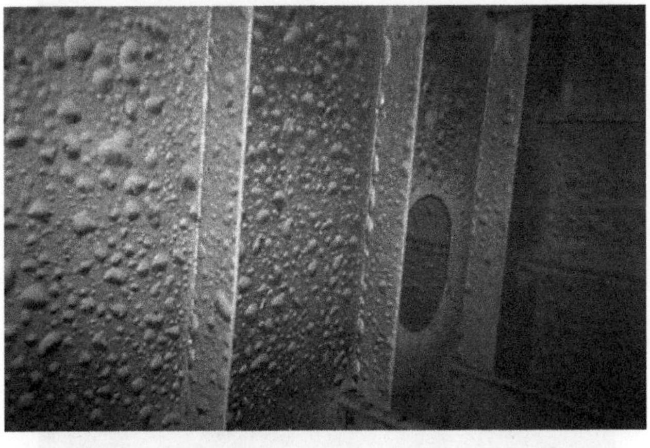

Figure 6.44 Osmotic blisters on a riser near the air–soil interface. The arrow indicates a location where a blister was broken for this photograph. Note the lack of corrosion beneath the blister.

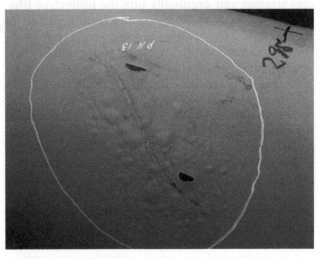

Figure 6.45 Coating blisters forming around scratches on a pipeline coating. *Source*: Photo courtesy R. Norsworthy, Polyguard Products, Inc.

Figure 6.43 Extensive blistering on continuously wetted surfaces indicating inadequate surface preparation [69, 70]. *Source*: Reproduced with kind permission of Brendan Fitzsimons, FITZ-COATINGS Ltd.

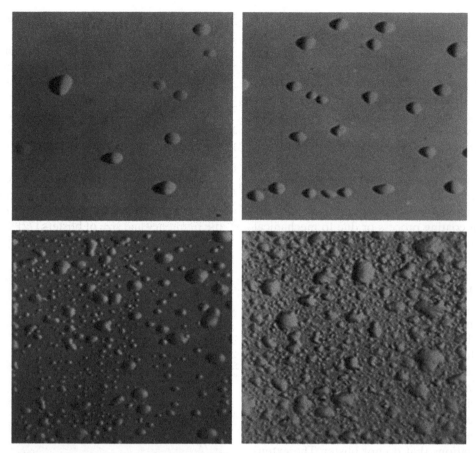

Figure 6.46   Blisters from ASTM D714. From the upper left, these pictures show Size 2 blisters with few, medium, medium dense, and dense ratings. *Source*: ASTM D714 [72]. Reproduced with permission of ASTM International.

Lifting and disbonding occur after corrosion starts, usually at holidays or cracks. As corrosion progresses the increased volume of the corrosion products compared to the metal from which they are formed creates stresses and eventually lifts the protective film. This is shown at the bottom of Figure 6.36 and in Figures 6.47 and 6.48. Figure 6.48a and b shows corrosion and undercutting progressing from deliberate scribe marks on the coatings being tested at an atmospheric corrosion test site. These scribe marks are placed on coating test panels before atmospheric corrosion exposures to determine how well the coating system will resist undercutting. All coatings eventually have holidays (holes in the coating), and the coatings testing industry uses these scribe marks in a deliberate attempt to replicate the service conditions expected in the field.

The term "edge corrosion" is used to describe corrosion occurring at edges or corners of metal surfaces. Figures 6.49 and 6.50 show this phenomenon.

Impact damage can produce holidays in coatings, which then lead to corrosion. Reverse impact damage (impact on the opposite side of the metal member producing mechanical damage to the coating) can lead

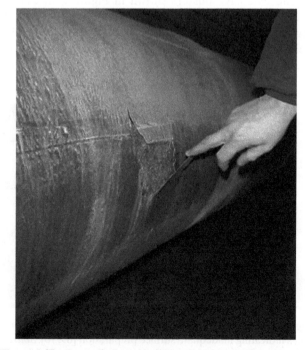

Figure 6.47   Protective coating lifting from the surface of a pipe due to corrosive undercutting. *Source*: Reproduced with kind permission of Brendon Fitzsimons, FITZ-COATINGS Ltd.

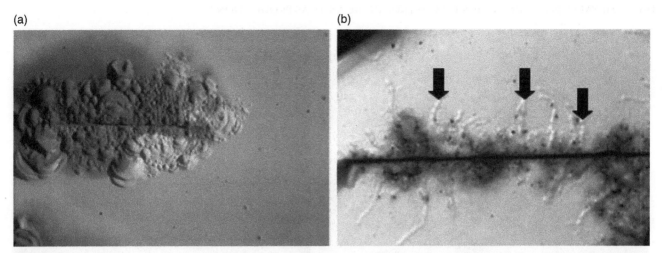

Figure 6.48 Undercutting and corrosion at deliberate scribe marks through the coating systems applied to atmospheric corrosion test panels. (a) Extensive undercutting with no apparent filiform corrosion at the edge of blisters. (b) The arrows in (b) indicate locations where filiform corrosion is progressing at the coating/steel interface.

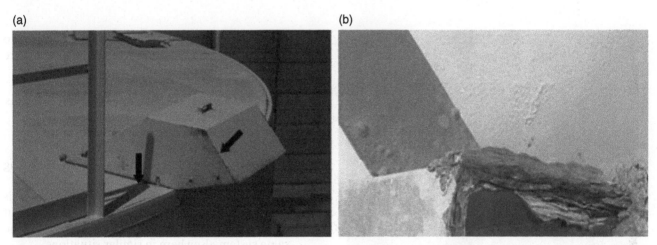

Figure 6.49 Edge corrosion on (a) bent sheet metal and (b) the end of a structural component on a large aboveground liquid storage tank in a coastal location.

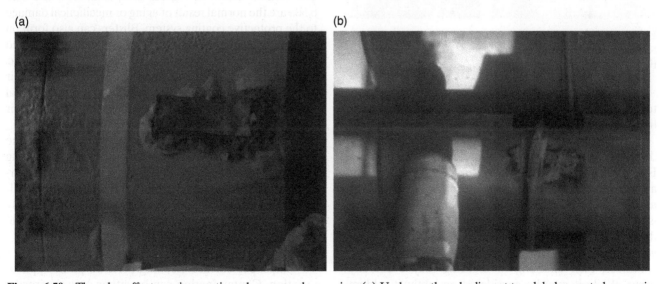

Figure 6.50 The edge effect causing coatings damage and corrosion. (a) Underneath and adjacent to a label mounted on equipment on a stationary offshore platform in a tropical location. Similar problems are apparent adjacent to the flange crevice on the left in this image. (b) Corrosion adjacent to a pipe support restraining strap.

(a) (b)

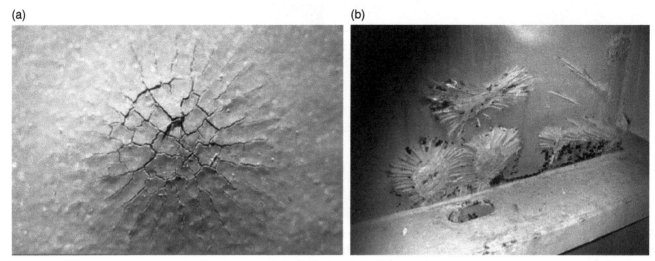

Figure 6.51 Protective coating cracks that radiate from the point of reverse impact and will eventually lead to corrosion [69, 70, 73].

Figure 6.52 Superficial chalking on a coating surface. *Source*: Reproduced with kind permission of Brendon Fitzsimons, FITZ-COATINGS Ltd.

to cracking and corrosion (Figure 6.51). This is more likely to happen with brittle coatings [69, 70, 73].

Chalking is the formation of a loose powder due to UV degradation of organic coating binders. This is shown in Figure 6.52. Epoxies are prone to this problem, and this is a reason why epoxies used for atmospheric exposure usually have non-epoxy overcoats. Another means of control is the use of inorganic pigments, which are immune to UV degradation. Chalking is a sign of UV degradation, but it is normally only a cosmetic problem. If UV degradation continues, then checking or cracking may eventually occur. While it may be considered unsightly, it seldom indicates a lack of coating protectiveness.

Failure Modes for New Coatings It is important to recognize these forms of coating failure as they are indications of one or more of the following [74]:

- Improper surface preparation – organic solvents, salt contamination, or improper surface profile.
- Incorrect application or curing temperature.
- Paint application either too thick or too thin.
- Incompatible coatings for the substrate (including primers and undercoats).
- Improper coating for the service conditions. This usually takes time for indications to develop, but identification of this problem can prevent using the same system elsewhere in similar situations.

The blisters shown in Figures 6.36, 6.42, 6.44, 6.45 and 6.50a are the normal result of aging or mechanical damage to the protective coating system. Blisters can also form on newly applied coatings when gases or liquids are trapped underneath the coating at the coating–substrate interface. This is shown in Figure 6.53, where the light-colored repair coating has extensive blisters, mostly on the bare metal surface, as opposed to the locations where this same repair coating has been applied over the existing coating.

These blisters form due to soluble material in the coating leaching out of the paint film and becoming trapped at the paint–substrate interface. The outer surface hardens and become less permeable and the solvents become trapped. The presence of blisters on newly applied coatings is an indication of one or more of the following conditions:

- Improper surface cleaning prior to paint application. Contamination can be either organic greases and oils or soluble salts.

Figure 6.53 Blisters formed on light-colored repair coating due to improper surface cleaning. *Source*: Reproduced with permission of NACE International. Brodar presentation.

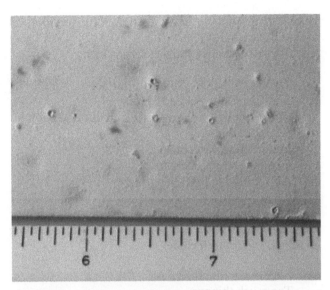

Figure 6.55 Pinholing on a coating surface. *Source*: Reproduced with permission of NACE International.

Figure 6.54 Wrinkled paint surface. *Source*: Reproduced with kind permission of Brendon Fitzsimons, FITZ-COATINGS Ltd.

- Paint applied too thick to allow evaporation of the solvent or suspension vehicle.
- Solvents evaporating when the temperature increases. This can be either organic contaminants or soluble salts.

Wrinkling is the result of paint being applied too thick or too hot. The surface of the coating expands more rapidly as it dries than the inner portions of the wet paint film. Figure 6.54 shows a wrinkled surface.

Pinholes are small visible holes in coating caused by:

- The spray gun being too close to the surface, which can force bubbles into the coating.

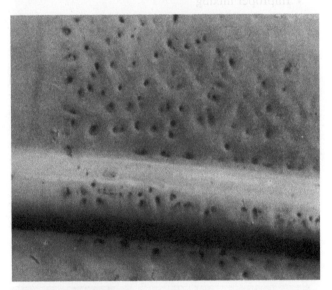

Figure 6.56 Fisheyes in a coating. *Source*: Reproduced with permission of NACE International.

- Incorrect solvent balance in the coating.
- Too volatile solvents.
- Hot weld spatter.

They are caused by a collapse of air or solvent vapor bubbles. Note how many of the pinholes in Figure 6.55 have craters around the edges.

Fisheyes look similar to pinholes; the difference is that they are caused by a lack of adhesion to the substrate. They are usually caused by improper surface cleaning or a lack of wetting of the substrate or particles in the paint (e.g. dust) by the paint film. The paint film pulls away from the contaminant leaving a tiny hole in the coating. The fisheyes shown in Figure 6.56 do not have the mounded rims shown for the pinholes in Figure 6.55.

Conditions that lead to fisheyes include:

- Water or oil on the substrate surface
- Improperly formulated coating
- Converter not properly dispersed into the paint
- Incorrect thinning
- Incorrect spray technique
- Excess wet film thickness
- Application at low temperature

The heaviness of liquid coatings can cause sagging or runs, especially on vertical surfaces. Conditions leading to sagging and runs include:

- Excess wet film thickness
- Too much thinner
- Low temperatures
- Improper mixing

Delamination between paint layers and the loss of substrate adhesion in new coatings, where underfilm corrosion has not had a chance to initiate and progress, is due to incompatibilities between the overcoating and the substrate, whether the substrate is metal or another coating. This is a common problem with recoating projects and is shown in Figure 6.57.

Figure 6.58 shows mud cracking on a flat surface (Figure 6.58a) and near a weld (Figure 6.58b). These cracks can form in fairly new coatings as they dry and set. They occur in thick films and are a result of the wet film thickness being too great or of excessive thinning of thick-film coatings. As the film dries and shrinks the cracks form. Early corrosion is a frequent result.

Overspray can look like abrasive dust on the surface of a coating and is poorly bonded to the coated surface. While unsightly, it is not a corrosion problem unless the oversprayed area is painted.

Pinpoint rusting, shown in Figure 6.59, is the result of too little paint or of a too rough surface profile [67–70].

Final Comments on Failed Organic Coatings Failure modes for aged coatings are to be expected and are an indication that the coating systems have reached their useful life. In new coatings the only reason to identify the failure mode is to avoid repeating the mistakes associated with the failure. No matter what the cause of the new coating failure, the repair procedure is the same – proper surface preparation followed by coating application in accordance with manufacturer's

Figure 6.57 Coating disbonding. *Source*: Reproduced with permission of NACE International. NACE CIP 1–5.

(a)

(b)

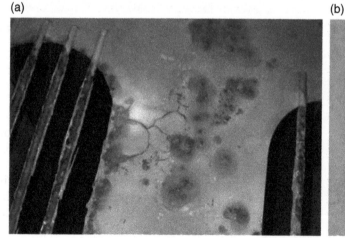
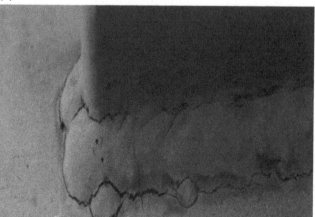

Figure 6.58 Typical mud cracking in thick coatings [73]. (a) Cracking starting at edges. (b) Cracking of thick coating at welded connection. *Source*: Reproduced with permission of NACE International. NACE CIP 1–5.

Figure 6.59 Pinpoint rusting. *Source*: Reproduced with permission of NACE International. NACE CIP 1–5.

recommended procedures including humidity and temperature restrictions.

Metallic Coatings

Metallic coatings find limited but important uses in oil and gas operations. The most commonly used metallic coatings are zinc, thermally sprayed aluminum (TSA), and corrosion-resistant metals and alloys such as chromium and electroless nickel. Zinc and aluminum are anodic to carbon steel in most environments, and they are more widely used. Chromium and other corrosion-resistant metals are limited to smaller applications because of their relatively higher cost and because any coating holidays would produce an unfavorable area ratio and lead to accelerated galvanic corrosion.

Zinc Coatings Galvanizing is a term usually reserved for zinc coatings applied to steel substrates by dipping the cleaned steel into molten zinc. The liquid zinc is sometimes alloyed to increase the fluidity of the liquid metal and to reduce the thickness of the zinc coating. Zinc is also applied by electroplating. The somewhat thinner zinc coatings produced by electroplating are preferred for threaded fasteners and other applications where close dimensional tolerances must be maintained. A third method of applying zinc is by Sheradizing, a method that deposits high-temperature zinc vapors onto the surface. This process produces zinc–iron intermetallic compounds on the surface, and this use is usually restricted to complicated parts with interior geometries that are difficult to electroplate and cannot tolerate the somewhat thicker and less precise geometries obtained

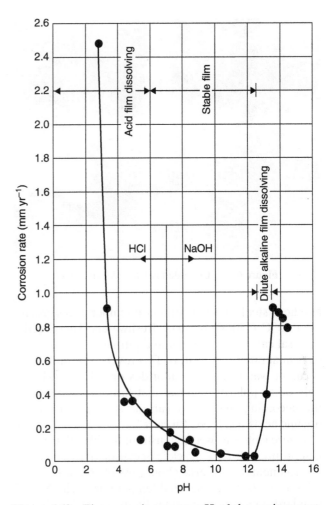

Figure 6.60 Zinc corrosion versus pH of the environment. *Source*: R. Baboian [61]. Reproduced with permission of NACE International.

with hot dipping. Both electroplating and Sheradizing find their main uses on threaded fasteners and other close-tolerance applications.

Zinc is an amphoteric metal, which means that it corrodes at unacceptable rates in both acids and bases. This is shown in Figure 6.60. The low corrosion rates in neutral atmospheres and in some neutral waters mean that zinc coatings, which can be applied in factories during manufacturing processes, are attractive alternatives to painted coatings for many applications. If the zinc coating is breached, the nearby zinc corrodes to protect the nearby exposed steel. Eventually the zinc is depleted and corrosion proceeds as shown in Figure 6.61. Note how the zinc coating is missing near the edges of the holes and where the metal is bent – two locations where coating holidays are likely to occur and corrosion will be accelerated.

While zinc is normally anodic to carbon steel, polarity reversals sometimes happen where zinc becomes

Figure 6.61 Corroded galvanized sign pole.

Figure 6.62 Disbonded coating on a galvanized pole.

anodic to steel. This only occurs at temperatures greater than 60 °C (140 °F) in some freshwaters. The polarity reversal can lead to accelerated pitting at coating holidays in aerated freshwaters. This reversal is unlikely to occur in waters high in chlorides or sulfates [2]. The only other polarity reversal that has been reported is tin becoming anodic to carbon steel in deaerated organic acids, e.g. food "tin cans."

Figure 6.60 was used to illustrate the idea that zinc is inappropriate for use in acidic or caustic (basic, high-pH environments). Concerns about possible polarity reversal also limits use at elevated temperatures. The NACE standard for offshore coatings limits use of hot-dipped galvanizing to temperatures below 60 °C (140 °F), presumably due to concerns about polarity reversal [26].

Atmospheric corrosion of zinc coatings can also lead to hydrogen embrittlement of high-strength steels, and zinc coatings are not used on high-strength fasteners. ASTM F3125 grade A490 bolts, which have high hardness, should not have zinc coatings of any type because of this concern. Lower strength (hardness) bolts manufactured to ASTM F3125 grade A325 and ASTM A193 Grade B7 are acceptable for galvanizing according to most authorities, although this practice was under review in 2017 and may change [75, 76]. Both electroplating and pickling (cleaning in acids prior to electroplating) can cause hydrogen embrittlement. International standards provide guidance on postplating hydrogen bakeout procedures to minimize this concern [77, 78].

Galvanized structural steel is usually used with no topcoating. If topcoating is necessary, e.g. for color coding or other purposes, special surface preparation precautions are necessary. If these precautions are not followed, disbonding of the topcoat from the galvanized substrate can result. This is shown in Figure 6.62.

Figure 6.63 Premature rusting at coating defects on galvanized walkway.

Coating defects can lead to premature corrosion on some galvanized structures. This is shown on the edges of a galvanized walkway landing alongside a large aboveground storage tank under construction (Figure 6.63). The spangle is still on the zinc coating, indicating that this rusting is very premature and due to coating defects present from the time of manufacture. Most coating holidays of this size on new construction are due to improper surface cleaning before placing the steel in the molten zinc bath.

Galvanized steel corrodes quickly in moist atmospheres. Figure 6.64 shows corrosion of the hardware around relatively uncorroded aluminum jacketing on an insulated offshore platform piping system. Note that the galvanizing is gone on the hardware and the aluminum is relatively uncorroded in this same environment.

Metallic zinc coatings are generally considered to be too thin and corrosion susceptible for use on offshore

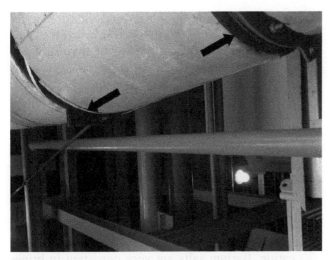

Figure 6.64 Corrosion of galvanized hardware around relatively uncorroded aluminum jacketing on an insulated offshore platform piping system. The arrows indicate the galvanized steel, which is exposed to a tropical marine atmosphere.

Figure 6.65 Corrosion of hot risers with improperly applied thermal sprayed aluminum coatings in the splash zone after only 2½ years [82]. *Source*: Reproduced with permission of NACE International.

structures (Figure 6.64), but they are widely used for onshore structures and on some process equipment. Possible liquid–metal embrittlement of offshore structural metals and piping due to melted zinc in fires is another reason why the use of metallic zinc coatings offshore has been limited.

Zinc and zinc-aluminum alloys can be thermally sprayed onto steel surfaces, and this normally provides a somewhat thicker coating that is considered to be more corrosion resistant [79].

Thermally Sprayed Aluminum Coatings TSA (also called flame-sprayed aluminum although there are non-flame processes as well) coatings are becoming more important for many applications. Tables 6.8 and 6.9 list several applications where thermally sprayed aluminum coatings are recommended. As experience with these coatings grows, it is likely that more applications will become apparent. They are currently recommended for various offshore applications and for corrosion under insulation in petrochemical piping [8, 26, 29, 79–81]. TSA coatings work best when they are sealed with organic sealers or semi-organic silicone sealers [82].

Conoco (now ConocoPhillips) has used TSA coatings for production risers since the mid-1980s. Most of this experience has been favorable, but problems have been reported, usually due to inappropriate TSA coating thickness, improper sealer application, or thermal cycling due to wave splashing and cooling [82]. Corrosion of hot risers and blisters due to inadequate sealing of TSA coatings are shown in Figures 6.65 and 6.66.

TSA coatings are also used to prevent environmental cracking of insulated stainless steel piping [83, 84].

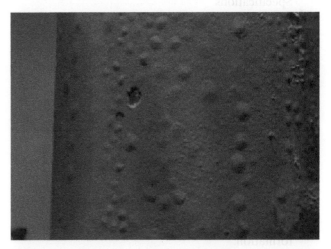

Figure 6.66 Blisters on heated unsealed TSA aluminum [82]. *Source*: Reproduced with permission of NACE International.

Cadmium Plating Electroplated cadmium was at one time the preferred metallic coating for bolts and other fasteners. The coatings are harder than similar electroplated zinc coatings and may be more corrosion resistant. Cadmium plating is still commercially available in North America, but environmental and occupational health concerns are limiting its use. Most organizations have stopped, or curtailed, the use of cadmium coatings, and the use of cadmium coatings is not recommended in recent NACE standards [26].

Amphoteric Coating Materials Zinc, aluminum, and cadmium are often termed the amphoteric coating metals. All of them have unacceptable corrosion rates in acidic and basic environments.

Chromium and Other CRAs   Chromium, nickel, and corrosion-resistant alloy (CRA) coatings are used for corrosion resistance and also as hard facing for erosion resistance in many applications including wellhead equipment, pumps, etc. In the absence of a reducible chemical, e.g. oxygen or acid ions, little galvanic corrosion with nearby carbon steel is likely.

Useful Publications

The following publications provide guidance on coatings selection and on preparing contracts for protective coatings [8, 9, 29]:

- NORSOK M-CR-501: Surface Preparation and Protective Coating
- NACE Publication 6J162: Guide to the Preparation of Contracts and Specifications for the Application of Protective Coatings
- SSPC Painting Manual Volume 2: Systems and Specifications

The Norwegian standard is directed toward offshore platform construction, whereas the NACE and SSPC publications also cover less harsh environments.

WATER TREATMENT AND CORROSION INHIBITION

The most common classifications of water into types used in oilfields are [82]:

- Connate (fossil) water – the original water trapped in the pores of a rock formation during its formation.
- Formation water – water present in the hydrocarbon-producing formation or related rock layers.
- Produced waters – these come from oil or gas wells and can be combinations of formation waters and condensates in various concentrations.
- Injection waters – these are surface waters injected into formations to maintain formation pressures. They contain dissolved solids and treatment chemicals. They may have been processed (filtered and chemically treated) in order to limit corrosion activity in the "down-hole" pipework.
- Condensed waters – these are waters that condense from the gas or oil well as temperatures and pressures change. They have low mineral contents and are often corrosive.
- Meteoric waters – these are waters that have come from surface sources and are normally in the upper layers of groundwater formations.

Connate, formation, produced and injection water are important to oil and gas production processes.

Surface waters are also classified by their salt contents into:

- Freshwater – low in salt content (<1000 ppm chlorides).
- Seawater – found in oceans and seas; this water is usually about 3½% sodium chloride plus significant concentrations of sulfate, magnesium, calcium, potassium, bicarbonate, and other ions. Scale deposits are always possible with seawater, and this can lead to corrosion.
- Brines – have higher salt contents than typical seawater. Most oilfield waters fit into this classification. Barium salts are very persistent in brines and seawater.
- Brackish waters – these are found in bays, estuaries, and where major rivers empty into the sea. They are too salty to be considered freshwater, and their composition is intermediate between freshwater and seawater.

Injection waters are necessary to maintain formation pressure and to properly dispose of subsurface waters that have been separated from produced hydrocarbons. Many different source waters are used for injection including seawater, freshwater, produced water, etc. It is important to properly treat injection waters, because any oxygen, bacteria, or scale-forming minerals from the surface can cause souring or plugging of formations.

Figure 6.67 shows a typical production profile for an oil field. The water production continues to increase for several years after the peak oil production [85]. Worldwide the water–oil ratio (WOR) averages about three barrels of water for every barrel of oil, but the

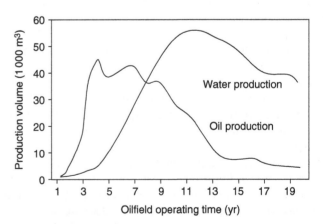

Figure 6.67   Typical production profile for an oil field. *Source*: M. Davies and P. J. B. Scott [85]. Reproduced with permission of NACE International.

figures for the United States, where fields are older and production rates have declined, are approximately seven barrels of water for each barrel of oil. Higher WORs are still profitable. The annual costs of produced water disposal is estimated at US$5–10 billion in the United States and as much as US$40 billion worldwide [85].

Oilfield waste water terminology has changed in recent years, and the following terms have been used since 2000 [85]:

- Water-based muds or fluids (WBM).
- Organic-phase drilling fluids (OPF), which refer to liquids based on drilling fluids used when water-based fluids have been replaced by fluids necessary for directional drilling, horizontal completions, and other high-technology drilling techniques.

Since 1960, US state and federal regulations has been requiring all processed wastewater to be reinjected into the oil reservoir for onshore production, unless the water is cleaned and used for secondary purposes [85]. Offshore wastewater is usually discharged into the ocean, and various governments have different standards on how clean the water must be before discharge.

Oil Production Techniques

Oilfield production can be classified as follows [85]:

- Primary production – uses reservoir energy to produce the oil and gas. The average recovery of the oil in place is 12–15%, but this varies depending on the viscosity of the oil.
- Secondary production – water or gas injection produces an additional 15–20% of the original oil in place.
- Tertiary production – an additional 10–15% of the original oil in place is recovered using enhanced recovery techniques.

Note that these three stages only recover approximately 50% of the original oil in a formation. This means that, as technology advances, more oil can be recovered from otherwise depleted reservoirs. Of course, the economics of this additional recovery depends on market conditions and availability of other oil resources.

Waterflooding   Waterflooding can lead to accelerated downhole corrosion and formation plugging due to scale formation. Injection water can mix with formation water. If this happens deep in the formation (below the water–oil boundary), this scaling may not have a significant effect on production rates [85, 86].

Enhanced Oil Recovery (EOR)   There are a number of methods of enhanced oil recovery. Two of the more common methods are steam injection and CO_2 injection.

The terminologies for these various techniques are not universally applied, and it is common to use the term enhanced oil recovery (EOR) for any recovery method that utilizes chemicals other than water. Secondary recovery is a term often used for repressurization of the reservoir with water or hydrocarbon-based gas to force oil out once the reservoir pressure has dropped.

Thermal processes usually involve heating the formation to reduce the viscosity of the unrecovered oil. The most common method is steam assisted gravity drainage (SAGD), where hot steam is injected into the higher levels of the producing formation and horizontal wells drilled at lower levels collect the oil and pump it to the surface. If steam is injected at excessive rates, the underground pressure and temperature can become too high and cause blowouts. Conversely, low temperatures will lower recovery rates [85].

Steam injection often involves injecting low-quality steam (approximately 80% vapor and 20% water) into the formation to lower the viscosity of heavy oil in the formation. Provided the downhole formation is maintained below about 400 °F (200 °C), downhole corrosion is usually not a problem. Higher temperatures in the range of 500–700 °F (260–370 °C) create CO_2 and H_2S problems, and the use of alloy tubular goods may become necessary [85].

CO_2 injection, often termed miscible flooding, often involves injecting water after the gas in a process known as water-alternating gas (WAG) recovery. This often produces aggressive corrosive environments with wet CO_2 and H_2S and may require stainless steel or nickel alloys in selected locations [85].

Water Analysis

Complete water analysis is seldom necessary. Table 6.13 shows commonly performed water analyses for different purposes. Many of these determinations are performed by water treatment companies that also provide chemicals for scale and corrosion control.

Several of these determinations, e.g. all dissolved gas determinations and pH – which is influenced by dissolved gases – are pressure and temperature sensitive. This is why most downhole pH determinations are calculated and why some samples must be collected in pressure-maintaining devices. It is also important that analyses be done in a timely manner, as water chemistry may change significantly after samples are collected. Storage conditions cannot replicate the dynamic conditions of flowing fluids. Field measurements are especially important for pH, dissolved oxygen, and alkalinity [85].

TABLE 6.13   Common Water Analysis Determinations

Determination	Produced Water and Other Waters	Injection Water	Cooling Water	Boiler Water
Alkalinity	X	X	X	X
Microbiological	X	X	X	
Barium		O		
Calcium	X	X	X	O
Carbonate	X	X	X	X
Carbon dioxide	X	O	O	X
Chloride	X	X	X	X
pH	X	X	X	O
Hydrogen sulfide		O		
Iron	X	X	X	O
Magnesium	X	X		O
Manganese	O			
Oxygen		O	O	O
Phosphate			O	O
Silica	X	X		X
Specific gravity	X	X	O	
Specific resistivity	X	X	O	X
Strontium		O		
Sulfate	X	X	O	X
Sulfite		O		O
Total dissolved solids	X	X	X	X
Zinc			O	

X, determination usually made.
O, determination occasionally made, e.g. manganese counts to correlate with iron counts for corrosion monitoring.
Source: From M. Davies and P. J. B. Scott [85]. Reproduced with permission of NACE International.

Gas Stripping and Vacuum Deaeration

Gas stripping implies that dissolved gases are removed from liquids using pressure reduction, heat, or an inert gas (stripping vapor – usually natural gas). Some processes use all three of these principles.

Most topside corrosion is due to the presence of oxygen. Vacuum deaerators and other thermal–mechanical means are used to remove dissolved gases, including oxygen, from liquids. These systems can effectively reduce the dissolved oxygen levels to 20–50 ppb. Further oxygen removal is then possible using oxygen scavengers, a form of corrosion inhibitors. This can reduce the oxygen level to the 10 ppb range. Issues with mechanical removal of dissolved gases include the initial capital costs and maintenance. Fouling with solids and bacteria can reduce efficiency, and defoamers may become necessary [87].

Corrosion Inhibitors

Corrosion inhibitors are substances which, when added to an environment, decrease the rate of attack by the environment [87–93]. Removal of oxygen, if present, with oxygen scavengers and adjustment of the pH to levels above 10 usually substantially reduces corrosion rates. While these approaches work in many aqueous environments, they are not practical for many production fluids, and the use of corrosion inhibitors, chemicals added to the environment in small concentrations, will often become necessary. These corrosion inhibitors will often reduce the corrosion rate to approximately 5–10% of the corrosion rate with no inhibitors.

The use of corrosion inhibitors was the main means of internal corrosion control in oil and gas production until the 1980s, when production from deeper, and consequently hotter, formations led to the increasing use of CRAs for environments where corrosion inhibitors will not work [87, 93].

Corrosion inhibition can be started or changed in situ without disrupting a production process. This is a major advantage over other corrosion control techniques, and it also means that the inhibitor chemistry or dosage rate can be changed as a field ages and sours or other conditions alter the corrosivity of the environment. Any change should be preceded by an extensive series of tests to ensure compatibility with the process fluid, the pipe work metallurgy, and the existing inhibitor. It has been observed that different inhibitors can "gum up" when mixed together, and obviously this has considerable implications for corrosion management.

There are many other chemical treatments used for oilfield production fluids, and corrosion inhibitors must be compatible with them. The most common compatibility problems are associated with hydrate inhibitors. Other chemicals used for scale and paraffin control, antifoaming agents, emulsions breakers, etc., also affect corrosion inhibitor performance, but they will be discussed only as they relate to corrosion control.

Types of Inhibitors   Corrosion inhibitors have been classified many ways, but one of the most common is into the following groups, based on how they control corrosion [94]:

- Adsorption or film-forming inhibitors
- Precipitation inhibitors
- Oxidizing or anodic passivation inhibitors
- Cathodic corrosion inhibitors
- Environmental conditioners or scavengers
- Volatile or vapor-phase inhibitors

These groupings and others are shown in Figure 6.68 [94].

Another possible classification is into organic and inorganic inhibitors. Most corrosion inhibitors used for oilfield applications are film-forming organic chemicals, but commercial multicomponent inhibitor packages often contain oxygen and H_2S scavengers and oxidizing agents in addition to the film-forming organic components.

Inhibitors do their work at low relative dosages (often expressed in ppm or quarts per 1000 barrels).

Most oilfield inhibitors work by forming hydrophobic films on metal surfaces. Filming amines, the first of these inhibitors to be widely used in oil and gas production, were developed in the 1930s. Many other organic corrosion inhibitors have been developed since that time. There are a wide variety of commercially available proprietary adsorbing inhibitors on the market.

They typically have hydrocarbon chains of C12–C18 with amine groups on the hydrophobic end and some other group on the opposite end [95, 96].

These thin films do not form new compounds on the surface and are considered to be chemisorbed or physisorbed – attached to the surface by relatively weak bonds having less energy than would be associated with chemical compound formation. These inhibitors work because one end of the relatively long-chain organic molecule is attracted to electrically conductive surfaces such as bare metals. The other end of the same molecule is either hydrophobic (it repels water) or oleophilic (it attracts oil). This means that the adsorbed inhibitor repels water and avoids water-wetting of the metal surface. This is shown schematically in Figure 6.69.

Halides, present in most oilfield waters, tend to increase the efficiency of these inhibitors by increasing adsorption of the slightly positive nitrogen groups present on the hydrophobic ends of these molecules [95]. Oxygen is an enemy of organic inhibitor films and can both penetrate films and interfere with film formation.

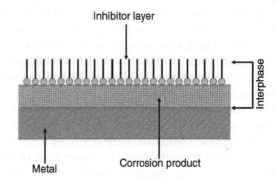

Figure 6.69   Adsorbing corrosion inhibitor with hydrophobic molecular tails away from the metal surface.

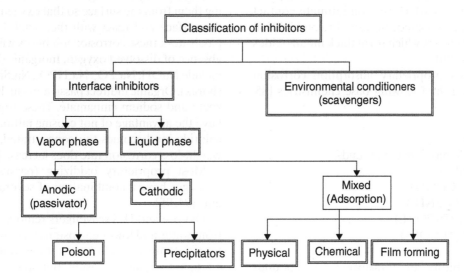

Figure 6.68   Corrosion inhibitor classifications [94]. *Source*: Reproduced with permission of John Wiley & Sons.

For this reason, oxygen is generally removed (or prevented from entering) oilfield waters that require inhibition with organic adsorbing inhibitors. Most types will not perform well in the presence of more than 0.5 ppm O_2, or, in some cases, as little as a few parts per billion. Because oxygen is the most important environmental chemical in determining corrosion rates, it is common to rely on oxygen removal, leak controls, and oxygen scavengers for topside corrosion control instead of the use of organic corrosion inhibitors.

Adsorbed inhibitor films are very thin and can be removed by mechanical shear forces if the fluid transport past the surface is too fast. The nature of these filming organic inhibitors is such that they will attach to most solid surfaces, and this means that fluid streams with sand or other solid particles will have reduced inhibitor efficiencies, because the inhibitor will also attach to sand and other particulate matter in the fluid stream [97]. Adsorbed inhibitors will also attach to any scale or corrosion products on the surface, and this also diminishes the corrosion-inhibiting effect by increasing the surface area for inhibitor attachment [97]. In older systems that have already corroded, it is essential to clean the surface, mechanically or chemically, before applying inhibitors. If rust or mineral scales are present, acid cleaning may be required. If acid cleaning is attempted, the equipment must be thoroughly rinsed and neutralized before returning the equipment to service.

Adsorbed corrosion inhibitors usually cover both anodes and cathodes. Because these inhibitors are based on organic chemicals, they normally cannot be used at elevated temperatures. The upper limit of their use depends on the chemical involved, but 200 °C (approximately 400 °F) is a common upper limit for the higher-temperature inhibitors, and most filming inhibitors lose effectiveness at much lower temperatures. These inhibitors, which rely on intimate contact with metallic surfaces, cannot be used in combination with oxidizing inhibitors, which form thick metal oxides on the surface [94–96].

There are many proprietary adsorption corrosion inhibitors based on the following base chemistries [85, 89–91, 96]:

- Imidazolines
- Quaternary ammonium compounds
- Amines (R-NH_2)
- Carboxyls (R-$COOH$)
- Thiourea (NH_2CSNH_2)
- Phosphonates (R-PO_3H_2)
- Benzonate ($C_6H_5COO^-$)

Precipitating inhibitors are film-forming compounds that form precipitates and cover the metal surface with mineral films that prevent water from reaching the metal surface. Silicates, phosphates, and molybdates fall into this category. They are used in process water and find limited use in oilfield fluids and production streams. Silicate inhibitors have the unusual property of being effective in already-corroded systems where most other corrosion inhibitors lose their effectiveness [88, 95]. Other precipitating inhibitors include calcium salts (calcium carbonate and calcium phosphate) and zinc salts (zinc hydroxide and zinc phosphate). Calcium compounds are widely used in potable water systems to maintain the pH of water at a high level (typically around pH 8–9) and with a slight oversaturation of calcium in the water so that any exposed surfaces will be covered with thin carbonate scales. This has been standard potable water treatment practice since the 1920s [2, 4, 89].

Passivating inhibitors that oxidize metal surfaces are commonly used in steam and water systems, but they are seldom used before effective hydrocarbon–water separation has occurred. They also tend to be ineffective in high-chloride waters like the majority of produced water systems.

Chromates are the most effective passivating inhibitors, but environmental concerns have limited their use, especially for any application where water discharge is possible. Alternatives to chromates are not as effective, although research continues on their development. At present, most non-chromate-oxidizing inhibitors are based on nitrites, which are considered to have fewer environmental problems than either chromates or phosphates. Bacterial decomposition of nitrites limits their use in open recirculating water systems. Molybdates and tungstates are also available. None of these oxidizers work in the presence of H_2S [94, 95].

Indirect passivators are alkaline chemicals that increase pH by reacting with hydrogen ions and removing them from the surface so that oxygen can adsorb onto the surface and react with the metal. Unlike the direct passivators, these corrosion inhibitors will not work in the absence of dissolved oxygen. Inorganic direct passivators include $NaOH$, Na_3PO_4, Na_2HPO_4, Na_2SiO_3, and $Na_2B_4O_7$ (borax). Organic indirect passivators include sodium benzoate and sodium cinnamate. These organic passivators have the advantage of not causing pitting corrosion if the chloride ion becomes too concentrated, but the general weight-loss corrosion rate does increase [94, 95].

Most proprietary oxidizing (passivating) inhibitor packages have a combination of several active ingredients [94, 95].

Oxygen and H_2S scavengers remove aggressive gases from water and lower corrosion rates. pH control is used to maintain water pH levels at controlled levels – high enough to limit corrosion but low enough to avoid unwanted scale deposits.

Oxygen scavengers do not work in acids and have no effect on pH. For this reason they are often used in conjunction with some form of pH adjustment, which is also necessary for both corrosion and scale deposition control. Sodium sulfite, ammonium bisulfite, and sulfur dioxide are examples of commercial oxygen scavengers, but others are also available [85]. Nitrites are often used for H_2S scavenging, which limits corrosion and also inhibits sulfate-reducing bacteria problems [85].

While oxygen scavengers are often combined with mechanical deaeration for large systems, the use of chemicals alone is sometimes justified for smaller systems [85].

At one time the boiler industry used hydrazine (N_2H_4) as the primary oxygen scavenger for boiler feedwater. Hydrazine had several advantages, including the fact that the by-products of its use were nitrogen gas and water. Unfortunately, hydrazine is carcinogenic, and the use of hydrazine has diminished in recent years.

Seawater and other water injection systems frequently use oxygen scavengers to control corrosion and, equally important, to minimize the possibility of microbial fouling of subsurface formations.

Most scavengers used in the oilfield are based on sulfites, bisulfites, or nitrites, but they are usually sold as proprietary chemical packages, with minimal identification of their chemistries, as either oxygen or H_2S scavengers. The H_2S scavengers will often raise the pH of water, and, if calcium carbonate scaling is a potential problem, they must be used in conjunction with scale inhibitors.

Batch processing of scavengers is possible, e.g. for drilling fluids, but continuous injection is more common.

Various chemicals are used to neutralize and buffer the pH of liquids. If an acid condenses from a liquid, e.g. gas condensate in wells or pipelines, the neutralizer must condense at the same temperature and pressure. This pH control is often necessary to prevent corrosion at low pHs and to prevent scale deposition, which can also lead to microbial corrosion, at higher pHs.

In boiler water systems it is common to use morpholine, an organic compound $O(CH_2CH_2)_2NH$, at ppm concentrations for pH adjustment. Morpholine is used for this purpose because its volatility is similar to water, and once it is added to water, the morpholine concentration becomes relatively evenly distributed in both the liquid and vapor phases. Hydrazine or ammonia oxygen scavengers are often used in conjunction with morpholine treatment [95, 96].

It is important to note that oxygen scavengers and oxidizing or passivating treatments work on opposite principles and the chemicals for these two purposes should not be used concurrently in the same system [95, 96].

Other environmental conditioning normally involves keeping the pH in an acceptable range. Low pHs promote corrosion, while high pHs lead to scaling.

Most vapor-phase corrosion inhibitors are low-molecular-weight amines that condense and form adsorbed films on metal surfaces. While some of these inhibitors have nitrites, which work as oxygen scavengers, most are merely amines. One example would be diethylamine, which, when used in sour gas, produces iron sulfide films on the surface that are protective in low-temperature, relatively dry gaseous environments [88].

Most commercial corrosion inhibitor packages are complex blends of many different chemicals, only a portion of which are the nitrogen-containing materials considered to be the primary film-forming chemicals. The "complex blends of many different chemicals" is the solvent system. In most cases there is only one, or two at most, active inhibitor ingredients. The solvent system is necessary to retain the inhibitors in uniform solutions, while at the same time giving the oil-soluble, water-soluble, or oil-soluble-water-dispersible characteristics needed for the particular system. It is the solvent systems that can cause damage to elastomers and man-made rubbers, not the active inhibitor chemistry. This is also the case with other organic based specialty chemicals such as demulsifiers, paraffin inhibitors, and asphaltene dispersants. The solvent system chemicals can be damaging to elastomeric seals and similar polymeric components of the system. NACE has issued a report on this problem, but testing to ensure that the problem does not exist is often necessary [98]. As a general recommendation, gaskets and seals made from tetrafluoroethylene (Teflon®) are usually impervious to any solvent attack [99].

Other compatibility problems are related to the use of hydrate inhibitors and other chemicals added to the system. It is common to test mixtures of these proposed chemicals to determine if one or the other chemical package will interfere with the performance of the other chemicals.

Inhibitor breakdown, e.g. at unforeseen elevated temperatures, can also lead to additional corrosion problems [100–103].

Application Methods Inhibitors are injected continuously (the preferred method) or in batch treatments, which may be necessary for some systems [93]. Most inhibitor injection systems are manufactured from 316L stainless steel, although some lines of PTFE or nylon are used if the pressures and temperatures are low.

Continuous injection is almost universally used except for downhole and pipeline applications where injection sites are difficult to establish and maintain. It is recommended that the corrosion rates be determined upstream of the inhibitor injection location so that the effectiveness of the inhibitor can be determined. Inhibitor injection rates are then adjusted so that acceptable corrosion rates are obtained at the end of the line. If corrosion rates are unacceptable high, then the injection rate can be changed. These changes are likely as a field ages and the corrosivity of produced fluids change.

The initial corrosion inhibitor dosage will usually be very high to satisfy inhibitor demand for the exposed metal and to ensure complete filming. Once this initial filming has been accomplished, the dosage rate is dropped to a minimum level necessary to maintain the film under operating conditions.

Corrosion inhibitor batch treatments are used for downhole tubing and for subsea pipelines. They must have low solubility in the system fluids. Batch treatments are characterized by short periods of high inhibitor dosages followed by long periods where the inhibitor level is relatively low. This affects the way these chemicals react with seals and other polymeric materials in the system [99].

Batch treatments involve a relatively short period of inhibitor feeding followed by a long period of nonfeeding, where one of the following systems operates:

- The inhibitor film is persistent and lasts for a relatively long time.
- A reservoir of inhibitor slowly feeds into the system needing corrosion control.

Batch treatments will last from one week to several months between treatments.

Tubing displacement batch treatments are used in wells where a batch of corrosion inhibitor is pushed down into the tubing to the bottom of the well. The well is shut in for several hours and then returned to production. This technique is used in wells with packers and with gas-lift wells.

Squeeze treatments are similar to tubing displacements except the inhibitor is displaced beyond the bottom of the tubing and into the geological formation. During this displacement the system seals are subjected to high concentrations of inhibitor. Once the well is returned to production, the inhibitor concentration slowly lowers as inhibitor is washed from the formation by the produced fluids. Squeeze treatments are usually done monthly or semiannually.

Batch treatments are also common in pipelines, especially subsea pipelines where access is limited. Pigs force inhibitor into the pipeline and this coats the pipe wall [104, 105]. This can be done with spiral foam pigs, gel pigs, or special pigs designed to spray inhibitor onto the top of the pipeline interior (Figure 6.70). A typical pig run in a subsea pipeline is intended to provide corrosion inhibitor that will last for a month.

Testing and Monitoring   Laboratory and field testing are commonly used to determine which of many possible corrosion inhibitor systems should be used in any given application. Testing in the laboratory can reduce the number of inhibitor packages under consideration by quickly eliminating those deemed to be unsuitable. Important parameters to be tested include shear testing – the ability of the inhibitor to "stick" to the metal surface when liquids are moving parallel to the surface at high velocities – which helps determine the inhibitor persistency [106]. Other tests are intended to measure shear testing at higher flow rates, partitioning of inhibitors between water and hydrocarbon phases, etc. [107–110]. It is best if the tests are conducted using the actual fluids from the field, because the presence of minor variables, e.g. organic acid contents, in crude oil will affect the ranking of prospective corrosion inhibitors.

While a variety of standardized laboratory tests are available, no consensus exists on their relevance to actual field performance [111, 112]. The lack of confirmation

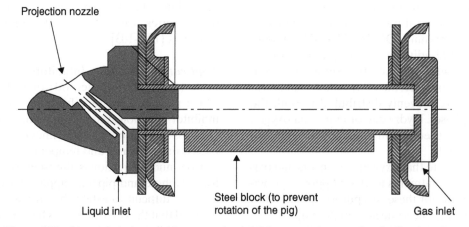

Figure 6.70   Venturi pig for spraying corrosion inhibitor onto the top of a pipeline interior to control top of line corrosion [104, 105]. *Source*: Reproduced with permission of NACE International.

of results between different laboratories conducting supposedly comparable tests means that laboratory screening tests cannot be relied upon to produce definitive answers on the best corrosion inhibitors for any given field. This must be determined by field tests.

Once preliminary laboratory tests have narrowed the possibilities, prospective inhibitor packages can be compared by field testing. In the past this has been done by using exposure coupons. Unfortunately this is a very expensive and time-consuming process. Reports indicate that the total time to conduct laboratory and field testing can be substantially reduced using enhanced electrical resistance probes to produce data on corrosion inhibitor performance in approximately 96h for each inhibitor tested in contrast to the 36 days previously required. Much of the reduction in time for field testing is due to the introduction of modern, quick-response electrical resistance (ER) probes that can indicate changes in corrosion rates due to upsets in a matter of hours instead of days [113].

It is important to note that all of these tests, both laboratory and field, require the use of replicate samples and testing at various inhibitor dosage levels, e.g. steps at 15, 30, and 50 ppm of inhibitor. An inhibitor that works well at one dosage might not be demonstrably better at a higher dosage.

Corrosion inhibitor performance monitoring is necessary to confirm that the corrosion inhibitors and dosage rates that have been selected are appropriate for the field in question. Corrosion coupons pulled every 90 days have become the standard method of ensuring that adequate corrosion control has been established. Unfortunately, these coupons cannot identify when in the 90 days most of the corrosion has occurred. ER probes should be used to supplement the coupon data, and it is recommended that two access fittings, one for flush-mounted coupons and one for flush-mounted ER probes should be mounted, normally at the six o'clock position, for each monitoring station in crude oil lines. In multiphase and gas lines, the 12 o'clock position is also vulnerable and should also be monitored.

It is very important that coupons be located in locations where maximum corrosion rates can be expected. Suggested locations for coupon locations along a wet gas piping system are shown in Figure 6.71 [113]. The ideas represented in this figure should be considered in placing corrosion coupons, ER probes, and other monitoring devices – they should be placed at locations where water accumulation and accelerated corrosion are most likely. For gas and multiphase pipelines, the possibility of top-of-line corrosion must also be considered.

Unfortunately, the exact locations where corrosion rates are likely to be the highest are hard to predict. The most aggressive locations will change, especially for multiphase systems, as temperatures and pressures change and fields age, altering the composition and production rates of produced fluids. Monitoring with coupons and ER probes is no substitute for inspection – the two techniques are complementary, and one cannot substitute for the other. NACE provides suggestions on the corrosion rates to be expected, as shown in Table 6.14 [113].

Unfortunately, too many organizations spend so much time collecting coupons and reporting weight loss data that they forget to question whether the data has been collected at the correct locations and what it means. Organizations with thousands of coupons showing that corrosion is under control have still had unfortunate leaks due to corrosion at locations where corrosion could have been predicted and inspections could have identified the problems. This is especially true as fields age and production rates decline, leading to more corrosive conditions, especially at locations like those shown in Figure 6.71. The practice of having the same organization that applies corrosion inhibitors also conduct monitoring on the effectiveness of the inhibitor program can lead to unnecessary difficulties.

CATHODIC PROTECTION

Cathodic protection is an electrical means of corrosion control where the structure to be protected is made the cathode in an electrochemical cell. Oxidation of the electrochemical cell is shifted to anodes leaving the structure to be protected as a cathode with a net reduction reaction, which suppresses the corrosion rate. While many oilfield structures are cathodically protected, this discussion will emphasize pipelines, the most common in oil and gas production. The principles discussed here for pipelines apply to any cathodically protected structure.

Figure 6.72 shows a simple cathodic protection system for a buried pipeline [114]. The pipeline is connected by a lead wire to a buried magnesium anode, which corrodes at an accelerated rate, thereby providing protective cathodic current to the pipeline.

The combination of protective coatings as the primary means of corrosion control and cathodic protection as a supplemental secondary means of corrosion has proven most economical for most pipelines and similar buried or submerged structures. The electrical current demands of the cathodic protection system are determined by the effectiveness of the protective coating, and they increase as the protective coating ages and degrades [115, 116].

Cathodic protection allows carbon steel structures, which have limited natural corrosion resistance in many

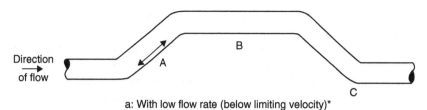

a: With low flow rate (below limiting velocity)*

A. Water oscillates – corrosion accelerated
B. Corrosion not accelerated
C. Water impinges at C – corrosion accelerated with higher flow rate
   (above limiting velocity)*
   Corrosion most severe at impingements

*Limiting Velocity – velocity above which erosion damage can be expected.

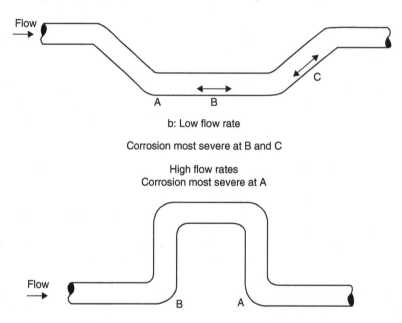

b: Low flow rate

Corrosion most severe at B and C

High flow rates
Corrosion most severe at A

c: Vertical rleer in gas line carrying small volume of water

A. In high-velocity flow, water impinges on Points A and B, accelerating corrosion
B. At low velocity, water accumulates in upstream leg of loop, cascades down in
   downstream loop, Impinging at Point A

Figure 6.71   Locations for coupon installation in a wet gas piping system. *Source*: NACE RP0775 [113]. Reproduced with permission of NACE International.

TABLE 6.14   Qualitative Characterization of Carbon Steel Corrosion Rates for Oil Production Systems [113]

	Average Corrosion Rate		Maximum Pitting Rate	
	mm yr^{-1}	mpy	mm yr^{-1}	mpy
Low	<0.025	<1.0	<0.13	<5.0
Moderate	0.025–0.12	1.0–4.9	0.13–0.20	5.0–7.9
High	0.13–0.25	5.0–10	0.21–0.38	8.0–15
Severe	>0.25	>10	>0.38	>15

mm yr^{-1}, millimeters per year; mpy, mils per year.
Source: Reproduced with permission of NACE International.

oilfield environments, to perform with little or no corrosion, provided the cathodic protection system is designed, installed, and maintained correctly. It was discovered in the nineteenth century and used on British naval vessels. Its use on pipelines dates to the early 1900s work of R. Kuhn and coworkers [117, 118], who used cathodic protection to lower the corrosion rates of buried onshore pipelines in Louisiana.

The corrosion (oxidation) reaction on a buried steel structure is:

$$Fe \rightarrow Fe^{+2} + 2e^- \tag{6.1}$$

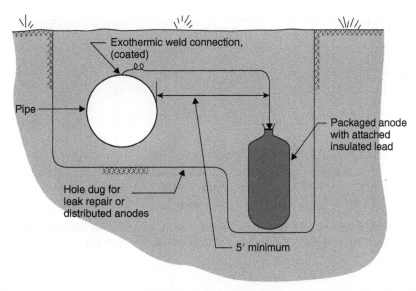

Figure 6.72 Single packaged anode buried in soil to protect a buried pipeline. *Source*: R. Bianchetti [114]. Reproduced with permission of NACE International.

The buried anode undergoes a similar reaction:

$$Mg \rightarrow Mg^{+2} + 2e^- \qquad (6.2)$$

Reduction reactions depend on the pH of the water in the environment, but are usually:

In acids:

$$2H^+ + 2e^- \rightarrow H_2 \qquad (6.3)$$

Or

$$O_2 + 4H^+ + 4e^- \rightarrow 2H_2O \qquad (6.4)$$

In neutral or basic solutions:

$$O_2 + 2H_2O + 4e^- \rightarrow 4OH^- \qquad (6.5)$$

Except in strong acids (pH < ~3), the concentration of oxygen is most likely to predominate, and most of the reduction reaction on a cathode will be oxygen reduction.

In order for cathodic protection to work, all components of an electrochemical cell – anode, cathode, electrolyte, and return circuit – must be present. The absence of any one of these will prevent successful cathodic protection. Sometimes people forget this, for example, with attempts to protect high-temperature pipelines where the environment is so hot that water evaporates and no electrolyte is present to transmit electric current.

How Cathodic Protection Works

The effectiveness of cathodic protection can be expressed in many ways. The first arguments for its use on pipelines and other oilfield structures emphasized the reduction in leaks due to external corrosion. This type of data was used by R. Kuhn, who presented data in the 1930s, showing a major reduction in leaks on natural gas pipelines due to the use of cathodic protection [117, 118]. While cathodic protection had been described in the 1800s and used to protect nails holding copper sheathing to the bottom of British ship hulls, Kuhn is generally recognized as the first engineer to use cathodic protection in the United States, where he applied it to controlling corrosion of cast iron natural gas pipelines starting in 1913 [119].

Figure 6.73 shows the reduction in leaks on a major pipeline system due to the application of cathodic protection [120]. This figure shows data from the 1940s and later, when the idea of cathodic protection started to gain widespread attention along the Gulf Coast of the United States. Discussions among Gulf Coast pipeline operators led to the formation in 1943 of the organization that has become NACE International, the largest organization devoted to corrosion control [119, 120].

The Evans diagrams (potential vs. log current) in Figures 6.74 and 6.75 illustrate the principles of cathodic protection. The corrosion (oxidation) reaction is shown by the line that goes up and to the right in Figure 6.74. The intersection of this oxidation reaction with the reduction reaction for dissolved oxygen in the electrolyte determines the corrosion rate. In Figure 6.74 the corrosion rate is reduced by over two orders of magnitude by the application of cathodic protection. Note that the

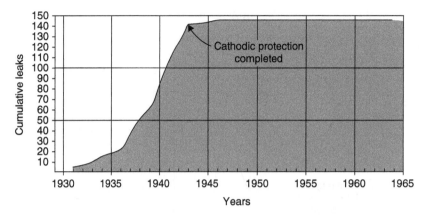

Figure 6.73 Effectiveness of cathodic protection in stopping the development of pipeline leaks. *Source*: J. Beavers [120]. Reproduced with permission of NACE International.

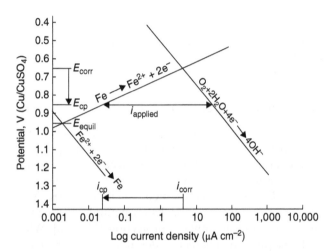

Figure 6.74 Evans diagram showing the principles of cathodic protection. *Source*: J. Beavers [120]. Reproduced with permission of NACE International.

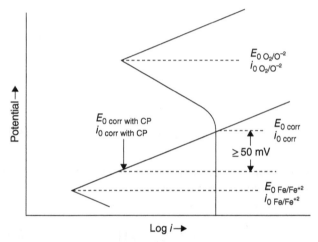

Figure 6.75 Evans diagram showing cathodic protection with the reduction reaction limited by oxygen diffusion control.

potential of the cathodically protected iron, shown at $-0.85\,V$ ($Cu/CuSO_4$), is above the equilibrium potential and that the corrosion rate, while reduced by more than two orders of magnitude, is not zero.

Figure 6.74 is used in a standard reference work on pipeline corrosion to explain how cathodic protection works. The same idea has been published previously [116]. It is deliberately simplistic and slightly unrealistic, because the corrosion rate of most buried or submerged steel is controlled by the diffusion of oxygen to the metal surface and not by hydrogen ion reduction. This means that the reduction reaction, shown as a slanting straight line in Figure 6.74, is more likely to be a vertical line indicating that the reaction is under oxygen diffusion control (concentration polarization) as shown in Figure 6.75.

Both figures make the same points:

- Cathodic protection substantially reduces the oxidation current (corrosion) of the structure being protected.

- Cathodic protection does not stop corrosion – it reduces the corrosion rate, hopefully to a negligible, or at least an acceptable, rate.

Note that neither diagram suggests that the potential after cathodic protection is below the equilibrium potential where the current for oxidation of iron is equaled by the current for reduction of iron ions.

While Evans diagrams, like those shown above, are used to explain cathodic protection, they were not used by the people who developed these techniques in the early mid-twentieth century.

In the oilfield, cathodic protection is applied to pipelines in soil and water environments, to offshore structures, and to process and storage vessels.

Cathodic protection is usually used in conjunction with protective organic coatings. The protective coating is considered the primary means of corrosion control, and the cathodic protection system is sized to provide corrosion control at defects in the coating. As the coating ages and becomes less protective, the demands for electrical current from the cathodic protection system increase. This combination of protective coatings and cathodic protection has become standard on most oilfield equipment. The exceptions are most offshore structures and some process equipment, which are often used in the uncoated state. Cathodic protection causes a pH shift, shown in Equations (6.3)–(6.5), to higher pHs, where most minerals are less soluble. The pH shift produces precipitates of calcareous deposits, usually calcite but sometimes other minerals, on the protected surface. These mineral deposits reduce the exposed metal surface and act as protective coatings on offshore structures [121, 122]. Figure 6.76 shows calcareous deposits caused by cathodic protection on the node of an offshore platform in a warm shallow sea. Whitish deposits cover most of the surface, but even the darker areas are covered with scale so hard that it is difficult to remove it with a hammer and chisel.

The oxidation at anodes in cathodic protection systems alters the potential of the protected structure and shifts it in a cathodic direction. While several reduction reactions are possible on cathodically protected surfaces, the most common reaction is the reduction of dissolved oxygen or, if the pH is low or the negative potential is large, the evolution of hydrogen gas.

Note the logarithmic slope of the oxidation (corrosion) rates in Figures 6.74 and 6.75. Neither of these figures implied the total elimination of corrosion. Small shifts of potential produce drastic reductions in corrosion rates, and it has been reported that a cathodic potential shift of −70 to −100 mV will reduce the corrosion rate to 10% of the original corrosion rate [123]. Cathodic protection *reduces* but does not eliminate corrosion. "A major activity of a CP engineer is to determine the actual level of CP required to reduce the corrosion rate to an acceptable level" [121].

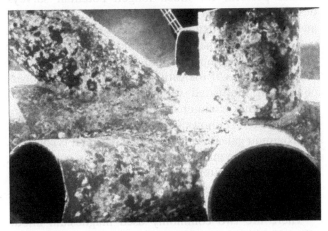

Figure 6.76 Calcareous deposits formed by cathodic protection on an offshore platform node. *Source*: Photo courtesy J. Smart.

Types of Cathodic Protection

There are two types of cathodic protection, galvanic or sacrificial anode cathodic protection and impressed current cathodic protection.

Galvanic Anode Cathodic Protection (Also Called Sacrificial Anode Cathodic Protection) Figure 6.72 showed a simple galvanic cell using a buried magnesium anode to protect a buried steel pipeline. Table 6.15 shows the potentials of selected metals in soil. Carbon steel and cast iron are naturally cathodic to most other structural metals. They are, however, cathodic to magnesium, aluminum, and zinc – the metals used for galvanic anodes.

Figure 6.72 showed a simple single anode attached to a pipeline. The anode, which corrodes to protect the structure, was located at a "remote" location, as far away from the pipeline as is practical. The purpose of this remote location was to ensure that the current from the anode was distributed to "ground," so that current was not wasted near the anode wire-lead connection location. Current then came from the "ground" to the holidays in the protective coating on the pipeline instead of being concentrated near the anode wire-lead connection to the pipeline. Modern pipeline coatings are much better, and it is now common to bury the galvanic anodes in the same trench as the new pipeline, because it is unlikely that significant coating holidays will be located near the anode when it is first buried. The remote anode

TABLE 6.15   Galvanic Series of Metals in Soil

Material	Potential (Volts CSE)[a]
Carbon, graphite, coke	+0.3
Platinum	0 to −0.1
Mill scale on steel	−0.2
High silicon cast iron	−0.2
Copper, brass, bronze	−0.2
Mild steel in concrete	−0.2
Lead	−0.5
Cast iron (not graphitized)	−0.5
Mild steel (rusted)	−0.2 to −0.5
Mild steel (clean and shiny)	−0.5 to −0.8
Commercially pure aluminum	−0.8
Aluminum alloy (5% zinc)	−1.05
Zinc	−1.1
Magnesium alloy (6% Al, 3% Zn, 0.15% Mn)	−1.6
Commercially pure magnesium	−1.75

[a] Typical potential normally observed in neutral soils and water, measured with respect to copper sulfate reference electrode.
Source: Beevers [124]. Reproduced with permission of NACE International.

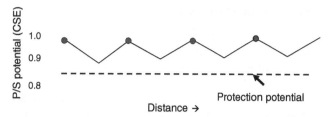

Figure 6.77 Potential plot along a pipeline with galvanic anode cathodic protection.

approach is still appropriate for repair construction, where the coating is assumed to have aged except at the small repaired location.

Galvanic anodes are typically supplied with approximately 3–5 m (10–15 ft) of lead wire, which is sufficient to locate the anodes at "remote earth" in most environments.

Galvanic anodes are often installed in "distributed anode" configurations. One anode protects a given length of pipe and, where the IR drop down the pipeline is too much and inadequate protection is available, another anode is located. The critical location is midway between the two anodes. This is shown in Figure 6.77. Note that the potential varies from approximately 1 V to somewhat more than 0.85 V. Since all voltages are negative relative to copper/copper sulfate electrodes (CSE), the potentials are plotted with larger negative numbers on top. This is in accordance with convention for cathodic protection but seems backward compared with conventional engineering practice. As long as the potential remains more negative than the protection potential (above the dotted line in Figure 6.77) corrosion on the pipeline will be minimized and the structure will be cathodically protected.

The potential profile in Figure 6.77 shows that the pipeline is more protected than necessary near the anodes and that the potential decays as the distance from the anode increases. The spacing between anodes is determined by the IR drop down the pipeline and by the current demand on the pipeline exterior. A typical spacing between anodes is of the order of hundreds of meters (yards), but there are wide variations depending on protective coating quality, diameter of the pipeline, and corrosivity of the environment.

Because the voltage of galvanic anodes is limited, the spacing between anodes is often the design-limiting parameter. This leads to increased construction costs on long-distance pipelines, and galvanic anodes are seldom used onshore for long-distance pipelines. They are used on small structures, in low-resistivity environments – where they can be relied upon to work – and for protecting "hot spots" where corrosion is intensified.

Table 6.16 lists some advantages and limitations of galvanic anode systems. The typical design life of

TABLE 6.16   Advantages and Limitations of Galvanic Anode Cathode Protection Systems

Advantages	Limitations
No external power required	Limited driving potential
Easy to install	Lower/limited current output per anode
Simple – can be installed and maintained with minimally trained personnel	May not work in high-resistivity environments
Minimum maintenance	High cost per ampere-year of current generated
Installation can be inexpensive if installed during construction	Installation can be expensive on long pipelines – requires many installations
Relatively uniform distribution of current	
Not a source of stray current	
Cannot be turned off – always active until anodes consumed	

TABLE 6.17   Primary Uses of Galvanic Anodes in Oilfield Applications

Magnesium
On-shore buried structures
Process equipment
Zinc
Marine pipelines
Process equipment
Freshwater ballast tanks
Ship hulls
Aluminum
Offshore structures
Limited use in process equipment

onshore galvanic anodes is 5–10 years, although some anodes perform for much longer. Potential surveys, described below, are necessary to determine when the anodes have neared or reached the end of their useful life.

Three different metals are commonly used for galvanic anodes – magnesium, zinc, and aluminum. Carbon steel is sometimes used for cathodic protection on process equipment fabricated with CRAs, but carbon steel anodes are not commercially available from most suppliers. Table 6.17 shows typical applications for each of the common galvanic anode metals. While it is common to refer to these materials by their primary constituent, all these anode materials are alloyed to ensure that they will reliably corrode and produce the necessary current for cathodic protection.

There are two commonly used magnesium anode alloys. The high-potential alloys have a native potential in soil of approximately –1.80 V relative to copper/copper

TABLE 6.18 Electrochemical Properties of Magnesium

A-h lb^{-1} theoretical	1000
Current efficiency (based on ~30 mA ft^{-2})	50%
A-h lb^{-1} actual	500
Consumption rate, lb-A-yr	17.4
OCP	V to Cu/CuSO$_4$
AZ-63 (H-1) alloy	−1.50 to −1.55 V
High-potential alloy	−1.75 to −1.77 V

Source: From T. May [125].

sulfate, and the H1 or AZ-63 anodes have a potential of −1.55 V. Table 6.18 lists important properties of these alloys [124–127].

Most of the cost of galvanic anode installation is labor and excavation. Thus the onshore installation costs for galvanic anodes are essentially the same for all anode sizes. The 17-lb (7.7 kg) anode is the most commonly used size in North America.

Most applications use the high-potential magnesium–manganese alloy developed by Dow Chemical Company in the 1950s [124–126].

Quality control problems with magnesium anodes have occurred in recent years, and many anode suppliers have been forced to conduct quality control testing on magnesium anodes [124–126].

Anode efficiency for magnesium anodes is 50% under normal conditions. This means that half of the electrical current produced by the corrosion of the anodes will be available for cathodic protection. The efficiency is less at low pHs.

Magnesium anodes are the most reliable of all galvanic anode materials – they will corrode in almost any wet environment. Nonetheless they are normally supplied with prepackaged backfills. The most common magnesium anode, 17 lb (7.7 kg), will weigh about 45 lb (20 kg) when the weight of the prepackaged backfill is added. This backfill is intended to provide a low-resistivity and wet environment to the anode. Most backfills are a combination of a hygroscopic soil (gypsum and/or bentonite clay) and ionic salts (calcium chloride).

Zinc anodes were used as early as 1824 to protect the nails holding copper cladding to the bottom of wooden ships. Alloy additions of aluminum and cadmium increase the efficiency of modern zinc anodes and also produce more uniform corrosion.

The potential of zinc in most soils is assumed to be −1.1 V CSE. This voltage is much lower than that for magnesium, but the efficiency of zinc anodes is generally considered to be approximately 90%, so much more of the electricity generated by corrosion of the anode is available for cathodic protection.

Unlike magnesium, zinc will not corrode in many soils, and the use of zinc in soils has been restricted to low-resistivity soils (<1000–2000 Ω-cm depending on the authority in question). Recent quality control problems with magnesium anodes have caused many organizations to use zinc anodes onshore in applications where they would not have been considered in previous years. Proponents of the use of zinc anodes for pipeline cathodic protection argue that both zinc and magnesium can produce adequate current to polarize pipelines having the high-quality coatings that have been introduced in recent years. Magnesium is alleged to corrode too fast, wasting electricity, whereas zinc will provide enough current and last longer. The same types of prepackaged backfills that are used for magnesium are supplied for zinc anodes. The backfills produce wet soil environments having resistivities in the hundreds of Ω-cm (ohm-cm). This low-resistance environment should corrode both zinc and magnesium. This practice of substituting zinc for magnesium is controversial and should only be used with careful monitoring to ensure that the desired cathodic protection is achieved. Many operating companies continue to avoid the use of zinc anodes except in soils with naturally low resistivity (usually high moisture swampy or coastal soils).

Table 6.19 summarizes the characteristics of zinc anodes. Temperatures above 60 °C (140 °F) have been found to cause zinc to be cathodic to carbon steel in some freshwater environments. This should not be a problem in seawater and other high-chloride environments [2, 127, 128].

For marine applications zinc anodes, which last longer than magnesium, are less efficient than aluminum anodes. Zinc should only be used in brackish water when the chloride concentration falls below approximately 6–10 ppt (parts per thousand) compared with approximately 35 ppt in open seawater [128]. Under these conditions, aluminum may not corrode and produce the necessary current.

Aluminum anodes have become the standard galvanic anode material for use in offshore applications. Early aluminum anode alloys used mercury as an "activator," but environmental concerns have caused these mercury-activated anodes to be replaced with indium-activated anodes. Table 6.20 shows the two types of aluminum anodes most commonly used offshore. These anodes cannot be used in freshwater applications, because they will passivate and become inactive if the salt content (commonly expressed as chloride concentration) is too low. This is also why aluminum anodes are not used on ships – they passivate in harbors and will not work once they are back in the ocean. Aluminum anodes can also be used in oilfield process equipment where produced water has a high salt content. Special alloying modifications are also available from some suppliers for use in cold water.

TABLE 6.19   Zinc Anode Characteristics

Element	Mil Spec A-18001K	ASTM B-418-01 Type I	ASTM B-418-01 Type II
Al	0.10–0.50%	0.10–0.50%	0.005% max
Cd	0.025–0.07%	0.025–0.07%	0.003% max
Fe	0.005% max	0.005% max	0.0014% max
Pb	0.006% max	0.006% max	0.003% max
Cu	0.005% max	0.005% max	0.002% max
Si	0.0125% max		
Zn	Remainder	Remainder	Remainder
Use	Seawater and brackish water ($T < 50°$ C) [120 °F]		Soil and freshwater
Nominal potential	−1.10V CSE		
Efficiency	90%		
Capacity	738 A-h kg^{-1} (335 A-h lb^{-1})		
Consumption	11.9 kg (A-yr)$^{-1}$ (26.2 lb (A-yr)$^{-1}$)		

TABLE 6.20   Aluminum Anodes for Offshore Use

Chemical Composition

Element	Mercury Activated	Indium Activated
Zn	0.03–0.50	2.8–3.5
Si	0.14–0.21	0.08–0.2
Hg	0.035–0.048	—
In	—	0.01–0.02
Cu	<0.01	<0.01
Fe	<0.12	<0.12
Other each	<0.02	<0.02
Al	Remainder	Remainder

Electrochemical Properties

Use	Open Seawater	Seawater/Mud
Potential (Cu/CuSO$_4$)	−1.05	−1.15
Nominal efficiency (%)	95	85
Nominal A-h lb^{-1}	1280	1150
Capacity – Seawater	2830 A-h kg^{-1} 1280 A-h lb^{-1}	2530 A-h kg^{-1} 1150 A-h lb^{-1}
Consumption – Seawater	3.10 kg (A-yr)$^{-1}$ 6.83 lb (A-yr)$^{-1}$	3.48 kg (A-yr)$^{-1}$ 7.83 lb (A-yr)$^{-1}$
Capacity – Mud	—	2180 A-h kg^{-1} 990 A-h lb^{-1}
Consumption – Mud	—	4.02 kg (A-yr)$^{-1}$ 8.87 lb (A-yr)$^{-1}$

There have been isolated cases where aluminum did not work offshore. These instances have been traced back to freshwater flushing from rivers into the ocean. This has happened as far as 150 km (100 miles) offshore in the Gulf of Mexico due to the freshwater flow from the Mississippi River. Freshwater is less dense than saltwater, so the tops of the water column near rivers may be fresh, while the deeper locations (>30 m or 100 ft) may be salty enough for efficient use of aluminum anodes.

Aluminum anodes for offshore platforms are available in sizes up to 500 kg (1200 lb) and larger.

Most of them are cast with a steel core, which can be welded to the platform leg or other structure.

Carbon steel anodes are occasionally used for cathodic protection of CRAs in process equipment – heat exchangers with corrosion resistant alloy tubing and protective-coated carbon steel headers and water boxes. Carbon steel galvanic anodes are used to increase the

(a) (b)

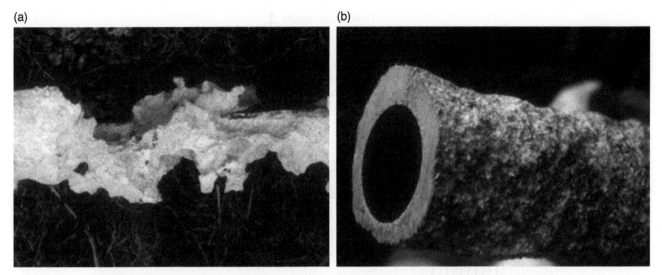

Figure 6.78 Corrosion of aluminum anodes. (a) Uneven corrosion of aluminum anode due to improper foundry procedures leading to segregation of alloying elements. (b) Uniform corrosion of aluminum anode as the result of proper foundry practice [121]. *Source*: Reprinted with permission of ASM International®. All rights reserved. www.asminternational.org.

area ratio of exposed carbon steel and lower the corrosion rates of the structural members. This is a relatively unusual application, and most anode suppliers do not carry carbon steel or iron anodes.

All three of the anode materials are used in process equipment, depending on the application and the conductivity of the environment.

Backfill materials are commonly used on shore to guarantee that galvanic anodes will corrode and provide the necessary current to protect the structure. These are usually supplied in water-permeable cloth bags with enough prepackaged backfill soil to more than double the weight of the metal anode. Most galvanic anode backfills contain gypsum, bentonite clay, and an ionic salt such as calcium chloride or sodium sulfate. The minerals in the backfill are hygroscopic and absorb moisture if it is available. They are also ionic and have low resistivity to ensure that the backfill will be corrosive whenever wetted. Anodes are sold with prepackaged backfills in cloth bags. Backfill is also sold in 50 lb (23 kg) bags for use with anodes shipped without backfill.

Prepackaged anodes are shipped with plastic wrapping to prevent them from becoming moist and corroding prior to installation. It is unfortunate that many prepackaged anodes are installed with this plastic intact, because the installation crews do not understand the purpose of the various plastic and cloth wraps.

Recent problems with the quality of magnesium anodes have caused many suppliers to institute quality control programs to ensure that the anodes will perform as expected [124–126]. NACE International and other organizations provide guidance on quality control test

procedures [129, 130]. Many organizations develop lists of quality-approved vendors, but this has not always worked, as different sources of anode materials come on the market and shortages in supplies from traditional vendors develop. Figure 6.78 shows the results of improper and proper foundry practice on the corrosion of aluminum anodes [121]. Similar patterns have been reported on other alloy systems.

Impressed Current Cathodic Protection (ICCP) When large currents are needed or high electrolyte resistivity prevents the use of galvanic anodes, the protective current for cathodic protection is supplied by an ICCP system similar to the one shown in Figure 6.79. The cathodic connection to the pipeline is identical to that shown in Figure 6.72 for galvanic anode cathodic protection. The buried pipeline and the anodes are both connected to an electrical rectifier, which converts alternating current to direct current, and imposes cathodic potentials on the structure and anodic potentials on the anode bed.

Unlike galvanic anodes, impressed current anodes need not be naturally anodic to carbon steel, and they usually are not. Most impressed current anodes are made from nonconsumable anode materials that are naturally cathodic to steel. The anodes are intended to serve as sites for oxidation of a component of the environment, usually oxygen from molecular water, and are not intended to oxidize the anode itself. Even though they are intended to be nonconsumable, these anodes do degrade with time. Because they are naturally cathodic to steel, they would accelerate corrosion if they were directly connected to the structure they are intended to protect.

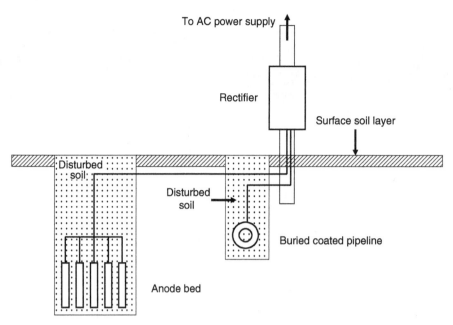

Figure 6.79   Impressed current cathodic protection of a buried pipeline.

It is important to always attach the leads from the rectifier to the proper terminals. The anode ground bed leads should always be attached to the positive terminal of the power supply. The negative terminal is always connected to the structure to be protected. Confusion on this point can result in impressed current cathodic protection systems being connected improperly. This causes increased corrosion on the structure rather than the intended reduction in corrosion rates.

The most common oxidation reactions on impressed current anodes are oxidation of oxygen from the water by one of the two following reactions:

In acids:

$$2H_2O \rightarrow O_2 + 4H^+ + 4e^- \tag{6.6}$$

In neutral or basic solutions:

$$4OH^- \rightarrow O_2 + 2H_2O + 4e^- \tag{6.7}$$

Note that the above two reactions are merely the reverse of Equations (6.4) and (6.5), the most common reduction reactions on a cathodically protected surface. All oxidation reactions lower the pH (acidify) the environment, so Equation (6.6) is the more likely oxidation reaction.

If any chloride ions are in the water, then chorine evolution can also happen:

$$2Cl^- \rightarrow Cl_2 + 2e^- \tag{6.8}$$

All of these reactions produce strong oxidizers, which can bleach or oxidize any organic materials nearby. The degradation of early impressed current anode lead wire insulation was once a problem, but modern lead wire insulation is much more resistant to this oxidation. Chlorine gas is also poisonous and care must be taken to vent this gas properly in cases of chlorine evolution.

The anodes used in ICCP are intended to be non-consumable, but oxidation of these materials does occur to a limited extent, and care must be taken to operate these anodes at recommended voltages and current densities to prevent premature degradation.

The wiring and connections of the ICCP system must totally isolate the system from the environment. Any exposed metal becomes part of the ICCP circuit and can lead to premature system failure.

Most ICCP uses electric current from a local power source connected through a rectifier, which changes alternating current (AC) to direct current (DC) [131]. In locations where conventional electric power is not available solar cells, batteries, thermoelectric generators, and other DC power sources have been used. At one time windmills were used in isolated locations, but the maintenance requirements on these mechanical systems have caused them to be replaced in many locations by solar cells [131–134].

Figure 6.80 shows a typical cathodic protection rectifier. Most rectifiers will have lead wire connections to the anode bed and to the protected structure as well as connections to an AC power source plus controls and displays to indicate power output, voltages, and current. They will also have lightning arrest capabilities and

Figure 6.80 A typical cathodic protection rectifier used for onshore cathodic protection. *Source*: Photo courtesy NACE International.

other safety features and be mounted in protective casings to protect them from the weather, wildlife, and vandalism. All of this costs money, and a typical rectifier, plus installation costs, will run into the thousands of dollars. For these reasons, ICCP is normally limited to situations where large amounts of current are needed; otherwise galvanic anodes would be cheaper.

Most of the cost of rectifier installation is for labor and installation; therefore it is common to use rectifiers somewhat larger than the measured or calculated current requirements for the installation would dictate. It is much cheaper to regulate the output of a rectifier than it is to reinstall a larger rectifier if the current demands cannot be met by the existing system. The total cost of installation also leads to the common use of an "anode bed" (also called a "ground bed") for ICCP systems. While ICCP anodes are intended to be nonconsumable, they have current density limitations. Anode beds with dozens, even hundreds, of anodes are not unusual. The purpose of these large anode beds is to allow for the use of higher-current output rectifiers, which minimizes the cost of cathodic protection by allowing one rectifier to supply cathodic protection current to large cathode surface areas. It is common for one rectifier/ground bed to provide cathodic protection for several miles or kilometers of buried pipeline. Galvanic anodes, which would need to be placed at hundreds of meters/yards intervals, would be much more expensive for this kind of application.

A typical ICCP system for a pipeline would include an AC-powered rectifier with a maximum rated DC output of between 10 and 50°A and 50V.

There are a number of anode materials used worldwide for ICCP. In relative order of importance, they are:

- High-silicon cast iron
- Graphite
- Mixed metal oxide
- Precious metal clad (platinum)
- Polymer
- Scrap steel
- Lead alloy

The above anodes are sometimes classified into massive anodes:

- High-silicon cast iron
- Graphite
- Scrap steel
- Lead alloy

and dimensionally stable anodes, which tend to be much smaller and less robust:

- Mixed metal oxide
- Precious metal
- Polymer

Each of these materials is discussed in separate sections below.

With the exception of scrap steel the materials used for ICCP anodes are naturally cathodic to carbon steel and would accelerate the corrosion of steel structures if they were connected directly to the structure. The purpose of the anodes in ICCP is to serve as a surface for the oxidation of either oxygen or chlorine gases, the two intended reaction products at the anode surface.

High-silicon cast iron anodes for cathodic protection became popular in the 1950s and are still the most commonly used impressed current anode materials. The Duriron Company in Dayton, Ohio, developed and marketed the first widely accepted anodes of this type, and they also developed a more corrosion-resistant anode with chromium additions in the 1970s. The patents on these alloys have expired, and both of these alloys are available worldwide. The original grade, ASTM A518 Grade 1, is still specified for some environments, but the more corrosion resistant alloy, ASTM A518 Grade 3 with chromium additions, is usually specified, because it is more widely available and more corrosion resistant [135–138]. The most common shapes for these anodes are cylindrical tubes or solid rods up to 210 cm (8 ft) long and weighing up to 127 kg (280 lb).

Buried applications of these anodes usually include a carbonaceous backfill (coke breeze), which increases efficiency by shifting most of the oxidation reaction to the backfill. This backfill prolongs anode life [137, 138].

Table 6.21 compares the properties of high-silicon cast iron anodes with those of graphite anodes, the next

TABLE 6.21 Properties of High-Silicon Cast Iron and Graphite Anodes

	Graphite	High-Silicon Cast Iron
Nominal Current Density		
Soil/freshwater, A m^{-2}	2–10	2–5
(A ft^{-2})	(0.2–1)	(0.2–0.5)
Soil/freshwater, A m^{-2}	5–10	5–10
(A ft^{-2})	(0.5–1)	(0.5–1)
Soil/freshwater, A m^{-2}	5–10	10–50
(A ft^{-2})	(0.5–1)	(1–5)
Consumption Rate		
Soil/freshwater, kg (A-yr)$^{-1}$	0.5–0.9	0.1–0.5
(lb (A-yr)$^{-1}$)	(1–2)	(0.2–1.2)
Soil/freshwater, kg (A-yr)$^{-1}$	0.1–0.2	0.05–0.3
(lb (A-yr)$^{-1}$)	(0.2–0.5)	(0.1–0.7)
Soil/freshwater, kg (A-yr)$^{-1}$	0.1–0.3	0.3–0.5
(lb (A-yr)$^{-1}$)	(0.2–0.7)	(0.7–1)
Comments/limitations		
	Avoid:	Avoid:
	Low pH	Dry soils
	High sulfate	High pH
	$T > 50\,°C$	High sulfate

most commonly used ICCP anode material for buried soil applications.

Graphite anodes contain particulate graphite held together with a light oil impregnation. They were developed in the 1940s and were the most popular ICCP anodes until the development of high-silicon cast iron. Graphite anodes are very fragile, and some users report up to 50% breakage between shipping and construction damage. They are, nonetheless, the preferred anode material for many buried soil applications, and, in some parts of the world, their use is greater than any other ICCP anode material. The most common size for these anodes is 7.6 cm diameter by 150 cm long (3 in. diameter by 60 in. long).

Graphite anodes are almost always used with a carbonaceous backfill, which prolongs anode life. The backfill moves the oxidation reaction to the backfill and both prolongs anode life and increases the relative contact area of the anode with the soil environment. Carbonaceous backfill is often supplied with embedded anodes in prepacked perforated steel cylinders that greatly reduce breakage during shipping and construction. The perforations allow the release of gaseous oxidation products, and eventual corrosion of the steel cylinders is acceptable, because they have served their purpose once the anodes are in place in the ground bed.

Table 6.21 compared the properties of graphite anodes with high-silicon cast iron, their primary competitor in most buried in soil applications. One limitation on graphite anodes is that they will disintegrate if the current density is too high. This is caused by a loss of the binder material due to gas evolution within the anode [137, 138].

Mixed metal oxide (MMO) anodes were originally developed and marketed in the 1960s for the chemical process industry [137, 138]. They were quickly adapted to cathodic protection. They are the third most commonly used ICCP anode material for buried soil applications and, in some markets, are the leading ICCP anode for marine applications, although high-silicon cast iron remains competitive. Their primary advantage in marine applications is their relatively light weight in comparison with that of high-silicon cast iron. The anodes consist of commercially pure niobium or titanium substrates with proprietary oxide mixtures on the surface. They are available as discs, tubes, and wire. For onshore applications they are often supplied as tubes with a prepackaged backfill, although they can also be used, e.g. in deep well applications (described later in this chapter) with loose carbonaceous backfills. They are available as wire, tubing, rod, strip, and mesh shapes. Life is limited by the oxidation of the substrate metal, and they are usually limited to temperatures less than 40 °C (140 °F).

MMO anodes have been used for buried applications, but with limited success. The titanium substrates do not work in dry soils, and niobium is not much better. Niobium-substrate MMO anodes do find use for tank bottom protection, where they are inserted between the tank bottom and a liner in a 0.3 m (12 in.) gravel/sand environment, but this is a very limited application. While titanium substrates have problems in many high-resistivity environments, they have become the standard for use in offshore applications [139]. The breakdown of MMO substrates seems to be dependent on the applied current density.

Table 6.22 summarizes some of the properties of mixed metal oxide anodes.

While other precious metals can also work as anode surface materials, virtually all precious metal-clad anodes use platinum as the surface material. Precious metal anodes were developed in the 1960s and were initially used in offshore applications where their weight advantages over high-silicon cast iron made them economically competitive. Platinum has a very high exchange current density, approximately 10 000 times higher than high-silicon cast iron [140]. This results in substantial weight savings for the installation of these anodes. The major problems with platinum and other precious metals, which all have similarly high exchange current densities, are their costs. For this reason, most platinum anodes are made by coating a thin layer of platinum onto either a titanium or niobium substrate.

Platinum anodes have been used in a wide variety of applications. Problems with anode breakdown (buildup

TABLE 6.22 Mixed Metal Oxide Anode Properties

	Carbon Backfill		Freshwater	Brackish Water	Seawater	Mud Saline
	High Current	Special				
Current Density, A m^{-2}	83–140	35–40	83–170	83–260	480–610	83–240
(A ft^{-2})	(7.7–13)	(3.3–3.8)	(7.7–16)	(7.7–24)	(45–57)	(7.7–22)
Life (yr)	20	20	20	15	15	15
Comments	Above ratings do not apply to Expanded Mesh Anodes					
	Current densities must be derated at temperatures below 5–10 °C					
	Electrolyte impurities can affect ratings					
	Mixed metal oxide surface is susceptible to abrasion damage					
	Attenuation should be considered in long, thin wires, and rods					

Source: Data from Slide 101 NACE CP Technologist chapter 2, January 2005.

of a high-resistance oxide film at the platinum–titanium interface) have diminished the use of titanium-substrate anodes in buried applications, and niobium-substrate anodes have similar, but lesser, problems. Platinum anodes remain the fourth most popular anode material for buried onshore applications, and they find other uses in process equipment and marine environments. For buried applications they are typically used with either prepackaged backfills or inserted vertically into deep wells, which are then filled with carbonaceous backfills.

Platinum and other precious metals are very soft, and these anodes should not be used in flowing water situations where abrasion can remove the very thin and fragile platinum surface layer.

Polymer anodes have a very limited market in oil and gas production operations. They are supplied as flexible wires with graphite embedded into the wire insulation. Their use is primarily as distributed anodes in low-current situations such as the ground side of storage tank bottoms and buried in parallel along relatively short well-coated pipelines, e.g. in industrial areas and tank farms where conventional ICCP anodes would require more expensive deep wells to avoid stray current problems. They have been in use since the 1980s, and there are reports of premature anode failure due to changes in the resistivity of their environments. Most buried applications call for the anodes to be buried in carbonaceous backfills.

The first applications of ICCP used scrap steel for anodes. This practice diminished with the development of graphite and high-silicon cast iron anodes in the 1940s and 1950s. Scrap steel is still used on occasion for ICCP anodes. The most common situation would be where an abandoned-in-place structure, usually a pipeline, is used as an anode for ICCP of a replacement or newer pipeline in the same right of way. Scrap steel is inexpensive, but, unlike the materials discussed above, it is consumed as an anode and has a limited life. Nonetheless this approach is still used, especially as the original uncoated

pipelines constructed in the 1930s through the 1950s are replaced with new parallel pipelines having modern coatings. The small exposed surface area of the new pipelines means that the average applied current density on the abandoned pipelines or gathering lines is very small, and the scrap steel anodes should last for many years [137, 138].

At one time lead anodes with silver or other precious metal additions were used as very heavy precious metal anodes. These anodes were used before precious metal cladding techniques were developed. The lead would corrode leaving an enriched silver or other precious metal surface. These enriched surfaces had high exchange current densities. The heavy weight was useful in locations such as Cook Inlet in Alaska where high tidal currents would damage less robust anodes. The anodes were mounted on sleds that sat on the sea bottom at remote anode bed locations. Like most anode sled arrangements, mechanical damage to the lead wires was a concern. While several NACE and other standards still list these materials [141], they have not been specified in North America for many years and are not listed by most cathodic protection anode suppliers.

Tables 6.23 and 6.24 summarize information on the use of ICCP anodes in seawater and underground service – the two most common ICCP environments. The advantages of MMO and platinized titanium anodes for seawater service are apparent. For onshore applications, the weight savings realized by the use of these lighter anodes is much less important, but the reduced cost of handling and transport of these much lighter anodes remains one of the reasons for their continued and increasing popularity, especially in remote areas where transportation to construction sites is restricted. The relative market share of the top three anode materials for onshore and natural water service is summarized in Table 6.25.

Most ICCP anode materials used in soil applications require, or at least benefit, from the use of backfill

TABLE 6.23 Impressed Current Anode Material Consumption Rates [141]

Impressed Current Anode Material	Typical Anode Current Density in Saltwater Service A m^{-2} (A ft^{-2})	Nominal Consumption Rate g (A-yr)$^{-1}$ (lb (A-y)$^{-1}$)
Platinum (on titanium, niobium, or tantalum substrate) or titanium mixed metal oxide	540–3320 (50–300)	3.6–7.3 (0.008–0.016)[a]
Graphite	11–43 (1–4)	230–450 (0.5–1.0)
Fe-14.5%Si-4.5%Cr	11–43 (1–4)	230–450 (0.5–1.0)

[a] This figure can increase when current density is extremely high and in low-resistivity waters.
Source: Reproduced with permission of NACE International.

materials that surround the anode and provide a more electrically conductive environment. These backfills make the anode environment more electrically conductive, which lowers electric power requirements. They also prolong the life of the anode. Carbonaceous materials are almost universally used for this purpose.

TABLE 6.25 Relative Market Share of ICCP Anodes

Environment		
Soil	Freshwater	Seawater
High-silicon cast iron	Mixed metal oxides	Mixed metal oxide
Graphite	Pt/Nb or Pt/Ti	Pt/Nb or Pt/Ti
Mixed metal oxides to a limited extent, but not in dry soil	High-silicon cast iron use is declining	High-silicon cast iron

TABLE 6.24 Summary of ICCP Anode Properties for Underground Cathodic Protection Systems [135]

	Graphite	Si–Cr Cast Iron	Mixed Metal Oxide	Platinum-Coated	Polymeric	Scrap Steel
Consumption rate	0.1–1 kg (A-yr)$^{-1}$ (0.2–2 lb (A-yr)$^{-1}$)	0.1–0.5 kg (A-yr)$^{-1}$ (0.2–1 lb (A-yr)$^{-1}$)	Coating/titanium bond determines anode life	8–16 mg (A-yr)$^{-1}$	Projected 20-yr life requires installation in carbonaceous backfill	9 kg (A-yr)$^{-1}$ (20 lb (A-yr)$^{-1}$)
Current density, maximum	5 A m^{-2} (0.5 A ft^{-2})	10 A m^{-2} (1.0 A ft^{-2})	100 A m^{-2} (9.3 A ft^{-2})	110 A m^{-2} (10 A ft^{-2})	52 mA m^{-1} (16 mA ft^{-1}) (1.3 A m^{-2} [0.12 A ft^{-2}])	Unknown
Common shapes	Cylindrical	Tubular and solid cylindrical	Tubular and wire	Wire	Wire	Pipe, rail, and casing
Handling precautions	Material is brittle	Material is brittle	Oxide can be damaged by abrasion	Platinum can be damaged by abrasion	Can be damaged by abrasion	None; anodes are heavy
Connections	Mechanical connections at the center or near end of anode	Mechanical connections at the center or near end of anode	Mechanical connections at the center or end of anode	Brazed or mechanical connections at the end of anode	Brazed or mechanical connections	Multiple connections, brazed or bolted
Packaging	Sold bare or in canisters	Sold bare or in canisters	Sold bare, with foam protectors, or in canisters	Sold bare or in canisters	Sold bare or packaged	No packaging
Environmental hazards	None known	None known	None known	None known	None known	None known
Date of first use	1940s	1950s	Early 1980s	1960s	early 1980s	1930s
Other notes	Typically fully impregnated with oil, wax, or resin	None	Connection seals for tubular anodes tested by the manufacturer	None	Typically installed in shallow, horizontal beds	Multiple sealed connections allow full use of anode

Source: Reproduced with permission of NACE International.

These materials, either metallurgical coke (manufactured from coal) or similar high-carbon particulate materials with hydrocarbon precursors extend the life of the anodes and move most of the oxidation reaction, which will eventually degrade the anode, to the backfill particle surfaces. Backfill materials are provided in pre-packaged format for some anodes or in bags similar to those provided for galvanic anodes [137, 138]. The relatively low-conductivity soil backfills used for galvanic anodes should not be used for ICCP anodes.

There have been fewer quality control problems with ICCP anode suppliers than with galvanic anodes. Most users rely on preapproved qualified vendors to ensure the quality of ICCP anodes.

Deep anode beds are sometimes used for galvanic anodes in locations where right-of-way restrictions or the lack of shallow groundwater dictates their use. They are much more common in ICCP systems where the stray current problems are more severe because of the relatively large currents normally associated with ICCP rectifiers and multiple-anode ground beds. Figure 6.81 shows a typical deep anode ground bed [139].

Deep wells require casings – liners for the hole that prevent contamination of groundwater and provide a means for venting the oxidation product gases. If the anodes are located in salty water, the chlorine gases that are liberated will be poisonous, but oxygen is the most common gas generated by most deep wells. Casings for deep wells are often made from PVC – a polymer that is resistant to oxidizing gases. Most cathodic protection deep wells are drilled using similar equipment to that is

used for drilling water wells. The drilling uses water-based drilling muds that must be thinned before installing the anodes and backfills, which are usually coke breeze or other carbonaceous materials.

The most efficient current distribution is achieved when anodes are located at electrically remote locations. If this is impractical, e.g. on floating hulls, then the use of flush-mounted anodes with dielectric shields – insulators placed between the anode and the structure to be protected – becomes necessary. Figure 6.82 shows a dielectric shield flush-mounted on a hull.

Flush-mounted anodes with dielectric shields (Figures 6.82 and 6.83) are relatively inefficient. They distribute more current than necessary near the dielectric

Figure 6.82 Dielectric shield flush mounted on a hull.

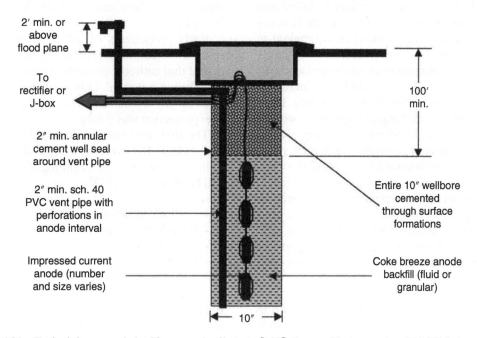

Figure 6.81 Typical deep anode bed in normal soil strata [139]. *Source*: Photo courtesy NACE International.

Figure 6.83 Coating disbonding caused by excessive current from a flush-mounted anode [135]. *Source*: Reproduced with permission of NACE International.

TABLE 6.26 Comparisons Between Galvanic Anode and Impressed Current Cathodic Protection Systems

System	
Galvanic Anode	Impressed Current
Low initial investment for small systems	Cheaper for large systems
Fixed voltage	Adjustable voltage
Small voltage	Small to large voltages
Fixed current	Adjustable current
Small current	Small to very large currents
Low maintenance	Higher maintenance
Stray currents unlikely	Stray currents possible
Reversed potentials impossible	Reversed polarity possible
No power source necessary	Requires external power
Excess current unlikely	Excess current can cause coating debonding

shield–structure interface in order to provide adequate current farther away – usually at the midpoints between two anodes. It is important to inspect and regulate the current to flush-mounted anodes in order to prevent coating disbonding caused by excessive cathodic current near the anodes. Figure 6.83 shows coating disbonding caused by excessive cathodic current on a flush-mounted anode [142].

The advantages of ICCP cathodic protection include the large electrical current available from one rectifier/anode bed installation, the low installation cost compared to galvanic systems requiring many anodes, and the long life of ICCP anodes if correctly installed and operated in the correct current density ranges. It is not unusual to have rectifiers and associated equipment in continuous operation for several decades with only routine maintenance and occasional anode replacement.

Limitations of ICCP systems include the increased possibility of hydrogen embrittlement of any high-strength (high hardness) steel, e.g. at improperly welded joints, and the increased likelihood of causing coating disbonding or stray current corrosion on nearby structures. Highly trained maintenance and inspection personnel (usually the same people) are also required because of the dangers associated with maintenance on rectifiers and to prevent reversed polarity connections at rectifiers. The relatively high maintenance and trained personnel requirements of ICCP systems are why most offshore cathodic protection systems use galvanic anodes.

Comparison of Galvanic Anodes and ICCP Table 6.26 compares galvanic anode and impressed current cathodic protection systems. The installation of a galvanic anode

is fairly inexpensive, and it is possible to install anodes for under $100 per anode if the structure is already exposed. This low initial cost plus the low maintenance/inspection costs make galvanic anode cathodic protection the option of choice for many oilfield applications.

Cathodic Protection Criteria

Operators of cathodic protection systems need to be able to determine if the structures are being adequately protected. Figures 6.74 and 6.75 showed that the corrosion rate is substantially reduced whenever the structure potential is shifted in a cathodic direction. The question then becomes how much cathodic protection is desired or necessary.

Early proponents of cathodic protection discussed criteria for cathodic protection. Some advocates suggested that cathodic protection needed to approach the equilibrium potential. Others suggested that a potential shift of any amount would yield prolonged life and overprotection was costly.

The first international standard on cathodic protection, NACE RP 0169 (now termed SP0169), listed the following means of determining if cathodic protection had been achieved [143]:

- −850 mV polarized potential
- 100 mV polarization
- 300 mV shift
- E-log i
- Net protective current

Recent revisions have eliminated some of these criteria, and the validity of the remaining criteria remains controversial.

-850 mV CSE Criterion   Over the years many authorities came to the conclusion that the −850mV CSE potential advocated by R. Kuhn and his colleagues in Louisiana was the easiest and most reliable way to determine if cathodic protection had been achieved. This idea was incorporated into the first international standard on cathodic protection, NACE RP0169 (since changed to SP0169) [143]. Kuhn's arguments were based on leak records that showed that if cathodically protected structures were kept at −850mV or more compared to CSE, and then leaks due to corrosion were substantially eliminated [115, 116, 119–121]. This approach was reinforced by Peabody, who published a practical galvanic series of metals in soil in 1967 (Table 6.27) [144]. This showed that carbon steel ("mild steel" in Peabody's terminology) would have a native, or unprotected, potential of somewhere between −0.2 and −0.8V CSE. Thus Kuhn's recommended potential of −0.85V (−850mV) is at least a 50mV shift in the cathodic direction, and usually much more. Peabody and Parker were the two standard references on cathodic protection of pipelines in 1969 when NACE RP169 was first published, and both books advocated the −850mV criterion, although they do discuss other criteria for determining if cathodic protection has been achieved [144, 145].

Many authorities pointed out, and still do, that it is unnecessary to have steel at −850mV CSE in order to achieve cathodic protection. While this has always been the case, most owner operators choose to use the −850mV criterion in NACE SP0169 and similar standards, because it is easy to measure and to train inspectors on how to perform the necessary measurements. Electricity is generally cheaper than trained labor, which is necessary to inspect according to the other, more complicated, criteria.

In cases where microbially influenced corrosion is suspected or at elevated temperatures, the protection potential is considered to be −950mV CSE [119, 123, 143]. Little controversy has appeared over the idea of a similar criterion for locations where MIC is suspected. The change of potential to −950mV due to temperature is not controversial and can be understood by anyone

who considers the Nernst equation, developed long before cathodic protection was common, that clearly explains why electrode potentials for any reaction will be affected by temperature. This is in contrast to the continuing controversy over the necessity to use an "instant off" or similar IR compensation technique to identify the "true" potential of a structure.

All of the above discussion has related to buried structures, primarily pipelines. Other reference electrodes are used in different applications. The corresponding voltage for silver–silver chloride electrodes, which are used in seawater, is −805mV, although this is usually rounded to −800mV.

-100 mV Shift Criterion   Advocates of the −100mV shift criterion point out that −850mV CSE is not necessary to achieve cathodic protection (an acceptable reduction in corrosion activity) in many, perhaps most, cases.

They also claim that in some circumstances −850mV CSE, however determined, may not produce protection. This latter claim is very controversial, and, except in the cases of elevated temperature, parallel zinc anodes, or microbial activity, it has not been unequivocally documented.

The −100mV shift criterion assumes that unshifted potentials can be determined. This is impractical for galvanic anode systems. It also assumes that the unshifted (or native) potential of the structure does not change with time. Areas with changing ground water levels due to seasonal wet and dry seasons are one example of where native (unprotected) potentials are likely to change.

Turning off ICCP systems to determine the native potential requires up to 48h for the potential to decay to the unshifted potential.

The difficulties and limitations discussed above have led most operators to prefer to use the −850 criteria for determining cathodic protection.

E-log i Criterion   There are structures where it is inconvenient or impossible to place reference electrodes along the structure being protected. Well casings are an excellent example of this situation. While the top of the well casing is available for electrical connections, the bottom of the casing is inaccessible. The *E*-log *i* criterion (Tafel curve method) is used in these situations to measure the current necessary to provide cathodic protection [146, 147].

Figures 6.84 and 6.85 show how this is done. A cathodic current source is connected to the casing and current is applied. The potential at the casing head is measured before the current is applied, and the change in potential is monitored as additional cathodic current is applied to the casing.

TABLE 6.27   Practical Galvanic Series

	Metal	Volts (CSE)
Noble or cathodic	Copper, brass, bronze	−0.2
	Mill scale on steel	−0.2
	Mild steel (rusted)	−0.2 to −0.5
	Mild steel (clean and shiny)	−0.2 to −0.8
Active or anodic	Zinc	−1.1
	Magnesium	−1.75

Source: A. W. Peabody [144] Reproduced with permission of NACE International.

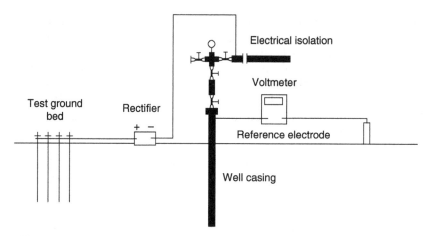

Figure 6.84   Test setup for E-log$i$ testing to determine the necessary cathodic protection current for a well casing [146]. *Source*: Reproduced with permission of NACE International.

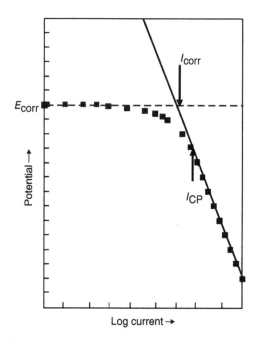

Figure 6.85   Plot of E-log$i$ data for a well casing.

At the corrosion potential no current is being supplied from the external power supply, and the corrosion potential, E_{corr}, is due to the natural oxidation and reduction reactions on the structure. The total oxidation and reduction currents are unknown. As current is supplied from an external power source, the potential versus log of applied current plot begins to curve downward when the applied current is similar in magnitude to the natural current. Increased applied current leads to a situation where virtually all of the cathodic current is coming from the power supply. The plot then becomes linear or straight. The applied current where the potential log current plot becomes "straight" or linear is assumed to be the current necessary to provide adequate cathodic protection. At one time it was suggested that the

"linear" portion of the E-log i plot should extend over one decade (or order of magnitude) of current. In recent years a two-decade (100-fold change in current) linear region has been recommended [147–149].

The current requirement from the E-log i method is considered conservative, and leak records seem to confirm that idea [149–156].

The E-log i or Tafel extrapolation method can also be used to determine the corrosion current. This is discussed in Chapter 7.

Inspection and Monitoring

The continued operation of cathodic protection systems requires monitoring to ensure that the system is working adequately. Third-party damage, coating degradation leading to increased current demands, changes in environment, and aging of cathodic protection components can all cause systems to degrade. There are many inspection and monitoring methods used in conjunction with cathodic protection. This section only discusses some of the more important methods. More complete discussions are available [114–116, 157–159]. Inspection and monitoring is required on at least an annual basis for most pipeline systems in the United States, and more frequent intervals are sometimes necessary. Rectifier operations must be monitored every 2½ months in the United States.

Potential Surveys   The most common means of inspecting a cathodically protected structure is by means of a potential survey. In any potential survey it is necessary to measure the potential of the structure in question relative to a standard potential. The most commonly used reference electrode is the saturated copper–copper sulfate electrode (CSE), which is used onshore and in freshwater applications. This is shown in Figure 6.86.

Silver–silver chloride electrodes are used in marine applications, and the conversion from one standard to the other is fairly simple. The –850 mV CSE standard theoretically becomes –805 mV with the silver–silver chloride electrode, but it is usually rounded to –800 mV. Zinc is sometimes used as a robust reference anode for permanently mounted test stations on offshore structures.

Reference electrodes can degrade and must be maintained [142]. It is common for inspectors to carry three electrodes with them. Two are used and checked against each other. If they do not produce the same result, they are then checked against a "less weathered" electrode in the hopes that one or the other electrode will still be in calibration.

In order to measure the potential of a structure, it must be connected through a high-impedance voltmeter to a reference cell in direct electrical contact with the same electrolyte. This is shown in Figure 6.87.

At one time it was common for pipeline surveyors to make electrical connections with the buried pipeline by driving a pointed rod into the soil over the pipeline. This caused unnecessary coating damage. It is now more common to locate test points along the right of way. These test points are electrical connections to the pipeline and allow the surveyor to make electrical connections to the pipeline without damaging the coating. A secondary advantage of using test points is that they are permanent locations and ensure that connections on subsequent surveys will be made at the same location. A typical flush-mounted test point is shown in Figure 6.88.

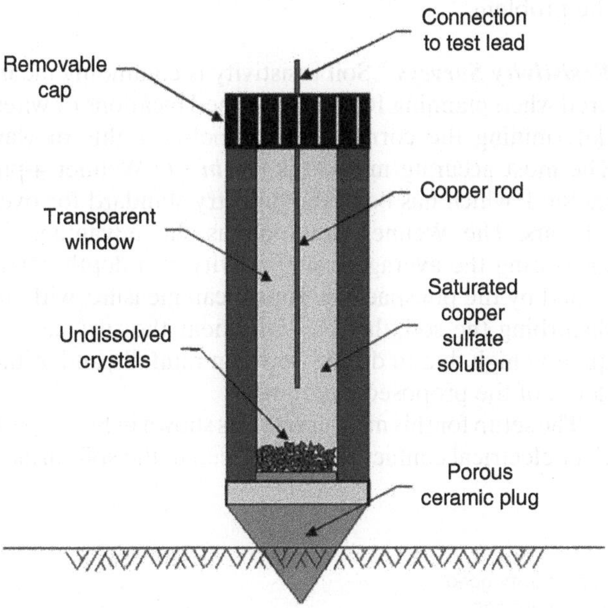

Figure 6.86   Saturated copper–copper sulfate electrode.

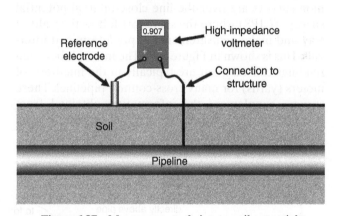

Figure 6.87   Measurement of pipe-to-soil potential.

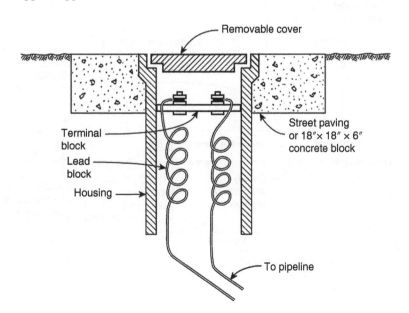

Figure 6.88   Typical at-grade test station. *Source*: R. Bianchetti [160]. Reproduced with permission of NACE International.

Test stations of this type are available from most cathodic protection equipment suppliers. The at-grade design has the advantage of being less likely to suffer vandalism or other third-party damage. It is, however, hard to find and subject to being covered over by soil erosion. Aboveground designs are also available. The minimum spacing for test stations is at the midpoint between anode locations, the most likely location for the pipeline potential to be unprotected. These test stations are often required by regulatory agencies, and their locations are recorded on maps of the cathodic protection system. It is possible to instrument these test points and relay the readings to remote locations.

Test points on cross-country transmission pipelines are typically located at intervals up to several kilometers (miles) and at cased road crossings, wherever they cross another utility, and at buried insulated joints.

Pipelines run for long distances, and the most common surveys are over-the-line close-interval potential surveys (CIPS) where the surveyors follow the right of way and make measurements at predetermined intervals. This is shown in Figure 6.89. The intervals between readings can vary, but are typically in the hundreds of meters (yards) for many cross-country pipelines. These surveys supplement the information obtained from readings at the test points, which usually are spaced much farther apart.

A typical pipe-to-soil potential profile is shown in Figure 6.90. Virtually all cathodic potentials are negative, and it is common to plot the larger negative voltages higher on the vertical axis. As long as the potential is more negative than −850 mV CSE, most authorities will consider the pipeline to be protected. The problem areas identified by this survey are shown near the center of Figure 6.90 where the negative voltages are less than −850 mV. The problem could be caused by a disconnected magnesium anode or by other factors. Once the unsatisfactory potential survey results are available, it is usually necessary to inspect the pipeline in these locations in greater detail to determine the source of the problem.

Resistivity Surveys   Soil resistivity is commonly measured when planning ICCP ground bed locations or when determining the corrosivity of pipeline rights of way. The most accurate method is the *in situ* Wenner 4-pin method, which has been the industry standard for over 50 years. The Wenner method has the advantage of measuring the average soil resistivity at a depth determined by the pin spacings. Thus it can measure, without disturbing the soil, the resistivity near the surface, frequently high due to drying between rainfalls, and at the depth of the proposed structure.

The setup for this measurement is shown in Figure 6.91. Four electrical contact pins are placed in the soil surface.

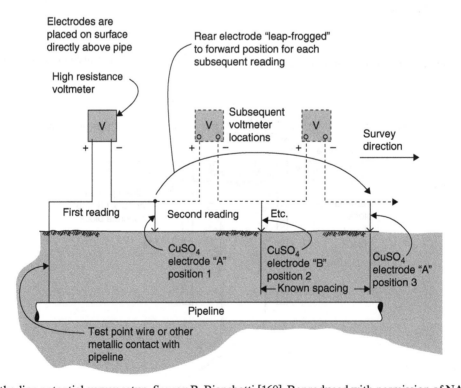

Figure 6.89   Over-the-line potential survey setup. *Source*: R. Bianchetti [160]. Reproduced with permission of NACE International.

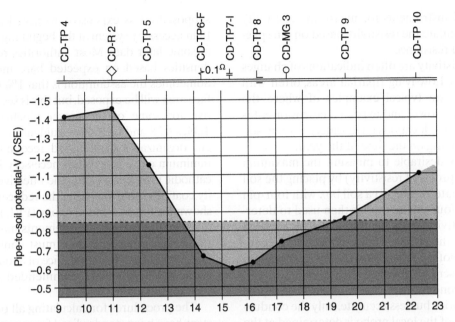

Figure 6.90 Protective potential profile indicating a lack of protection near the center of the plot. *Source*: R. Bianchetti [160]. Reproduced with permission of NACE International.

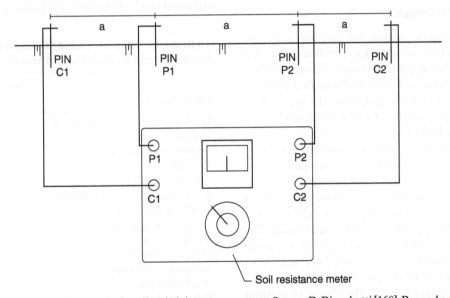

Figure 6.91 Wenner 4-pin soil resistivity measurement. *Source*: R. Bianchetti [160]. Reproduced with permission of NACE International.

An AC electrical current is applied between the outer pins to produce current flow through the soil. The voltage measured between the inner pins is used to calculate resistivity. The four pins are arranged in a straight line, and the distance between pins is adjusted to reflect the depth of the soil of interest. Once the measurements have been made, it is easy to calculate the average resistivity of the soil at the depth equal to the pin spacing. Adjusting the pin spacing allows determination of changes in resistivity with depth [144, 145, 160, 161].

The soil resistivity is calculated from the following formula [157–161]:

$$\rho = 2\pi A R \tag{6.9}$$

where

ρ = soil resistivity (ohm-centimeters).
A = distance between pins.
R = resistance measured with the ohmmeter.

Most commercial instruments for measuring resistivity automatically calculate the resistivity based on pin spacing and measured resistance.

Changes in resistivity are often indications of changes in moisture levels. Low-lying riparian areas, often with more vegetation, are typical examples of where the resistivity would be lower and expected corrosion rates would increase. Some locations have widely varying soil resistivities depending on the time of the year.

It is sometimes desirable to measure the maximum conductivity (reciprocal of resistivity) by placing the soil from the appropriate depth into a soil box with four-pin connections. The four-pin method is then used to determine the conductivity of the wetted soil. This method cannot determine the effects of soil compaction. on conductivity and is not as reliable as the *in situ* four-pin measurements described above.

Single probe conductivity measurement is also possible. This is less work but less accurate, only the conductivity at the depth of the local probe is determined at the precise location where the probe is placed.

Cathodic Protection Design

Most of the following discussion will emphasize design of cathodic protection systems for pipelines. The US Department of Defense publication on cathodic protection and the Handbook of Cathodic Protection contains complete chapters on cathodic protection design for storage tanks, buried pipelines, ships (applicable to designs for spar platforms and FPSOs), marine structures, well casings, water tanks and boilers, and process equipment [158, 162].

The first step in any cathodic protection design is to determine the total electrical current demand. For existing structures this can be done by measuring the current necessary to produce the desired potential shift. This is done by connecting the structure to a temporary DC power source and varying the current until the necessary polarization, determined by either the E-log i method or by simple measurement of the potential at remote locations from the temporary anodes. The choice of method depends on whether or not the remote location is accessible for potential measurement. In either case it is necessary to wait until changes in the applied current have produced steady-state potentials before increasing the current to the next level. This can take minutes to hours, depending on the size of the structure involved.

New pipeline cathodic protection design is often based on the current expected to be necessary after 20 years of service. The coating degradation is not expected to worsen after that period of time. Designs provide more current than is necessary during the early life of the system and are intended to last indefinitely. For this reason, it is common to overdesign the system, because it is easier, and

supposedly less expensive, to install a somewhat larger than necessary system at the beginning than it is to retrofit at some later date. Most authorities recommend current densities based on expected bare metal exposed area. Sometimes the assumption is that 1% of the possible surface area will be exposed, but this is seen as very conservative for some of the newer pipeline coating systems. Tables 6.28 and 6.29 show guidelines from several different organizations' published recommendations for the minimum current densities necessary for buried pipeline cathodic protection [158]. The effectiveness of high-quality coatings is obvious from the reduced current demands shown for fusion-bonded epoxy (FBE) and polyethylene.

Once the current requirements have been identified, the design procedure then must consider a number of alternatives based on the choice of anode type. Figure 6.92 is one of a number of recommended design procedures that are available.

The procedures for calculating all of the above design steps have been standardized for many years. Many of the necessary formulas, e.g. for calculating ground-bed resistance, are based on work done in the 1930s. While the calculations can be done by hand (as they were originally), it

TABLE 6.28   Recommended Minimum Cathodic Protection Design Current Densities for Different Soils

Environment	Milliamperes per square foot (mA ft^{-2})
Soil with resistivity <1 000 Ω-cm	6.0–25.0
Soil with resistivity 1 000–10 000 Ω-cm	3.0–6.0
Soil with resistivity 10 000–30 000 Ω-cm	2.0–3.0
Soil with resistivity >30 000 Ω-cm	1.0–2.0
Highly aggressive soil with anaerobic bacteria	15.0–40.0
Still freshwater	2.0–4.0
Moving freshwater	4.0–6.0
Turbulent freshwater	5.0–15.0
Hot freshwater	5.0–15.0
Still seawater	1.0–3.0
Moving seawater	3.0–25.0

Source: From Kroon et al. [151].

TABLE 6.29   Recommended Minimum Cathodic Protection Design Current Densities for Different Coatings

Coating	Design Current Density (mA m^{-2})
None	20
Tape wrap	1.25
Coal tar epoxy	0.75
FBE	0.1
Polyethylene	0.1

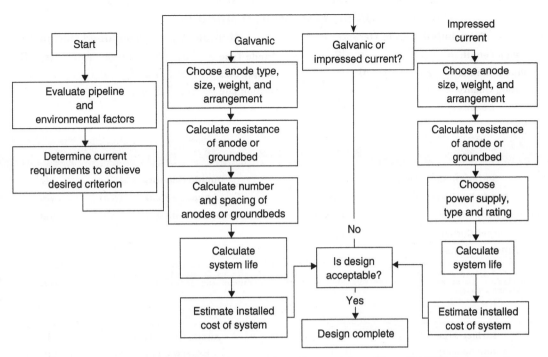

Figure 6.92   Cathodic protection design procedure. *Source*: Adapted from NACE International CP Technology Course Slide5/140 chapter 4.

is more common to do them using computer software. Most of this software is based on spreadsheets, and many cathodic protection contractors have developed their own in-house software for this purpose. In recent years several websites have become available that do these calculations. Figures 6.93 and 6.94 are examples of what is available. As the use of computers increases, this kind of user-friendly software will become more common.

For many years the 1960s book, Control of Pipeline Corrosion by A. W. Peabody was considered to be one of the premier reference materials on cathodic protection [144]. The book was updated in 2001 and 2018, and the compact discs that accompany the updated books have spreadsheet-based software included. Figure 6.95 is one example of the screens used by the compact disc that accompanies this widely used handbook.

The compact disc also contains sample problems showing the following calculations:

- Determining protective current requirements
- Anode resistance to earth
- Conventional ground bed design
- Deep anode bed design
- Cathode resistance to earth
- Total DC circuit resistance
- Current attenuation
- System life of galvanic anode systems

They allow the user to input the data in metric or US conventional units.

Temperature Effects on Cathodic Protection   Chemical reactions associated with corrosion are highly temperature dependent. Many design guidelines contain advice on increasing current density for above-ambient temperatures. The consumption rate of anodes depends on temperature, and this must be considered in cathodic protection design and replacement scheduling. The increased consumption rates of anodes can be minimized by using remote anode locations in cooler environments, but this leaves some designs more prone to mechanical damage due to soil movement and other causes.

Computer-Aided Cathodic Protection Design   The spreadsheet-based calculations discussed in conjunction with Figures 6.93–6.95 are simple arithmetic calculations using the same formulas that were calculated by hand in previous years. In recent years computer-aided cathodic protection designs for offshore structures have been tried by several organizations. These computer-aided designs are of two types [121, 163–166]:

- Personal computers used to make the types of calculations (such as the wetted surface area calculations discussed above) that have commonly been used for cathodic protection design. The computer is a time saver in these calculations and allows a greater number of alternatives to be considered, but the actual design methodology is not changed.

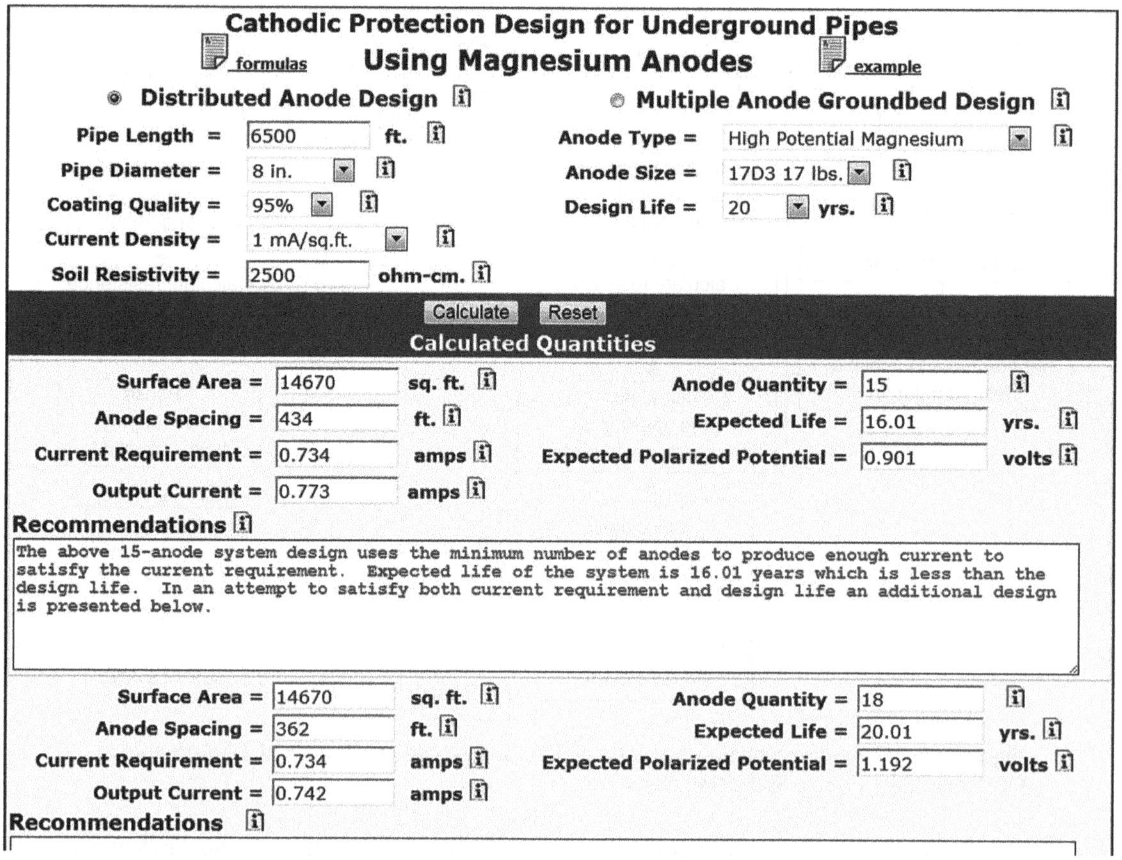

Figure 6.93 Online cathodic protection design screen based on spreadsheet formulas. *Source*: Image courtesy Mesa Products, Inc. http://www.mesaproducts.com and http://www.cpdesigncenter. com/private/galvanicdesign/galvanicdesignframeset.htm (accessed 28 January 2017).

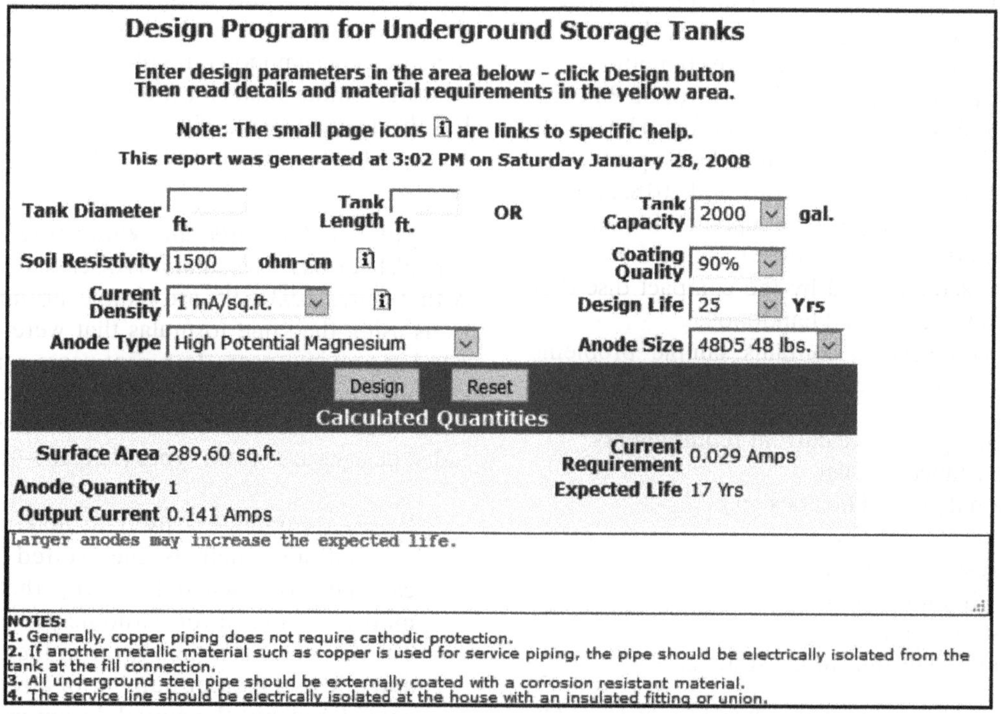

Figure 6.94 Online screen for calculating cathodic protection for underground storage tanks. *Source*: Image courtesy Mesa Products, Inc. http://www.mesaproducts.com and http://www. cpdesigncenter.com/private/ust/ustdesignframeset.htm (accessed 28 January 2017).

NACE
Companion to the Peabody Book
October 26, 2000
Revision 1.1M

Dwight's Equation for Single Vertical Anode Resistance to Earth - millimeters

$$R_V = \frac{1.59\,\rho}{L}\left(\ln\frac{8L}{d} - 1\right)$$

ρ = Soil resistivity in ohm-cm	ρ =	10,000 ohm-cm
L = Rod length in mm	L =	2134 mm
d = Rod diameter in mm	d =	203 mm
R_V = Resistance of vertical rod in ohms	R_V =	25.6 ohms

Dwight's Equation for Single Vertical Anode Resistance to Earth - meters

$$R_V = \frac{0.00159\,\rho}{L}\left(\ln\frac{8L}{d} - 1\right)$$

ρ = Resistivity of backfill material (or earth) in ohm-cm	ρ =	10,000 ohm-cm
L = Length of anode in meters	L =	2.13 m
d = Diameter of anode in meters	d =	0.203 m
R_V = Resistance of one vertical anode to earth in ohms	R_V =	25.6 ohms

Figure 6.95 Screen from CD accompanying *Peabody's Control of Pipeline Corrosion*, 2e. *Source*: R. Bianchetti [160]. Reproduced with permission of NACE International.

- The use of numerical techniques, such as finite element, finite difference, or boundary integral, to model the potential current distribution around a structure. Initial efforts to use these techniques found limited acceptance because of the time delay caused by communications difficulties between the operator, the cathodic protection designer, and the computer expert.

The increased memory capabilities of personal computers now allow design engineers to make calculations once requiring mainframe computers. Figure 6.96 shows a sample plot of the cathodic protection on an offshore platform node. The various color arrangements or shadings allow quick assessment of areas that might be inadequately protected [163–166]. Comparisons of plots for different anode arrangements allow the designer to quickly determine which anode locations are the most effective and where inspection points should be located to determine if adequate cathodic protection is being achieved at high-stressed node welds and other critical locations.

Additional Comment About Cathodic Protection Design   The US Department of Defense manual on cathodic protection offers the following caution [158]:

The use of computers has greatly aided in the design of CP systems (e.g. drawing programs, word processors, and spreadsheets). Many designers use previously complete designs as templates for new designs, and in their haste, forget to change some of

Figure 6.96 Computerized model of the cathodically protected region around a node on an offshore platform [142]. *Source*: Reprinted with permission of ASM International®. All rights reserved. www.asminternational.org.

the parameters from the previous design. Review all design documents to ensure that the information is accurate.

This problem predates the use of computers.

Additional Topics Related to Cathodic Protection

Some of the problems associated with cathodic protection include stray current corrosion, hydrogen embrittlement and stress corrosion cracking of high-strength (or high-hardness) steel, and cathodic disbondment of coating. In addition to these well-documented problems, many authorities have questioned the standards used over recent decades for determining if a structure is cathodically protected.

"Instant Off" Potentials   In recent years the biggest controversy in cathodic protection has been over the idea of measuring "instant off" potentials to determine if a structure is protected from corrosion. There are many publications with pros and cons on this subject, and the ideas behind "instant off" potentials are the subject of continuing debates [166].

The first advocates of "instant off" potentials cited the need for accounting for IR drops between the structure and the electrolyte. This was based on the mistaken assumption that the –850 mV CSE potential was the "equilibrium potential" for carbon steel in soil. This is not the case because:

- The –850 mV criterion came from leak records and measurements of "current on-potentials" on cathodically protected pipelines, primarily on the Gulf Coast of the United States. R. Kuhn, from Louisiana, was the most prominent early advocate of this idea, and he based his reasoning on leak records that showed that structures held at potentials at least as negative as –850 mV CSE had much lower leak records than unprotected steel. This standard was considered acceptable for most situations although a –950 mV criterion was generally recommended when microbial activity was likely [115–118, 123].
- The –850 mV criterion is at a lower (smaller) negative number than the equilibrium potential. This was shown in Figure 6.74 where the equilibrium potential is shown at –950 mV CSE in an acid environment. The equilibrium potential will vary with pH in accordance with the Nernst equation discussed in earlier chapters, but the equilibrium potential will always be at a greater negative potential than the –850 CSE protection potential used for cathodic protection. As stated previously, cathodic protection does not eliminate oxidation or corrosion on a protected structure, but it has been shown to significantly reduce corrosion.

In the decades after the NACE RP 0169 standard was adopted, many authors discussed errors in potential measurement, but these errors were considered to be insignificant in most cases [115, 116]. This may be true for the wet, swampy soils that were common in Louisiana where Kuhn and coworkers first developed the –850 mV CSE criterion, but some recent authors suggest that this may not be the case in drier soils [166].

The IR drops that were originally considered to be insignificant for bitumastically coated pipelines protected with galvanic anodes were questioned for impressed current systems and for measurements directly over galvanic anodes. This led to the development of NACE RP0169-92, which mandated that IR drops must be compensated for using an "instant off" criterion. Difficulties in defining how this "instant off" should be done, and questions on whether or not this "instant off" potential is necessary, continue as of this writing (2017). Since the requirement to determine an "instant off" negative potential means that the negative "current on" potential will be greater than the negative "instant off" potential, the requirement is conservative. It is, nonetheless, questioned by many and unpopular with field personnel [167–171].

Figure 6.97 illustrates the concept of the "instant off" method of determining IR errors in cathodic protection. When the cathodic protection current is applied, the structure assumes a potential that is intended to reduce corrosion. In order to measure the potential, a voltmeter is connected to the system and the current-on potential is measured. The measured potential includes the potential of the structure and any current resistance (IR) drops in the circuit, e.g. the reference electrode–electrolyte IR drop and the structure–electrolyte IR drop. By turning off the current, the voltage supposedly "instantaneously" drops to the potential of the structure in the electrolyte. The potential then decays to the unprotected potential, a process that can take anywhere from minutes to days. Proponents of the "instant off" method suggest that the "instant off" potential must

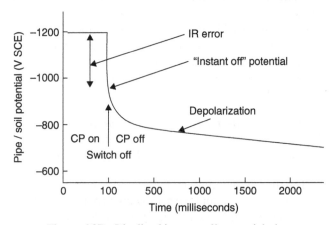

Figure 6.97   Idealized instant-off potential plot.

satisfy whatever criterion, either −850 mV CSE or −100 mV potential shift, is used for the system. The IR drop can be as large as 1 V in some instances.

Proponents cite electrochemistry textbooks that identify several IR drops in an electrochemical circuit, and the supposed failures of the "current on" criterion that had been in use for several decades. Opponents of this concept argue that measuring the potential with the current applied has worked for decades. The original "current on" criterion was developed for onshore pipelines along the Gulf Coast of the United States, a region where soils generally have low resistance. The IR drop of concern is more likely to be of concern with high-resistivity environments. The length of time (from microseconds to several seconds) of the "instant decay" depends, at least in part, on the environment and must be empirically determined.

Stray Current Corrosion Caused by Cathodic Protection Systems The potential field around a cathodically protected structure can cause stray current corrosion on nearby structures. This is a significant problem with ICCP systems that share rights of way with nearby structures/utility systems. Galvanic anodes, with their much lower driving potentials, are unlikely to cause this problem.

Figure 6.98 shows two pipelines crossing each other. The cathodic protection system is causing corrosion where current from the cathodic protection system leaves the unprotected line.

Stray, or interference, current is detected by turning the cathodic protection system on and off and monitoring the potential of the unprotected line. If the potential of

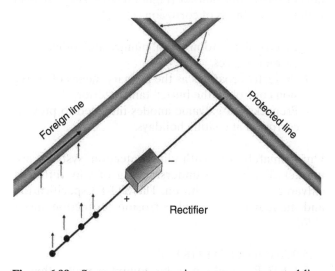

Figure 6.98 Stray current corrosion on an unprotected line crossing a cathodically protected line. The arrows show current paths from the anode bed through the nearest portion of the foreign line and then leaving the foreign line near the protected line and causing corrosion.

the unprotected line varies with the cycle of turning the cathodic protection system on and off, then stray current corrosion will occur.

Stray current corrosion can usually be handled by bonding a short section of the unprotected line to the protected line and using current from the cathodic protection system to protect both structures. "Hot spot" galvanic anodes can also be attached near the location where interference is occurring. These anodes can have the effect of shifting the potential to a protected level. It is also possible to use insulating joints or to adjust the output of the "offending" rectifier to a level where the stray current is reduced. Improvements in the coating on the foreign line are also helpful [172].

Stray current effects may also be due to natural electromagnetic phenomena. Telluric currents can be identified by recording the pipe-to-soil potential for 24 h. If no recognizable pattern is identifiable, then the currents are probably not man-made and are telluric. These "telluric currents" can have a number of causes, but they normally have minimal effect on corrosion because they usually do not last very long [163]. Corrosion-related problems from telluric currents are more likely to be a concern in high-latitude (nearer to the Poles) locations [173, 174].

Disbonded Coatings Coating disbonding can be caused by excessive cathodic protection or by inadequate coating-to-metal adhesion, which is usually due to poor surface preparation prior to the coating process.

Excessive cathodic protection can cause hydrogen gas evolution at the metal–coating interface. Gas pressure buildup eventually leads to coating disbonding. This was shown in Figure 6.83. Some organizations try to avoid hydrogen blistering by limiting the negative potentials allowed for cathodic protection, but the current density at the metal surface is more important. Water is only stable over a 1.23 V range between oxygen evolution at anodes and hydrogen evolution at cathodes. International standards for testing coatings for cathodic disbonding resistance use galvanic anodes to test for this phenomenon, and these galvanic anodes can achieve high current densities near coating holidays but very limited potentials. Cathodic protection also shifts the pH at cathodes to more alkaline (higher pH) values, and some coatings are not resistant to disbonding in the presence of alkalis.

Coating disbonding due to poor surface preparation is shown in Figure 6.99, where a fusion-bonded epoxy coating has lifted from the surface. The metal underneath the disbonded coating is discolored, but no significant corrosion has occurred. The whitish deposits underneath this disbonded coating show that cathodic protection has reached the metal surface, increased the

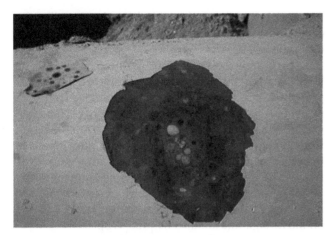

Figure 6.99 Whitish deposits underneath disbonded fusion-bonded epoxy coating. *Source*: Photo courtesy R. Norsworthy, Polyguard Products, Inc.

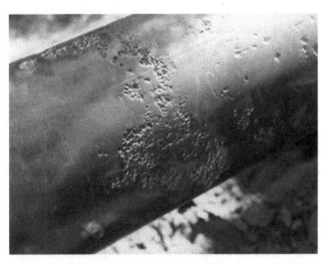

Figure 6.100 Pitting due to disbonded coating, which shielded cathodic protection current. *Source*: Photo courtesy R. Norsworthy, Polyguard Products, Inc.

pH of the moisture at the metal/environment interface, and caused these mineral deposits, similar to those shown in Figure 6.76.

Disbonded coatings lifted away from the metallic substrate can act as dielectric shields preventing cathodic protection currents from reaching the shielded metal surface, and the unprotected metal surface can corrode. Fusion-bonded epoxy coatings have the reputation for not causing this kind of shielding. Other commonly used pipeline coatings do shield the metal surface and allow corrosion to occur as is shown in Figure 6.100. This pipeline, which had substantial areas of disbonded coatings, was found to be corroding along much of the pipeline, even though cathodic protection was applied and aboveground pipe-to-soil readings indicated that the structure was cathodically protected.

Figure 6.101 STI P3 tanks in storage prior to installation.

Dielectric shielding masks the problems of corrosion underneath the coating, and this unprotected area cannot be identified by potential measurements using close interval surveys or other commonly used over-the-ground pipeline inspection techniques. This type of coating disbonding can increase cathodic current demand by increasing the total surface area of uncoated metal [175].

Misapplication of Galvanic Anodes on Small Underground Storage Tanks Many owners of underground storage tanks are small operations with no professional corrosion staff or experience. The Steel Tank Institute in the United States developed STI p3 tanks to meet this market (Figure 6.101). They are sold with three methods of corrosion control:

- Electrical isolation from pumps and other stray current sources.
- Protective coatings as the primary means of corrosion control on the buried tank exterior.
- Prepackaged galvanic anodes intended to prevent corrosion at coating holidays.

Unfortunately, the cathodic protection systems are marketed with a misunderstanding of why different galvanic anodes are chosen. The STI-P3 specification and manual for external corrosion protection states [176]:

5.0 ANODE INTEGRITY
5.1 STI-P3® tanks may be equipped with either zinc or magnesium anodes. Whereas magnesium anodes are designed only for installation in soil resistivities of 2000 ohms-cm or greater, zinc anodes are effective in all soil resistivities.

Most managers, having no background in cathodic protection, will choose the zinc anodes based on the recommendation that they are "effective in all soil resistivities."

This recommendation on the use of zinc anodes is the reverse of most recommendations on the use of zinc anodes. These tanks are usually not sold with anode backfills, so the natural resistivity of the local soil will determine the corrosion rate of the anodes, and zinc anodes cannot be effective in high resistivity soils.

The lessons to be learned from this situation are:

- STI P-3 tanks are sold with inappropriate cathodic protection systems.
- What appears to be an industry standard may be a marketing organization recommendation.
- Industry standards and marketing organization publications can contain mistakes.

Aboveground storage tanks have many advantages, including ease of inspection, for many oil and gas production operations.

Summary of Cathodic Protection

Cathodic protection is a widely used means of lowering corrosion rates. The early advocates of cathodic protection usually did not claim perfect protection, merely a reduction in leaks. While cathodic protection has lowered corrosion rates on oil and gas production structures for many years, it cannot stop all corrosion, and all cathodic protection systems must be periodically inspected to ensure that they are working correctly.

Standards for Cathodic Protection

The list below shows some of the international standards relevant to cathodic protection. These are consensus standards based on industrial practice at the time of publication and are reviewed and revised on a periodic basis. The current version of the standard should always be used. As one example NACE SP 0169-2013 is the 2013 version of a standard that originated in 1969. Significant changes have been introduced into this and many other standards. While earlier versions of these standards are sometimes difficult to obtain, it is important to recognize that existing equipment and systems may reflect earlier versions of standard industrial practice. Earlier versions of these standards can be obtained from technical libraries and from the issuing organization. Most current standards can be downloaded from the Internet.

NACE Standards

- SP0169, Control of External Corrosion on Underground or Submerged Metallic Piping Systems.
- SP0176, Corrosion Control of Submerged Areas of Permanently Installed Steel Offshore Structures Associated with Petroleum Production.
- SP0177, Mitigation of Alternating Current and Lightning Effects on Metallic Structures and Corrosion Control Systems.
- SP0186, Application of Cathodic Protection for External Surfaces of Steel Well Casings.
- SP 0193, External Cathodic Protection of On-Grade Carbon Steel Storage Tank Bottoms.
- SP0285, Control of External Corrosion on Metallic Buried, Partially Buried, or Submerged Liquid Storage Systems.
- SP0286, Electrical Isolation of Cathodically Protected Pipelines.
- SP0388, Impressed Current Cathodic Protection of Internal Submerged Surfaces of Steel Water Tanks.
- RP0675, Control of External Corrosion on Offshore Steel Pipelines.
- TM 0497, Measurement Techniques Related to Criteria for Cathodic Protection on Underground or Submerged Metallic Piping Systems.

ASTM Standards

- ASTM G 8, Cathodic Disbonding of Pipeline Coatings.
- ASTM G 19, Cathodic Disbonding of Pipeline Coatings by Direct Soil Burial.
- ASTM G 42, Standard Test Method for Cathodic Disbonding of Pipeline Coatings Subjected to Elevated Temperatures.
- ASTM G 95, Test Method for Cathodic Disbondment Test of Pipeline Coatings (Attached Cell Method).

DNV Standards

- DNV-RP-B401, Cathodic Protection Design.
- DNV-RP-F103, Cathodic Protection of Submarine Pipelines by Galvanic Anodes.

British Standards

- BS 7361-1, Cathodic Protection.
- EN 12068, Cathodic Protection. External Organic Coatings for the Corrosion Protection of Buried or Immersed Steel Pipelines Used in Conjunction with Cathodic Protection. Tapes and Shrinkable Materials.

- EN 12473, General Principles of Cathodic Protection in Sea Water.
- EN 12474, Cathodic Protection for Submarine Pipelines.
- EN 12495, Cathodic Protection for Fixed Steel Offshore Structures.
- EN 12499, Internal Cathodic Protection of Metallic Structures.
- EN 12696, Cathodic Protection of Steel in Concrete.
- EN 12954, Cathodic Protection of Buried or Immersed Metallic Structures. General Principles and Application for Pipelines.
- EN 13173, Cathodic Protection for Steel Offshore Floating Structures.
- EN 13174, Cathodic Protection for Harbour Installations.
- EN 13509, Cathodic Protection Measurement Techniques.
- EN 13636, Cathodic Protection of Buried Metallic Tanks and Related Piping.
- EN 14505, Cathodic Protection of Complex Structures.
- EN 15112, External Cathodic Protection of Well Casing.
- EN 50162, Protection Against Corrosion by Stray Current from Direct Current Systems.

REFERENCES

1 Hudson, R. (2000). *Coatings for the Protection of Structural Steelwork*. Teddington, UK: National Physical Laboratory.

2 Revie, R.W. and Uhlig, H.H. (2008). *Corrosion and Corrosion Control*. New York, NY: Wiley.

3 Trethewey, K.R. and Chamberlain, J. (1995). *Corrosion for Science and Engineering*, 2, 302–335. London, UK: Longman Scientific and Technical.

4 Roberge, P.R. (2006). *Corrosion Basics: An Introduction*. Houston, TX: NACE International.

5 Tator, K.B. (2003). Organic coatings and linings. In: *Metals Handbook, Volume 13A – Corrosion: Fundamentals, Testing, and Protection*, 817–833. Materials Park, OH: ASM International.

6 Schweitzer, P.A. (2006). *Coatings: Applications and Corrosion Resistance*. Boca Raton, FL: CRC Press.

7 Tator, K.B. (2003). Zinc-rich coatings. In: *Metals Handbook, Volume 13A – Corrosion: Fundamentals, Testing, and Protection*, 834–836. Materials Park, OH: ASM International.

8 NORSOK M-CR-501. *Surface Preparation and Protective Coating*. Oslo, Norway: Norwegian Technology Standards Institution.

9 NACE Publication 6J162. *Guide to the Preparation of Contracts and Specifications for the Application of Protective Coatings*. Houston, TX: NACE International.

10 NACE No. 2/SSPC-SP 10. *Near-White Metal Blast Cleaning*. Houston, TX: NACE International.

11 NACE WAB-2/SSPC-SP 10 (WAB). *Near-White Metal Wet Abrasive Blast Cleaning*. Houston, TX: NACE International.

12 Schreiner, T. (February 2016). NACE, SSPC publish joint wet abrasive blast standards, NACE WAB-2/SSPC-SP 10 (WAB), near-white metal wet abrasive blast cleaning. *Materials Performance* 55 (2): 14–15.

13 Appleman, B.R. (1987). Painting over soluble salts: a perspective. *Journal of Protective Coatings & Linings* (October): 68–82.

14 Trimber, K. 2017. New SSPC/NACE Wet Abrasive Blast Cleaning Standards. https://ktauniversity.com/new-sspc-wet-abrasive-blast-cleaning-standards (accessed 6 January 2017).

15 NACE 6G198/SSPC-TR 2. *Wet Abrasive Blast Cleaning*. Houston, TX: NACE International.

16 SSPC-VIS 1. *Guide and Reference Photographs for Steel Surfaces Prepared by Dry Abrasive Blast Cleaning*. Pittsburgh, PA: SSPC.

17 A. Beggs and M. Damiano, SSPC visual standards update 2003: a picture is worth a thousand words. In: *Ultra-High Pressure Waterjetting: A JPCL eBook*, 5–13, Technology Publishing Company, Pittsburgh, PA. http://www.paintsquare.com/store/assets/JPCL_uhp_ebook1.pdf (accessed 16 April 2018).

18 Soltz, G.C. (1991). *The Effects of Substrate Contaminants on the Life of Epoxy Coatings in Sea Water*. NSRP 0329 (June 1991). San Diego, CA: National Steel and Shipping Company.

19 NACE Publication 6G186. *Surface Preparation of Soluble Salt Contaminated Steel Substrates Prior to Coating*. Houston, TX: NACE International.

20 ISO 8502. *Preparation of Steel Substrates Before Application of Paints and Related Products – Tests for the Assessment of Surface Cleanliness*. Geneva: ISO.

21 SSPC-Guide 15. *Field Methods for Analysis of Soluble Salts from Steel Substrates*. Pittsburgh, PA: SSPC.

22 Meade, B.W., Palle, S., and Hopwood, T. (2012). Effects of chloride contamination on coatings performance. *SSPC Conference Proceedings*, Tampa, FL (1 January 2012).

23 Alland, K., Vendenbossche, J.M., Vidic, R., and Ma, X. (2013). *Evaluation of Bridge Cleaning Methods on Stell Structures*, FHWA-PA-2013-007-PIT WO 2 December 2013. Harrisburg, PA: The Pennsylvania Department of Transportation.

24 Stiner, H. (2016). Soluble Salts and Coating Performance. https://www.corrosionpedia.com/2/6502/soluble-salts/soluble-salts-and-coating-performance (accessed 16 April 2018).

25 Brodar, J. (2007). Maintaining large diameter water line internals: new directions for repairing coal tar enamel, coal tar epoxy coatings, and other options. *NACE International Western Area Conference*, Phoenix, AZ (6–9 November 2007).

26 NACE SP 0108. *Corrosion Control of Offshore Structures by Protective Coatings*. Houston, TX: NACE International.

27 Smart, J.S. and Heidersbach, R. (1987). (Marine) organic coatings. In: *Metals Handbook, Volume 13 – Corrosion*, 912–919. Materials Park, OH: ASM International.

28 Webb, A., Verborgt, J., Martin, J.R. et al. (2007). Reducing corrosion control costs with rapid-cure coatings. *NRL Review* 63–69.

29 SSPC (2008). *Painting Manual, Volume 2: Systems and Specifications*. Pittsburgh, PA: SSPC.

30 NACE TM0174. *Laboratory Methods for the Evaluation of Protective Coatings and Lining Materials on Metallic Substrates in Immersion Service*. Houston, TX: NACE International.

31 Pinney, S.G. (2006). *Coating Inspection, What's Changed?*, NACE 00611. Houston, TX: NACE International.

32 Vincent, L.D. (2001). *Surface Preparation Standards*, NACE 01659. Houston, TX: NACE International.

33 ISO 8501-1. *Preparation of Steel Substrates Before Application of Paints and Related Products – Visual Assessment of Surface Cleanliness*. Geneva: ISO.

34 ASTM D610. *Evaluating Degree of Rusting on Painted Steel Surfaces*. West Conshohocken, PA: ASTM International.

35 SSPC SP-1. *Solvent Cleaning*. Pittsburgh, PA: The Society for Protective Coatings.

36 ISO 8502-6. *Extraction of Soluble Contaminants for Analysis – The Bresle Method*. Geneva: ISO.

37 ISO 8502-9. *Field Method for the Conductometric Determination of Water-Soluble Salts*. Geneva: ISO.

38 National Physical Laboratory (2000). *Surface Preparation for Coating*. Teddington, UK: National Physical Laboratory.

39 ISO/TR 15231:2001 (2001). *Collected Information on the Effect of Levels of Water-Soluble Salt Contamination*. Geneva: ISO.

40 NACE SP0508. *Methods of Validating Equivalence to ISO 8502-9 on Measurement of the Levels of Soluble Salts*. Houston, TX: NACE International.

41 ASTM D4417. *Field Measurement of Surface Profile of Blast Cleaned Steel*. West Conshohocken, PA: ASTM International.

42 NACE SP0287. *Field Measurement of Surface Profile on Abrasive Blast-Cleaned Steel Surfaces Using a Replica Tape*. Houston, TX: NACE International.

43 ISO 8502-2. *Method for the Grading of Surface Profile of Abrasive Blast-cleaned Steel – Comparator Procedure*. Geneva: ISO.

44 ISO 8503-1. *Specifications and Definitions for ISO Surface Profile Comparators for the Assessment of Abrasive Blast-cleaned Surfaces*. Geneva: ISO.

45 ISO 8503-3. *Method for the Calibration of ISO Surface Profile Comparators and for the Determination of Surface Profile – Focusing Microscope Procedure*. Geneva: ISO.

46 ISO 8503-4. *Method for the Calibration of ISO Surface Profile Comparators and for the Determination of Surface Profile – Stylus Instrument Procedure*. Geneva: ISO.

47 SSPC-VIS 2. *Evaluating Degree of Rusting on Painted Steel Surfaces*. Pittsburgh, PA: SSPC.

48 SSPC-VIS 3. *Photographs for Steel Surfaces Prepared by Hand and Power Tool Cleaning*. Pittsburgh, PA: SSPC.

49 NACE VIS7/SSPC-VIS 4. *Guide and Reference Photographs for Steel Cleaned by Waterjetting*. Houston, TX: NACE International.

50 NACE VIS9/SSPC-VIS 5. *Guide and Reference Photographs for Steel Surfaces Prepared by Wet Abrasive Blast Cleaning*. Houston, TX: NACE International.

51 ISO 2808. *Determination of Film Thickness*. Geneva: ISO.

52 SSPC PA-2. *Measurement of Dry Film Thickness with Magnetic Gages*. Pittsburgh, PA: SSPC.

53 ASTM D7091. *Nondestructive Measurement of Dry Film Thickness of Nonmagnetic, Nonconductive Coatings Applied to Non-Ferrous Metals*. West Conshohocken, PA: ASTM International.

54 ASTM D4138-07a. *Standard Practices for Measurement of Dry Film Thickness of Protective Coating Systems by Destructive, Cross-Sectioning Means*. West Conshohocken, PA: ASTM International.

55 ASTM D3359. *Measuring Adhesion by Tape Test*. West Conshohocken, PA: ASTM International.

56 ASTM D4541. *Pull-Off Strength of Coatings Using Portable Adhesion Testers*. West Conshohocken, PA: ASTM International.

57 ASTM D6677. *Evaluating Adhesion by Knife*. West Conshohocken, PA: ASTM International.

58 NACE SP0188. *Discontinuity (Holiday) Testing of New Protective Coatings on Conductive Substrates*. Houston, TX: NACE International.

59 NACE RP0274. *High Voltage Electrical Inspection of Pipeline Coatings Prior to Installation*. Houston, TX: NACE International.

60 DNV-RP-B101. *Corrosion Protection of Floating Production and Storage Units*. Oslo, Norway: DNVGL.

61 Baboian, R. (2002). *NACE Corrosion Engineer's Reference Book*, 3, 74–381. Houston, TX: NACE International.

62 NACE SP0178. *Design, Fabrication and Surface Finish Practices for Tanks and Vessels to Be Lined for Immersion Service*. Houston, TX: NACE International.

63 SSPC-PA Guide 11. *Protecting Edges, Crevices, and Irregular Steel Surfaced by Stripe Coating*. Pittsburgh, PA: SSPC.

64 NACE RP0304. *Design, Installation, and Operation of Thermoplastic Liners for Oilfield Pipelines*. Houston, TX: NACE International.

65 NACE No. 11/SSPC-PA 8. *Thin-Film Organic Linings Applied in New Carbon Steel Process Vessels*. Houston, TX: NACE International.

66 Goerz, K., Simon, L., Little, J., and Fear, H. *A Review of Methods for Confirming Integrity of Thermoplastic Liners – Field Experiences*. NACE 04703. Houston, TX: NACE International.

67 Mills, G.D. (2015). Chapter 69: The analysis of coatings failures. In: *Paint and Coating Testing Manual, 15th Edition of the Gardner-Sward Handbook* (ed. J.V. Koleske), 830–848. Conshohocken, PA: ASTM International.

68 Vincent, L. (2004). *The Protective Coatings User's Handbook*. Houston, TX: NACE International.

69 Fitzsimons, B. and Parry, T. (2015). Coating failures and defects. In: *ASM Handbook 5B, Protective Organic Coatings* (ed. K. Tator), 502–512. Materials Park, OH: ASM International.

70 Fitzsimons, B. and Parry, T. (2011). *Fitz's Atlas 2 of Coating Defects*. Surrey, UK: MPI Group.

71 Tator, K.B. (2015). Soluble salts beneath coatings. In: *ASM Handbook, Volume 5B, Protective Organic Coatings* (ed. K.B. Tator). Materials Park, OH: ASM International.

72 ASTM D714. *Evaluating Degree of Blistering of Paints*. West Conshohocken, PA: ASTM International.

73 Amtec Corrosion and Coatings Consultants. Introduction to paint problems. In: *Amtec Guide to Coating Failures & Coating Breakdown*. http://www.amteccorrosion.co.uk/coatingfailuresguide.html (accessed 18 January 2017).

74 Cavallo, J. (ed.) (2017). *Coating Failure Analysis*. Houston, TX: NACE.

75 ASTM F3125/F3125M. *High Strength Structural Bolts, Steel and Alloy Steel, Heat Treated, 120 ksi (830 Mpa) and 150 ksi (1040 Mpa) Minimum Tensile Strength, Inch and Metric Dimension*. ASTM International, West Conshohocken, PA, www.astm.org (accessed 5 April 2018).

76 ASTM A193/A193M. *Standard Specification for Alloy-Steel and Stainless Steel Bolting for High Temperature or High Pressure Service and Other Special Purpose Applications*, ASTM International, West Conshohocken, PA, www.astm.org (accessed 5 April 2018).

77 ISO 9588. *Post-coating Treatments of Iron or Steel to Reduce the Risk of Hydrogen Embrittlement*. Geneva: ISO.

78 ASTM B850. *Post-Coating Treatments of Steel for Reducing Risk of Hydrogen Embrittlement*. West Conshohocken, PA: ASTM International.

79 API RP583. *Corrosion Under Insulation and Fireproofing*. Washington, DC: API.

80 Papavniasam, S. (2014). *Corrosion Control in the Oil and Gas Industry*. Houston, TX: Gulf Publishing.

81 Fitzgerald, B.J., Lazar, P., Kay, R.M., and Winnik, S. (2003). *Strategies to Prevent Corrosion Under Insulation in Petrochemical Industry Piping*, NACE 03029. Houston, TX: NACE International.

82 Thomason, W.H., Olsen, S., Haugen, T., and Fischer, K.P. (2004). *Deterioration of Thermal Sprayed Aluminum Coatings on Hot Risers Due to Thermal Cycling*, NACE 04021. Houston, TX: NACE International.

83 NACE SP0198. *Control of Corrosion Under Thermal Insulation and Fireproofing Materials – A Systems Approach*. Houston, TX: NACE International.

84 van Roig, J. and de Jong, J. (2009). *Prevention of External Chloride Stress Corrosion Cracking of Austenitic Stainless Steel with a Thermal Sprayed Aluminum Coating*, NACE 09348. Houston, TX: NACE International.

85 Davies, M. and Scott, P.J.B. (2006). *Oilfield Water Technology*. Houston, TX: NACE International.

86 Smith, J. and Cobb, W. (1997). *Waterflooding*. Houston, TX: Midwest Office of the Petroleum Technology Transfer Council.

87 Shankardass, A. (2004). Corrosion control in pipelines using oxygen stripping. *Oilsands Water Usage Workshop*, Edmonton, AB, Canada (24 February 2004).

88 Kane, R. (2006). Corrosion in petroleum production operations. In: *Metals Handbook, Volume 13C – Corrosion: Environments and Industries, Volume 13C – Environments and Industries*, 922–966. Materials Park, OH: ASM International.

89 Frenier, W. (2017). *Chemical Methods for Controlling Corrosion in Oil and Gas Activities*. Houston, TX: NACE International.

90 Frenier, W. (2017). *Chemical and Mechanical Methods of Pipeline Integrity*. Richardson, TX: Society of Petroleum Engineers.

91 Kelland, M. (2014). *Production Chemicals for the Oil and Gas Industry*, 2. Boca Raton, FL: CRC Press.

92 Palmer, J.W., Hedges, W., and Dawson, J.L. (2004). *A Working Party Report on the Use of Corrosion Inhibitors in Oil and Gas Production*. London, UK: European Federation of Corrosion.

93 Byars, H. (1999). *Corrosion Control in Petroleum Production*, 2. Houston, TX: NACE International.

94 Revie, R.W. (ed.) (2011). *Uhlig's Corrosion Handbook*, 3, 1000. New York, NY: Wiley.

95 Bradford, S. (2001). Chapter 11: Corrosion inhibitors. In: *Corrosion Control*, 2, 345–371. Edmonton, AB, Canada: CASTI Publishing, Inc.

96 Foss, M., Gulbrandsen, E., and Sjoblom, J. (2009). Effect of corrosion inhibitors and oil on carbon dioxide corrosion and wetting of carbon steel with ferrous carbonate deposits. *Corrosion* 65 (1): 3–14.

97 Kolts, J., Joosten, M., and Singh, P. (2006). *An Engineering Approach to Corrosion/Erosion Prediction*. NACE 06560. Houston, TX: NACE International.

98 NACE TM0187. *Evaluating Elastomeric Materials in Sour Gas Environments*. Houston, TX: NACE International.

99 NACE 1G286. *Oilfield Corrosion Inhibitors and Their Effects on Elastomeric Seals*. Houston, TX: NACE International.

100 Rasmussen, K., Wright, A., Schofield, M., and Lund, J. (2004). *Investigation of a Corrosion Failure in an Offshore Gas Compression System: The Role of H_2S Scavenger and Organic Acids*. NACE 04360. Houston, TX: NACE International.

101 Powell, D.E. (2004). *Methodology for Designing Field Tests to Evaluate Different Types of Corrosion Inhibitors in Crude Oil Production Systems.*, NACE 04367. Houston, TX: NACE International.

102 Gulbrandsen, E., Nesic, S., Strangeland, A. et al. (1998). *Effect of Precorrosion on the Performance of Inhibitors for CO_2 Corrosion of Carbon Steel*, NACE 98013. Houston, TX: NACE International.

103 Miksic, B., Furman, A., and Kharshan, M. *Effectivenets of the Corrosion Inhibitors for the Petroleum Industry Under Various Flow Conditions*, NACE 2009-09573. Houston, TX: NACE International.

104 Gunaltun, Y. and Payne, L. (2003). *A New Techique for the Control of Top of the Line Corrosion: TLCC-PIG*, NACE 03344. Houston, TX: NACE International.

105 Freeman, E.N. and Williamson, G.C. (2007). New technology effective in reducing top-of-the-line corrosion rate. *Pipeline & Gas Journal* (January): 45–47.

106 NACE 1D182. *Wheel Test Method Used for Evaluation of Film Persistant Inhibitors for Oilfield Applications.* Houston, TX: NACE International.

107 NACE 5A195 (1995). *State-of-the-Art Report on Controlled-Flow Laboratory Corrosion Tests.* Houston, TX: NACE International.

108 NACE 1D196 (1996). *Laboratory Test Methods for Evaluating Oilfield Corrosion Inhibitors.* Houston, TX: NACE International.

109 ASTM G170. *Standard Guide for Evaluating and Qualifying Oilfield and Refinery Corrosion Inhibitors in the Laboratory.* West Conshohocken, PA: ASTM International.

110 Papavniasam, S., Revie, R.W., and Bartos, M. *Testing Methods and Standards for Oil Field Corrosion Inhibitors*, NACE 04424. Houston, TX: NACE International.

111 Papavinasam, S., Revie, W.R., Attard, M. et al. (2009). Comparison of laboratory methodologies to evaluate corrosion inhibitors for oil and gas pipelines. *Corrosion* 59 (10): 897–912.

112 Bieri, T.H., Horsup, D., Reading, M., and Woolham, R.C. *Corrosion Inhibitor Screening Using Rapid Response Corrosion Monitoring*, NACE 06692. Houston, TX: NACE International.

113 NACE SP0775. *Preparation, Installational, Analysis, and Interpretation of Corrosion Coupons in Oilfield Operations.* Houston, TX: NACE International.

114 Bianchetti, R. (2001). Chapter 9: Cathodic protection with galvanic anodes. In: *Peabody's Control of Pipeline Corrosion*, 2 (ed. R. Bianchetti), 177–199. Houston, TX: NACE International.

115 Parker, M.E. and Peattie, E.G. (1999). *Pipeline Corrosion and Cathodic Protection*, 3, 100. Houston, TX: Gulf Publishing.

116 Toncre, A.C. (1981). The relationship of coatings and cathodic protection for underground corrosion control. In: *Underground Corrosion, ASTM STP 741* (ed. E. Escalante), 166–181. Philadelphia, PA: American Society for Testing and Materials.

117 Kuhn, R.C. (1953). Cathodic protection of underground pipelines against soil corrosion. *American Petroleum Institute Proceedings IV* 14: 153.

118 Kuhn, R.J. (1933). Cathodic protection of underground pipe lines from soil corrosion. *API Proceedings* 14 (Section 4): 153–167.

119 Beavers, J. (2001). Chapter 3: Cathodic protection – how it works. In: *Peabody's Control of Pipeline Corrosion*, 2 (ed. R. Bianchetti), 21–47. Houston, TX: NACE International.

120 Beavers, J. (2000). Chapter 16: Fundamentals of corrosion. In: *Peabody's Control of Pipeline Corrosion*, 2 (ed. R. Bianchetti), 297–317. Houston, TX: NACE International.

121 Heidersbach, R. (2003). Cathodic protection. In: *Metals Handbook, Volume 13A, Corrosion – Fundamentals, Inspection and Testing*, 855–870. Materials Park, OH: ASM International.

122 England, H. and Heidersbach, R. (1981). Morphology of calcareous deposits produced on cathodically protected steel in the deep ocean. *Oceans* 13: 652–655.

123 Baeckmann, W. and Schwenk, W. (1997). Fundamentals and practice of electrical measurements. In: *Handbook of Cathodic Protection* (ed. W. Baeckmann, W. Schwenk and W. Prinz), 79–138. Houston, TX: Gulf Publishing.

124 Beevers, J. (2001). Chapter 1: Introduction to corrosion. In: *Peabody's Control of Pipeline Corrosion*, 2 (ed. R. Bianchetti), 1–6. Houston, TX: NACE International.

125 May, T. (2004). Magnesium anodes – a quality crisis? *Materials Performance* 43 (1): 2–6.

126 May, T. (2008). Magnesium anode quality – part 2. *Materials Performance* 47 (8): 36–39.

127 Hoxeng, R. and Prutton, C. (1949). Electrochemical behavior of zinc and steel in aqueous media. *Corrosion* 5 (10): 330–338.

128 Britton, J. and Baxter, R. (2017). The Design and Application of Deep Water Offshore Cathodic Protection Systems – Some Practical Considerations. https://stoprust.com/technical-library-items/24-cp-design-in-deep-water (accessed 15 January 2017).

129 NACE SP0387. *Metallurgical and Inspection Requirements for Cast Sacrificial Anodes for Offshore Requirments.* Houston, TX: NACE International.

130 ASTM G97. *Laboratory Evaluation of Magnesium Sacrificial Anode Test Specimens for Underground Applications.* West Conshohocken, PA: ASTM International.

131 Beavers, J. (2001). Chapter 10: Cathodic protection with other power sources. In: *Peabody's Control of Pipeline Corrosion*, 2 (ed. R. Bianchetti), 201–210. Houston, TX: NACE International.

132 B. Laoun, K. Niboucha, and L. Serir (2008). Cathodic protection of a buried pipeline by solar energy. *Revue des Energies Renouvelables* 12 (1): 99–104. http://www.cder.dz/download/Art12-1_10.pdf (accessed 27 January 2017).

133 Peidmont, E. and Tehada, T. (2010). *Solar Powered Cathodic Protection System*, NACE 100092. Houston, TX: NACE International.

134 Bailey, D., Marsh, C., Stephenson, L. et al. (2014). Demonstration of Photovoltaic-Powered Cathodic Protection System with Remote Monitoring System. ERCE/CERL TR-14-3 (February 2014), available from DTIC as report ADA597424.

135 NACE 10A196. *Impressed Current Anodes for Underground Cathodic Protection Systems.* Houston, TX: NACE International.

136 ASTM A518/A518M. *Corrosion-Resistant High-Silicon Iron Castings.* West Conshohocken, PA: ASTM International.

137 Bianchetti, R. (2001). Chapter 8: Impressed current cathodic protection. In: *Peabody's Control of Pipeline Corrosion*, 2 (ed. R. Bianchetti), 157–176. Houston, TX: NACE International.

138 Morgan, J. (1987). *Cathodic Protection*, 2. Houston, TX: NACE International.

139 Baxter, R. (2009). Deepwater corrosion services, Houston, TX, private communication (January 2009).

140 Fontana, M. (1998). *Corrosion Engineering*. New York: McGraw-Hill.

141 NACE SP0176. *Corrosion Control of Submerged Areas of Permanently Installed Steel Offshore Structures Associated with Petroleum Production*. Houston, TX: NACE International.

142 Heidersbach, R. (1987). Cathodic protection. In: *Metals Handbook, Volume 13 – Corrosion*, 466–477. Materials Park, OH: ASM International.

143 NACE SP0169. *Control of External Corrosion on Underground or Submerged Metallic Piping Systems*. Houston, TX: NACE International.

144 Peabody, A.W. (1967). *Control of Pipeline Corrosion*. Houston, TX: NACE International.

145 Parker, M. (1954). *Pipeline Corrosion and Cathodic Protection*. Houston, TX: Gulf Publishing.

146 Bauman, J. (2004). *Well Casing Cathodic Protection Using Pulse Current Technology*, NACE 04049. Houston, TX: NACE International.

147 Bianchetti, R. (2001). Chapter 5: Survey methods and evaluation techniques. In: *Peabody's Control of Pipeline Corrosion*, 2 (ed. R. Bianchetti), 65–100. Houston, TX: NACE International.

148 Bianchetti, R. (2001). Chapter 12: Construction practices. In: *Peabody's Control of Pipeline Corrosion*, 2 (ed. R. Bianchetti), 237–260. Houston, TX: NACE International.

149 Bauman, J. (2004). *Well Casing Cathodic Proection Using Pulse Current Technology*, NACE 04049. Houston, TX: NACE International.

150 Klechka, E., Koch, G., Kowalski, A. et al. (2006). *Cathodic Protection of Well Casings Using E-log I Criteria*, NACE 06071. Houston, TX: NACE International.

151 Kroon, D.H., Williams, G.D., and Moosavi, A.N. (2004). Cathodic protection of well casings in Abu Dhabi. *Materials Performance* 18 (8): 18–23.

152 NACE SP0186. *Application of Cathodic Protection for External Surfaces of Steel Well Casings*. Houston, TX: NACE International.

153 Mendoza, D., Aguilar, A., and Perez, R. (2011). *Cathodic Protection Behavior of API X-52 and API X-65 Steels Buried in Natural Soil*, NACE 11052. Houston, TX: NACE International.

154 Appendix D to Part 192. Criteria for cathodic protection and determination of measurements. https://www.law.cornell.edu/cfr/text/49/appendix-D_to_part_192 (accessed 16 April 2018).

155 NACE RP0186.2001 (2001). *Application of Cathodic Protection for External Surfaces of Steel Well Casings*. Houston, TX: NACE International.

156 National Iranian Gas Company (2010). Buried Pipelines Cathodic Protection. http://igs.nigc.ir/STANDS/IPS/e-tp-740.PDF (17 April 2018).

157 Bianchetti, R. (ed.) (2018). *Peabody's Control of Pipeline Corrosion*, 3. Houston, TX: NACE International.

158 UFC 3-570-1 (2016). Cathodic protection. US Department of Defense document (28 November 2016).

159 Holtsbaum, H.B. (2016). *Cathodic Protection Survey Procedures*, 3. Houston, TX: NACE International.

160 Bianchetti, R. (ed.) (2001). *Peabody's Control of Pipeline Corrosion*, 2 (CD-ROM)e. Houston, TX: NACE International.

161 NACE TM0497. *Measurement Techniques Related to Criteria for Cathodic Protection on Underground or Submerged Metallic Pipeline Systems*. Houston, TX: NACE International.

162 Baeckmann, W., Schwenk, W., and Prinz, W. (eds.) (1997). *Handbook of Cathodic Protection*. Houston, TX: Gulf Publishing.

163 Strommen, R. (1982). Computer modeling of offshore cathodic protection utilized in cp monitoring. *Proceedings, Offshore Technology Conference*, Paper OTC-4367-MS, Houston, TX.

164 Adey, R.A. and Baynham, J.M. *Design and Optimisation of Cathodic Protection Systems Using Computer Simulation*, NACE 00723. Houston, TX: NACE International.

165 Froome, T. and Baynham, J. (2013). *Assessing Interference Between Sacrificial Anodes on Anode Sleds*, NACE-201302344. Houston, TX: NACE International.

166 Gummow, R. (November 2010). Examining the controversy surrounding the –850 mV CP criteria. *Pipeline and Gas Journal* 237 (11): 85.

167 Norsworthy, R. (2010). Revising CP criteria requires careful consideration. *Oil and Gas Journal* 108 (33): 134–139.

168 B. Cherry (2006). How instant is instant? *Journal of Corrosion Science and Engineering* 9: 1–9. https://www.researchgate.net/publication/289427865_How_instant_is_instant (accessed 17 April 2018).

169 Wyatt, B. (2003). Advanced systems of overline assessment of coatings and cathodic protection. *UMIST Cathodic Protection Conference*, Manchester, UK (10–11 February 2003).

170 Brenna, A., Beretta, S., Uglietti, R. et al. (2017). Cathodic protection monitoring of buried carbon steel pipeline: measurment and interpretation of instant-off potential. *Corrosion Engineering, Science and Technology* 52 (4): 253–260.

171 Ondak, E. and Rizzo, F. (2008). *Electrochemical Analysis of Pipeline CP Critieria*, NACE 08608. Houston, TX: NACE International.

172 Szeliga, M.J. (2001). Chapter 11: Stray current corrosion. In: *Peabody's Control of Pipeline Corrosion*, 2 (ed. R. Bianchetti), 216–241. Houston, TX: NACE International.

173 Rix, B. and Boteler, D. (2001). *Telluric Current Considerations in the CP Design for the Maritimes and Northeast Pipeline*, NACE 01317. Houston, TX: NACE International.

174 Hesjevik, S. (2001). *Telluric Current Effect on Short Gas Pipelines in Norway: Risk of Corrosion on Buried Gas Pipelines*, NACE-01313. Houston, TX: NACE International.

175 Nishikawa, A., Nonaka, H., and Fujimoto, S. (2016). Evaluation of cathodic protection under disbonded coating on buried steel structures by laboratory and field tests. *Corrosion* 72 (10): 1311–1322.

176 STI-P3 (2017). Specification and Manual for External Corrosion Protection of Underground Steel Storage Tanks. Revised February 2017. Lake Zurich, IL: Steel Tank Institute.

7

INSPECTION, MONITORING, AND TESTING

The concepts of inspection, monitoring, and testing often overlap, and many organizations have slightly different definitions of the terms. For the purposes of this book, the following ideas will be used to differentiate between the various terms:

- Inspection is used to determine the condition of a system.
- Monitoring is used as a tool for assessing the need for corrosion control or the effectiveness thereof.
- Testing has two oilfield definitions. Hydrostatic testing involves filling a structure with liquid to determine if it has an adequate strength to withstand the desired stresses or pressures, which often involve code-mandated safety factors. Other tests are performed to determine the suitability of equipment, materials, chemicals, etc. for use in field applications. These tests are often performed in laboratories, but may also involve field trials.

The equipment involved in oilfield production is so large and complicated that inspections and monitoring procedures must be selected in a cost-effective manner to ensure safe and efficient operation. It is important to recognize that not every process can be inspected, monitored, or tested before or during operation. The Pareto principle, often expressed as the idea that approximately 80% of all problems come from 20% of the equipment involved, or the concept of the "insignificant many and the mighty few," has been used by many organizations to prioritize inspection, monitoring, and testing [1–4].

Figure 7.1 shows a simple illustration of these ideas and comes from a standard that originated in 1975 [5]. The horizontal piping shown in Figure 7.1 has three areas of differing corrosion risks. Organizations are likely to concentrate both inspection and monitoring efforts in locations A and C of this structure, because they are the locations where corrosion and/or erosion damage is more likely to occur. Testing for the effectiveness of corrosion inhibitors should also emphasize the flow conditions in these areas, and it is likely that corrosion control efforts effective in these high-susceptibility locations will also be effective in low-susceptibility areas like location B in Figure 7.1.

For many years oil and gas production organizations spent significant portions of their efforts inspecting and monitoring corrosion at locations where the consequences of failure were relatively low. The idea of risk-based inspection introduced the relative importance, or consequences, of equipment failures and allows organizations to concentrate their inspection, monitoring, and maintenance efforts where they will produce the most improvement in operational reliability by considering both the likelihood of failure and the consequences thereof. API and other organizations now have risk-based inspection procedure standards [1, 6–12]. These documents also introduce the following concepts:

- Probability of failure.
- Consequence of failure.
- As low as reasonably practicable (ALARP), which indicates that the systems cannot be 100% perfect.

Metallurgy and Corrosion Control in Oil and Gas Production, Second Edition. Robert Heidersbach.
© 2018 John Wiley & Sons, Inc. Published 2018 by John Wiley & Sons, Inc.

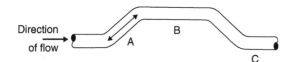

A. Water oscillates – corrosion accelerated

B. Corrosion not accelerated

C. Water impinges at C – corrosion accelerated with higher flow rate (above limiting velocity)*

* Limiting velocity – velocity above which erosion damage can be expected.

Figure 7.1 Areas of increased corrosion susceptibility in a horizontal piping system. *Source*: NACE SP0775 [5]. Reproduced with permission of NACE International.

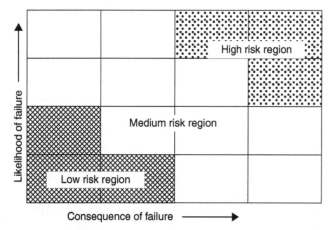

Figure 7.2 Example of an RBI risk matrix.

Figure 7.3 Ultrasonic inspection port installed near a walkway instead of a location likely to produce useful data.

Figure 7.2 shows the concept behind RBI as discussed in several standards and reports. It is common to place the consequences of failure on the horizontal axis and the probability of failure on the vertical axis. Inspection, monitoring, and maintenance efforts are concentrated on those items that fall in the upper right of Figure 7.2 (high probability of failure with significant consequences). The lower left receives the least attention.

RBI guidance has problems in implementation and is an active area of technical discussion and refinement [12]. The difficulty in assessing risks is a major problem with RBI implementation, but the concepts of concentrating inspection and monitoring efforts on those portions of a complex system that are both high consequence and high likelihood of occurrence have gained general acceptance. The difficulty in assessing the risks leads to the concept of ALARP – the idea of as low as reasonably practicable [13]. This idea justifies attempts to concentrate inspection and maintenance efforts on the most likely and consequential scenarios and minimizing time and effort on those situations with low consequences or likelihood of occurrence.

At the time of this writing, API risk-based inspection standards have concentrated on downstream operations, but they also have been applied to production and pipeline operations for many years. The API standards have guidelines for pressure vessels, heat exchangers, atmospheric storage tank shell courses and bottom plates, compressors, pumps, and pressure relief devices. All of this equipment is common to upstream as well as downstream operations, and the standards are divided into sections on inspection planning, determination of the probability of failure, and modeling consequences [6, 7].

It is very important that inspection and monitoring occur in appropriate locations not chosen for the convenience of inspection or monitoring personnel. Figure 7.3 shows an ultrasonic inspection (UT) port on insulated piping located near a walkway. This port was not installed in a location where it would produce the useful information and serves no useful purpose. Instances like this are far too common.

Manufacturing and construction details also need to be considered when prioritizing inspections. Figure 7.4 shows weld locations on piping, which should be considered in inspection. Construction details that cause problems such as the mounting details (and holes in thermal insulation jacketing) shown in Figure 7.5 are also important.

Identification of these high-priority areas is the reason why design, operations, and inspection personnel need to understand the different kinds of corrosion and other degradation mechanisms likely to occur on specific types of equipment.

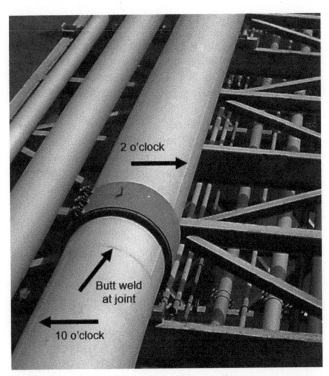

Figure 7.4 Weld locations where inspections should be concentrated.

Figure 7.5 Torn thermal insulation jacketing due to motion of the structural supports on the piping.

INSPECTION

Inspections can be planned and scheduled or occur during unplanned shutdowns, construction modifications, etc. The primary purpose of inspections is to assess the fitness for intended service of the equipment in question. Most inspections are intended to be nondestructive, and industry tends to use the terms nondestructive

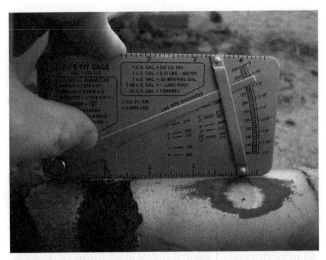

Figure 7.6 The use of a manual pit gage to measure the depth of external pitting on a pipeline.

testing (NDT), nondestructive inspection (NDI), and nondestructive evaluation (NDE) interchangeably [4].

Fabrication-related NDT tends to emphasize welding, locations where manufacturing defects are more likely to occur, and in-service inspections also concentrate on welds, which tend to be locations more likely to be associated with structural failure, and on corrosion.

Choosing where to inspect depends on an understanding of the structural loading of the equipment concerned as well as the flow patterns of any fluids involved and any other parameters likely to affect long-term equipment reliability and corrosion degradation.

Risk-based inspection procedures are intended to concentrate efforts on high-risk, high-consequence (e.g. safety or expense) failure locations within in a system. Projected results of using risk-based inspection priorities are likely to result in increased inspection intervals for most inspection locations, and some high-consequence locations where increased inspection frequency will be indicated. The purpose of these procedures is to move from mandated or calendar-based inspection to inspections based on logic, data, and documented experience [5, 6].

The following discussion covers some of the NDT inspection methods most likely to be used in upstream oilfield operations.

Visual Inspection (VT)

The most common form of inspection is visual inspection (VT) to identify surface abnormalities that can affect system performance. Optical aides such as pit gages, borescopes, fiber-optic cameras, and TV-camera remote-operated underwater vehicles (ROVs) extend VT beyond locations where inspectors can have personal access [14, 15]. Figure 7.6 shows a manual pit gage being used to measure the depth of pitting on the exterior surface of a pipeline.

Surface cleaning, not necessary with some other techniques, is very important for VT, especially when checking for cracks, e.g. from stress corrosion or corrosion fatigue. Once marine growth was removed, the pit shown in Figure 5.30 was visible. Fracture mechanics analysis then determined that underwater welding was necessary to prevent this pit from becoming a dangerous stress riser that could lead to corrosion fatigue of the platform node [15].

The primary advantage of VT is the curiosity and integrity of the individual inspector. While checklists and preplanned inspection procedures are important, this technique, more than others, allows for the inspector to observe unanticipated phenomena. This is the reason why all inspectors need some training in all forms of degradation, mechanical and corrosion related, that they are likely to encounter. Visual inspectors can also identify situations where other inspection techniques are appropriate.

Recent experience with pipeline corrosion has led to increased emphasis on VT, and external corrosion direct assessment and internal corrosion direct assessment techniques (including VT) have received increased emphasis in recent years [16–19].

Benefits of Visual Inspection [4, 14]

- Large areas quickly scanned.
- Pit depths and pitting rates can be identified.
- Video techniques can be used if personnel access is denied.
- Does not require extensive training or equipment.

Limitations of Visual Inspection [4, 14]

- Internal inspection usually requires shutdown.
- Borescopes and cameras only work during operation if medium is transparent.
- Limited to surface defects.

Penetrant Testing (PT)

Penetrant testing, often called dye penetrant testing, can be applied to virtually any nonporous surface and is a common method of inspecting for cracks of all types. The procedure is fairly simple and inexpensive [20]:

- The surface is cleaned.
- A surface-wetting liquid with a colored dye is applied to the surface and allowed to seep into defects through capillary action.
- Excess penetrant is removed from the surface.
- A powdered developer is applied to pull the trapped penetrant from the defect and spread it on the surface so it can be seen.
- VT then determines the location of the defects.

Figure 7.7 shows dye penetrant indications of pitting corrosion and of stress corrosion cracking on stainless steel components.

Penetrants come in two basic types, fluorescent and visible penetrants, and are selected based on penetrating capability and contrast of the dyes. Fluorescent penetrants are primarily used in construction and manufacturing and require the surfaces being inspected to be visually inspected using ultraviolet light. It is common for field inspectors to have portable aerosol cans of dye penetrant, cleaner/remover, and developer.

Benefits of Penetrant Testing [4, 14, 20]

- Simple and rapid.
- Makes surface defects easier to be seen.
- Works on all nonporous materials.

Limitations of Penetrant Testing [4, 14, 20]

- Requires skilled inspectors.
- Limited to surface defects.

(a)

(b)

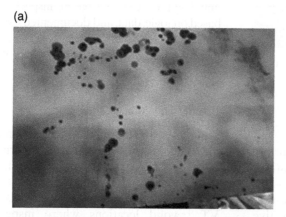

Figure 7.7 Dye penetrant indications. (a) Exterior of stainless steel pipe corroded from the inside. (b) Stress corrosion cracking on a stainless steel component.

- Requires direct access to the surface being inspected.
- Chemical cleaning and disposal is necessary.
- Paint or coatings may mask defects.

Magnetic Particle Inspection (MT)

Magnetic particle inspection (MT) serves most of the same purposes as dye penetrant testing, but it is considered to have two advantages. It can detect near-surface flaws (e.g. hydrogen blisters or weld defects) that would be missed by surface-specific inspection methods such as visual or dye penetrant inspection. For magnetic materials, it can sometimes indicate smaller defects that would not be detected by penetrant inspection. It is used for detecting cracks and similar defects on welds, drill tools, pipelines, and any other iron-based or ferromagnetic components [21].

The process involves applying a magnetic field, either with permanent magnets or with an AC coil, to the area to be inspected. Then a suspension of iron particles is sprayed onto the surface, and the residual magnetic fields "decorate" surface or near-surface flaws – defects that interrupt the magnetic field in the magnetized part being inspected. It is common to apply these particles over a contrast-enhancing temporary coating and to use colored particles to enhance visibility. MT usually involves dry powder application of magnetic powders and visible light inspection, but wet sprays and ultraviolet light inspection is also possible [4]. MT can even be performed underwater to inspect subsea pipelines [22]. Figure 7.8 shows magnetic particle stress corrosion cracking indications on the exterior of an in-service crude oil pipeline.

Benefits of Magnetic Particle Inspection [4, 21, 23]

- Relatively simple and rapid method of inspection.
- May detect fine cracks missed by eye and dye penetrant.

Limitations of Magnetic Particle Inspection [4, 21, 23]

- Extensive training necessary.
- Only ferromagnetic materials inspected.
- Requires smooth, clean surface.
- Paint or coatings may reduce sensitivity.
- May need to demagnetize surface after inspection.

Ultrasonic Inspection (UT)

UT uses high-frequency sound waves to measure the distance from a source transducer to a reflection source such as a defect or metal surface, e.g. the opposite side of the metal being inspected. Any change in density of the material through which the sound wave is traveling will produce an echo that can be detected by ultrasonic detectors. This technique is usually used as a portable technique where the transducer (Figure 7.9) is placed on the metal surface to be inspected. The pulse (sending sound waves)–echo (receiving sound waves) technique is the most common UT technique [4]. High-frequency sound waves are introduced into a material (pulse), and reflected sound (echo) measurements indicate the distance from the material surface that the reflections (echoes) are coming from. Figures 7.9 and 7.10 show a typical setup where a transducer (sound source and receiver) is placed on the surface and echoes from defects within the material and from the far side of the material produce three different return signals – the original pulse at the surface A, an echo from beyond the center of the sample B, and an echo from the far

Figure 7.8 Magnetic particle crack indications on the exterior of a petroleum pipeline.

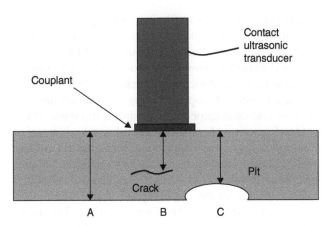

Figure 7.9 Ultrasonic pulse–echo transducer on a plate with a crack and a corrosion pit.

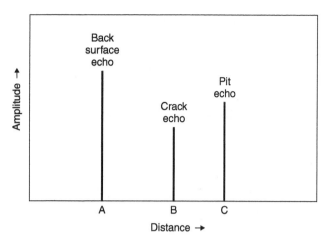

Figure 7.10 Pulse–echo display of the sample shown in Figure 7.9.

Figure 7.11 Ultrasonic inspection every meter at the 12 o'clock position on an exposed pipeline. Note the dark areas on the pipeline where dirt and debris have been removed so that good sensor contact can be maintained with the pipeline surface.

surface. The pulse–echo technique is used for determining remaining wall thicknesses due to corrosion and for weld flaw detection during construction and repairs.

It is common to monitor corrosion by inspecting in the same location at locations marked on the outside of pressure vessels and topside piping. Ultrasonics are also used for one-time inspections, as shown in Figure 7.11.

One limitation of conventional pulse–echo UT is that they only indicate the condition of the structure at locations near the ultrasonic transducer. Other UT techniques have been developed to measure larger areas, although they are less commonly used [14].

It is common to mark the locations where UT measurements have been made (Figure 7.12a). This allows the inspector to return to the same location for subsequent

inspections, and this becomes a means of monitoring if corrosion has progressed or not. Variations in UT readings from nearby locations can be indications that internal pitting corrosion is occurring. Figure 7.12b shows UT wall thickness monitoring points that, even if carefully chosen, are unlikely to identify the precise location where metal loss is greatest. Figure 7.12b is from a crude oil marine terminal. The wall loss due to CUI varies from less than 20% to over 70% on this piping, and if the operator were to rely on only one UT probe, they would unlikely have the probe in the most critical location. UT wall thickness *monitoring* is *not* a substitute for inspection.

The metal in Figure 7.12b is carbon steel. Wall losses of up to 75% have occurred, but the wall loss depths, as indicated by the markings, are variable around the piping. Variations of wall loss at nearby locations are indicators of pitting corrosion.

Benefits of Ultrasonic Inspection [4]

- Only requires direct access to one side of the test piece.
- Accurate thickness and flaw depth measurement.
- Can penetrate thick materials.
- Analytic techniques, based on ANSI/ASME B31G, API 653 and 510, and similar codes, can be used for determining maximum allowable operating pressures and estimated remaining service life [4, 14, 23–27].

Limitations of Ultrasonic Inspection [4]

- Training is relatively extensive and may require several years of experience to produce skilled inspectors.
- Limited use on very thin material.

Radiography (RT)

The use of X-rays and gamma rays has been applied to industrial inspection for many years. The techniques are very similar to those used for medical radiography, and many advances in medical radiography have been adopted for industrial applications [4].

Common oilfield uses for radiography include weld quality inspection, weight-loss corrosion inspections, and the measurement of the extent of scale, hydrate, and paraffin buildup inside pipelines. A typical radiographic exposure using film as a radiation detector is shown in Figure 7.13. The radiation is absorbed by any material between the source and the detector. Figure 7.13 shows that thin cracks are normally missed by radiography, but volume defects, e.g. pitting corrosion and internal porosity, are readily detected.

(a)

(b)

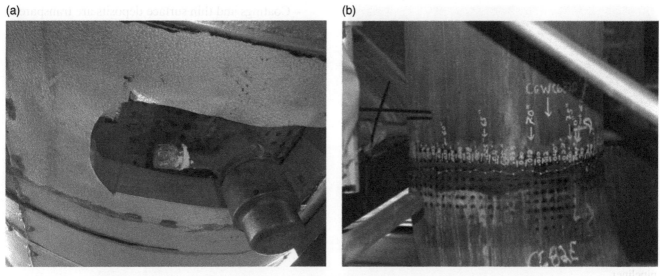

Figure 7.12 Markings where ultrasonic inspection readings have been made on crude oil piping. (a) At a corrosion inhibitor injection port. (b) After insulation has been removed on overwater piping at a marine terminal.

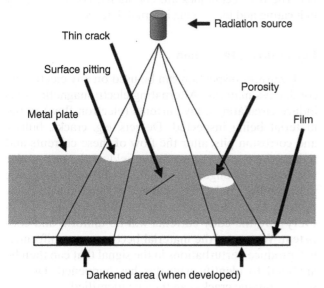

Figure 7.13 Schematic of film radiography of a metal with a corrosion pit, an internal crack, and internal porosity defects.

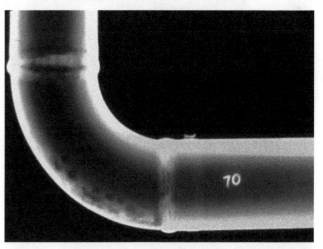

Figure 7.14 Radiograph showing erosion–corrosion at piping bend. *Source*: Courtesy of NACE International.

Oilfield radiography uses X-ray generators, which require high-voltage electricity, or radioisotopic gamma rays. Isotopic radiation sources are limited in the radiation flux they can produce, and they often require longer exposure times than X-rays. A number of radioactive isotopes are available, but iridium-192 is the most commonly used gamma source for oilfield use [28].

At one time image capturing was with film, but advances in electronic radiation detection are capturing much of the market. These electronic methods use similar equipment, e.g. image plates, but they can be processed much quicker, present no chemical disposal problems, and are more amenable to automated image transfer and analysis. These advances have led to significant cost and time reductions [28].

Radiography detects differences in mass between the source and the imaging device. Heavier or thicker materials require longer exposure times, and lighter materials may not be detected. This can be an advantage, as radiography of coated pipelines does not require coating removal before examination. This is shown in Figure 7.14, which shows the effects of erosion–corrosion on the inside of a coated pipe.

Radiography allows inspection and imaging of materials lighter than metals. This is shown in Figure 7.15, where a region of disbonding has been located in a nonmetallic

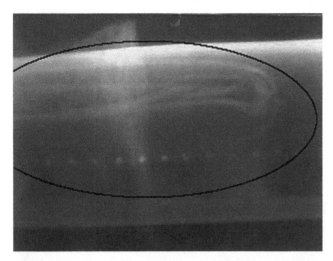

Figure 7.15 Radiograph showing damage to nonmetallic pipeliner.

Figure 7.16 Flange near damage shown in Figure 7.15.

liner using radiography. A nearby flange showing the lining is shown in Figure 7.16. Despite this capability, the great majority of oilfield uses for radiography remain in weld quality and metal loss imaging.

Radiation safety is a major concern when using radiography, and dosage monitoring of all nearby personnel, exclusion of non-necessary personnel from the exposure area, and appropriate radiation shielding are all necessary.

Benefits of Radiographic Inspection [4, 14, 28]

- Rapid use of electronic cameras instead of film.
- Image as permanent record.

- Coatings and thin surface deposits are transparent, which allows minimal surface preparation.
- Can be used on most materials.
- Shows fabrication errors (e.g. misalignments), weld defects, and weight-loss corrosion.

Limitations of Radiographic Inspection [4, 14, 28]

- Only local areas can be inspected.
- Only 2D image – no info on depth of defect.
- Access to both sides of the structure is necessary.
- Radiation safety precautions necessary.
- Free access necessary for radiation source.
- Orientation of crack-like defects means they may be missed.
- Expensive.

Because radiography cannot determine the depth of internal defects and will miss tight cracks like those shown in Figure 7.13, it is common to combine radiography with UT. The two techniques are considered complementary in locating and identifying internal defects.

Eddy Current Inspection

Eddy current inspection can be used on any electrically conductive material. Alternating electromagnetic fields induce circulating eddy currents (electron flow) in the material being inspected. Defects, e.g. cracks, bulges, and corrosion pits, alter the flow of these currents and reduce the secondary magnetic field of the part under inspection. Changes in conductivity and magnetic permeability can be analyzed and correlated with flaws. As long as the material being tested is very uniform in every way, the eddy currents will be uniform and consistent. Whenever the material becomes nonuniform, it will produce perturbations in the signal that can then be analyzed to identify why this has happened. Defects such as fatigue cracks can then be identified.

The basic equipment consists of an alternating electrical current source, a connected coil (probe) that can be passed near the part being inspected, and a voltmeter to measure the voltage change across the coil. The examination process involves moving the probe across the part being inspected and noting where the current changes. Figure 7.17 shows inspection of heat exchanger tubing. In this type of inspection, it is common to draw the probe through the tubing and note locations where indications appear. Further inspection, e.g. with ultrasonics, can identify the extent of the irregularity detected by eddy current. Heat exchanger inspections can analyze many tubes very quickly. It is common to block off any tubes with eddy current irregularities until the total number of blocked tubes affects equipment performance.

Figure 7.17 Eddy current inspection of heat exchanger tubes. *Source*: From Ref. [3].

Benefits of Eddy Current Inspection [4]

- Detects both surface and slight subsurface defects.
- Probes do not have to contact the part.
- Works through paint and some coatings.

Limitations of Eddy Current Inspection [4]

- Relatively extensive training is required.
- Limited to conductive materials.
- Limited depth of penetration.

Magnetic Flux Leakage (MFL) Inspection

Whenever a magnetic field is applied to a nonhomogeneous steel structure, the flux lines induced by the magnetic field source "leak" from the metal at discontinuities such as pits. MFL inspection is a method of electronically detecting this leakage and recording the locations where this occurs. MFL inspection is commonly used to detect wall loss corrosion (pitting) on both sides of storage tank bottoms and both interior and exterior corrosion on pipelines. Figure 7.18 shows how magnetic flux anomalies are detected by sensors located between the two pole sources of magnetic fields.

Figure 7.19 shows MFL inspection of storage tank bottoms. MFL scanning can quickly identify locations where corrosion is likely to have occurred both on the upper and lower surfaces of tank bottoms. It provides information on the location and relative severity of pitting corrosion, which is then usually verified by more time-consuming UT.

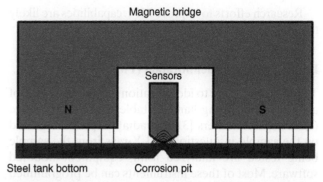

Figure 7.18 MFL detection of pitting corrosion on the far surface of a metal.

Figure 7.19 MFL inspection of a large aboveground storage tank floor. *Source*: Image courtesy Rosen Inspection.

MFL is also the most common NDT technique used in smart pigs for pipeline inspection.

Benefits of MFL Inspection

- Relatively insensitive to surface conditions.
- Rapid inspection of large surfaces.
- Equipment does not need to touch metal surface.

Limitations of MFL Inspection

- Cannot determine magnitude (depth of penetration) of corrosion or wall thinning.
- No discrimination between internal versus external corrosion.
- Limited/no detection of cracks or crack-like defects using conventional MFL equipment.
- Limited detection of gouges and laminations.
- Surface conditions and/or floor buckling alters sensitivity. Rough surfaces generate electrical signal noise and reduce sensitivity.
- Usually supplemented by UT inspections to confirm locations and extent of corrosion.

Research efforts to improve MFL capabilities are likely to remove some of the limitations listed above [29, 30].

Positive Material Identification (PMI)

This term is applied to identification and confirmation of alloy materials using hand-portable X-ray fluorescence (XRF) spectrometers [31]. A radiation probe is placed on the sample in question, and X-rays from the sample being tested are analyzed using equipment-mounted software. Most of these instruments can be programmed for dozens of alloys, and the typical readout tells the operator the alloy with the closest match to the detected X-rays. Early versions of portable XRF devices used radioactive isotopes to excite the X-ray spectra, but in recent years these radioactive isotopes have been replaced by small X-ray tubes, which mean much less documentation of radioactive materials. Once the power source is turned off, X-ray emission tubes emit no radiation and are not a health and safety hazard.

Figure 7.20 shows one of these XRF analyzers in operation. All the operator needs to do is clean the surface so that the bare metal is exposed and turn on the detector. The machine will analyze the sample in a matter of seconds and compare it with a series of preloaded alloy possibilities before providing the nearest match.

Handheld PMI detectors have become the industry standard for PMI of corrosion-resistant alloys (CRAs) [31–37]. The portable XRF spectrometers used for this purpose cannot detect carbon and other light elements, and hardness testing is the traditional way of sorting carbon steel samples.

Figure 7.20 Handheld X-ray fluorescent spectrometer being used for positive materials identification. *Source*: Photo courtesy of Thermo Fisher Scientific.

Benefits of PMI [31]

- Fast and accurate alloy identification.

Limitations of PMI [31]

- Radioactive, requires secure storage and dosage monitoring.
- Cannot analyze carbon steels and elements lighter than magnesium.
- Requires direct access to cleaned surface for analysis.
- Initial equipment cost is relatively high compared with some other techniques.

While PMI using handheld instruments has become routine for handling CRAs, it is important to note that most misidentification of alloys in oil and gas industries is associated with low-alloy steels that are often used instead of carbon steels. PMI instruments can be used to check for the presence and concentration of Cr and Mo in low-alloy steels.

Optical emission spectroscopy is another possible PMI technique. It can identify light elements, but it leaves a small burnt area on the metal surface, which can affect corrosion and cracking resistance. For this reason it is seldom used in upstream field applications, although it is a standard technique in manufacturing.

Thermography

Thermography, also called infrared (IR)/thermal testing, uses IR cameras to detect temperature differences in equipment. It is often used as a remote inspection technique to determine fluid levels in storage tanks, fluid leaks on insulated piping, losses in wall thickness due to erosion–corrosion, and a variety of other purposes [38, 39]. The technique cannot identify the reason for the detected temperature differences. It is used as a quick means of determining locations where closer inspection using other means is appropriate. Figure 7.21 shows a thermal image of piping on an offshore platform. Locations of possible leaks in the piping are readily identified.

Benefits of Thermography [38, 39]

- Identification of hot spots, e.g. due to scale buildup in furnaces or leaks in thermal insulation.

Limitations of Thermography [38, 39]

- Cannot determine corrosion or wall thinning.

Figure 7.21 Thermal imaging of piping on an offshore platform. http://diydrones.com/profiles/blogs/pipeline-inspection-with-uav-thermal-diagnostics, 28 February 2017.

Figure 7.22 Slumping storage tank due to wall thinning.

Additional Remarks About Inspection

There are a number of other inspection techniques used in oilfield applications including alternating current field measurement (ACFM) for external crack detection.

Inspectors need guidance on how to collect and analyze, or protect for shipment for laboratory analysis, biological samples and surface deposits.

One of the biggest problems with inspection is that it can become a routine, and organizations may not pay attention to the results of inspection. This is shown in Figure 7.22, which shows an acid storage tank built from carbon steel that experienced gradual wall thinning due to corrosion. The use of carbon steel for the storage of concentrated sulfuric acid is an accepted industry practice [40–42]. This tank was inspected on a regular

basis to determine the extent of the expected gradual wall thinning, and the original design called for replacement once the thinning reached a certain prescribed extent. Unfortunately, the inspection reports, which proved to be accurate, were filed and not brought to the attention of the appropriate decision makers. The tank was not replaced at the appropriate time, and a subsequent filling of the tank produced the slumping shown in this picture.

Upstream oil and gas operations have many routine inspections, but in far too many cases, the inspection process becomes a routine, and organizations do not realize the implications of reports available somewhere in their organization. Preventable leaks and damaged equipment are sometimes the unfortunate results.

Another problem with inspection is that many useful observations indicating problems with operating equipment are made by personnel whose primary job focus is something else. Figure 7.23 shows images from an insulated piping system at a major marine crude oil loading terminal. Contractor personnel from two different service organizations, responsible for corrosion inhibitor injection or for corrosion coupon monitoring, saw the water and corrosion in Figure 7.23a for several years, but because it was not on their checklists of assigned duties, they did not report their observations to the owner/operator of the terminal. Numerous external moisture leak indications like those shown in Figure 7.23b and c were also visible to many operators, inspectors, and engineers working for the owner/operator, but, once again, no one reported these observations to the appropriate decision makers, until most of the wall thickness of the carrier pipe was corroded away in several locations.

Situations like those shown in Figures 7.22 and 7.23 are not unusual in oil and gas production. The complexity of equipment associated with oil and gas production means that many people are likely to observe situations that can be corrected before they become serious, but the same personnel are so busy doing their primary responsibilities that they tend to overlook other contributions to the safety and reliability of the equipment in their organizations. Mechanical integrity programs cannot rely solely upon contract inspectors and need adequate involvement of plant personnel familiar with plant operations. These personnel also need to be familiar with any changes in operating conditions. It is not unusual for corrosion to be most serious during periods of equipment shutdown or layup.

Many organizations would benefit from recognizing the Pareto principle that most problems are associated with a small fraction of their equipment. Risk-based inspection procedures are attempts to apply this thinking [6]. Having more inspections is often counterproductive. Conducting the right inspections, in the most important

(a)   (b)   (c)

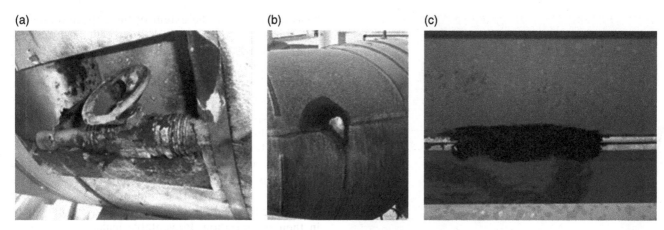

Figure 7.23 (a) Corrosion and moisture indications inside an access hatch for corrosion inhibitor injection on an insulated piping system. (b, c) Moisture leakage indications at seams on the thermal insulation jacketing of the same piping system.

high-risk locations, is preferable to having so many inspections that the organization spends too much time conducting inspections and too little time thinking about what these inspections mean in terms of safety and reliability. The problem is compounded when most, if not all, of the inspections are contracted to outside inspection organizations leaving few, if any, operator personnel spending significant portions of their time doing on-site inspection.

Many organizations have maintenance and inspection budgets related to production volumes. As production fields age and production volumes decrease, equipment ages, and many production fluids become more corrosive. The need for increased inspection and maintenance suffers from decreasing budgets at times of increasing needs for inspection and maintenance.

Inspector training on various NDT techniques can lead to American Society for Nondestructive Testing (ASNT) certification. Other certification organizations are also available.

Inspection intervals are established using a variety of API and ASME guidelines for guidance [6, 7, 25–27, 43].

As mentioned in Chapter 5, oil and gas production and pipeline organizations must concentrate on failure mechanisms most likely to produce leaks (pitting corrosion) and sudden failure (embrittlement and or fatigue cracking). Those inspection techniques associated with pitting corrosion and with brittle failure (linear or narrow cracks) should receive major emphasis.

MONITORING

Inspection is used to determine the condition of equipment at the time of inspection, while monitoring allows operators to determine if conditions and corrosion rates are changing. The two techniques are complementary, and both are necessary. While this monitoring is usually called *corrosion* monitoring, it would be more accurate to call it *corrosivity* monitoring.

Corrosion monitoring is used to determine changes in the corrosivity of environments and to determine the effectiveness of corrosion control techniques such as chemical inhibitor injection [44]. Most oilfields become more corrosive as fields age, production rates decrease, and water cuts increase. Souring of formations, often caused by inadequate injection water treatment, can also cause increased corrosivity. In low-temperature, low-pressure situations where corrosion inhibitors are used to minimize and control corrosion rates of carbon steel, the proper application of corrosion monitoring becomes the principal means of determining and maintaining corrosion control.

Monitoring Probes

Most monitoring techniques require the insertion of metal samples of some type into corrosive production fluids. Two typical arrangements are shown in Figure 7.24. The corrosion coupon shown on the left is exposed to produced water at the bottom of the pipe and to oil that is flowing above the denser water. The flush-mounted probe on the right is only exposed to the produced water on the bottom of the pipe. It is obvious that the two different probes will be exposed to different corrosion environments and return different information.

High-velocity gas streams in pipes can cause problems. Any aqueous phases are usually restricted to thin layers on the surface of piping, and probes that protrude into the piping may miss this corrosive liquid. Surface probes are more appropriate for these situations [44].

It is important to monitor corrosion in the appropriate location. Common locations for internal corrosion are near the bottom or the top, depending on whether corrosion is expected from the water or the gas phase. Oil is generally noncorrosive.

A major limitation of monitoring systems is the limited size and shape of the probes used to monitor corrosion. Probes are manufactured from wire, sheet, or plate having chemistries close to, if not the same as, plate and tubular products used in oilfield applications. The crystal orientation and relative grain boundary areas of the exposed probe samples are different than the flat, as-rolled or as-drawn surfaces of most equipment (Figure 7.25). For these reasons, most probes will have slightly higher corrosion susceptibilities than the actual structures in which they are placed. This produces slightly increased corrosion rates, which are conservative and to be desired. The statistical nature of pitting and other forms of localized corrosion also mean that monitoring with small corrosion probes cannot replicate the corrosion rates of larger metallic structures. It is also difficult, and often impossible, to place probes in the most corrosion-susceptible locations in a complicated piping system.

While the true corrosion rates of large-scale equipment cannot be determined from small corrosion probes, "awareness of *changes* in the corrosion rate is often the major requirement of monitoring with the absolute value of the corrosion rate being less important" [4]. Figure 7.26 shows how corrosion rate monitoring can be correlated with corrosion inhibitor usage. Numerous examples of these correlations between corrosion rates and associated events are possible. They can be associated with mass-loss corrosion coupons, and, with the use of electrochemical corrosion monitoring, it is often possible to identify specific events (e.g. a pump that turns on at a certain time every day) with changes in corrosion rates.

If the efficiency of corrosion inhibitor application is being monitored, it is often advisable to place the monitoring devices as far downstream (away) from the inhibitor injection point as possible. This is intended to detect if enough inhibitor is reaching the downstream locations of the system.

Figure 7.24 Intrusive and flush-mounted corrosion probes inserted into a three-phase production system. *Source*: Hedges [4]. Reproduced with permission of NACE International.

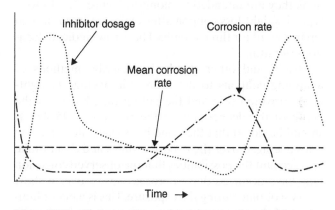

Figure 7.26 Changes of corrosion rates correlated with corrosion inhibitor applications.

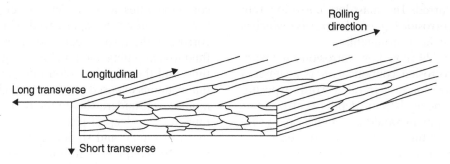

Figure 7.25 Rolling direction and resulting grain structures [44]. *Source*: Reproduced with permission of John Wiley & Sons.

Figure 7.27 A typical weight-loss coupon after exposure to an oilfield environment for three months. Arrows indicate the three areas with pitting.

Mass-loss (Weight-loss) Coupons and Probes The most common method used for corrosion monitoring is the insertion of mass-loss coupons into the environment of interest. This is the simplest form of corrosion monitoring, and it can be used in any environment. The results are easily understood. Coupons are exposed for a period of time (weeks or months) and then removed from the environment [5].

Fittings such as those shown in Figure 7.24 are often used for this purpose. It is important that the coupons be inserted in the appropriate location, usually top or bottom, depending on whether gas-phase or water-phase corrosion is being monitored. Unfortunately, it is all too common for the fittings and probes to be placed in the most convenient access location, and this means that many coupons are not exposed to the corrosive conditions they are intended to monitor. Figure 7.27 shows a typical weight-loss coupon after exposure in an oilfield environment for three months. The arrows indicate locations of pitting.

NACE and other standards prescribe methods of analyzing coupons to determine the average (weight-loss) corrosion rate and the pitting rate, based on the depth of the deepest pit on the coupon [5, 45–47]. It should be noted that these methods calculate corrosion rates averaged over the exposure times and do not take into account the possibilities that the observed corrosion may have occurred due to process upsets over a short interval of time during the exposure. This is a major limitation of the way most coupon exposure testing is conducted. Pitting initiation takes time, and there is no way of determining from mass-loss exposure coupons when during the exposure interval pitting, or any other form of corrosion, has occurred. The mass-loss corrosion rate also assumes that corrosion is general corrosion, which is seldom true for oilfield environments.

Figures 7.28 and 7.29 show how corrosion rates vary with time. Most corrosion coupon analysis assumes a linear, constant corrosion rate. In other words exposure for ½ year would produce 1/10 the depth of corrosion or mass loss that equipment would experience in 5 years. This is seldom the case, but fortunately the "rate decreases with time" situation is most common for general attack corrosion in many environments. In other words,

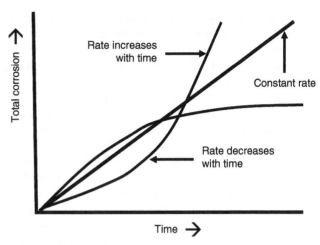

Figure 7.28 Differing corrosion rates found on industrial equipment and structures.

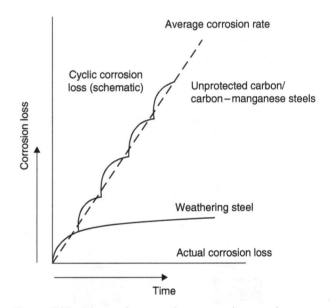

Figure 7.29 Linear (apparently constant) corrosion rates resulting from oxide spalling on carbon steels. Weathering steel produces a protective film and has lower total corrosion weight losses than result from the nonprotective rust that forms on carbon steels [48].

short-term mass-loss exposures tend to overestimate corrosion rates and predict shorter useful equipment lives than is often experienced. Unfortunately pitting corrosion, the form of corrosion most likely to produce fluid loss in piping and pipelines, requires an incubation time, and then the corrosion rate increases. This means that coupons exposed for short periods of time, typically months, will underestimate the overall pitting corrosion rates for the deepest pits and overestimate useful lifetimes of piping and similar equipment. To summarize, mass-loss coupons will often overestimate corrosion rates for general, overall corrosion and underestimate

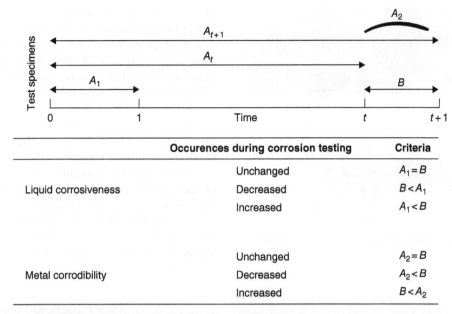

	Occurences during corrosion testing	Criteria
	Unchanged	$A_1 = B$
Liquid corrosiveness	Decreased	$B < A_1$
	Increased	$A_1 < B$
	Unchanged	$A_2 = B$
Metal corrodibility	Decreased	$A_2 < B$
	Increased	$B < A_2$

Figure 7.30   Planned interval testing. *Source*: Ref. [49]. Reproduced with permission of NACE International.

the corrosion rates for pitting corrosion. Because the purpose of corrosion monitoring is to determine if corrosion rates are changing or not, the actual corrosion rate is seldom important. Unfortunately, too many organizations do not understand the limitations of corrosion monitoring and assume that corrosion rates determined by monitoring small samples indicate the actual corrosion rates of their equipment. This results in unnecessary product losses, downtime on equipment, and environmental contamination.

One means of determining if corrosivity has changed during the exposure interval is to use planned interval corrosion testing, a procedure that has been available since the 1940s. The principles of this method are shown in Figure 7.30. The method requires exposing numerous samples for different periods of time and comparing the corrosion rates for the various exposures.

This method, which has been available since the 1940s [42], has been generally supplanted by online electrical corrosion monitoring techniques that can determine changes in corrosion rates in much shorter times.

Corrosion coupons are normally supplied by companies that specialize in the preparation of these samples from standard alloys in accordance with UNS and similar alloy chemical-content standards. These alloys are similar to API-grade OCTG alloys, but they are not the same. Differences in minor constituent chemistry, thermal and mechanical processing history, nature of inclusions and other imperfections, etc. are likely in these coupons. These differences seldom affect the results of weight-loss and pitting determinations, but they can produce unintended results in some environments.

Because monitoring is intended to determine the changes in corrosivity of the environment, it is not necessary to try and replicate the exact alloy in the system for most applications. The use of simple carbon steel samples, which are generally the most corrosion-susceptible alloys in a system, is usually appropriate. Any changes in corrosivity to carbon steel will probably have similar effects on other alloys, which are usually somewhat more corrosion resistant. An important exception to this suggestion is in microbially influenced corrosion (MIC) monitoring of stainless steels. These alloys are generally more susceptible to MIC in hydrotesting systems than are carbon steels in similar environments.

Figure 7.31 shows a typical carbon steel coupon purchased from a major coupon supplier and exposed in an H_2S-containing environment. The arrows show locations of hydrogen-induced cracking (HIC). The carbon steel in this coupon was harder (HRC27) than the OCTGs used in the equipment where the coupon was exposed. While the coupon had cracking, none is likely to occur on the piping system being monitored.

Coupon exposure time must be considered. Short-term exposures (15–45 days) are likely to indicate higher corrosion rates than longer-term exposures. It takes time for biofilms, scale deposits, and pitting to develop. Longer exposures (60–90 days) are sometimes necessary to detect and define pitting attack [49]. Pitting is normally analyzed using optical microscopes at relatively low magnifications. Figure 7.32 shows corrosion coupons from an offshore gas field analyzed using a scanning electron microscope. These instruments are able to

Figure 7.31 Hydrogen-induced cracking on a carbon steel mass-loss coupon.

observe surfaces at much higher magnifications and detect pitting that would be missed at lower magnifications [50]. This allows for faster corrections to corrosion inhibitor treatments. Many oilfield service companies have access to these microscopes, but the use of scanning electron microscopes for coupon analysis remains unusual.

The purpose of all monitoring, to include using mass-loss coupons, is to measure changes in corrosivity. Having monitoring coupons with the same surface finish is important, but it is impossible to match the surface of the equipment. Various authorities recommend that the coupon surface finish should match the equipment, but this is impossible for coupons inserted into equipment after several months, or years, of service, which will no longer have replicable surface finishes.

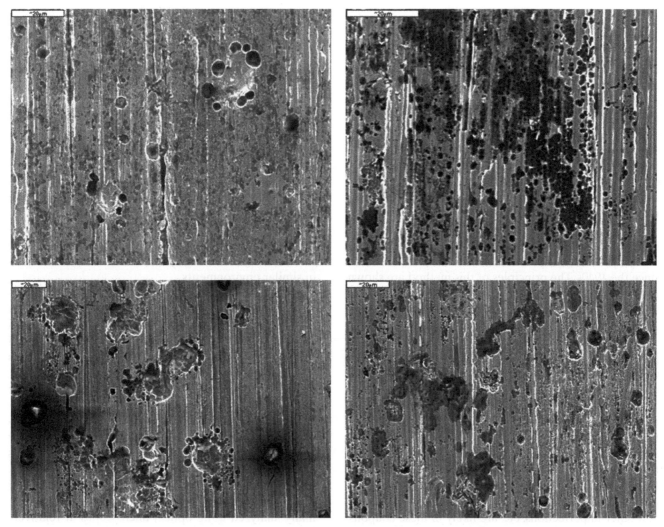

Figure 7.32 Scanning electron micrographs showing early indications of pitting on corrosion coupons from an offshore gas field.

Benefits of Mass-loss Coupons [4, 5, 14]

- Can be used in any corrosive environment.
- Relatively simple procedure, easily understood, and widely accepted.
- Works for general attack and localized corrosion mechanisms.
- Mass-loss coupons are relatively inexpensive.

Limitations of Mass-loss Coupons [4, 5, 14]

- Coupons must be inserted into the fluid, exposing personnel to potential hazards.
- Practical limitations usually mean that data is only available 1–6 times per year.
- Cannot be automated.
- Only determines average corrosion rate, cannot determine effects of upsets or unusual occurrences.
- Short exposure times overestimate general corrosion rates and may miss the onset of pitting or the results of microbial or other films that produce underdeposit corrosion.
- Labor-intensive technique.

Mass-loss coupons are the most common means of monitoring the effectiveness of corrosion inhibitors for corrosion control. Unfortunately many operators have the same contractor responsible for inhibitor application and for monitoring the effectiveness of the inhibitors. This conflict of interest can lead to problems. Another problem with mass-loss coupons is that it is often impossible to place them in the most corrosive locations. Organizations that rely on this data, instead of inspections of the equipment involved, can experience unanticipated equipment failures. Coupons, like all other monitoring techniques, can only indicate whether the corrosion rates are changing. They cannot identify what the corrosion rates are on the most corroded equipment in a system, often in dead legs, the bottom of upward inclines like shown in Figure 7.1, or other inaccessible locations.

Electrical Resistance (ER) Probes   Whenever possible at least two different monitoring techniques should be used. The combination of mass-loss coupons with ER probes is the most common combination in use in upstream oil and gas operations.

Electrical resistance probes allow continuous online monitoring of corrosion and are the second most common monitoring technique for oilfield corrosion. Figure 7.33 shows typical commercially available ER probes. The probes are based on the principle that as corrosion or erosion of the probe occurs, the reduced metal has increased resistance to electrical current. Monitoring the changes in resistance provides an indication of the corrosion inside process equipment. Figure 7.33 shows a number of different geometries including several flush-mounted probes that do not extend into the fluid and measure corrosion at the vessel wall level.

The output from ER probes can be transmitted to any desired location. These probes will work in any environment, and that is a major advantage over other online monitoring techniques. Because all that is measured is the resistivity of the remaining metal in the probe, it can even be used in situations where the environment alternates, e.g. gas bubbles or liquid slugs in piping and pipelines.

The resistivity of metals changes with temperature. Modern ER probe systems have temperature compensation built into the probes, so this is no longer a problem. Electrical probes can be shorted by sulfide deposits, and this is a major limitation to ER probe use [4]. Figure 7.34 shows iron sulfide (FeS) deposits on an ER probe surface (Figure 7.34a) and extensive underdeposit corrosion on another ER probe (Figure 7.34b).

Response times for ER probes depend on the corrosivity of the environment and the metal cross section of the probes. While thinner probes could be manufactured, their service life would be too short making them

Figure 7.33   Typical ER probes. *Source*: Hedges [4]. Reproduced with permission of NACE International.

(a)

(b)

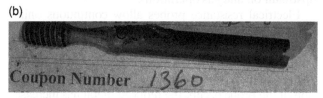

Figure 7.34 Iron sulfide (FeS) deposits on ER probes. (a) Probe with intact scale on the surface. (b) Different probes with scale removed to show extensive underdeposit corrosion.

impractical. At present response times range from ½ to 1 day for very corrosive environments to 5–10 days at the lower end of typical corrosion rates [51].

Benefits of ER Probes [4, 14, 51]

- Continuous online monitoring possible.
- Can be used in almost any environment (conducting or nonconducting).
- Useful in monitoring inhibitor persistence.
- Sensitive to both corrosion and erosion.
- Can be used to monitor sand erosion as well as corrosion.

Limitations of ER Probes [4, 14, 51]

- Results are indicative of general corrosion or erosion; technique does not measure localized corrosion. Specialized probes can be designed to sense crevice corrosion.
- Probes require insertion into the corrosive fluid.
- Response time is slower (hours to days) than for other electrochemical monitoring techniques.

- Iron sulfide deposits can produce misleading results.
- Temperature compensation techniques are not sensitive to rapid temperature changes.

Electrochemical Corrosion Rate Monitoring Techniques

Electrochemical corrosion monitoring techniques include [4, 14, 44, 52]:

- Linear polarization resistance (LPR)
- Tafel extrapolation
- Galvanic monitoring
- Electrochemical noise
- AC impedance spectroscopy

The first three techniques are appropriate for use in oilfield monitoring applications. The latter two, while they have many laboratory- and research-oriented advocates, cannot at the present time produce better results than LPR and Tafel extrapolation. They also require much more expensive, and delicate, instrumentation, and will not be discussed, even though they do appear in the corrosion literature and in standards.

Most of these techniques are based on Faraday's law (Chapter 2), which shows a direct relationship between electric current and the mass of metal lost or deposited in an electrochemical cell. The determination of corrosion rates also depends on knowledge of the valency (oxidation state) of the corrosion reactions [14]. Most instruments are factory calibrated based on the assumption that the corroding metal is iron or carbon steel and corrosion produces Fe^{+2} (instead of Fe^{+3}) ions. This is a conservative approach and appropriate for corrosion monitoring, where changes in corrosion rates are more important than the determination of the true corrosion rates.

Linear Polarization Resistance (LPR) Linear polarization probes (Figure 7.35) are sold with electrodes made from the material being monitored; in most cases this is carbon steel. The probes, which are sold with

(a)

(b)

Figure 7.35 Commercial LPR probes. (a) Two electrode probes protruding into the fluid stream. (b) Three electrode flush-mounted probe.

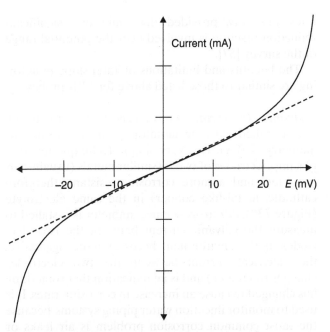

Figure 7.36 Voltage versus corrosion current plot at potentials near the equilibrium (corrosion) potential.

$$i_{corr} = \frac{\beta}{R_p} \qquad (7.2)$$

where

R_p = polarization resistance, Ω (ohms)
i_{corr} = corrosion current, amps
β = the Stern–Geary constant

The Stern–Geary constant can be calculated from theoretical considerations or directly measured in separate experiments [4, 51–57]. As a practical matter, most instrument suppliers sell their instruments calibrated based on the "average" Stern–Geary constant for iron and carbon steel and on the assumption that corrosion of iron produces Fe^{+2} ions (instead of Fe^{+3} ions). These assumptions are justified for monitoring purposes, because the purpose of electrochemical monitoring is to determine if the corrosion rate is changing or remaining steady. The absolute corrosion rate is not determined, but that is not the purpose of monitoring. It is unfortunate that most users of the equipment do not realize that the corrosion rates are not "real."

LPR instruments can determine changes in corrosion rates within minutes, sometimes even seconds. This real-time indication that corrosion rates are changing is the main advantage of this technique over the simpler and more widely used electrical resistance technique, which may take hours or days to respond to changes in corrosion conditions.

Like all electrochemical monitoring techniques, the electrodes must be kept free of biofouling and of oily deposits. This is a major limitation and the reason why this technique is not more widely used in production monitoring before separation processes remove hydrocarbons from water.

either two or three electrodes depending on the manufacturer, are small and can be inserted into the fluids of interest in the same manner as electrical resistance probes. The technique is based on the observation that at potentials very near (±20 mV) the corrosion potential, the voltage versus current plot is frequently linear [14]. This is shown in Figure 7.36. The voltage/current slope is directly proportional to the mass-loss corrosion rate. The method is restricted to aqueous solutions and is less accurate in high-resistivity waters. Because the purpose of LPR monitoring is to determine changes in corrosion rates, e.g. the results of corrosion inhibitors, the inaccuracy of the reported corrosion rates is seldom a concern.

The polarization resistance, R_p, is the slope of the voltage versus current line near the corrosion potential [4, 14, 51, 52]:

$$R_p = \frac{\Delta E}{\Delta i} \qquad (7.1)$$

where

R_p = the polarization resistance, Ω (ohms)
ΔE = the change in potential
Δi = the change in current

The polarization resistance is then converted to a corrosion current using the Stern–Geary equation, Equation (7.2) [4, 14, 51, 52]:

Benefits of LPR Probes [4, 14, 51–55]

- The probes do not have to be removed to obtain corrosion rate data allowing real-time measurements to be collected.
- Corrosion rate data can be obtained as often as measurements can be made.
- The probes can be remotely controlled to record and transmit data to the corrosion engineer's office.

Limitations of LPR Probes [4, 14, 51–55]

- Probes require insertion into the fluid.
- In multiphase systems the electrodes can become covered in oil or condensate, which blocks off part of the electrode area so that the actual area is not known.

- Cannot be used for measuring localized corrosion rates (e.g. pitting).
- Cannot be used in sour systems (H_2S) since conductive iron sulfide deposits can short circuit the electrodes.

Tafel Extrapolation This technique uses the same instrumentation as used in LPR monitoring, and most instruments are sold with the option of operating in either the LPR or the Tafel extrapolation mode. At potentials greater than a few millivolts from the equilibrium (corrosion) potential, potential–current plots frequently become linear on a log-linear plot when the potential is plotted on a linear basis and the current on a logarithmic scale. This is shown in Figure 7.37.

Prior to the application of applied current, the voltmeter reads the corrosion potential relative to a reference electrode. As applied current is increased, the applied current versus potential curve shows no change in potential when most of the reduction current on the working electrode is due to the corrosion reaction. Eventually the effects of the applied cathodic current become apparent, and the curve slopes downward. Once most of the current is due to the applied current, the slope becomes linear, and the original current becomes negligible. The log-linear portion of a polarization curve is called the "Tafel region" in recognition of the German chemist who first described this behavior. The Tafel slope is then extrapolated back to the original potential to determine the oxidation (corrosion) current before added current was applied [57].

This technique can measure low corrosion rates at equal or greater accuracy than weight-loss measurements. It is possible to measure extremely low corrosion

rates this way, provided that only one significant reduction reaction is involved over the potential range of the survey [57].

The benefits and limitations of Tafel slope monitoring are similar to those listed above for LPR probes.

Galvanic Monitoring Galvanic corrosion monitoring is also called galvanic monitoring or zero-resistance ammetry (ZRA). This very simple technique involves placing electrodes of two dissimilar metals (usually carbon steel and a more corrosion-resistant, therefore cathodic, metal-like copper) in the same electrolyte (Figure 7.38). A zero-resistance ammeter is installed to measure the galvanic current between the two electrodes. If the environment becomes more aggressive, the electrical current between the two electrodes changes (increases) and is an indication that something has changed to cause an increase in corrosion rates. It is used to monitor injection water piping systems, because the most common corrosion problem is air leaks or bacteria, which depolarize the cathode and increase current flow between the two electrodes. The instrumentation for this technique is relatively simple, and, like other electrochemical techniques, the results from many electrodes can be monitored at a central location [4, 14, 55, 58, 59]. The response time is as rapid as for LPR probes.

Benefits of Electrochemical Monitoring Techniques [4, 14, 55]

- Faster response time than other techniques.
- Real-time monitoring is possible.
- Corrosion rate data can be obtained as often as measurements can be made.
- Probes do not need to be removed to obtain data.

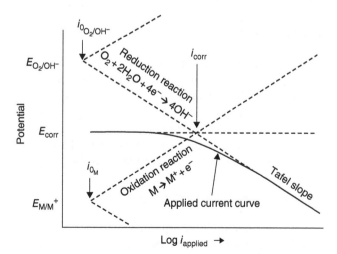

Figure 7.37 Applied current cathodic polarization curve of a corroding metal showing Tafel extrapolation.

Figure 7.38 Galvanic current monitoring probe.

- Probes can be remotely controlled to record and transmit data to a central location anywhere in the world.
- Can provide information about the stability of a system such as the persistence of corrosion inhibitor films.

Limitations of Electrochemical Monitoring Techniques [4, 14, 55]

- Electrodes are subject to fouling by scale or biofilms; sulfide scales can also lead to erroneous results.
- Probes require insertion into the fluid of interest.
- Electrodes can become covered with oil or condensate blocking off part of the electrode in multiphase systems.
- Cannot measure localized corrosion.
- Conductive sulfide deposits prevent the use in sour systems.
- Erosion cannot be measured.

Additional Comments on Electrochemical Monitoring

Electrochemical monitoring has very fast response times, and some organizations have coupled electrochemical monitoring systems with automated control systems for corrosion inhibitor injection. This is not recommended for most oilfield situations, because of the complicated interactions between corrosion inhibitors, scale and hydrate inhibitors, etc.

The LPR and Tafel extrapolation techniques should only be used in water systems after the majority of hydrocarbons have been removed. Galvanic monitoring is less sensitive and can work in multiphase systems, but it can only monitor corrosion in water-based liquid systems and will not monitor corrosion in top-of-line condensate locations.

Electrochemical monitoring instrumentation is often oversold. Many vendors claim that these techniques can be used for monitoring localized pitting and crevice corrosion and other phenomena. While this may be true in the laboratory, field use for identification and monitoring of pitting corrosion is not available and unlikely considering the limitations of small electronic probe size compared with the large size of oilfield systems. These techniques are complementary to inspection techniques, but they cannot substitute for well-planned inspections.

Electrochemical noise and AC impedance spectroscopy are not suitable for use on oilfield corrosion monitoring situations with the single possible exception of using AC impedance spectroscopy to monitor coating systems for incipient breakdown in field tests of competitive coating systems.

Hydrogen Probes

The reduction reaction associated with corrosion in acids and in sour service environments is hydrogen gas evolution. Some of the hydrogen atoms migrate into the metal and can cause HIC and other problems. The use of hydrogen probe monitors is a relatively simple and inexpensive means of monitoring corrosion activity in these situations [14, 45, 59–63].

There are three types of hydrogen probes [14, 45, 59–63]:

- Hydrogen pressure (or vacuum) probes
- Electrochemical hydrogen patch probes
- Hydrogen fuel cell probes

Figure 7.39 shows a schematic of a hydrogen pressure probe that can be externally mounted on the outside of pipelines or storage tanks. The seal between the probe and the structure must be gas tight, and this sometimes requires welded patches, although temporary probes that can be removed and used in other locations are also available. This type of probe can be used to monitor changes in corrosion activity due to internal corrosion inhibitor treatments. The results of one field study are shown in Figure 7.40.

Response times for hydrogen probes are typically several hours, and the probes can only sense corrosion activity for localized areas [63].

These probes have also been used to monitor hydrogen gas permeation into interior components of submerged offshore structures due to cathodic protection of the submerged exterior surfaces.

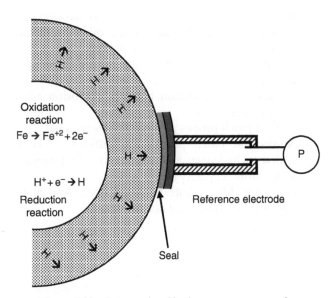

Figure 7.39 Schematic of hydrogen pressure probe.

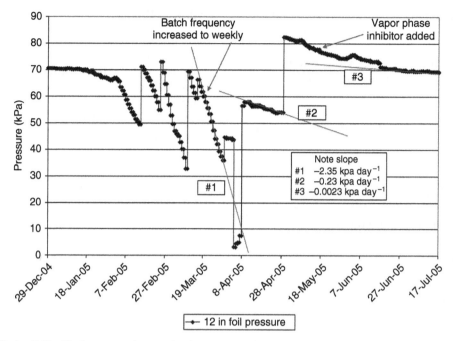

Figure 7.40 Hydrogen probe monitoring of corrosion activity on inhibitor-treated pipeline. *Source*: Mabbott [63]. Reproduced with permission of NACE International.

Benefits of Hydrogen Probe Monitoring [14, 52, 60–62]

- Useful to correlate with HIC and H blistering to warn off further events or increased damage.
- Some hydrogen probes are easy to relocate to other areas of interest.
- Online monitoring possible.

Limitations of Hydrogen Probe Monitoring [14, 52, 60–62]

- The correlation of hydrogen flux and corrosion varies, especially in situations where other chemicals can be involved in reduction reactions.
- Not useful in oxygen reduction in neutral or base environments.
- No accurate correlation between corrosion rates and hydrogen flux.
- Welded patch probes may require stress relief.
- Nonwelded patch probes are difficult to keep sealed to hydrogen and limited to the maximum temperature capacity of the seals.
- Does not differentiate between steels subject to HIC and steels that are not susceptible.

Sand Monitoring

Sand monitoring is necessary for a number of reasons to include prevention of unexpected erosion failures on pipelines, wells, and topside piping systems [64, 65].

There are three major types of sand-monitoring probes. One form is an ER probe, as described above. The probes are usually made of a corrosion-resistant alloy, and as the probes wear away, the resistance increases. This technique is simple and reliable. The probe must be located in an appropriate area where erosion is to be expected to occur.

Another probe design involves a CRA tube with a vacuum. If the tube wears to the point that a leak develops, the vacuum is lost, and an electric signal is generated. Once again, this technique is simple and reliable.

Acoustic sand monitors are also available. These monitors can be mounted externally on piping, and the signal from them provides an indication of the volume of sand moving through the piping. An alternative version of this design involves inserting a probe into the fluid stream and monitoring the acoustic signal from the sand striking a sensing element. This system has the advantage of providing real-time data and warnings when problems may develop. Figure 7.41 shows how the acoustic signal on a topside monitoring station indicated when a sand event (a large "slug" of sand) due to a gravel pack failure caused problems on an offshore platform.

The efficiency of acoustic sensors depends on the relative velocity of the fluids involved. Figure 7.42 shows a sensitivity regime map for a commercial acoustic sand detector. Detection is much more efficient in low liquid fluid streams, provided the superficial gas velocity is high enough (approximately $10\,m\,s^{-1}$ [$40\,ft\,s^{-1}$ or more]).

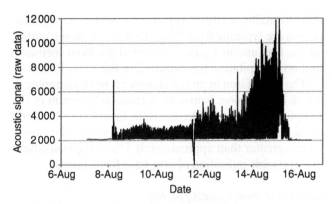

Figure 7.41 Acoustic data indicating the progressive sand production due to a gravel pack failure.

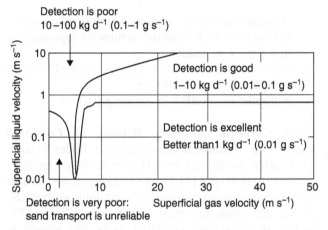

Figure 7.42 Sensitivity regime map for a commercial acoustic sand detector.

Location of sand sensors is critical. The best places are in locations where erosion is likely to occur, either immediately downstream from piping bends or downstream from flow restrictions like valves or chokes [64, 65].

Whenever sand production is encountered, the presence and proper disposal of naturally occurring radioactive materials (NORM) must also be considered. These materials must be disposed of in accordance with local regulations for hazardous waste disposal.

Fluid Analysis

The chemical and suspend solid contents of production fluids can be monitored to analyze corrosion problems upstream of the sampling point. Chemical monitoring is also used to ensure that treated fluids, e.g. steam and injection water, do not cause corrosion, biological growth, or scaling problems.

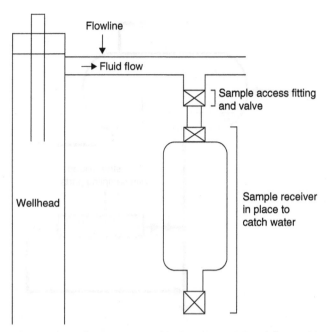

Figure 7.43 Sample receiver for collecting water samples. *Source*: NACE SP0192 [66]. Reproduced with permission of NACE International.

Sampling Procedures Fluid sampling affects the validity of any chemical monitoring system. Various test probes and sampling ports can be placed in topside piping systems to monitor a variety of parameters. Figure 7.43 shows a typical sampling receiver for collecting water samples for further analysis. These receivers are usually located below a flow line or similar piping system, but horizontal collection locations are sometimes used to avoid the collection of sand, silt, or microbiologically created material [67].

Multiple sensors can be placed in sidestream devices like the one shown in Figure 7.44. These sidestream devices should be located after oil–water separators to avoid hydrocarbon fouling of probes, coupons, and sensors. Devices like this have been used in power plants since the 1920s and are commercially available for use in both high- and low-pressure models for use in oil and gas production. Sidestream devices have the limitation that they do not reproduce the fluid flow patterns of the main flow channels, so they should only be used for sampling and monitoring those fluid parameters that will not be affected by diversion into the device (water chemistry) [68–70].

Iron Counts This corrosion monitoring technique dates back to the 1950s and is still widely used to provide a simple indication of corrosion activity upstream of the sampling point. Figure 7.45 shows the response of iron counts to corrosion inhibitor injection into a system.

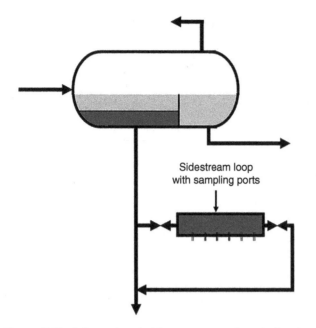

Figure 7.44 Schematic of sidestream sampling device for drawing water samples or for continuous monitoring of water quality.

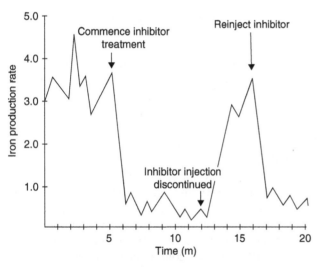

Figure 7.45 Iron count rate history showing the effects of corrosion inhibitor injection. *Source*: NACE SP0192 [66]. Reproduced with permission of NACE International.

NACE SP0192 contains detailed information on water collection and analytical procedures [66]. The samples can be analyzed in the field using a variety of commercially available colorimetric tests or shipped to a laboratory. Iron count analysis needs to distinguish between iron associated with corrosion and naturally occurring iron present in the formation water. A common way of doing this is to compare the iron content of the water sample with the manganese content. All carbon

steel contains some manganese, and the manganese counts are typically from 0.5 to 1.5% of the iron counts if all the iron and manganese detected are from corrosion and no precipitation has occurred in the water [66].

The lack of iron in produced water is not a guarantee of a lack of corrosion. Iron-scale formation upstream of the sampling point is possible, especially in waters with neutral or higher pH and high carbonate contents. Manganese counts greater than approximately 1% of the iron counts are a possible indication of iron-scale formation.

Benefits of Iron Counts [14, 66]

- Quick, inexpensive, and easy field analysis.

Limitations of Iron Counts [14, 66]

- Assumption is made that corrosion is proportional to iron content, but upstream mineral deposits may lower the iron counts.
- Representative sampling may be difficult due to complicated flow conditions.
- Not reliable in sulfide-containing systems

Other Chemical Analyses Related to Corrosion Monitoring   Relatively clean water, e.g. boiler feedwater, is often monitored for pH, electrical conductivity, and dissolved oxygen [14]. These measurements are usually conducted using commercially available probes. The probes are often mounted, along with ER or LPR probes, in sidestream sampling loops like shown in Figure 7.44, and online monitoring relayed to central control stations is common.

pH Monitoring   pH measurements provide direct evidence of changes in fluid parameters that affect corrosion. Low pHs are corrosive to carbon steel, while neutral or higher pHs are relatively benign, especially in the absence of dissolved oxygen.

Benefits of pH Monitoring [14]

- Simple.
- Probes have rapid response.

Limitations of pH Monitoring [14]

- Interference from sodium, lithium, and potassium ions.
- Frequent probe maintenance is necessary.

Electrical Conductivity Monitoring   The electrical conductivity of water is a direct measurement of the presence or absence of dissolved ions. Conductivity measurements are routinely used to monitor steam condensate systems. Any increases in conductivity are evidence of leaks into the system. These leaks are often from condensers, and it is common to monitor the conductivity immediately downstream from condensers. Many gas fields produce low-conductivity condensate, and conductivity monitors can also be used to monitor changes from the production of condensate to the onset of high-conductivity (high salt) formation water contamination.

Benefits of Conductivity Monitoring [14]

- Simple and rapid response.
- Early warning of leak, e.g. in steam or condensate return system.

Limitations of Conductivity Monitoring [14]

- Routine cleaning necessary to avoid bridging the electrodes.
- Temperature sensitive.

Oxygen-level Monitoring   Oxygen is the most corrosive gas commonly found in topside oilfield fluids. The presence of dissolved oxygen is an indication of leaks into production fluids, which normally have very low oxygen concentrations. Oxygen scavengers can remove oxygen down to 1 ppb or less, although some boiler water treatments deliberately keep oxygen levels at 0.02–0.2 ppm in order to provide enough dissolved oxygen to passivate piping systems. Online oxygen probes are commonly used to monitor dissolved oxygen in topside production water and injection water systems. The presence of high levels of oxygen is an indication of leaks in the system and an indicator of the need for additional oxygen scavengers or other oxygen control procedures. Electronic oxygen probes must be periodically replenished of the chemicals used to detect oxygen [14].

Benefits of Oxygen Monitoring [14]

- Changes in oxygen indicate problems in the system.

Limitations of Oxygen Monitoring [14]

- Electrode poisoning in hydrocarbon processes.
- Other dissolved gases (H_2 and CO_2) can interfere with measurements.
- Regular maintenance of electrodes.

Bacterial Growth Monitoring   MIC can occur on the exteriors of tank bottoms and on exteriors of buried or submerged piping. The presence of bacteria in these situations is unavoidable. Corrosion control by a combination of protective coatings and cathodic protection effectively minimizes this problem, and microbial growth monitoring is not necessary.

On the insides of piping systems, monitoring for the presence and growth of bacteria has been proven necessary in many circumstances. NACE TM0194 and other industry standards address this problem by suggesting methods for monitoring bacteria and the effectiveness of biocide treatments [71–77]. The NACE standard provides guidance on effective sampling and culture procedures for both planktonic, freely floating bacteria and sessile, surface-attached bacteria. A number of commercially available field testing technologies are available to determine bacterial populations and activity.

Planktonic bacteria, bacteria freely floating in the liquid, are collected using sampling devices similar to those shown in Figures 7.43 and 7.44. It is important that any liquid samples be contained in clean glass or plastic containers. Samples should be analyzed as soon as possible, and, if delays of more than one hour are unavoidable, the samples should be kept in airtight glass containers. Refrigeration of samples kept more than four hours is also recommended.

Sessile bacteria grow in biofilms on metal surfaces. The standard discusses coupons for collecting these biofilms, often at the six o'clock position in oil and gas piping.

Sampling is recommended just prior to and after biocide treatments. Bacteria can then be cultured and assessed for responses to biocides.

Naturally Occurring Radioactive Materials (NORM)

Oil and gas production will also bring to the surface some NORM, which may concentrate in corrosion products and scales. Removal of these products and scales necessitates handling with appropriate protective equipment and disposal in accordance with established government restrictions and guidelines. Scales containing barium or strontium (usually but not always carbonates) tend to have measurable NORM levels. Production tubulars and scrap metal from topside processing equipment may have NORM contamination that results in disposal problems. A recent report titled "Managing Naturally Occurring Radioactive Material (NORM) in the oil and gas industry" does not mention corrosion, although it does discuss NORM in produced waters, scales, sludge, and pigging debris [78].

Health and safety standards associated with NORM are regulated by individual states in the United States and by various governments worldwide. A detailed discussion of this subject is beyond the scope of this book.

Additional Comments on Monitoring

Table 7.1 compares several corrosion monitoring techniques and shows the forms of corrosion for which they are suitable [52].

Additional monitoring methods are available and discussed in NACE and other publications.

The purpose of monitoring is to identify when changes in corrosion rates are occurring and to correlate these changes with corrosion control procedures [4]. No monitoring method can identify actual corrosion rates. Oilfield systems are too complicated, the sizes of samples used to monitor corrosion are much smaller than the surface areas of exposed equipment and piping, and the metallurgical conditions of probe materials are different from the conditions on complicated structures that have welds, stresses, and other complications not replicated in monitoring samples. Once changes in corrosion rates are noted, causes can be identified, e.g. the need for additional corrosion inhibitors or changes in the corrosivity of produced fluids or injection water. Corrective actions can then be taken if necessary.

Many organizations suffer from too much monitoring. It is common to have chemical and inhibitor suppliers responsible for the application of the appropriate chemicals and also for the application and analysis of monitoring coupons or probes used to determine the effectiveness of these treatments. Aside from the obvious conflicts of having suppliers monitor their effectiveness, it is often the case that too many coupons or other sources of data are collected. This has been known to lead to organizations spending too much time and effort on the insertion, removal, and collection of field monitoring data and not enough time analyzing the meaning of the data collected. Widely publicized oilfield failures have been associated with systems where hundreds of thousands of coupons had been collected, yet unexpected leaks on major equipment still occurred. The unfortunate consequences of reduced production rates are that oilfields become more corrosive at the same time that aging equipment and increased corrosivity require more corrosion control and monitoring.

Monitoring cannot replace inspection. The two procedures are complementary. Appropriate corrosion monitoring can reduce the need for inspections and indicate potential locations where problems are occurring and additional inspections are warranted.

Probes can only work when they are activated. Many data collection units often rely on alkaline batteries for power. These are less reliable than hardwired data collection units.

TESTING

Testing is used in two senses in oilfield applications. Hydrostatic testing is commonly used to "proof" newly constructed or altered equipment to ensure that the equipment will be safe to operate under the intended temperature and pressure conditions. The other use of the term is for relatively short-term laboratory or field trials to determine materials compatibility, the effectiveness or corrosion inhibitors, etc.

Hydrostatic Testing

Hydrostatic testing is required on pipelines and storage tanks after construction and major repairs. Incomplete removal of the hydrotest water after piping and pipeline testing can result in major corrosion problems. While it would be desirable to use clean water for these tests, as a practical matter river water or seawater is often used. If this water is not removed and the pipeline dried, then

TABLE 7.1 Comparison of Corrosion Monitoring Methods [52]

Corrosion Phenomenon	Corrosion Monitoring Method			
	Ultrasonic	Coupons (Mass Loss)	ER Probes	LPR
General (uniform) corrosion	Excellent	Excellent	Excellent	Good
Pitting corrosion	Fair	Excellent	NA	NA
Galvanic corrosion	NA	Excellent	NA	NA
MIC	Fair	Good	NA	NA
Erosion–corrosion	Good	Excellent	Excellent	NA
SCC	Fair	Good	NA	NA
Intergranular corrosion	NA	Good	NA	NA
Hydrogen-induced corrosion	Fair	Fair	NA	NA
Crevice and under deposit corrosion	NA	Good	NA	NA

ER, electrical resistance; LPR, linear polarization resistance; MIC, microbiologically influenced corrosion; SCC, stress corrosion cracking; and NA, not applicable.
Source: Reproduced with permission of Springer.

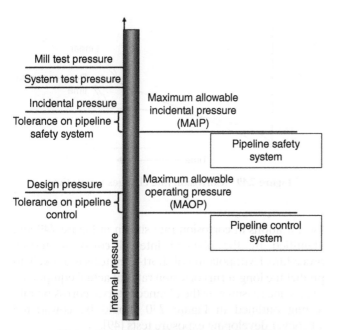

Figure 7.46   Pipeline pressure testing [86].

Figure 7.47   Crack and leak at seam weld during hydrostatic testing. *Source*: Photo courtesy J. Smart.

(a)

(b)

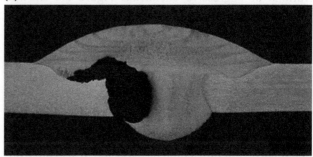

Figure 7.48   MIC pitting of stainless steel after hydrotesting in freshwater. (a) 304 stainless steel pitting at low spots that did not drain after only six weeks, and (b) weld in 316 stainless steel one year after freshwater hydrotesting.

microbial colonies can form, and corrosion can be noted in a matter of days. Treatment of the test water with biocides can minimize this problem, but disposal of the biocide-containing test water can become a problem. Industry standards on hydrostatic pressure testing and on the treatment and disposal of hydrotest water are available [79–86]. Both liquid and gas lines are tested with water, because the energy release from compressed gases would be dangerous.

Figure 7.46 shows various levels of pipeline pressure testing and compares these levels with design pressures [86].

If a newly constructed pipeline successfully passes a hydrostatic pressure test, it can be assumed that no hazardous defects are present in the tested pipe. Older pipelines, especially those manufactured prior to 1970 using low-frequency electric resistance welding (LFERW) and lap welding (LW) of the longitudinal seam, are known to be susceptible to failure, and these older pipelines also need to be tested, even if there is no evidence of corrosion or other in-service deterioration. Some of the factory-welded seams in these types of pipe can be susceptible to failure [82, 83]. Similar tests are done on aboveground storage tanks (ASTs) after construction or repairs, but the pressure levels are limited to the pressure exerted by the height of the water column [26, 87]. Figure 7.47 shows a leak along a seam weld after hydrostatic testing during construction of aboveground piping.

Most problems with corrosion following hydrotesting are associated with MIC, although other forms of corrosion are also possible [87–90]. Figure 7.48 shows

pitting corrosion after hydrotesting of stainless steel piping at a weld. Welds are relatively rough surfaces where biofilms are more likely to form and biocide treatments are less likely to be effective. Contrary to other environments, stainless steels are considered more susceptible to post-hydrotesting MIC than carbon steels. The reasons for this are not clear, but may be due to the relatively thin passive films formed on stainless steels compared with those formed on carbon steel. Stainless steel piping also tends to be thinner than carbon steel piping.

The choice of water for hydrotesting is important, and the cleaner the water the less likely that post-testing corrosion problems, especially MIC, will occur. Unfortunately, in many oilfield applications, the large size of pipelines and similar equipment means that river or ocean water is used [88]. It then becomes even more important to treat the water with biocides, oxygen scavengers, and corrosion inhibitors [87–90].

Laboratory and Field Trial Testing

Laboratory and field trial testing procedures have been developed to provide short-term evaluation methods for new or replacement materials or chemical treatments prior to their adoption for field use. The tests tend to concentrate on potential weaknesses or vulnerabilities of the materials being tested, e.g. H_2S compatibility for metals, decompression resistance of polymeric liners, etc.

Test Duration Accelerated laboratory tests rely on one of two approaches to produce short-term accelerated results. Either the environment is made more aggressive, or the methods of determining incipient failure are improved over field conditions. Common means of making environments more aggressive include increasing temperatures, pressures, or concentrations of aggressive chemicals. Figure 3.8 showed how varying salt concentrations altered the corrosivity of brines by reducing the oxygen solubility. These interactive effects must always be kept in mind whenever developing corrosion testing protocols.

Testing and sampling evaluations need to recognize the time dependence of various kinds of corrosion. Most exposure testing involves weight-loss measurements or electrochemical monitoring of corrosion rates. Figure 7.49 shows two idealized corrosion rate plots. The linear corrosion rate would be expected in acid environments where the corrosion is dependent on transport of reducible chemicals to the metal surface. This is relatively rare in oilfield applications. The parabolic, decreasing with time, kinetics are far more common. Exposures for very short periods would predict very high corrosion rates, whereas longer test exposures might reveal minimal increases in overall corrosion rates after the system equilibrates.

Reporting corrosion rates based on the weight loss or pitting depth after one time interval does not recognize

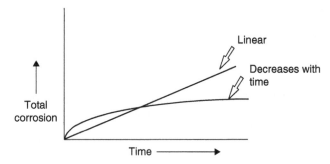

Figure 7.49 Corrosion rate changes versus time.

the variations in corrosion rate shown in Figure 7.49 and identified by the planned interval corrosion testing procedure. Extrapolation of short-term tests is unlikely to predict the long-term corrosion rates of actual equipment. This is one reason why the planned interval corrosion rate testing outlined in Figure 7.30 should be considered whenever developing exposure tests [49].

Actual corrosion rates are relatively unimportant in laboratory screening testing. It is more likely that the testing will allow ranking of materials, chemical treatments, etc. by the order of corrosion resistance, and this will allow choices of materials or chemicals for further field tests.

Laboratory Test Environments There are at least eight different types of environments in which oilfield materials might be tested [91–93]:

- Sour water environments.
- Sweet water environments.
- HCl environments (associated with acidizing treatments of formations or scale removal from downhole components).
- Drilling mud acid environments.
- Organic acid environments.
- Hydrocarbon environments.
- Supercritical CO_2 environments.
- Atmospheric environments of various types (e.g. marine, desert sunlight, etc.).
- Synthetic seawater.

Materials compatibility should be tested in environments as close to the actual operating environment as possible, but this is not always possible. Standardized screening environments are contained in some standards. Tables 7.2 and 7.3 show two sets of these suggested environments.

Notice the high pressures and temperatures in Tables 7.2 and 7.3. This type of testing requires specialized high-pressure test chambers, called autoclaves, which are typically limited to several liters

TABLE 7.2   Recommended Hydrochloric Acid Environments for Testing Materials Compatibility [91]

A – Concentrated Acid	B – Partially Spent Concentrated Acid	C – Partially Spent Dilute Acid
15 wt% HCl	15 wt% HCl	1.5 wt% HCl
	16.5 g l^{-1} CaCO$_3$	1.65 g l^{-1} CaCO$_3$
100 mol% N$_2$	100 mol% N$_2$	100 mol% N$_2$
80 ± 3 °C (175 ± 5 °F)	80 ± 3 °C (175 ± 5 °F)	80 ± 3 °C (175 ± 5 °F)
7.0 ± 0.3 MPa (1000 ± 50 psig) total pressure	7.0 ± 0.3 MPa (1000 ± 50 psig) total pressure	7.0 ± 0.3 MPa (1000 ± 50 psig) total pressure

Source: Reproduced with permission of NACE International.

TABLE 7.3   Suggested Sour Water Environments for Testing Materials Compatibility [91]

A – High Temperature	B – Low Temperature
30 g l^{-1} NaCl	30 g l^{-1} NaCl
1 mol% H$_2$S	1 mol% H$_2$S
14 mol% CO$_2$	14 mol% CO$_2$
Balance CH$_4$	Balance CH$_4$
80 ± 3 °C (175 ± 5 °F)	50 ± 3 °C (120 ± 5 °F)
7.0 ± 0.3 MPa (1000 ± 50 psig) total pressure	7.0 ± 0.3 MPa (1000 ± 50 psig) total pressure

Source: Reproduced with permission of NACE International.

capacity. Laboratory testing at under these conditions requires specialized equipment, and many service companies are available to do this kind of testing on a contract basis.

Analysis of Samples After Exposure Post-exposure analysis of metal samples usually involves weight-loss determinations and low-magnification inspection for signs of pitting or crevice corrosion. The incubation times for these forms of corrosion may be longer for field exposures than laboratory test exposures. One means of compensating for relatively short exposure times in laboratory or field testing is by examining the samples with high-powered microscopes, which may detect pitting at levels not discernible to ordinary visual examination. The laboratory use of scanning electron microscopes has become routine for this purpose. These instruments are usually combined with X-ray spectrometers for chemical analysis, similar to those described in the section on positive metal analysis. The combination of high magnifications with chemical analysis of local areas is a major advantage of the use of scanning electron microscopes for laboratory and field failure analysis.

The pitting corrosion coupons shown in Figure 7.32 were from an offshore gas field. It is relatively easy to identify which of these coupons had the most corrosion. Laboratory screening tests, where exposure times are often limited, may require analysis and comparison of

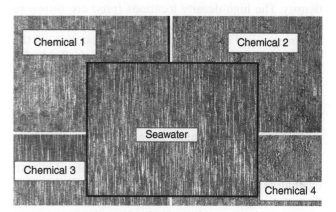

Figure 7.50 High-magnification scanning electron microscope images of pitting corrosion.

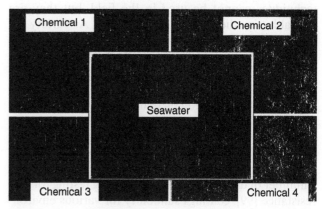

Figure 7.51 Automated image analysis images of the same coupons shown in Figure 7.50.

samples having less apparent corrosion. This is shown on the coupons in Figure 7.50. It seems to most viewers that the samples labeled chemical 1 and chemical 3 have less corrosion than those labeled chemical 2 and chemical 4. How much difference is noted depends on the technician doing the evaluation. Figure 7.51 shows images analyzed by a standard automated image processor. The instrument was set to identify locations where the images in Figure 7.50 had a certain predetermined

TABLE 7.4 Pitting Results Determined by Automatic Image Analysis

Chemical	ASTM Rating	Area (mm^2)
Chemical 1	A-3	0.018
Chemical 2	A-4	0.385
Chemical 3	A-2	0.085
Chemical 4	A-5	0.785
Seawater	A-3	0.132

density. The high-density locations (pits) are shown in Figure 7.51 as white images, somewhat like a negative of the images in Figure 7.50. Automated counting of the images revealed the information shown in Table 7.4, which clearly shows that chemicals 2 and 4 are less effective in preventing corrosion than the other two chemicals and produce more corrosion than applying seawater to the same alloys.

The use of automated image analysis techniques eliminates differences of interpretation between different evaluators and also removes the tendency of human evaluators to respond differently at various times, e.g. due to being tired, overworked, or distracted. While automated image analysis results depend on instrument settings, e.g. the density of an image that will be read as a pit, once these settings are in place, the instrumentation will always interpret in the same manner. These techniques cannot be used in the field, but they are available in the laboratory [93–98]. A variety of techniques are now available that can survey pitting profiles, determine pit density, etc. Most metallurgical uses of automated image analysis have been for grain size, inclusion density, and phase analysis, but there is no reason why corrosion pit density cannot be routinely automated as well.

Types of Standardized Test Procedures NACE and other organizations have prescribed test procedures for a number of different purposes including testing for H$_2$S resistance [99], coatings suitability for various environments, elastomer and polymer performance, corrosion inhibitor effectiveness, etc. Some of these tests, e.g. for coatings performance, require exposure times as long as 5000 h (approximately 7 months). These long-term laboratory tests are often used after short-term screening tests reduce the number of candidate materials or procedures.

Laboratory tests are useful for screening candidate materials, but field exposure is the ultimate test. Field trials of various materials do not have NACE standards, but the laboratory testing procedures described in the various NACE test manuals provide guidance on how the results of field trials can be evaluated.

REFERENCES

1 ASME PCC-3-2007 (reaffirmed 2012). *Inspection Planning Using Risk-Based Methods*. New York: ASME.

2 Konz, S. (2001). Methods engineering. In: *Handbook of Industrial Engineering: Technology and Operations Management*, 3 (ed. G. Salvendy), 1353–1391. New York: Wiley.

3 NDT Resource Center (2017). *NDT and Eddy Current Testing*. http://www.ndt-ed.org/EducationResources/High School/Electricity/eddycurrenttesting.htm (accessed 24 February 2017).

4 Hedges, B., Chen, H.J., Bieri, T.H., and Sprague, K. (2006). *A Review of Monitoring and Inspection Techniques for CO$_2$ and H$_2$S Corrosion in Oil and Gas Production Facilities: Location, Location, Location*, NACE 06120. Houston, TX: NACE International.

5 NACE SP0775. *Preparation, Installation, Analysis, and Interpretation of Corrosion Coupons in Oilfield Operations*. Houston, TX: NACE International.

6 API RP580 (September 2008). *Risk-Based Inspection*, 2. Washington, DC: API.

7 API RP581 (April 2016). *Risk-Based Inspection Methodology – Documenting and Demonstrating the Thinning Probability of Failure Calculations*, 3. Washington, DC: API.

8 DNV-RP-G101. *Risk Based Inspection of Offshore Topsides Static Mechanical Equipment*. Oslo, Norway: DNVGL.

9 CEN (European Committee for Standardization) (April 2008). *Risk-Based Inspection and Maintenance Procedures for European Industry (RIMAP)*, CWA 15740. Belgium: CEN.

10 ISO 31000. *Risk Management Standard*. Geneva: ISO.

11 ISO 17776:2016. *Offshore Production Installations – Guidelines on Tools and Techniques for Hazard Identification and Risk Assessment*. Geneva: ISO.

12 Helle, H. (2012). Five fatal flaws in API RP 581. *14th Middle East Corrosion Conference and Exhibition*, Manama, Kingdom of Bahrain (February 2012).

13 HSE UK R2P2 (2001). *Reducing Risks, Protecting People*. Sudbury, UK: HSE Books.

14 NACE 3T199. *Techniques for Monitoring Corrosion and Related Parameters in Field Applications*. Houston, TX: NACE International.

15 Smart, J. (1980). Corrosion failure of offshore steel platforms. *Materials Performance* 41–48.

16 NACE SP0206. *Internal Corosion Direct Assessment Methodology for Pipelines Carrying Normally Dry Natural Gas (DG-ICDA)*. Houston, TX: NACE International.

17 NACE SP0208. *Internal Corrosion Direct Assessment Methodology for Liquid Petroleum Pipelines*. Houston, TX: NACE International.

18 ANSI/NACE SP0502. *Pipeline External Corrosion Direct Assessment Methodology*. Houston, TX: ANSI/NACE International.

19 NACE SP0204. *Stress Corrosion Cracking (SCC) Direct Assessment Methodology*. Houston, TX: NACE International.

20 ASTM E165. *Standard Practice for Liquid Penetrant Examination for General Industry*. West Conshohocken, PA: ASTM International.

21 ASTM E1444. *Standard Practice for Magnetic Particle Testing*. West Conshohocken, PA: ASTM International.

22 Palmer, A.C. and King, R. (2006). *Subsea Pipeline Engineering*. Tulsa, OK: PennWell.

23 ASTM A342/A342M. *Standard Test Methods for Permeability of Feebly Magnetic Materials*. West Conshohocken, PA: ASTM International.

24 ANSI/ASME B31G. *Manual for Determining the Remaining Strength of Corroded Pipelines*. New York: ASME.

25 API 510. *Pressure Vessel Inspection Code: Maintenance Inspection, Rating, Repair, and Alteration*. Washington, DC: API.

26 API 653. *Tank Inspection, Repair, and Reconstruction*. Washington, DC: API.

27 API 570. *Piping Inspection Code*. Washington, DC: API.

28 Galbraith, J.M., Williamson, G.C., and Creech, M. (2008). *Advances in Pipeline Radiography*, NACE 08140. Houston, TX: NACE International.

29 API 1160. *Managing System Integrity for Hazardous Liquid Pipelines*, Table 9.1. Washington, DC: API.

30 Feng, Q., Li, R., Nie, B. et al. (2017). Literature review: theory and application of in-line inspection technologies for oil and gas pipeline girth weld defection. *Sensors (Basel, Switzerland)* 17 (1): 50. doi: http://doi.org/10.3390/s17010050.

31 API RP578. *Guidelines for a Material Verification Program for New and Existing Alloy Piping Systems*. Washington, DC: API.

32 ASTM E1916. *Standard Guide for Identification and/or Segregation of Mixed Lots of Metals*. West Conshohocken, PA: ASTM International.

33 ASTM E 1476. *Standard Guide for Metals Identification, Grade Verification, and Sorting*. West Conshohocken, PA: ASTM International.

34 ASTM E 572. *Standard Test Method for Analysis of Stainless and Alloy Steels by X-ray Fluorescence Spectrometer*. West Conshohocken, PA: ASTM International.

35 ASTM E 322. *Standard Test Method for X-Ray Emission Spectrometric Analysis of Low-Alloy Steels and Cast Irons*. West Conshohocken, PA: ASTM International.

36 MSS SP-137. *Positive Material Identification of Metal Valves, Flanges, Fittings, and Other Piping Components*. Vienna, VA: Manufacturers Standardization Society (MSS) of the Valve and Fittings Industry.

37 PFI ES-42. *Positive Material Identification of Piping Components Using Portable X-ray Emission Type Test Equipment*. New York: Pipe Fabrication Institute.

38 NACE SP0198. *Corrosion Control Under Thermal Insulation and Fireproofing Materials – A Systems Approach*. Houston, TX: NACE International.

39 ASNT NDT Handbook (2001). *Infrared and Thermal Testing (IR)*, 3, vol. 3. Columbus, OH: American Society for Nondestructive Testing.

40 NACE SP0294. *Design, Fabrication, and Inspection of Storage Tank Systems for Concentrated Fresh and Process Sulfuric Acid and Oleum at Ambient Temperatures*. Houston, TX: NACE International.

41 NACE RP0391. *Materials for the Handling and Storage of Commercial Concentrated (90 to 100%) Sulfuric Acid at Ambient Temperatures*. Houston, TX: NACE International.

42 Baboian, R. (2002). *NACE Corrosion Engineer's Reference Book*, 3. Houston, TX: NACE International.

43 API 579-1/ASME FFS-1. *Fitness for Service*. New York: American Society of Mechanical Engineers.

44 Roberge, P. (2007). *Corrosion Inspection and Monitoring*, 192. New York: Wiley.

45 Dean, S. and Sprowls, D. (1987). In-service monitoring. In: *Metals Handbook, Volume 13, Corrosion*, 197–203. Metals Park, OH: ASM International.

46 NACE RP0497. *Field Corrosion Evaluation Using Metallic Test Specimens*. Houston, TX: NACE International.

47 ASTM G4. *Corrosion Coupon Tests in Field Applications*. West Conshohocken, PA: ASTM International.

48 Dolling, C. and Hudson, R. (2003). Weathering steel bridges. *Proceedings of the Institution of Civil Engineering, Bridge Engineering* 156 (1): 39–44.

49 NACE TM0169. *Laboratory Testing of Metals*. Houston, TX: NACE International.

50 Eckert, R., Aldrich, H., Edwards, C., and Cookingham, B. (2003). Microscopic differentiation of natural gas initiation mechanisms in natural gas pipeline systems. In: *Corrosion*. Houston, TX: NACE International Paper No. 03544.

51 Bieri, T.H., Horsup, D., Reading, M., and Woollam, R.C. *Corrosion Inhibitor Screening Using Rapid Response Corrosion Monitoring*, NACE 06692. Houston, TX: NACE International.

52 Groysman, A. (2010). *Corrosion for Everybody*. New York: Springer.

53 Revie, R.W. and Uhlig, H.H. (2008). *Corrosion and Corrosion Control*, 71–73. Hoboken, NJ: Wiley-Interscience.

54 Dean, S.W. (2003). Corrosion monitoring for industrial processes. In: *Metals Handbook, Volume 13A, Corrosion: Fundamentals, Testing, and Protection* (ed. S.D. Cramer and B.S. Covino), 697–702. Materials Park, OH: ASM International.

55 NACE Publication 3D170 (1984). *Electrical and Electrochemical Methods for Determining Corrosion Rates*. Houston, TX: NACE International.

56 Stern, M. and Geary, A.L. (1957). Electrochemical polarization 1. A theoretical analysis of the shape of polarization curves. *Journal of the Electrochemical Society* 104 (1): 56–63.

57 Fontana, M. (1986). *Corrosion Engineering*, 499–502. New York: McGraw-Hill, Inc.

58 NACE Publicaton 1C187. *Use of Galvanic Probe Corrosion Monitors in Oil and Gas Drilling and Production Operations*. Houston, TX: NACE International.

59 Papavinasam, S. (2014). *Corrosion Control in the Oil and Gas Industry*. Watham, MA: Gulf Professional Publishing.

60 NACE 1C184 (2008). *Hydrogen Permeation Measurement and Monitoring Technology*. Houston, TX: NACE International.

61 ASTM G148. *Evaluation of Hydrogen Uptake, Permeation, and Transport in Metals by Electrochemical Technique.* West Conshohocken, PA: ASTM International.

62 Dean, F., Fowler, C., Farnel, R., and Mishael, S. (2010). *Hydrogen Flux and Corrosion Rate Measurements on Hydrogen Induced Cracking Susceptible and Resistant A516 Steels in Various Sour Environments,* NACE 10179. Houston, TX: NACE International.

63 Mabbott, K. (2006). Corrosion monitoring validation – case study. *NACE Northern Area Western Conference,* Calgary, Alberta, Canada (6–9 February 2006).

64 Hedges, B. and Bodiington, A. (2004). *A Comparison of Monitoring Techniques for Improved Erosion Control: A Field Study,* NACE 04355. Houston, TX: NACE International.

65 MacKinnon, A., Brown, J., and Brown, G. (2011). *Keeping Acoustic Sand Monitoring Simple,* NACE 11396. Houston, TX: NACE International.

66 NACE SP0192. *Monitoring Corrosion in Oil and Gas Production with Iron Counts.* Houston, TX: NACE International.

67 NACE RP0189. *On-Line Monitoring of Cooling Waters.* Houston, TX: NACE International.

68 NACE International Publication 1D199. (1999). *Internal Corrosion Monitoring of Subsea Production and Injection Systems.* Houston, TX: NACE International.

69 Singh, R. (2016). *Corrosion Control in Offshore Structures,* 43–44. Waltham, MA: Elsevier-Gulf Publishers.

70 API RP45. *Analysis of Oilield Waters.* Washington, DC: API.

71 NACE TM194. *Field Monitoring of Bacterial Growth in Oil and Gas Systems.* Houston, TX: NACE International.

72 API RP 38. *Biological Analysis of Subsurface Injection Waters.* Dallas, TX: API.

73 NACE RP0173 (1973). *Collection and Identification of Corrosion Products.* Houston, TX: NACE International.

74 NACE TM173-2015. *Methods for Determining Quality of Subsurface Injection Water Using Membrane Filters.* Houston, TX: NACE International.

75 NACE TM0212-2012. *Detection, Testing, and Evaluation of Microbiologically Influenced Corrosion (MIC) on Internal Surfaces of Pipelines.* Houston, TX: NACE International.

76 Jensen, M., Jensen, J., and Lundgaard, T. *Improving Risk-Based Inspection with Molecular Microbiological Methods,* NACE-2013-2247. Houston, TX: NACE International.

77 Skovus, T.L., Enning, D., and Lee, J.S. (eds.) (2017). *Microbiologically Influenced Corrosion in the Upstream Oil and Gas Industry.* Boca Raton, FL: CRC Press.

78 International Association of Oil & Gas Producers (2016). *Managing Naturally Occuring Radioactive Material (NORM) in the Oil and Gas Industry,* Report 412 (March 2016). London, UK: International Association of Oil & Gas Producers.

79 API Publication 1157 (1998). *Hydrostatic Test Water Treatment and Disposal Options for Liquid Pipeline Systems.* API.

80 ASTM E1003. *Hydrostatic Leak Testing.* West Conshocken, PA: ASTM International.

81 DNV-OS-F101. *Submarine Piping Systems.* Oslo, Norway: DNVGL.

82 49 CFR 192. Subpart J, Pressure Testing for Natural Gas Pipelines.

83 49 CFR 195. Subpart E, Pressure Testing of Liquid Pipelines.

84 API 650. *Welded Steel Tanks for Oil Storage.* Washington, DC: API.

85 API RP1110. *Pressure Testing of Steel Pipelines for the Transportation of Gas, Petroleum Gas, Hazardous Liquids or Carbon Dioxide.* Washington, DC: API.

86 AS/NZ 2885.5.2002 (2002). *Pipelines – Gas and Liquid Petroleum, Part 5: Field Pressure Testing.* Sydney, NSW, Australia: Standards Australia.

87 Kelland, M. (2014). Chapter 18: Chemicals for hydrotesting. In: *Production Chemicals for the Oil and Gas Industry,* 2, 397–402. Boca Raton, FL: CRC Press.

88 Darwin, A., Annadorai, K., and Heidersbach, K. (2010). *Prevention of Corrosion in Carbon Steel Pipelines Containing Hydrotest Water – An Overview,* NACE 10401. Houston, TX: NACE International.

89 Heidersbach, K. (2017). Microbiologically induced corrosion – detection, prevention, and treatment. *7th API Biennial Inspection Summit,* Galveston, Texas.

90 Javaherdashti, R. and Akvan, F. (2017). *Hydrostatic Testing, Corrosion, and Microbially Influenced Corrosion: A Field Manual for Control and Prevention.* Boca Raton, FL: CRC Press.

91 NACE TM0298. *Evaluating the Compatibility of FRP Pipe and Tubulars with Oilfield Environments.* Houston, TX: NACE International.

92 NACE TM0183. *Evaluation of Internal Plastic Coatings for Corrosion Control of Tubular Goods in an Aqueous Flowing Environment.* Houston, TX: NACE International.

93 ASTM D1114. *Preparation of Substitute Ocean Water.* West Conshohocken, PA: ASTM International.

94 DeHoff, R.T. and Rhines, F.N. (1968). *Quantitative Microscopy.* New York: McGraw-Hill Book Co.

95 ASTM E1245. *Determining the Inclusion or Second-Phase Constituent of Metals by Automatic Image Analysis.* West Conshohocken, PA: ASTM International.

96 ASTM G46. *Examination and Evaluation of Pitting Corrosion.* West Conshohocken, PA: ASTM International.

97 ASTM E45. *Methods for Determining the Inclusion Content of Steel.* West Conshohocken, PA: ASTM International.

98 ASTM E112. *Methods for Determining Average Grain Size.* West Conshohocken, PA: ASTM International.

99 NACE TM0103. *Laboratory Test Procedures for Evaluation of SOHIC Resistance of Plate Steels Used in Wet H_2S Service.* Houston, TX: NACE International.

8

OILFIELD EQUIPMENT

Previous chapters have covered the fundamentals of metallurgy, corrosion control, inspection, and corrosion monitoring. This chapter discusses materials and corrosion control concepts associated with specific types of oilfield equipment. No single book can possibly cover all of the subjects necessary for a complete understanding of this very complex subject. This chapter is an attempt to indicate some of the more important concepts, from both economic and safety/reliability standpoints, associated with materials selection and corrosion control in oilfield operations.

Drilling operations, wells, pipelines, and flowlines present problems unique to oil and gas production and are discussed in detail. Other subjects, e.g. facilities and process equipment, are covered in less detail, because the materials and corrosion issues are not unique to oil and gas production and are discussed in great detail in other publications. As an example, floating production storage and offloading (FPSO) vessels are very similar to moored ships, and much of the materials and corrosion control technology associated with them is derivative from that used for vessels that move cargo instead of being moored and stationary for long times. Similar comments would apply to newer forms of offshore production platforms, e.g. spars and tension-leg platforms.

DRILLING AND EXPLORATION

Drill strings are composed of high-strength materials, normally carbon steel with low-alloy connections. The most common forms of material failure in drilling are fatigue and corrosion fatigue [1–6]. Other forms of corrosion, to include environmental cracking, are also important [5–8]. Most problems in drilling are associated with the drill string, at least in part because downhole size restrictions necessitate the use of high-strength materials at stresses close to their operating limits. At one time drill string failures due to torsion failures ("twist offs") and tensile overload were common [8], but the incidence of these failures has been reduced in recent years [5]. One in seven wells experience problems, and fatigue is the leading cause of drill string incidents, which can cost hundreds of thousands of dollars in downtime costs [3].

Drill Pipe

One of the practices leading to fatigue problems is the widespread use of directional drilling, which places high stresses at several locations within the drill string (Figure 8.1). Within a single joint the locations most likely to produce fatigue cracking are near the ends where cross-sectional areas change (Figures 8.2 and 8.3).

Figure 8.4 shows a typical drill pipe failure. The rounded holes in Figure 8.4 are caused by erosion from the drilling fluid mud solids as they progress through narrow fatigue cracks. These holes are often termed "washouts" and are a common failure mode in drill pipe.

At one time there were major problems with cracks and washouts located near the last engaged threads on joints (Figure 8.5). Improved designs of both the pin and box ends of drill pipe have minimized these problems (Figure 8.6). Many proprietary connection designs are currently available.

Metallurgy and Corrosion Control in Oil and Gas Production, Second Edition. Robert Heidersbach.
© 2018 John Wiley & Sons, Inc. Published 2018 by John Wiley & Sons, Inc.

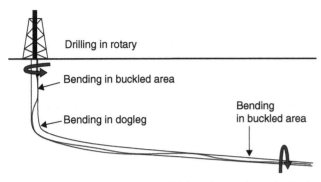

Figure 8.1 Locations where drill string fatigue is likely. *Source*: Vaisberg et al. [3]. Reproduced with permission of Oil & Gas Science and Technology.

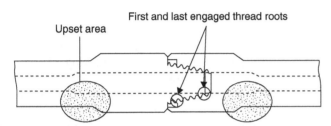

Figure 8.2 Critical drill pipe fatigue areas. *Source*: Vaisberg et al. [3]. Reproduced with permission of Oil & Gas Science and Technology.

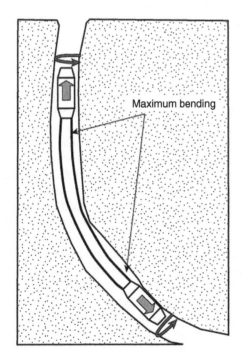

Figure 8.3 Locations of maximum bending stresses on drill pipe in doglegs (bends) in drilling holes. *Source*: Vaisberg et al. [3]. Reproduced with permission of Oil & Gas Science and Technology.

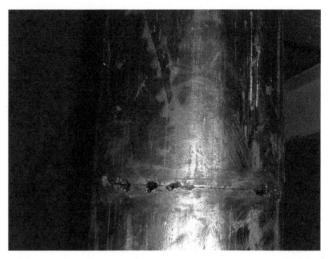

Figure 8.4 Drill pipe corrosion fatigue cracking and washout by drilling mud.

Figure 8.5 Washout near the threads on an oilfield drill pipe.

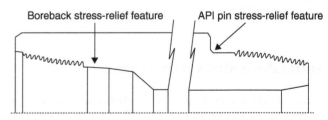

Figure 8.6 Stress reduction by elimination of sharp features and cross-sectional changes. *Source*: Vaisberg et al. [3]. Reproduced with permission of Oil & Gas Science and Technology.

It is common for suppliers to list drill pipe by API standard designations along with descriptions of the type of connections available [9, 10]. The necessity for drilling to greater depths plus horizontal drilling requirements, especially offshore, has led to the development of high-strength drill pipe and connectors that are not covered by API or other international standards. These are reported in the technical literature, and a variety of proprietary, experimental high-strength and H_2S-resistant materials are commercially available.

Cumulative damage to drill pipe is important in determining remaining fatigue life [11]. Early efforts concentrated on records showing the number of trips (downhole insertions) of pipe, but recent efforts also recognize differences based on where in the drill string the pipe has been located. The top of the string normally has the highest tensile loading, but other locations, as shown in Figure 8.1, can also experience very high loads.

Drill pipe strings are withdrawn from the hole for various reasons. It is recommended to break different joints on each trip out of the hole. Drill strings are normally broken in a manner that each pin and box connection can be inspected on every third trip. These visual inspections look for signs of wear, galling, and washouts (indications of fatigue cracks or pitting being enlarged from the inside out). Periodic inspections for fatigue cracks use magnetic particle inspection enhanced by the use of fluorescent particles and ultraviolet (UV) (black light) illumination [1]. It is important that these inspections be concentrated near the locations shown in Figure 8.3.

The above discussion has concentrated on mechanical fatigue loading. It is important to recognize that most fatigue failures in drilling operations are actually corrosion fatigue problems, and any means of lessening the corrosivity of the environment will serve to prolong the life of drilling components [1, 4, 11, 12].

The primary means of controlling drilling fluid corrosion is by control of the drilling mud pH. Sodium hydroxide is usually used to maintain pH levels, and pHs are maintained at levels higher than necessary for optimal drilling mud performance. Typical pH ranges for mud-based drilling fluids are between 8.5 and 11, and higher pHs serve to reduce corrosion. Most reasons for keeping drilling mud pH under control are for fluid rheology purposes, but corrosion control is also maintained, and it is common to maintain the pH somewhat higher than necessary for rheology reasons into order to control corrosion. The ideal pH for many clay-based drilling fluids is 9, but corrosion control often results in pHs as high as 11. These high pHs are normally less corrosive, and they also tend to reduce the concentrations of CO_2 and H_2S dissolved in drilling fluids.

Other means of controlling corrosion in drilling operations include the use of chemical oxygen and H_2S scavengers and the use of internal coatings on drill pipe [1, 4, 5, 13, 14]. Air, and thus oxygen, is often introduced into drilling fluids during topside operations when downhole materials are separated from the fluids before they are recycled into the wellbore.

Lower-pH drilling fluids, e.g. polymers and clear brines, require greater attention to oxygen scavengers and other corrosion-inhibiting chemicals than the use of clay-based muds. The use of filming corrosion inhibitors has been recommended in the past [13], and these inhibitors, often necessary in polymer or clear drilling fluids, tend to attach to most solid surfaces, so their efficiency in mud-based drilling fluids is questionable. They are often applied to freshwater rinses that are often used to clean drill pipe after use prior to storage.

The effectiveness of drilling fluid additives in controlling corrosion (and thus corrosion fatigue) is monitored using ring-shaped corrosion coupons like those shown in Figure 8.7. These rings are inserted between the tool joint (Figure 8.8) at the top of the first stand above the

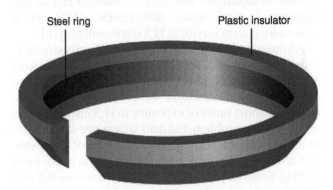

Figure 8.7 Drill pipe corrosion ring coupon fabricated in accordance with API RP 138-1B. *Source*: Brondel et al. [4]. Reproduced with permission of Schlumberger, Ltd.

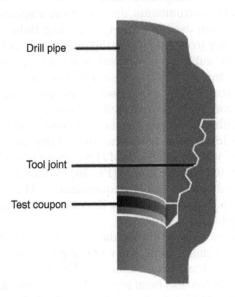

Figure 8.8 Drill pipe corrosion coupon inserted in tool joint. *Source*: Brondel et al. [4]. Reproduced with permission of Schlumberger, Ltd.

drill collar and should usually be left in place for at least 40 h (usually 100 h). The ring is then sent to a laboratory for weighing and analysis [1, 15–17]. Initial corrosion rates in the first few hours are usually faster and may give misleading results; this is the reason for waiting at least 40 h before pulling coupons. As in all corrosion monitoring, coupons can only indicate changes in corrosivity and cannot indicate true corrosion rates. They are used by drilling operators to determine the efficiency of chemical corrosion control measures, which can be adjusted as necessary.

Various inspection techniques have been reported for drill pipe, but their cost effectiveness is questionable [3]. The best approach seems to be to track the usage history of the pipe in question and to retire it before fatigue problems become likely.

The high strength levels necessary for drill pipe and associated connections mean that they cannot be manufactured in accordance with the guidelines for H_2S service suggested in ANSI/NACE MR0175/ISO 15156, which address long-term service in H_2S environments [18].

Aluminum drill pipe, sold primarily for its strength/weight advantages, does not have H_2S cracking problems, and other attempts to develop H_2S-resistant drill pipe and connections have been reported. The relatively short times of exposure to H_2S environments minimize this problem, but drill strings are still subject to environmental cracking. Most control of this problem is due to pH adjustments that keep the pH of most drilling fluids in pH ranges (typically between 8.5 and 10) where H_2S levels are minimized in bentonite clay drilling muds. Polymer and clear brine drilling fluids require the use of more aggressive H_2S scavengers, usually zinc compounds. Zinc carbonates work well in high-pH environments, and the more expensive zinc chelates are used in brine-type drilling fluids. In both cases they work by causing the precipitation of insoluble zinc sulfides. Iron-based magnetite (iron oxide, approximately Fe_3O_4) is also used, but the efficiency is lower except at low pHs, which are not desirable for both corrosion control and rheological reasons [1]. The ability to withstand relatively long-term exposure to low levels of H_2S does not impart immunity to H_2S "kicks" where high levels of H_2S may occur for periods of up to several hours [19, 20]. While no NACE standard covers drilling equipment in H_2S environments, a Canadian Industry Recommended Practice is available [21].

Corrosion of drill pipe during storage is a concern. This is normally handled by washing the inside of drill pipe with fresh water after use. The use of corrosion inhibitors has also been reported [11, 14], and the widespread use of internally coated pipe also helps.

External wear of drill pipe is controlled by the use of hard bands, wear-resistant metal deposits on the outside of the box ends of tool joints. New drill pipe is available with factory-applied hard bands, and hardbanding can also be applied to used drill pipe. Water cooling is necessary to prevent degradation of internal coatings during reapplication of hard facing to used drill pipe. Hard facing limits wear on the external diameter of the drill pipe, and it also reduces wear on casing [14].

Tool Joints

External wear of drill pipe is controlled by the use of hard bands, wear-resistant metal deposits on the outside of the box ends of tool joints. New drill pipe is available with factory-applied hard bands, and hardbanding can also be applied to used drill pipe. Water cooling is necessary to prevent degradation of internal coatings during reapplication of hard facing to used drill pipe. Hard facing limits wear on the external diameter of the drill pipe, and it also reduces wear on casing [14]. The hard facing, frequently containing tungsten carbide, chromium, molybdenum, nickel, tin, or boron particles or alloying additions, is harder than the drill pipe, and thus it is often more subject to stress corrosion cracking (SCC) [6].

Blowout Preventers (BOPs)

Blowout preventers (BOPs) are large devices used to mechanically seal, control, and monitor oil and gas wells and avoid blowouts – sudden releases of downhole fluids caused by sudden changes in downhole formation pressure [22]. They are also intended to prevent drill pipe and well casing from being ejected from the well when a blowout threatens. Because they are safety devices, they are typically overdesigned and fabricated from high-strength low-alloy steel castings or forgings, typically iron–nickel–chromium–molybdenum alloys chosen for their ability to be strengthened (hardened) by heat treatment. The fluid-wetted surfaces of these high-strength components are often corrosion-resistant alloys (CRAs) applied over high-strength steel substrates. They are usually installed redundantly in stacks like shown in Figure 8.9.

BOPs are critical safety equipment and are usually very conservatively designed, but they can have corrosion and degradation problems. Elastomeric components need to be inspected and replaced on a regular basis, and high-strength parts, e.g. shears, are subject to environmental cracking and also need inspection and possible replacement.

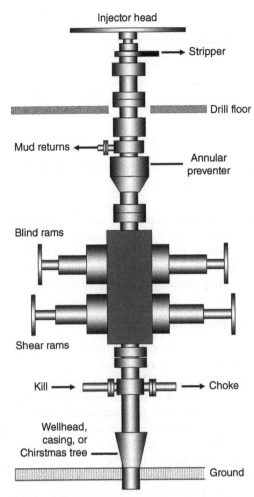

Figure 8.9 A typical blowout preventer.

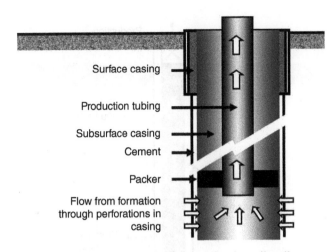

Figure 8.10 Simplified schematic of an oil well.

WELLS AND WELLHEAD EQUIPMENT

Wells are the oilfield equipment with the most corrosion problems. Reasons for the corrosivity of downhole environments include relatively high downhole formation temperatures, the effects of high pressure on solubility of corrosive gases (primarily CO_2 and H_2S), and the tendency of many newer wells to be developed at greater depths and temperatures [23–25]. This is coupled with the difficulty of monitoring downhole corrosivity. Most corrosion monitoring is done near the wellhead, and conditions downhole may not be recognized. To cite just one example, a frequently used means of monitoring wellbore corrosion is by iron analysis (iron counts) in the production fluids. This relatively insensitive technique is unlikely to detect pitting or other localized corrosion occurring at isolated locations on production tubing strings that may be thousands of meters (or feet) long.

Figure 8.10 is a simplified schematic of a typical oil well. The produced fluid, normally a combination of crude oil, natural gas, and water, is produced up the center production tubing. The well is separated from the downhole environment by metallic casing or liners. Casings are normally cemented to the adjacent formation and are considered permanent installations in the well. Liners serve a similar purpose, separating the production tubing from the wellbore, but they are not cemented in place and can be removed. It is common to have packers to separate the formation liquids from the annular spacing between the production tubing and the larger-diameter casing, but many wells, e.g. gas wells and gas-lift wells, may be completed with no packers. Packers prevent produced liquids from rising too high in annular spaces. It is common to have downhole pumps to bring oil to the surface, and the production string is hung in the well from high-strength tubing hangers. Tubing hangers are suspended at the wellhead, which also has valves, controls, and other devices. Pumps and wellhead equipment are discussed at the end of this section of wells.

Locations where downhole oil well components are likely to corrode are shown in Figure 8.11 and include [26]:

- The interior surfaces of production tubing strings, as well as tubing hangers, wellhead and Christmas tree components, at locations where they are in contact with corrosive produced or injection fluids. Production tubing interiors are the most likely locations for downhole corrosion. Tubing and wellhead component corrosion can be minimized by proper materials selection (usually CRAs), the use of internal coatings, or the use of corrosion inhibitors.
- Internal surfaces of the casing and the exterior of the production tubing exterior if the fluid in this space is filled with corrosive liquids or gases. If air leaks into the top of the annulus, corrosion can

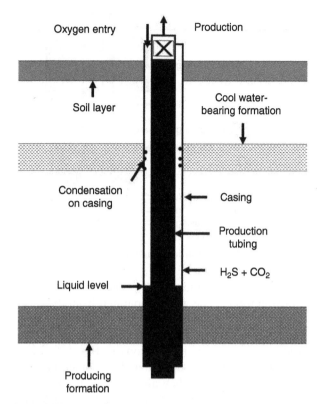

Figure 8.11 Locations where downhole components of an oil well are likely to corrode.

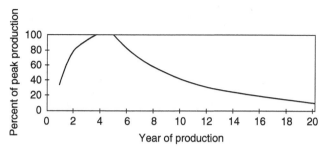

Figure 8.12 Production profile for a typical oilfield.

occur in the upper regions of the casing–production tubing annulus even in the presence of corrosion-inhibited packer fluids. While the strict definition of fluids includes liquids, vapors, and supercritical fluids, the term packer fluid is understood to mean a liquid or mud used to fill the casing–tubing annulus to shut off the pressure of the formation fluids and to prevent them from rising in the annulus.

• External casing corrosion, especially where external formations produce changes in temperature, causing condensation in wells having no packer fluid liquids.

• The produced fluid interface between the vapor-filled annular space and the bottom-hole liquid can become corrosive.

The drawings in Figures 8.10 and 8.11 are simplified, and many more components are found in most wells. It is also simplistic to show wells having vertical boreholes. Deviated wells like those shown in Figure 8.1 are becoming increasingly common, not only offshore but also in environmentally sensitive areas and in hydraulic fracturing operations. Downhole conditions in deviated wells are sometimes similar to deep pipelines, and many of the corrosion problems in pipelines will also be found in deviated wells. The principal differences between

deviated wells and pipelines are the much higher temperatures and pressures common to well strings. Internal pipeline environments can be controlled to a greater extent than in wells, where the downhole temperatures and pressures cannot be changed.

History of Production

Figure 8.12 shows the production profile for a typical oilfield. Peak production normally takes several years to develop, and this is followed by gradual declines in volume. It is common to develop oilfields based on the idea that they will be in production for approximately 30 years [27]. As production goes down, money available for maintenance decreases at times when the needs for maintenance and inspection increase. There are many fields that have been profitably producing oil for more than 50 years, and as the prices of hydrocarbon fuels increase, the extension of field life to even longer production cycles is likely.

As the prices of hydrocarbons change, many fields considered uneconomical become profitable again. It is not unusual to keep wells in production, especially oil wells, e.g. offshore, because this is less expensive than removing them from service and dismantling the associated production equipment, offshore production platforms, etc. When production fields age, downhole pressures degrade, especially for gas wells. The resultant changes in pressure-dependent produced fluid compositions alter corrosivity. Oil wells produce increasing formation water cuts, and this changes many oilfields from noncorrosive to corrosive at times when income levels, which often determine maintenance and inspection budgets, decline. Injection water breakthrough also changes the corrosivity of produced fluids. All of these trends have led many operators to alter their design procedures to emphasize materials selection and maintenance programs that will require higher capital costs for more CRAs instead of relying, as in the past, on the use of corrosion inhibitors to minimize the corrosion of carbon steel downhole components [27, 28].

Downhole Corrosive Environments

Oil is not corrosive to carbon steel in the absence of liquid water, so crude oils with low water cuts, where water is suspended as small droplets surrounded by oil (emulsified) are generally not corrosive. This was shown in Figure 3.1, which shows how corrosion rates of oil well production tubing change with water cut, and in Figure 3.2, which shows the idea of emulsified water surrounded by oil in deviated (nonvertical) oil wells.

Most oil wells also produce varying amounts of formation waters that are typically very high in dissolved minerals to include chlorides and other ionic salts. Formation waters are corrosive, because CO_2 and other acid-forming organic decomposition products are always present, at least in small amounts. In the producing formation the formation water is saturated with dissolved minerals. Changes in temperature and pressure as these fluids move up can lead to dissolved gas breakout, which may cause corrosion, and to the precipitation of mineral scales, which may be protective but can also lead to localized corrosion at breaks in the scale and to scale plugging of tubing (Figure 3.38). The natural buffering action of dissolved minerals in formation waters often prevents or minimizes corrosion, and dissolved chemicals often buffer the water, preventing the formation of acidic water by dissolved gases such as CO_2 and H_2S, which may also be in the fluid stream. In summary, many oil wells are noncorrosive or minimally corrosive.

The presence of high levels of CO_2 in oil wells can cause corrosion (sometimes called "sweet corrosion") under conditions where mineral-deposit scales are not protective [26]. Increasing corrosion rates after a critical water cut appears (approximately 40–50% depending on the field) can be detected by monitoring corrosion rates on the surface or by downhole caliper surveys. For many oil wells, this is when downhole corrosion inhibitor injection is started. Unfortunately, when corrosion is detected, e.g. by caliper surveys indicating approximately 10% wall loss, the tubing surface has been roughened and covered with corrosion products. This can make the introduction of successful corrosion inhibitor programs difficult, especially without prior downhole cleaning, although it has been reported that many corrosion inhibitors work better on corrosion product or scale-covered surfaces than on clean fresh metal. Some operators are reluctant to wait for corrosion to happen before starting corrosion inhibitor programs [29].

Natural gas is not corrosive, provided it remains in the vapor state, but most gas fields are considered corrosive from the beginning of production. Wells may not be corrosive if the production streams reach the surface under appropriate temperature and pressure conditions.

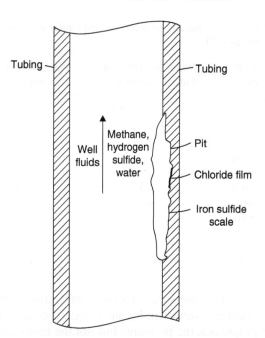

Figure 8.13 Barnacle corrosion in production tubing [4, 24]. *Source*: Reprinted with permission of ASM International®. All rights reserved. www.asminternational.org.

Once lowered temperature and pressure conditions allow condensation of higher-end organic molecules and water, gas condensates usually become very corrosive. Corrosion often appears at intermediate levels in production tubing (Figures 8.11 and 8.13) as localized corrosion at imperfections in iron carbonate or other protective films [4, 24, 30–33]. These imperfections can occur anywhere on the tubing, but they are more likely to happen at joints or other areas of flow disturbance and can be minimized by using premium low-turbulence joint connections.

The reason for corrosion in CO_2-rich condensate environments is that the condensed liquids are low in mineral content, so they do not form the thicker protective scales likely to be formed in oil wells, where mineral-saturated formation waters are being coproduced. Condensates have little buffering capacity. They will be acidic and corrosive to a far higher degree than the high-mineral-content formation waters that wet the interiors of oil well tubing. Condensates of CO_2 and water are highly corrosive, and the presence of any organic acids (acetates, formic acids, etc.) greatly increases their corrosivity [27]. Temperature zones and their resulting rate-determining factors in CO_2 environments at various temperatures were illustrated in Figure 3.33. Note how the protectiveness of iron carbonate scales changes at approximately 60 °C (140 °F) and increases with increasing temperature at deeper locations. This means that many gas wells do not need corrosion inhibitors in their lower, high-temperature locations.

TABLE 8.1 Solubility of Water in Gas Under Reservoir Conditions [27]

Reservoir Pressure	Water Solubility in Gas, bbl (mmscf)$^{-1}$	Water Solubility in Gas, bbl (mmscf)$^{-1}$
Bara (Psia)	127 °C (260 °F)	138 °C (280 °F)
239 (3460)	2.4	3.2
207 (3000)	2.6	3.6
172 (2500)	3	4
138 (2000)	3.5	4.7
103 (1500)	4.3	5.9
68 (1000)	6	8.3
34.5 (500)	11.2	
27.6 (400)	13.8	

Source: Reproduced with permission of NACE International.

Gas well pressures change with time. Initial formation pressures can be fairly high, but, as production progresses, the pressures, but not the temperatures, of formations change and become lower. Table 8.1 shows the changes in water solubility in natural gas at two different temperatures, only 20° F (11° C) apart. As production proceeds and formation pressures decrease, the relative amount of water in produced natural gas increases. This means that, like oil wells, the corrosivity of gas wells is likely to increase [27].

Chemical influences on CO_2 corrosion were discussed in detail in Chapter 3. Figures 3.27–3.33 illustrate various aspects of CO_2 corrosivity. There are no universally accepted standards or models for CO_2 corrosivity. This is a subject of continuing research, and consensus is unlikely in the foreseeable future. Various CO_2 corrosivity prediction models are available [1, 27, 28, 30–32]. Figures 3.31 and 3.32 are associated with CO_2 corrosion models developed by K. DeWaard and coworkers, which are some of the first mathematical models to have been proposed. These early models have been widely adopted and discussed. Many operators find that the corrosion rates predicted by the DeWaard and associates approaches tend to predict corrosion rates higher than experienced in field exposures. Most of the newer models do an excellent job of predicting corrosion rates under the controlled laboratory testing conditions used for their development, and they have been verified by laboratory experimental data. Unfortunately unidentified or nonincluded field conditions, often minor production fluid constituents, are not recognized in these models, and predicted corrosivity in the design process must be confirmed by monitoring and testing once production starts [27]. Changes in produced fluid characteristics also necessitate continuous monitoring, especially for gas wells [27, 28, 31]. As produced fluid corrosivity increases, the need for more diligent monitoring and inspection also increases. This often happens at times when decreased production rates suggest to management that lowered inspection and maintenance budgets, which are often based on field production income, tend to also decrease.

CO_2 or sweet corrosion is the most common environmental problem causing weight-loss corrosion in oil and gas production. H_2S is substantially less corrosive, as was shown in Figure 3.3, which compares the influence of dissolved oxygen, CO_2, and H_2S on water corrosivity. H_2S is much less aggressive than the other gases. Unfortunately H_2S, or "sour" weight-loss corrosion as it is often termed [26], is not the only problem associated with H_2S production. H_2S also serves as a hydrogen-entry promoter, promoting the absorption of monatomic hydrogen formed by reduction reactions. The resulting absorbed hydrogen can lead to various forms of hydrogen-related cracking.

The term "sweet corrosion" has historically been used to indicate oilfield corrosion under conditions where dissolved CO_2 in the aqueous phase causes corrosion. It is also used to differentiate between conditions where the downhole fluids have enough H_2S to come under the guidance of ANSI/NACE MR0175/ISO 15156. The parallel term "sour corrosion" implies corrosion in fluids with enough H_2S for ANSI/ANS/NACE MR0175/ISO 15156 to apply. Both terms, sweet and sour corrosion, are not used as frequently as they once were.

Most new oil and gas wells are now completed under the assumption that they will eventually become "sour" and produce undesirable levels of H_2S. The appropriate guidance for these designs should usually be ANSI/NACE MR0175/ISO 15156 with the publication date specified [18, 24, 27, 34–36]. This standard, like many others, undergoes periodic updates and changes, and it is important that the appropriate version of any standard be understood by all parties concerned, often several decades after the design and installation of original equipment has been accomplished [37]. Readers are cautioned that ANS/NACE MR0175/ISO 15156 is primarily concerned with hydrogen-related cracking phenomena. It does not address other forms of corrosion, often considered to be "weight-loss" (mass-loss) corrosion to distinguish these forms of corrosion from the environmentally induced embrittlement cracking discussed in ANSI/MR0175/ISO 15156.

Weight-loss corrosion can also occur under high-temperature high-pressure downhole conditions. The models for H_2S corrosion are less widely known and not often used. The scales formed by downhole H_2S corrosion are often hard and compact, unlike the more porous carbonate scales formed in CO_2-rich environments. Relatively thick iron sulfide corrosion products can restrict flow in a manner similar to the scales formed from carbonate or other mineral-rich fluids. Unlike the scales shown in Figure 3.38, which come from precipitation of produced fluids as temperature and pressure conditions change in wells, this plugging is due to corrosion, but it can have the same undesired effect of restricting oil well production rates [34].

A recent report discusses weight-loss corrosion and proposes explanations for corrosion in H_2S-containing waters. The parameters associated with when weight-loss corrosion occurs in wells containing H_2S and/or CO_2 are illustrated in Figure 8.14 [34]. This model for corrosion prediction is very complicated, only recently presented, and has not been confirmed by other operators. It only serves to illustrate how complicated an understanding of downhole corrosivity is likely to be. Predictive models are no substitute for monitoring and

inspection once production begins, and changes in corrosivity as fields age are to be expected.

Most of the long-recognized environmental variables on corrosivity of oil and gas production are associated with three gases – oxygen, CO_2, and H_2S – and were discussed in detail in Chapter 3. Less attention has been paid to the influence of organic acids. These relatively small organic molecules have similar molecular weights, and volatility, to the heavier components in natural gas. Examples of organic acid terms appearing in the oilfield corrosion literature include carboxylic acids (formic acid, acetic acid, propionic acid, etc.), formates, acetates, propionates, fatty acids, oxalic acid, etc. All of these organic acid formers have similar properties. As organic chemicals, they tend to be less ionic than mineral acids (hydrochloric acid (HCl), nitric acid (HNO_3), etc.) but

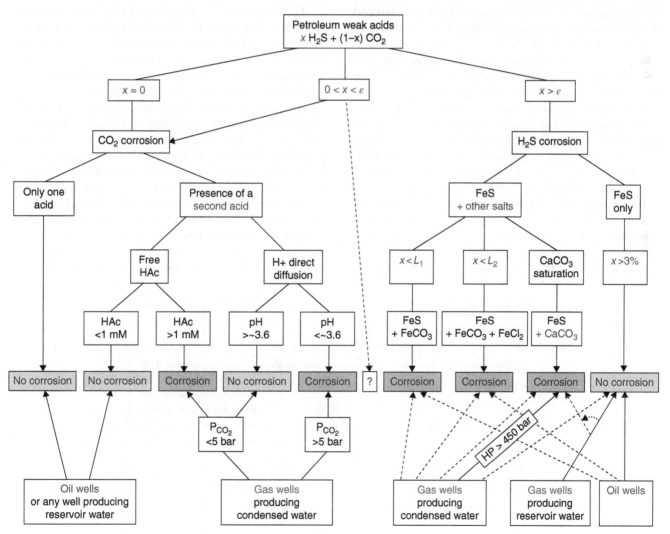

Figure 8.14 Logigram for predicting corrosion in oil and gas wells in the presence of CO_2 and/or H_2S as weak acids. *Source*: Crolet and Bonis [34], NACE 10365. Reproduced with permission of NACE International.

they can be significantly ionic and corrosive in some fluids, such as condensates, especially at elevated temperatures when they will tend to become more ionic than at lower temperatures. CO_2 pitting corrosion in the absence of organic acids may not occur. "There is no record of CO_2 corrosion in a producing well in the absence of acetic acid" [38]. This may only mean that whenever acetic acid (the second smallest of these organic acids) was present in natural gas streams, it was detected and reported, while formic acid, which is smaller, and propionic acid and other organic acids likely to have been in the same produced fluids were not detected or reported. Virtually all Gulf of Mexico gas condensate wells produce detectible amounts of organic acids in the condensate, and organic acids are also reported to be a problem worldwide [37–41].

Figure 8.15 shows the effect of pH on dissociation/ionization of acetic acid at room temperature. The ionization of organic acids would be even more pronounced at the elevated temperatures found in condensate-producing gas well production streams. While these acids are not strong and corrosive in many low-temperature applications, downhole conditions in gas wells with minimal buffering minerals in the condensate water can lead to very aggressive corrosive environments [43, 44].

Formation water is usually high in mineral content. Once it starts to be produced in gas wells, the mineral content of the water in gas wells may increase and the pH change and increase. The produced gas stream may become less corrosive when this happens [26].

Injection fluids – water, steam, natural gas, CO_2, or H_2S – may eventually become part of production well fluid streams. Injection waters, even if they are reinjected formation waters, will have different chemistries and scaling tendencies than the original formation waters.

The timing of injection water breakthrough will often be different for various wells in the same field. Changes in corrosion rates detected by electrical resistance probes or corrosion coupons are usually among the first indications that injection water breakthrough has occurred.

Inadequate biocide treatment of surface waters will often result in souring of formation waters due to inadequate injection water treatment, and corrosion rates frequently increase when injection water breakthrough occurs.

Workover and formation treatment fluids may also affect corrosion of producing wells. This is usually controlled by monitoring corrosion rates after these operations and by adding increased corrosion inhibitors until the topside produced stream returns to the original, prior to treatment, corrosion rates. Coupons cannot detect these changes in corrosion rates, and electrical resistance probes are usually used [31].

This section has reviewed downhole corrosivity in various types of wells. New oil wells are often not corrosive, and it is common to monitor corrosion rates and to not start corrosion inhibitor injection until a predetermined amount of corrosion, usually associated with increased water cuts, has occurred. A common starting point is once caliper surveys or other downhole inspections indicate a wall loss of 10% [27]. The major drawback to this approach is that by this time the tubing surface may be roughened, and the effects of inhibitor treatments may be lessened and perhaps ineffective. Gas wells are usually considered corrosive from the beginning. While oxygen (usually due to leaks from the surface), CO_2, and H_2S are the primary contributors to corrosivity of produced fluids, the effects of organic acids are receiving increased attention in recent years. Stimulation fluids and injection water breakthrough

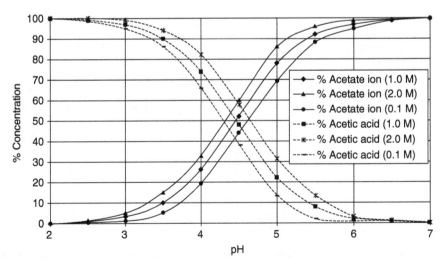

Figure 8.15 Dissociation of acetic acid at 25°C (77°F) [42]. *Source*: Garsany et al. [41], NACE 02273. Reproduced with permission of NACE International.

can also affect corrosivity. Many models are available to predict corrosivity in sweet or CO_2-controlled corrosion, and ANSI/NACE MR0175/ISO 15156 is used to prevent cracking-related problems in H_2S or sour environments. Unfortunately, these models are usually based on laboratory data and do not adequately account for many of the unrecognized variables in downhole produced fluids.

In the absence of produced fluid corrosivity data, the following guidance has been suggested for new wells [27]:

- **Gas wells** Assume condensed water pH based on fluid analysis and calculation of *in situ* pH.
- **Oil wells** Condensed water is too conservative; the produced fluids are unlikely to be that aggressive. Assume $200 \, mg \, l^{-1}$ alkalinity until produced fluid data is available.

The guidelines in the previous paragraph indicate the type of corrosion control that was common until quite recently. Most wells were completed with carbon steel tubing and casing [24]. Increased costs of component replacement, and the trend for wells to remain in production for even longer than the 30-year life that was once assumed to be the practical life limit of downhole tubing and equipment, have led many operators to install CRAs from the beginning, even in nominally nonaggressive conditions. The thinking behind this approach is that even one tubing string replacement at approximately 30 years is too expensive and the incremental costs of installing CRAs (usually 13Cr tubing) on initial completion is good insurance against the costs of more than one million dollars per well associated with retubing a well [24, 27].

Annular Spaces

The annular spaces between tubing and casing or liners are usually not used for production flow, but they are sometimes used for downhole treatments, e.g. inhibitor injection. Most annular spaces are filled with packer fluids, primarily water-based liquids, but sometimes oil or muds. These fluids serve to shut off production from downhole pressure and to minimize the pressure differences between the outside of the tubing and the casing interior. These fluids are usually much heavier than water. Solids-free packer fluids can have specific gravities from 0.83 to 2.3. This density is usually achieved by the addition of soluble salts. These salt solutions, which must be very concentrated in order to increase the density of the brines, should not be very corrosive, and usually they are not. The reasons for this were illustrated using the ideas shown in Figure 3.8, which showed how the corrosion rate declines due to decreasing oxygen

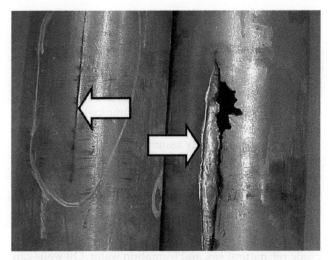

Figure 8.16 Cracks in duplex stainless steel downhole tubing caused by H_2S generated from degradation of the packer fluid. The wider crack on the right is caused by erosion of high-pressure fluids escaping into the tubing–casing annulus. This is a typical washout pattern.

solubility in salt water. Brines used for this purpose include sodium chloride, calcium chloride, calcium bromide, zinc bromide, and cesium formate [1].

Corrosion inhibitor packages are usually included in packer fluids to minimize any possibilities of corrosion. Organic film-forming inhibitors are useful up to approximately 250 °F (120 °C). Beyond these temperatures thiocyanate inhibitors have been used. Unfortunately, the thiocyanate ion (SCN^-) can break down and form H_2S gases. Figure 8.16 shows duplex stainless steel tubing that cracked due to H_2S generated from packer fluid degradation downhole. The literature discusses several instances of problems of this type, and alternatives to thiocyanate corrosion inhibitors are under development [42, 44–47]. No completely acceptable alternatives have been demonstrated to have long-term downhole stability.

Types of Wells

The previous section contained a detailed discussion of the environmental variables that affect corrosivity in well environments. Most wells will experience a variety of these environments. The discussions that follow are intended to highlight the environments associated with the primary fluids being handled in various types of wells. The main differences between corrosion in oil, gas, and injection wells occur on the production tubing interior surfaces. Annular space corrosion is approximately the same for all of these types of wells and is usually controlled by packer fluid corrosion inhibitors as discussed above.

Most wells and associated wellhead equipment are completed with materials that meet ANSI/NACE MR0175/ISO 15156 requirements. This is because operators have learned that even if the original formation and produced fluids do not contain H_2S, it is possible that they will "sour" and produce H_2S at some time in their production history. It is not unusual for wells to remain in production for much longer than the 30 years they have been typically assumed to be productive, and using H_2S-tolerant materials is much less expensive than retubing at some later date. Well casings, which are considered permanent installations, are seldom removed or replaced.

Oil Wells   Oil wells typically produce a combination of crude oil, natural gas, and formation water. Oil wells can be profitable when less than 10% of the produced liquid is oil and the rest is mineral-rich salty water. Corrosion control is not necessary in many wells until the water cut increases to the extent that tubing interiors become water wetted. Many operators wait until a predetermined corrosion rate or wall loss, typically 10% as determined by caliper surveys, before they begin corrosion inhibitor treatments [23]. The disadvantage of this approach is that once corrosion has started, corrosion inhibitors may be less effective due to their limited ability to penetrate beneath corrosion products and scale. Pitting corrosion may also be well advanced by this time. Most oil wells start out with natural flow to the surface, but after time downhole pressures subside and artificial lift devices become necessary.

Many operators have adopted the policy of starting all new completions with CRAs, typically 13Cr tubing. Downhole equipment, filters, pumps, etc., and wellhead equipment are typically made from CRAs. Galvanic corrosion is seldom a problem unless the downhole water becomes acidic due to injection water souring or the bottom-hole formation water has a pH less than approximately 3.6 as shown in Figure 8.14.

Gas Wells   Unlike oil wells, gas wells are usually corrosive from the start of their production. The reason for this is that most natural gas will contain varying amounts of water vapor as well as natural gas condensates, higher-molecular-weight organic chemicals that tend to condense from the natural gas, which is typically more than 90% methane with a molecular weight of 16 compared with water at 18, formic acid at 46, acetic acid at 60, etc. Liquid condensation is more likely as molecular weight increases, so it is inevitable that whenever water condenses on tubing interiors, it will be accompanied by higher-molecular-weight organic acids. This means that whenever water condenses in a production string, it is likely to be in the presence of organic acids, which lower

the pH of the water and make it acidic and corrosive. Natural gas may also contain other corrosive chemicals including elemental sulfur and mercury. While concentrations of these corrosive chemicals may be low and difficult to detect using normal analytical chemistry techniques, high-volume gas wells can eventually produce significant amounts of corrosive condensates. If these liquids are not suspended in the gas stream and removed from the well, they eventually cause downhole corrosion. At temperatures where organic filming inhibitors can work, the use of these inhibitors is the preferred method of corrosion control. For deep hot gas wells operating at temperatures where organic inhibitors break down, the use of CRAs is necessary.

While a number of software systems have been developed for the purpose of modeling gas well corrosion, they differ in their predicted corrosion rate outputs, and they are not at the stage of development where they can reliably predict corrosion rates prior to the start of production [27, 28, 30, 31, 48]. SOCRATES, a proprietary software package, gained wide use for selecting alloys for downhole applications [49], and other software is under development. Nyborg reviewed the software available and reported difficulties with all of the corrosion rate prediction models [30]. Most of the currently available models are not intended for use in the presence of H_2S or organic acids.

Injection Wells   Injection fluids are usually water or steam but sometimes gas, e.g. natural gas, CO_2, or H_2S, injected into reservoirs to either avoid pollution (disposal wells) or to maintain pressure. Enhanced oil recovery is also important using waterflooding, steam injection, and, in recent decades, miscible flooding with CO_2, which may be pipelined long distances to maintain production after less expensive means have neared or reached their limits.

Because injection water chemistry can be controlled, it is common to use carbon steel or 13% Cr for many injection wells, although fiber-reinforced plastic (FRP) and lined pipe are also used. If bottom-hole conditions lead to corrosive water accumulation, the tubing and equipment at the bottom of the hole may have CRA, usually 13Cr [50, 51].

Most corrosion control of injection water is by vacuum deaeration or gas stripping to remove dissolved oxygen. Oxygen scavengers are then added to the water, bringing dissolved oxygen levels down to approximately 5–10 ppb. Sodium bisulfite ($NaHSO_3$), ammonium bisulfite ((NH_4)$_2HSO_3$), and sulfur dioxide (SO_2) are some of the oxygen scavengers that are used for this purpose. Filming corrosion inhibitors are seldom used, because the large volumes of water involved mean that mechanical removal (vacuum deaeration or gas stripping) and scavenging are more economical [52].

Cracking associated with deaerator welds was one time a worldwide safety problem, and NACE has issued a standard on how to control this problem [53].

This discussion has concentrated on the use of surface waters for injection purposes. The handling of formation water prior to reinjection emphasizes maintaining positive pressures on all piping and vessels to minimize air entry. Oxygen scavengers are then added prior to reinjection.

Because injection water oxygen control is the primary means of corrosion control, it is necessary to monitor the oxygen levels at various stages in treatment and transport. This is done with online electronic oxygen sensors as well as galvanic probes and other means of corrosion monitoring. Galvanic corrosion probes that will respond quickly to changes in oxygen content are often the first indication that oxygen control has been compromised [54, 55].

Reservoir souring due to inadequate injection water treatment is a serious concern. This often happens several years after water injection has started and is usually due to inadequate water treatment to include the use of biocides that are not effective on anaerobic bacteria. Seawater treatment often includes some means of removing the natural sulfates present in seawater [52, 56]. This is done primarily for barium-sulfate scale control, but it also has an effect on reservoir souring.

Removal of the bulk of the sulfates by nanofiltration membranes brings the level of potential $BaSO_4$ precipitation down into levels at which scale inhibitors give sufficient control. The first commercial sulfate removal systems were used in the North Sea and subsequently in the US sector of the Gulf of Mexico. Now they are used all over the world.

Tubing, Casing, and Capillary Tubing

Oil country tubular goods (OCTG) include seamless drill pipe, tubing, and casing [9, 57, 58]. Most corrosion problems in oil and gas production are associated with tubing and casing. The reason for this is the large volume of these components used per well. While it is common to use equipment CRAs for wellhead, pumps, packers, etc., economic incentives and availability drive the trends for continued use of carbon steel and low-alloy tubing and casing. Concerns with eventual souring of fields have convinced most operators that all downhole equipment, to include tubing and casing, should meet the recommendations of ANSI/NACE MR0175/ISO 15156 [18, 35, 36, 50].

Tubing Corrosion Most well designs use tubing strings for upward production of produced fluids. Formation temperatures do not change during the lifetime of production, but production rates decrease as fields age. The resulting changes in temperature and pressure as fluids move up production tubing strings alter locations of gas breakout in oil wells and condensation formation in gas wells, and these altered locations are where most corrosion problems are likely to occur. Injection water breakthrough can also alter the corrosivity of both oil and gas wells. The cost of tubing replacement can be upward of $1 million per well, and many operators have adopted the policy of starting oil well production with 13Cr tubing in the hopes that they can avoid tubing replacement at some time in the future [24, 27]. Gas well tubing may possibly require even more CRAs (duplex stainless steels or higher grades of CRAs). The models for predicting when these alloys are necessary are still under development and controversial, but many operators do use software-based tools for tubing selection [24, 30, 31, 48, 49]. Table 8.2 shows the NORSOK standard recommendations for various well components to include tubing and liners [59]. Note that while recommendations for tubing and liners are identical for these removable and replaceable components, the table does not address casing, which is permanent and cannot be removed and replaced.

Erosion corrosion occurs downhole if production rates are too high, and either multiphase fluid flow or sand erosion results. This problem normally appears at the tubing joints (Figure 5.75).

As fields age and downhole temperatures and pressures change, scale formation and deposition can occur. This can reduce the cross section of tubing available for produced fluid flow (Figure 3.38) and cause losses of production. This downhole scaling is treated by using "pickling" acids, usually HCl or HF with added corrosion inhibitors [57, 58, 60–63]. The same acids are used to stimulate downhole geologic formations, either during initial production or later in the production cycle. Unfortunately, if the acids remain too long in contact with the scale, they will start to cause corrosion of the tubing, as shown in Figures 8.17 and 8.18.

Casing Corrosion Casing is the structural metal liner for the walls in oil and gas wells [64]. The purpose of casing is to separate the fluids within the well from other undesirable fluids at levels closer to the wellhead than the producing formation. The casing is normally cemented to the adjacent formations. Once the metal casing reaches a predetermined depth, a perforating gun is lowered to punch holes in the casing to start production. While most casing is carbon steel, 13Cr and higher grades of CRAs are also used. Casing is intended to last for the life of the wells and not be pulled, inspected, or replaced. Corrosion of casing can be from either the interior, in the annular spaces shared with tubing

TABLE 8.2   Materials Selection for Wells [59]

Well Type	Tubing and Liner	Completion Equipment (Where Different from Tubing and Liner)
Production	13Cr is base case	
	Low-alloy steel for systems with low corrosivity	13Cr
	13Cr and 15Cr alloys modified with Mo/NI, duplex and austenitic stainless steels, and nickel alloys are options for high corrosivity	
Aquifer water production	13Cr is base case	
Deaerated seawater injection	Low-alloy steel	UNS N09925, Alloy 718, 22Cr or 25Cr duplex stainless
Raw seawater injection	Low-alloy steel with GRP or other lining	Titanium or inhibitors in oxygen-free systems
	Low-alloy steel for short design life	Titanium or inhibitors in oxygen-free systems
	Titanium or inhibitors for oxygen-free systems	
Produced water and aquifer water injection	Low-alloy steel	13Cr with same limitations as for tubing in this service
	Low-alloy steel with GRP or other lining	13Cr with same limitations as for tubing in this service
	13Cr. Provided oxygen <10 mg m^{-3}	
	22Cr duplex, Alloy 718, N09925. Provided oxygen <20 mg m^{-3}	
Gas injection	Materials selection same as for production wells	
Alternating injection and combination wells	Materials selection shall take into account the corrosion resistance of different materials options for the various media	

Source: Data from table 2, NORSOK Standard M-001.

Figure 8.17 Corroded downhole tubing caused by leaving acid treatment in tubing for too long.

exteriors, or on the exterior (Figure 8.19). Sand can also cause erosion on both casing and tubing [65, 66].

Internal casing corrosion in the annular spaces is discussed above and applies to both the casing interior and tubing exterior. Most corrosion control of casing interiors is by appropriate alloy selection and by packer fluid inhibitor use, either filming inhibitors at lower temperatures or oxygen, CO_2, and H_2S scavengers at elevated temperatures.

The annular spaces in new wells should be considered corrosive until caliper or other borehole inspections indicate that corrosion is at low or negligible levels. Pumping

Figure 8.18 Month-old replacement tubing that needed replacement due to improper acidizing treatment.

wells should be tubed as close to the bottom of the wellbore as possible to minimize corrosion damage to the casing, which is often concentrated at the liquid/vapor interface. Casing pumps are generally not advisable [1].

Exterior corrosion protection is usually provided by cementing, which provides a high-pH noncorrosive environment for the casing exterior, and by electrical

Figure 8.19 External pitting corrosion on oilfield casing.

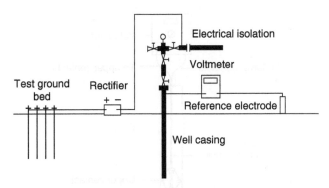

Figure 8.20 Experimental setup to determine required cathodic protection current for existing well casing exterior. *Source*: NACE SP0186 [70]. Reproduced with permission from NACE International.

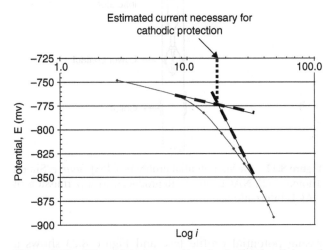

Figure 8.21 *E*-log *i* data for a well casing indicating the need for approximately 18 A of applied cathodic protection current. *Source*: NACE SP0186 [70]. Reproduced with permission from NACE International.

separation at the wellhead with insulated flange assemblies that are intended to isolate the well from flowlines that could produce stray current corrosion. At one time it was common practice to only cement the casing from the producing formation to seal the well annulus. Modern practice tends to cement the entire wellbore [67]. The reader is cautioned that many electrical isolation assemblies are shorted by maintenance and other work during the decades that wellheads are in service, and downhole corrosion due to stray electrical currents sometimes occurs. The only way to ensure that this has not happened is to conduct periodic downhole inspections. It is not always economically feasible to electrically isolate wells from flowlines and other nearby equipment.

External coatings are available that can withstand the mechanical damage associated with casing installation. These coatings, which will be damaged during installation, can reduce the cathodic protection current to approximately 10% of the current needed for casings with no external coating [68].

Cathodic protection is not normally installed on well exteriors until condition surveys or leak records indicate that corrosion is occurring due to external casing corrosion. The electric current requirements for cathodic protection on casings is then determined on existing well casing by applying electricity to the casing exterior and plotting the current–surface potential data on a voltage vs. log current (*E*-log *i*) scale [67–69] (Figures 8.20 and 8.21). For the data shown in Figure 8.21,

approximately 18 A of current is considered necessary to protect the casing exterior. It is common practice to determine the current necessary for either each well in the field, for small fields, or for selected representative wells in larger fields. The currents necessary for each individual well are then supplied from a central rectifier and ground bed. Individual controls on the current leads to each well are necessary to prevent too much current from being drained into the low-resistance wells, denying adequate protection to other wells [71].

One of the concerns expressed about determining the required current for cathodic protection using the *E*-log *i* test is whether or not enough current reaches the bottom of the casing string. Casing potential profiles determined with interior inspection devices have confirmed that this method applies cathodic protection current to casing exteriors as deep as 3960 m (13 000 ft) [68]. Figure 8.22 shows the instrument used for this

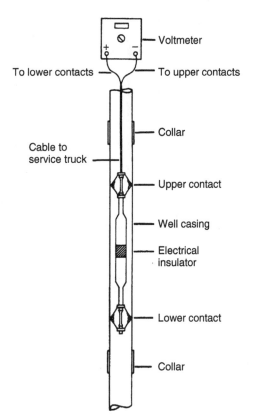

Figure 8.22 Casing potential profile tool [68]. *Source*: Simon Thomas [72], NACE 00056. Reproduced with permission of NACE International.

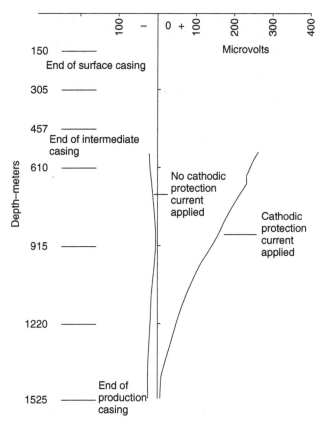

Figure 8.23 Typical casing potential profile plot [68]. *Source*: Simon Thomas [72], NACE 00056. Reproduced with permission of NACE International.

casing potential profile test, and Figure 8.23 shows a typical casing potential profile. The unprotected profile shown in Figure 8.23 shows locations where corrosion is to be expected, while the profile with cathodic protection current applied shows a steadily increasing potential from the bottom of the well, indicating that the casing is protected from external corrosion. A significant drawback to the casing potential profile method of determining if cathodic protection is needed or effective is the need to shut down the well for the measurements. The lost production can be significant.

Galvanic anodes have been used for well casing cathodic protection, but the use of impressed current systems is more common. The design and location of ground bed and rectifier systems for well casings follows the general procedures for other cathodic protection systems. Particular attention should be paid to keeping all anodes at a sufficient distance from the nearest well casing to prevent stray current corrosion [68–71].

Capillary Tubing Corrosion Capillary tubing is a very-small-diameter string of tubing usually run alongside the outside of the production tubing and banded to the outside of the production tubing. It is common for this tubing

to be made from CRAs, and the negligible galvanic effect to due contact with the production tubing is seldom a problem due to the lack of chemically reducible species in the casing–tubing annulus. Capillary tubing is used for a variety of purposes and is available in alloys ranging from austenitic stainless steel to high-nickel alloys. This tubing seldom has serious corrosion problems, and most alloys are H_2S, CO_2, and chloride resistant to various degrees depending on temperature and other variables.

Corrosion Inhibitors for Tubing and Casing in Production Wells

The use of corrosion inhibitors has been common in producing wells since the 1930s. It remains the most common means of corrosion control and is common in wells with carbon steel tubing. Wells finished with materials selected in accordance with ANSI/MR0175/ISO 15156 are not protected from weight-loss corrosion. That standard only addresses materials selection for resistance to environmental cracking.

Filming inhibitors are commonly used in flowing fluids. Chemical scavengers, e.g. oxygen and H_2S scavengers, are used in confined spaces like casing–tubing annular spaces.

The temperature of production fluids is a major consideration in choosing corrosion inhibitors. Up to approximately 65 °C (150 °F), inhibitor treatment is routine. At higher temperatures up to approximately 150 °C (300 °F), extensive testing is necessary to identify appropriate inhibitors. It is possible to inhibit corrosion as high as 220 °C (425 °F), but these applications are questionable and not generally accepted within the oilfield community.

In recent years environmental limitations on discharge of inhibitor-containing waters have become important and have limited the choices of chemicals used and dosage limits. Corrosion inhibitors must also be compatible with other chemicals, e.g. hydrate inhibitors.

Cleanliness is an important consideration in the use of corrosion inhibitors. Particles can clog injectors, and 10 μm filters are often specified. Inhibitors are usually diluted in solvents – crude oil, hydrate-inhibiting glycols, etc. – and the injection of high volumes of solvent, which requires larger orifices to maintain higher flow rates, can tolerate larger particles suspended in inhibitor fluids. Clean surfaces also produce better corrosion inhibitor efficiency. With lower surface areas for attachment, clean wells can be inhibited at lower inhibitor dosage rates. Periodic cleaning may lead to appreciable savings in overall inhibitor consumption, but this cleaning also means that inhibition dosage rates must be somewhat higher immediately after cleaning until the surfaces have been recoated with inhibitor films [24].

Sand and other particles reduce the efficiency of filming inhibitors, because films will form on these solid surfaces [72].

Gas Wells   Corrosion inhibitor treatments in gas wells can be fairly economical, because they produce relatively small amounts of liquid requiring inhibitors. Gas phases are noncorrosive. The high fluid velocity in gas wells means that filming corrosion inhibitors must be resistant to high shear stresses.

While continuous treatment produces more reliable corrosion control results, most gas wells have batch treatment of corrosion inhibitors. Approximately 10 ppm of inhibitor in condensate phase can yield corrosion control, but in practice the liquid production rate may not be known, so it is common to use rules of thumb to estimate the required injection rate, e.g. 1 pint (MMscf)$^{-1}$ (141 (million Sm)$^{-3}$) [72].

Most batch treatments attempt to apply approximately 3–5 mils (75–125 μm) of corrosion inhibitor, but the actual coverage will typically be only about one-tenth of the calculated coverage [73, 74].

Gas wells usually produce condensates in an annular dispersed pattern (Figure 5.70). The resultant mixing ensures that corrosion inhibitors will be transported to tubing surfaces where they can adhere and provide corrosion protection.

Oil Wells   Unlike gas wells, which are considered to be corrosive from the start of production, it is common practice to wait until oil wells start to corrode due to increased water cuts (Figure 3.1) before starting corrosion inhibition of wells and produced fluids. This approach has the disadvantage of starting corrosion control only after a predetermined amount of corrosion, e.g. 10% wall loss, which means the tubing will have surface roughening and corrosion product buildup. Mechanical or chemical cleaning of the tubing interior then becomes advisable before starting inhibitor programs [24].

The much higher volumes of liquid water produced in oil wells means that there are strong incentives to use the lowest possible effective doses of corrosion inhibitors. Some low doses of corrosion inhibitors are more harmful than beneficial, because this can produce relatively small anodic surfaces surrounded by larger inhibitor-protected cathodic surfaces. The users of these chemicals usually do not know the chemistry and operating principles of the commercial combined inhibitor packages [73].

It is common to start with dosages in the 5–15 ppm range for water-soluble inhibitors and approximately 25 ppm for oil-soluble inhibitors based on calculated barrels of water per day produced. Adjustments are then made based on topside monitoring of corrosion rates with coupons or electrical resistance probes.

Application Methods   Virtually all authorities claim that continuous injection is the most efficient way in corrosion inhibitor application, but the special equipment needs for continuous injection downhole mean that batch treatments are more common for production wells.

Batch treatments are most effective for gas wells. All methods require that producing wells be shut in for periods of up to one day depending on the depth of the well and the method used. The three common methods of batch treating wells are [29]:

- Batch and fall
- Tubing displacement
- Squeeze treatments

The batch and fall method is used when gas wells are confirmed to be corroding. The well must have low liquid levels and sufficient natural bottom-hole pressure to return the well to operation once the treatment is completed. Diluted liquid inhibitor is injected into the shut-in well and allowed to fall through the tubing for

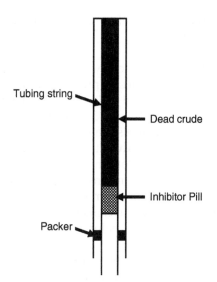

Figure 8.24 Tubing displacement method of lining production tubing with corrosion inhibitor.

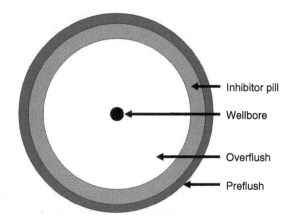

Figure 8.25 Diagram showing location off corrosion inhibitor in formation after squeeze treatment.

approximately six hours. The well is then returned to service, and the natural turbulence of the produced gas stream distributes the corrosion inhibitor to the tubing walls. This method is relatively simple but does not work well in deviated wells, some of which may have locations where the well slope is not steep enough to deliver the inhibitor to the bottom of the well.

The tubing displacement method works for both oil and gas wells. The well is shut in, and the liquid corrosion inhibitor "pill" of diluted inhibitor in a liquid carrier is placed in the well as shown in Figure 8.24. The well is back-pressured to force the pill to the bottom of the well using crude oil, nitrogen, or natural gas. Gas is used to back-pressure the well if the well will not produce with a full head of liquid. Once the well is returned to service, the concentrated inhibitor is diluted by the produced fluid and forms protective films on the tubing wall. The inhibitor film is diluted to spread the inhibitor and increase the likelihood of contact with all surfaces, but it should not be diluted to less than 1000 ppm (0.1%). The time to place the pill at the bottom of the well is typically around six hours, and this procedure is repeated at planned intervals, varying from every two weeks to quarterly [73].

Squeeze treatments are similar to tubing displacement treatments, but the inhibitor pill is displaced beyond the bottom of the well and out into the formation (Figure 8.25) [29, 75–79]. The liquid forcing the inhibitor into the formation is usually filtered produced water, which replicates the formation water chemistry. An alternative is to use fresh water with from 2 to 3% potassium chloride added to minimize any clay swelling in the formation. If low formation pressures are a concern, then gas, diesel, or condensate may be used to displace the inhibitor pill into the producing formation [29].

The inhibitor is allowed to soak into the formation before returning the well to operation. This "soak in" time may be up to 48 h.

During all forms of batch treatment, the elastomeric seals at the bottom of the well are exposed to high concentrations of the inhibitor and the carrier solvent. Once wells are returned to service, the inhibitor concentrations remain high for several weeks. The effects of these concentrated chemicals on the sealing materials in use should be evaluated before starting batch treatments [80, 81].

Concentrated inhibitor solutions returned into the produced fluids by batch treatments can cause problems in topside processing equipment. Their compatibility with existing process operations must also be checked [73].

Continuous inhibitor injection to downhole well locations can be by the use of capillary tubing or by insertion through the annulus. Either method requires specialized downhole equipment that must be installed during initial installation in the well or during extensive workovers.

Dedicated capillary tubing, usually strapped to the outside of the production tubing, is perhaps the most reliable means of inserting corrosion inhibitors into the produced fluid stream. But this method is not popular with production engineers, because this makes the production tubing string, with the attached capillary tubing, harder to place and remove. It slows workovers and other operations [73].

The alternative to continuous injection through capillary tubing is to fill the annulus with the corrosion inhibitor. This method is sometimes used when large volumes of chemicals must be injected into the production stream, e.g. if corrosion inhibitor and hydrate inhibitors are both being injected into the downhole fluids.

This method requires a clean annulus, and, if the injection rate is too slow, the inhibitor may lose

TABLE 8.3   Corrosion Inhibitor Treatment Methods for Producing Oil and Gas Wells

Method (Frequency)	Basic Description	Applications	Disadvantages	Additional Comments
Batch and fall (monthly)	Shut-in, pump into tubing, allow to fall	Gas wells	Cannot treat below liquid level	Wells >1000 psi, rate of fall depends on gas density
Tubing displacement (quarterly)	Shut-in, pump full tubing displacement minus safety factor	Use full volume diluted chemical or slug displaced with diesel, crude oil, water, or nitrogen	Overdisplacement can damage formation	Requires reservoir and subsurface evaluation to identify wells that can have displacement treatment
Squeeze (monthly or semianually)	Similar to displacement, chemicals forced into reservoir	Chemical returns into produced fluid from reservoir storage	High risk of formation damage	Not normally recommended due to risk of formation damage
Gas lift (continuous)	Pump small volume of chemical into gas lift gas at wellhead continuously	Gas lift wells only		
Downhole injection (continuous)	Chemical injection lines	Any new well that chemical injection lines can have installed	Plugging potential, initial capital cost	Injection valves adjusted to control injection rate
Dump bail	Use wireline bailer to spot chemical to location in well	May need several trips to deliver sufficient inhibitor using concentrated products	Mechanical risk	Aluminum bailers do not hold much chemical volume

effectiveness due to overheating in the annulus before it is injected. The method is most often used on oil wells with artificial gas lift. The pressure valve in a side-pocket mandrel opens and inserts inhibitor into the production stream. Intervals between injections can be as long as several hours. Disadvantages of this method include the fact that the inhibitor must be persistent and protect the tubing until the next injection of inhibitor, and high instantaneous concentrations of inhibitor may need to be handled in downstream processing facilities [73].

Valves for both methods of continuous inhibitor injection are prone to sticking and plugging.

Table 8.3 summarizes the various inhibitor treatment methods and practices for producing oil and gas wells for one operator in the Gulf of Mexico. Tubing displacement is probably the most common method of corrosion inhibition, because it has the fewest equipment requirements and drawbacks.

The economics of using corrosion inhibitors for production well corrosion control depends on the method of inhibitor application. The annual cost of corrosion inhibitors for one well is typically in the thousands of dollars. Continuous inhibitor injection systems cost in the tens of thousands of dollars. The biggest costs are for injection tubing and the increased costs of more complex and time-consuming workovers. For batch treatments, the largest cost is the lost production during downtime for injection [73].

The use of corrosion inhibitors for downhole corrosion control has been standard practice for many years, and most wells currently in production use this approach. The complexity of continuous injection and the lost production revenue associated with shutdowns for batch injection have led many operators to decide to use CRAs, usually 13Cr, on most new well completions [24]. The conversion to 13Cr martensitic stainless steels is appropriate, because any fluid conditions that would preclude the use of this series of CRA would also be so aggressive that corrosion inhibitors would not work and more CRAs would become necessary. One alternative to this approach is the use of internally coated tubing, which is discussed in the next section. Other more CRAs are also finding increasing use in new hot formations [24].

Internally Coated Tubing for Oilfield Wells

Internal coatings are often used on water injection wells as an economical means of enhanced corrosion control for low-cost carbon steel tubing. As discussed earlier in this chapter in the section on injection wells, the primary means of corrosion control for these wells is by the use of oxygen scavengers to minimize the corrosivity of the injection water. As an added means of corrosion control, it is common to use internally coated injection tubing for those wells that cannot use fiber-reinforced tubing. The usual reasons for using steel pipe are pipe diameter, fluid pressure, and the length of the injection tubing string. Internal coatings are also used on production tubing, both as a means of corrosion control and to reduce friction and increase flow rates [82–90]. The smoother surfaces of internally coated or lined tubing also reduce paraffin buildup and scale deposits (Figure 8.26).

Internal Coating Systems for Tubing in Wells There are a variety of coating systems available for tubing lining. Table 8.4 summarizes the relative cost of the resins. While cost is a consideration in resin selection, the maximum temperature that the tubing will see in downhole service is often the deciding factor in determining which resin system should be used. Maximum service temperatures are in the range of 120 °C (250 °F)–150 °C (300 °F).

Problems with Internally Coated Tubing Problems with internally coated tubing include wireline damage, deformation and cracking of coatings due to mechanical loads, and debonding of coatings due to sudden pressure releases.

Wireline damage in tubing is similar to the damage shown for drill pipe in Figure 8.27. The damage most likely to occur at joints and can be minimized by slow wireline speeds (<0.5 m s^{-1} or <100 ft min^{-1}). Downhole tools should be plastic coated, and wheeled centralizers or guides that keep the wireline near the center of the tubing also help. Corrosion inhibitor applications should be reintroduced immediately after running wirelines in order to refilm any exposed metal surfaces.

Figure 8.26   Paraffin deposits on the inside of Gulf of Mexico production tubing.

Figure 8.27   Wireline damage to the internal coating on drill pipe. *Source*: Byars [26]. Reproduced with permission of NACE International.

TABLE 8.4   Relative Costs of Resin Materials Used as Tubing Coatings and Liners

Material	Abbreviation	Relative Cost
Polyethylene	PE	1
Polypropylene	PP	1.1
Polyamide (Nylon) 11	PA11	6
Polyketone	PK	5
Polyvinylidene fluoride	PVDF	15
Polytetrafluoroethylene	TFE	50
Fluorinated ethylene propylene copolymer	FEP	65
Ethylene-tetrafluoroethylene copolymer	ETFE	65
Chlorotrifluoroethylene-ethylene copolymer	ECTFE	65

Rounded tubing connectors minimize stress concentrations that can lead to cracking. Of course these premium connectors cost extra.

Another problem associated with lined or internally coated tubing is rapid pressure release. All organic coatings are permeable to moisture and gases, and sudden pressure releases can lead to disbonding like shown in Figure 6.26.

Concluding Remarks About Internally Coated Tubing

Coatings are imperfect means of corrosion control and must be supplemented by inhibitors of some sort. For water injection wells the inhibition is usually by the use of oxygen scavenging after some other means of deaerating the water. For production wells most common corrosion inhibitor packages rely on thin-film formation of some sort of organic molecules attached to the surface.

International organizations have developed standards and guidance on internal coatings and lining to include the following NACE publications [83–89]:

- NACE SP0181, Liquid-Applied Internal Protective Coatings for Oilfield Production Equipment.
- NACE SP0188, Discontinuity (Holiday) Testing of Protective Coatings on Conductive Substrates.
- NACE SP0191, Application of Internal Plastic Coatings for Oilfield Tubular Goods and Accessories.
- NACE SP 0291, Care, Handling, and Installation of Internally Plastic-Coated Oilfield Tubular Goods and Accessories.
- NACE TM0185, Evaluation of Internal Plastic Coatings for Corrosion Control of Tubular Goods in an Aqueous Flowing Environment.
- NACE TM0187, Evaluating Elastomeric Materials in Sour Gas Environments.
- NACE TM 0192, Evaluating Elastomeric Materials in Carbon Dioxide Decompression Environments.

Wireline

Oilfield wireline is supplied for a variety of downhole measurement and control purposes. It is available made from both carbon steel (often termed "plow steel" to imply high-carbon steel with high yield strengths usually obtained by extensive cold working) and CRAs. Table 8.5 shows several CRAs available for wirelines. Many wireline suppliers use proprietary trade names for the alloys being supplied, but a quick Internet search can usually identify the UNS numbers associated with these alloys. Crevice corrosion of multistranded wireline is one possible problem with these lines, and high PRENs are sometimes considered to be indications of alloys appropriate for this service.

TABLE 8.5   Corrosion-resistant Alloys Available for Oilfield Wirelines

UNS Designation	Common or Trade Name	PREN[a]	Critical Pitting Temperature (CPT)	
			°F	°C
S31600	316	26	72	22
S32305/S31803	2205	36	108	42
S20910	XM-19	38	106	41
N08028	28	40	129	54
N08926	25-6MO	47	149	65
S31277	27-7MO	56	176	80
R30035	MP35N	53	183	84

[a] PREN, pitting resistance equivalent number = Cr + 3.3 Mo + 30 N.

Wirelines and their attached tools can cut internal coatings (Figure 8.27) and cause subsequent corrosion problems [26, 90].

High-strength wire is subject to hydrogen embrittlement in sour wells. Carbon steel wireline can be used in well fluids only because grease coatings on the metal surface prevent wetting of the wire and the absorption of hydrogen. Some hydrogen is inevitably absorbed into the metal, and allowing the hydrogen to bake out between trips is necessary. These bakeout times are normally several days long and must be used for all wireline metals. The use of CRAs reduces, but does not eliminate, the tendency for hydrogen damage.

Coiled Tubing

Coiled tubing is used for many of the same purposes as wireline. The principal advantage of coiled tubing is the stiffness of the tubing, which allows it to be "pushed" along horizontal legs in deviated wells. For workovers and other short-term applications, the exposure time is relatively short, hours or days, and corrosion can be a minimal concern. Fatigue becomes the major consideration in determining tubing life, because of the large strains induced during laying from reels. The number of trips for coiled tubing is limited to dozens of trips at the most. Unfortunately, the use of coiled tubing in H_2S environments can produce problems even with short-term exposures, and clear guidance on appropriate approaches to avoid system failures is lacking [91–94].

Coiled tubing is also used for velocity tubing when the natural pressure in gas wells has degraded. The insertion of smaller-diameter tubing is used to increase fluid velocity restoring production and the removal of liquids from the wellbore. Velocity tubing may be used for years, and conventional corrosion control methods, usually with inhibitors, are employed.

CRA tubing is available and used in some applications. These applications of coiled tubing are more likely to be used for subsurface controls and other long-term applications, e.g. power to hydraulic lift pumps.

Material and Corrosion Concerns with Artificial Lift Systems

Natural gas and most new oil wells flow due to downhole formation pressure. As oil wells age, it usually becomes necessary to add some form of artificial lift system, pumps, or other devices designed to bring the liquid crude oil to the surface. There are several options including:

- Beam-pumped wells
- Gas-lift systems
- Hydraulic lift systems
- Electrical submersible pumps (ESPs)
- Progressive cavity pumps

Each of these methods has unique materials and corrosion requirements. Most oil wells have some sort of artificial lift. Beam-pumped wells are the most common for onshore applications, and ESPs are most common in offshore locations where their reduced topside weight and increased flow capacity are important. Most operators try to stay "pumped off," maintaining a minimum liquid level over the pump intake. This results in little liquid in the annulus, and most of the liquid in the annulus will be oil that produces hydrocarbon condensation on the annulus interior with CO_2 or H_2S in the water, which tends to remain below the oil in the annulus [26].

CRAs are used for many components of artificial lift systems. This seldom creates galvanic corrosion problems with carbon steel and other alloys downhole, because the level of reducible chemicals (oxygen, hydrogen ions, etc.) is very low. At the worst, the corrosion rate on the larger carbon steel components will be twice the rate if no CRA cathodes were present, because the large anode sizes tend to reduce any effect of galvanic coupling.

Beam-pumped Wells Figure 8.28 shows the topside components of a beam-pumped well. The moving head of the well assembly causes a long string of sucker rods to move up and down inside the well. A chrome-plated polished rod on the surface passes through an elastomeric-sealed stuffing box and is connected to a long series of sucker rods that are strung inside the tubing to the bottom of the well. At the bottom of the well, a downhole pump moves the liquid oil and produced water up the tubing string. As downhole pressure

Figure 8.28 Topside of a beam pumping unit.

decreases, gas breakout will occur, and gas is usually removed from the tubing near the surface. Figure 8.29 is a drawing of a beam pumping system. These systems are commonly used on low-volume onshore wells. It is not uncommon to see hundreds of the pumping jacks shown in Figure 8.28 in an oilfield that may be several decades old but still economically productive.

The most common form of materials failure is corrosion fatigue of the sucker rods, long rods, usually of carbon or low-alloy steel (e.g. UNS G41420 – AISI 4142 alloy), with upset ends that are formed and machined for connecting to other rods. Fatigue cracking tends to occur at the upset ends where stresses are the greatest. Figure 8.30 shows damaged sucker rods, one with corrosion pits and the other with numerous fatigue cracks. Both rods eventually failed at shear lips like those near the bottom of Figure 8.30. Figure 8.31 shows greater detail on how fatigue cracking progressed to failure on a different sucker rod.

Another sucker rod failure mode is wear, which is most likely to occur near the midpoint of the rods at locations where the well deviates from horizontal. Factory-installed sucker rod guides made from abrasion-resistant injection-molded polymers are installed along the rod at selected locations. Up to three guides per rod are sometimes installed. These guides also reduce wear on downhole tubing, which is softer than the rod metal and more likely to wear, leading to leaks.

Sucker rods should be full-length normalized after quenching and tempering to eliminate undesired differences in microstructure and mechanical properties near the upset ends. Rods are commonly supplied in 25 ft (7.62 m) and 30 ft (9.14 m) lengths with diameters from 5/8 in. (15.88 mm) to 1⅛ in. (28.58 mm). Their relative length to diameter means that they are flexible and can bend around some high-radius downhole tubing bends.

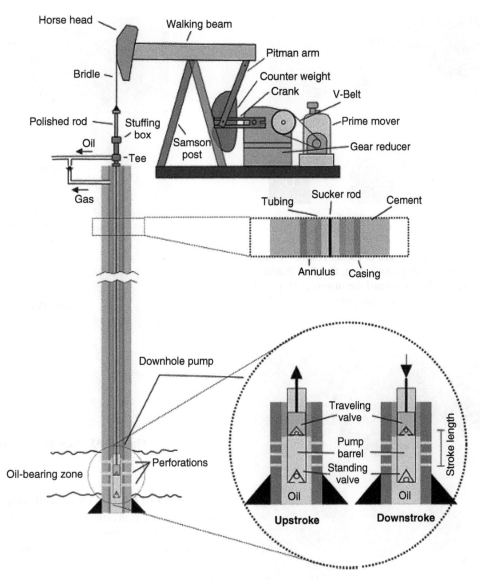

Figure 8.29 Beam-pumped well diagram.

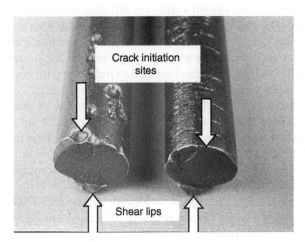

Figure 8.30 Broken sucker rods.

Most sucker rods are made from heat-treated low-alloy steel. Titanium, aluminum, and fiberglass sucker rods are also available. The primary advantage claimed for these alternate materials is their increased strength-to-weight ratios compared with steel. While these materials have been available for many years, low-alloy steel sucker rods are usually used.

Corrosion of sucker rod strings sometimes occurs near the wellhead. This is usually due to air leaking in at the polished rod-stuffing box seal. The most common means of preventing this corrosion is by maintaining a positive pressure on the downhole tubing, ensuring that any gas leakage is from inside the well to the atmosphere. The deep mesa-type pitting corrosion shown in Figure 5.96 was due to high concentrations of CO_2 in the

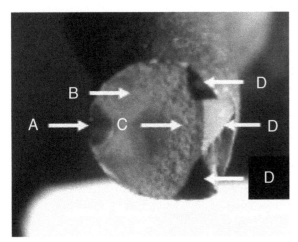

Figure 8.31 Fatigue fracture surface of sucker rod. A – Crack initiation site. B – Cyclic crack propagation. C – Fast fracture surface. D – Shear lips pointing toward the observer (top and bottom) and away from the observer (center).

Figure 8.32 Galvanic corrosion on a sucker rod pump.

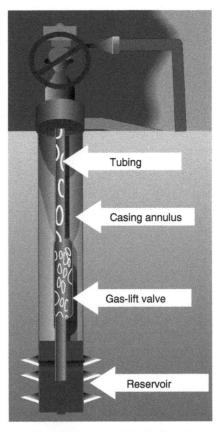

Figure 8.33 Gas-lifted oil well.

produced fluids. The 9-chrome material in use at the time proved inadequate for corrosive conditions, and 9-chrome alloys are no longer common for downhole applications.

A number of industry standards are available on the manufacture, care, and testing of sucker rods [95–100].

Corrosion and wear of downhole pumps is controlled by the use of appropriate alloys [31, 48]. Minimal corrosion is expected at the bottom of wells because of the absence of oxygen and other reducible species, but it can occur [101]. Figure 8.32 shows galvanic corrosion on a sucker rod pump. Note that corrosion has also occurred on the rod at the far left of the picture.

Gas-lifted Oil Wells   These wells usually have packers, and if the lift gas is corrosive due to the presence of oxygen, moisture, or corrosive gases (e.g. CO_2 or H_2S from unprocessed natural gas), corrosion can occur in the annular spaces.

Lift gas is usually transported downhole in the annulus, and one of two types of valves is installed downhole to inject the gas into the production string (Figure 8.33). Conventional gas-lift valves are installed as the tubing is placed in the well and must be removed with the tubing string. Side-pocket mandrels allow the installation and removal of valves by wireline while the mandrel remains in the well, eliminating the need to pull tubing for repair or replacement of valves.

Corrosion problems in gas-lifted wells may be due to wet lift gas or the absence of packers. Corrosive lift gas causes problems in the annulus as well as on tubing interiors. It is not uncommon for lift gas to be more corrosive than produced reservoir gas, which is sometimes reinserted as lift gas.

Corrosion control is achieved in gas-lifted wells through the use of noncorrosive lift gas and by the use of internal production tubing coatings, CRA tubing strings, and/or filming corrosion inhibitors [26, 50].

Hydraulic Lift Systems   These systems get power from high-pressure liquids injected into either the tubing that transports the high-pressure liquid to the bottom of the well. The downhole assemblies then transmit energy to the produced fluid, which is usually produced through the annulus. Downhole equipment can be either jet pumps (injecting the higher-pressure lift liquids into the production tubing using Venturi jet pumps with no moving parts) or reciprocating piston pumps to force produced fluids to the surface. Hydraulic lift fluids are either reinjected crude oil or produced water. If produced water is reinjected, it may be corrosive if it has been contaminated at the surface by oxygen-containing air. This corrosion is controlled by the use of corrosion inhibitors, including oxygen scavengers and film-forming inhibitors.

Venturi systems can be used in deviated wells where beam-pumped sucker rods or other mechanical systems will not work. Reciprocating piston pumps can operate until well depletion and handle low production rate wells. Either type of pumping system can also deliver chemicals, e.g. corrosion, paraffin, or emulsion inhibitors, to the produced fluid.

Drawbacks to hydraulic lift systems include their high energy consumption and relatively high cost. The use of coiled tubing to deliver hydraulic fluids downhole has reduced costs and enabled their use in additional wells.

Electric Submergible Pumps (ESPs)   These pumps, which are the most common oil well pumps for offshore use, have few corrosion problems. The pumps themselves are available in a variety of CRAs, and the electric power cables, which usually run in the annulus, have CRA (often nickel-based) cable sheaths. Metallic sheathing is necessary, because all polymers used for electrical insulation are permeable to gases, and corrosive gases, e.g. CO_2 or H_2S, could cause corrosion of the copper power wire. The downhole components of these pumps are made from materials that can withstand downhole temperatures of up to 150 °C (300 °F) and the associated high-pressure chemistries of the produced fluids. They lose efficiency with downhole gas contents greater than approximately 10%. Improved versions of these pumps are available with downhole gas separators that can remove gases and transport them to the surface in parallel tubing strings.

ESPs have tight dimensional tolerances and can have erosion and wear problems when producing sand.

Progressive Cavity Pumps   The moving parts of progressive cavity pumps that are exposed to production fluids are the helical screw rotor and drive shaft, which are usually chromium hard-faced (thicker than electrolytic chrome plating) heat-treatable carbon or low-alloy

Figure 8.34   Cross-sectional view of an elastomeric stator for a progressive cavity pump. Stators for oil wells usually have longer stators (more turns) because of relatively high downhole pressures.

steels, and the stator, made from a chemically resistant hard elastomer (Figure 8.34).

Problems with helical screws and drive shafts seldom occur. Stators are the components most likely to degrade and need replacement. Properly chosen materials should last three to five years, and most of the metallic components of the pump can be reused or refurbished after the elastomeric stator is replaced [1].

Friction between the stator and the rotor raises the temperature of both components above the downhole fluid/formation temperature. Candidate stator materials should be tested for fluid compatibility using established standards that measure swelling resistance, changes in hardness, and other parameters associated with the exposure of polymeric materials to downhole fluids at the estimated temperature of the operating pump [88, 89, 102, 103]. Table 8.6 presents the elastomeric materials most likely to be available for downhole progressive cavity pumps. The table only shows generic classifications of the main resin components, and all commercial elastomers will have minor constituents that affect their chemical and wear resistance. This is why fluid compatibility testing is recommended for any large-scale application, e.g. use in a major new field.

Wellheads, Christmas Trees, and Related Equipment
Wellheads are the components at the top of a well that provide structural and pressure-containing support for the drilling and production equipment. The primary purposes of wellheads are to provide suspension points for downhole equipment and pressure seals for the tubing and casing strings running from the bottom of the hole.

After completions are finished, Christmas trees (often abbreviated XT in many oilfield specifications and documents) are attached to most wells. "Trees" are assemblies of valves, chokes, spools, and other fittings used to control flow of fluids and equipment into and out of the well. Figure 8.35 shows a Christmas tree attached to a wellhead on an offshore production platform in the Gulf of Mexico. Some Christmas trees are

TABLE 8.6   General Elastomer Selection Guide for Progressive Cavity Pumps

	Elastomer Type			
Characteristics	Buna	High Nitrile	Hydrogenated	Fluoroelastomers[a]
Mechanical properties	Excellent	Good	Good	Poor
Abrasion resistance	Very good	Good	Good	Poor
Aromatic resistance	Good	Very good	Good	Excellent
H₂S resistance	Good	Good	Excellent	Excellent
Water resistance	Very good	Good	Excellent	Excellent
Temperature limit[b]	95°C (200°F)	95°C (200°F)	135°C (275°F)	150°C (300°F)

[a] Common trade names for fluoroelastomers include Viton, Hyflon, Halar, and Teflon.
[b] The internal operating temperature of pumps may be significantly higher than the reservoir fluid temperature due to friction between rotors and elastomeric stators.

Figure 8.35   Christmas tree attached to a wellhead.

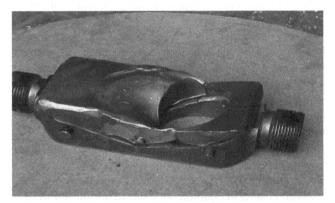

Figure 8.36   Erosion of a gate valve component producing loss of a sealing surface. *Source*: Photo courtesy of NACE International.

more complex. Subsea Christmas trees, which must operate for many years with no inspection or maintenance, are often machined from a single block of metal instead of being bolted together with flange connections [66, 104, 105].

Wellheads, which support downhole tubing, casing, and other components, are connected at the top of wells to Christmas trees, which control production rates and fluid flows out of the well and may also direct fluids and equipment into the well. The primary materials considerations for wellheads are strength, erosion resistance, and corrosion resistance. The relatively small size of wellhead and Christmas tree equipment compared with tubing, casing, pipelines, etc. means that materials costs are secondary considerations. It is common to use whatever material is necessary to ensure reliable production and minimal maintenance for wellhead components.

Christmas trees, which regulate flow rates and also change the primary direction of flow from vertical to horizontal, have fewer strength requirements, because they are not limited by downhole space limitations. The requirements for pressure containment suggest the use of low-alloy heat-treatable steels for the bodies of these components. Lining the fluid-exposed surfaces with hard, erosion-corrosion-resistant alloys provides the necessary environmental resistance, but all components must be made from H₂S-resistant alloys, because no coating or hard facing is without defects that can allow gas penetration. While a composite structure of high-strength steel with erosion-corrosion lining is possible for the main components, the smaller internal components (gates, springs, etc.) are not coated and are made from erosion-corrosion-resistant alloys. Figure 8.36 shows erosion on a Christmas tree gate valve. The use of hard corrosion-resistant materials and control of production to fluid flow rates that minimize erosion are means of controlling this erosion. The use of company guidelines has largely replaced API RP14E erosion recommendations for this application, because many companies find the API guidelines to be overly conservative

[106, 107]. This leads to perceived unnecessary reductions in production rates. The economic consequences of reduced production rates must be balanced against the costs of production shutdowns and possible reduced reliability.

Production wellheads have a variety of high-strength components necessary to support the weight of the casing and tubing. This places major limitations on the type of materials that can be used. The restrictions of ANSI/NACE 0175/ISO 15156 usually determine which materials can be used except in the most benign onshore applications [18, 35, 36]. API Spec 6A prescribes strength and impact properties for wellhead materials [25]. Required strength levels depend on the pressure ratings of the equipment, e.g. equipment utilized to 10000 psi (69 MPa) must be made from materials having a minimum yield strength of 60000 psi (414 MPa), and higher-pressure equipment must be made from materials using materials with a specified minimum yield strength of 75000 psi (517 MPa). Hardness levels and Charpy impact properties are also specified [25]. Other similar specifications cover subsea wellheads and Christmas trees [104]. The ISO standard for wellheads and Christmas trees is slightly different [108].

The wellhead and Christmas tree standards also specify various classes of service depending on pressure, corrosivity, and temperature. It is common for suppliers to supply valves with different materials selected for:

- Body and bonnets
- Flanges
- Wetted internals
- Valve stems
- Various types of seals
- Trim components including both wear and non-wear components

Valves are supplied depending on different API or ISO service classifications, which vary depending on pressure, corrosivity of the internal environment, and temperature.

Charpy impact toughness requirements in API specifications are intended to minimize brittle behavior. Many wellhead and Christmas tree components are massive pieces of metal that are heat-treated after forming, usually by forging. If the heat treatment is not correct, then brittle fracture can result [109]. Installations of subsea completions – and the tremendous expenses associated with their retrieval, maintenance, or repair – mean that subsea wellheads and Christmas trees are now intended to last for the life of the producing well. This can be 30 years or longer with no anticipated inspection, maintenance, or parts replacement. The tremendous expenses associated with subsea

completions means that materials selection and fabrication must be as careful and reliable as possible [105].

Springs in Christmas trees present difficult materials selection and fabrication choices. Most springs must have very high strength and hardness to work effectively. High strength and hardness makes them susceptible to chloride SCC and H_2S and not in compliance with the requirements of ANSI/NACE MR0175/ISO 15156. The acceptable high-strength alloys for this service are cobalt alloys, which are allowed to HRC 55 or 60 depending on the alloy, and precipitation-hardened nickel-based alloys, which can be used to a maximum hardness of HRC 50 [36].

FACILITIES AND SURFACE EQUIPMENT

Topside equipment and surface facilities are much more likely to have corrosion problems than downhole wells and wellheads. This is because they are exposed to air, and any air leaking into fluids can cause internal corrosion as well as atmospheric corrosion of the equipment exteriors.

Figure 8.37 shows a typical arrangement of topside equipment for a medium-sized offshore platform in the Gulf of Mexico. Most of the process equipment is listed by category in the labels associated with the various bays of the structures.

ISO 21457 and NACE Report 1F192 suggest a number of alloys for a variety of downhole and topside applications [50, 110]. Other standards will also suggest alloy selections. Most organizations will also ensure that production fluid-exposed equipment also meets the conditions of ANSI/NACE MR0175/ISO 15156 [18, 36, 37]. It should be noted that the latter standard only addresses H_2S-related cracking problems and does not address other forms of corrosion.

Piping

Most topside corrosion problems are associated with piping systems and their welded or flanged joints. Carbon steel is used for most piping, because it is inexpensive and has few internal corrosion problems in hydrocarbon systems. Other materials are used for corrosive environments, and this includes the use of CRAs for some hydrocarbon systems, e.g. if they are necessary due to high levels of CO_2 or H_2S, usually in gas handling lines. The alloy classifications in ANSI/NACE MR0175/ISO 15156 provide a list of possible alloys for this use. While the standard is only intended to describe the alloys acceptable for H_2S service, produced fluid piping systems should always be made from these alloys with the final selection depending on weight-loss or other

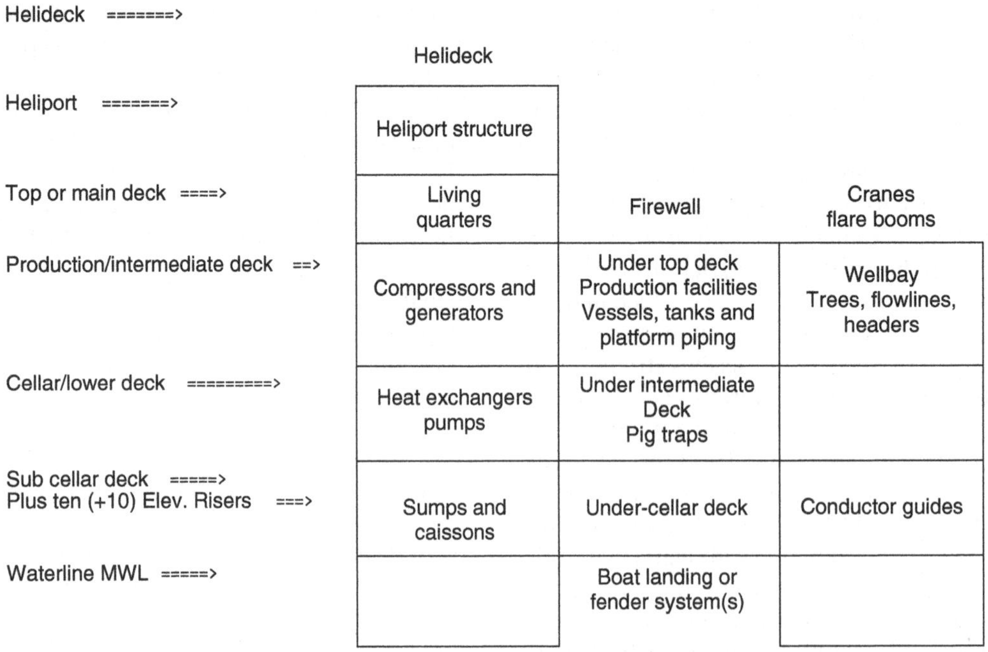

Helideck   =======>

Helideck		
Heliport structure		
Living quarters	Firewall	Cranes flare booms
Compressors and generators	Under top deck Production facilities Vessels, tanks and platform piping	Wellbay Trees, flowlines, headers
Heat exchangers pumps	Under intermediate Deck Pig traps	
Sumps and caissons	Under-cellar deck	Conductor guides
	Boat landing or fender system(s)	

Heliport   =======>

Top or main deck   ====>

Production/intermediate deck   ==>

Cellar/lower deck   =========>

Sub cellar deck   =====>
Plus ten (+10) Elev. Risers   ===>

Waterline MWL   =====>

Figure 8.37   Typical arrangement of topside equipment on an offshore platform.

corrosion resistance. Further guidance is available in ISO 21457 and NACE Report 1F192 [50, 110]. While these reports discuss a number of internally clad piping systems (CRA lining of carbon steel piping), several organizations have reported debonding problems with clad piping, and the use of internally clad piping is considered questionable by many authorities.

Titanium can be used in seawater and other systems with fluid velocities as high as 7.6 m s^{-1} (25 ft s^{-1}). Seawater velocity needs to be greater than 1.2 m s^{-1} (4 ft s^{-1}) to avoid biofouling.

Copper-based alloys (usually 90/10 copper nickel or 70/30 copper nickel) are used in seawater handling systems, fire suppression piping, etc. The liquid velocity limits for copper-based piping depend on size and vary from 3.5 m s^{-1} (11.5 ft s^{-1}) for 10 cm (4 in.) cupronickel piping to lower levels for other copper alloys [111].

It is important to keep the piping system design as simple as possible with limited dead legs and other locations where deposits can accumulate and lead to corrosion. It is also important to ensure inspectability and provide for extra tie-ins so that designs can be altered as necessary.

FRP piping is sometimes used for water and other systems. Problems associated with this piping are generally due to the increased support structures necessary for this piping. Other problems have been reported with joining the piping. A major advantage of the piping is the smooth interior surfaces that limit fouling attachment sites. This has led to the increase use of FRP piping for fire suppression systems.

Figure 8.38   Corrosion of pipe at neoprene support gasket.

Figure 8.38 shows corrosion of a carbon steel pipe protected from abrasion by a neoprene gasket. Unfortunately, the neoprene collects moisture, and the external pipe coating is not an immersion coating. Figure 8.39 shows a better approach to piping supports. A round U-bolt with a plastic sleeve holds the pipe against a nonabsorbent plastic half-round. Several suppliers of this kind of piping attachments are commercially available. Figures 8.38 and 8.39 are related to corrosion under pipe supports (CUPS), which is discussed in greater detail in Chapter 5.

Thermal expansion and contraction can be a major problem in piping systems. The vertical expansion loop shown in Figure 3.19 should normally be avoided. This loop on a steam injection line can lead to slugging when

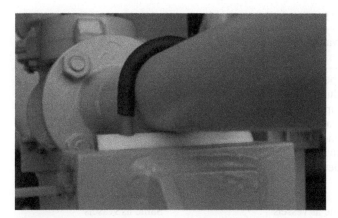

Figure 8.39 Insulated U-bolt assembly to avoid corrosion at pipe support.

Figure 8.40 Corrosion at a control valve in a "dry" fire sprinkler system. *Source*: Photo courtesy Mark Hopkins, Hughes Associates, Baltimore, MD.

condensate builds up in the upstream portion of the loop. Most authorities would recommend horizontal expansion loops whenever possible.

Firewater Systems   Firewater lines are stagnant except during monthly testing, so it is generally acceptable to use carbon steel, which corrodes for a small number of days after the testing until the dissolved oxygen in the water is used up. This intermittent service is much less corrosive than constant use. Other materials used for this piping include cupronickels, PVC, and austenitic stainless steel. Unfortunately all metallic firewater lines are subject to under-deposit corrosion due to biofouling or debris. This was shown in Figure 4.54, which shows corrosion at the six o'clock position in a copper-nickel firewater line from a Gulf of Mexico platform.

Many companies use FRP for this service, because metallic corrosion products can clog nozzles. FRP also has smooth surfaces and is less likely to promote biofouling communities. API RP14G, "Fire Prevention and Control," lists FRP as one of the acceptable piping options [112].

Most firewater systems are "wet" systems with water in all lines at all times. In cold climates, this can become an icing problem, and dry systems must be used. Unfortunately "dry" systems can often have water trapped at valves and low spots in the piping system, leading to corrosion like that shown in Figure 8.40. Dry systems require frequent maintenance and inspection, drying with inerting gases, etc. Because these systems are very difficult to dry, it is common to find some residual water at low spots. This can lead to microbially influenced corrosion (MIC) similar to the corrosion problems found after hydrotesting equipment [113–115].

Seawater Systems   Seawater lines are often titanium or cupronickel to minimize corrosion.

Pumps often have nickel-aluminum bronze (NAB) housings, pump bodies, shafts, and fasteners. Selective

phase attack of some NAB components is possible, but this problem, which is due to improper foundry practice, is relatively rare.

Cast stainless steel may be cheaper, but it is usually not as strong or cavitation resistant. Stainless steel shafts can also develop pitting or crevice corrosion in gland areas during shutdowns. Very large pumps may have cast iron bodies with CRA impellers. This leads to some galvanic corrosion, but the relatively large size of the cast iron body minimizes the problem.

Water Handling and Injection Equipment   Table 8.7 summarizes corrosion problems and control methods for oilfield water handling systems.

NACE SP0499 cites NORSOK guidance for injection water based on dissolved oxygen equivalents of

$$\text{Oxygen equivalent} = \text{ppb oxygen} + 0.3 \text{ ppb free chlorine}$$

For conditions where the oxygen equivalents are

50 ppb for 90% of operation time and
200 ppb for 10% of operation time, noncontinuous

and the temperature is 30 °C (86 °F) or less, NORSOK recommendations in Table 8.8 are suggested [59, 116].

Storage Tanks

Most of this discussion will deal with aboveground storage tanks (ASTs), which are much more common in oil and gas production than the, usually smaller, underground storage tanks (USTs). Protective coatings supplemented by cathodic protection are used for external corrosion of USTs, and these tanks present few unique problems.

TABLE 8.7   Internal Corrosion and Corrosion Control of Water Handling and Injection Equipment

System or Equipment (Environment)	Usual Problem Areas	Evaluation Methods (Detection and Monitoring)	Most Common Corrosion Control Methods
Vessels and tanks	Shell when water is very corrosive Bottom under deposits Vapor space when sour	Inspections Failure history Coupons, probes Occasionally iron content and/or bacterial activity Galvanic probes (oxygen detection)	Oxygen-free operation (exclusion and/or removal) Cathodic protection (CP) Coatings Periodic cleanout Occasionally: biocides or inhibitors Nonmetals – if air intrusion is tolerable
Filters	Connections when bimetallic At filter media (sand) level	Same as vessels	Same as vessels
Gathering and injection lines and plant piping	Along bottom under deposits	Same as vessels	Oxygen-free operation Coatings and linings (including cement) Periodic line pigging Occasionally: biocides or inhibitors
Injection and transfer pumps	All wetted parts Seals leaking air	Failure history Inspection Galvanic probe (oxygen detector)	Oxygen-free operation Metallurgy
Injection wells	Tubing interior, inhale valves, fittings, etc. Wellhead and Christmas tree Annulus above packer (depends on packer fluid)	Failure history Coupons Occasionally iron content Galvanic probes (oxygen detection)	Oxygen-free operation Coatings and linings Occasionally biocides or inhibitors

Source: Byars [26]. Reproduced with permission of NACE International. Appendix 1A, pp. 18–22.

TABLE 8.8   NORSOK Recommendations for Injection System Materials [59, 116]

Injection Water	Tubing and Liner	Completion Equipment (When Different from Tubing/Liner)
Deaerated seawater	Low-alloy steel	UNS N09925 (alloy 925), UNS N07718 (alloy 718), 22Cr or 25Cr duplex SS
Raw seawater	Low-alloy steel with glass-reinforced plastic (GRP) or other liner; unlined low-alloy steel for short design life; titanium (with design limitations)	Titanium (with design, limitations)
Produced and aquifer water	Low-alloy steel; low-alloy steel with GRP or other liner; 13Cr (provided oxygen <10 ppb); 22Cr duplex 55, UNS N07718, UNS N09925 (provided oxygen <20 ppb)	13Cr (with limits as for tubing for this service)

Figure 8.41 shows the exterior of the same AST shown in Figure 6.20. Exterior corrosion of the air-exposed portions of ASTs is controlled by the use of protective coatings and seldom presents unique challenges. The exterior surfaces have relatively simple geometries and, if properly coated, may perform well for many years. Maintenance and safety concerns are associated with personnel ladders, floating roofs, and other attachments.

Figure 8.42 shows the locations on a crude oil AST where corrosion is likely. Little corrosion is expected on the interior walls of the tank where metal surfaces will be hydrocarbon wetted. Corrosion problems occur in the vapor space above the stored liquid and at the bottom of the tank, where water and sludge deposits may accumulate (Figures 8.43 and 8.44). ASTs of this general design may store over ½ million barrels of liquid product. Many problems associated with these large tanks are due to their tremendous weight, especially the weight of the exterior walls, which are much heavier than the liquids inside. It is not unusual for tanks to settle unevenly and produce wrinkled bottoms like the one

Figure 8.41 The painted exterior of the same aboveground storage tank shown with stripe coating in Figure 6.20.

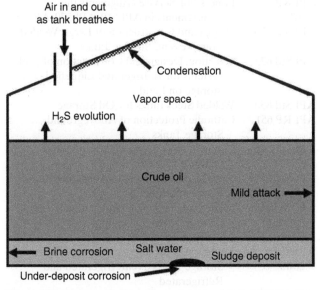

Air in and out
as tank breathes

Condensation

Vapor space

H₂S evolution

Crude oil

Mild attack

Brine corrosion   Salt water   Sludge deposit

Under-deposit corrosion

Figure 8.42 Cross section of a typical crude oil AST showing locations where corrosion is likely to occur.

Figure 8.43 Corrosion on the interior ceiling of a crude oil AST.

Figure 8.44 Bottom of crude oil AST showing solids buildup in the foreground and discoloration due to corrosion by water on the lower course of the steel wall in the background.

Figure 8.45 Ring bulging at outer wall of AST tank bottom. *Source*: Photo courtesy J. O'Hearn, Corrpro, reproduced with permission.

shown in Figure 8.45. These locations can distort and stress the metal so much that cracking and leaking can occur. The uneven bottom also prevents drainage and leads to corrosion. Repeated problems like those shown in Figure 8.45 have led to the development of API inspection standards for ASTs. Figure 8.46 shows the bottom profile of a large fuel-oil AST that was inspected in accordance with API 653 [117]. This tank was part of a tank farm less than 20 years old that needed to be replaced due to extensive leaks of hydrocarbon into the local groundwater. Similar tanks, with less settlement at the exterior rings (Figure 8.47), may last for many decades with only superficial staining of the tank bottoms.

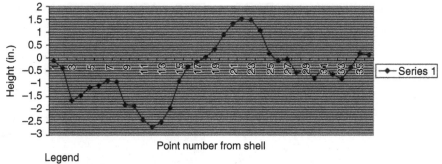

Legend
Series 1:   Differential between ideal slope and actual settlement, in inches.
X-axis: Ideal slope of bottom, in inches.

Figure 8.46   Inspection profile on the bottom interior of a large fuel-oil AST.

Figure 8.47   Concrete ring walls supporting the exterior of an AST.

The magnitude of leaks from some of these large storage tanks has led to improved AST design and inspection standards. These are summarized in Table 8.9. Many, if not most, corrosion problems come from uneven settling of the tremendously heavy walls. Improved guidelines on how to design exterior rings to support these walls has led to major improvements in design, but thousands of existing tanks with internal and external corrosion of the tank bottoms continue in use. Many leaks are not detected until long after they have produced environmental consequences.

The liquid on the interior bottom of a storage tank can be due to produced water, as indicated in Figure 8.42, but it can also be condensation from the vapor space at the top of the tank that finally settles at the bottom. The drawing in Figure 8.42 is simplified, and a number of vertical supports extend from the vapor space to the bottom of the tank. Water draining along these supports can produce corrosion of the vertical components as

TABLE 8.9   API Storage Tank Guidelines

API 510	Pressure Vessel Inspection Code: In-service Inspection, Rating, Repair, and Alteration
API RP 579-1	Fitness-For-Service (Augments the requirements in API 510, API 570, API 653)
API Std 620	Design and Construction of Large, Welded, Low-pressure Storage Tanks
API Std 625	Selection, Design and Construction of Tank Systems for Refrigerated Liquefied Gas Storage on Land
API Std 650	Welded Steel Tanks for Oil Storage
API RP 651	Cathodic Protection of Aboveground Storage Tanks
API RP 652	Lining of Aboveground Storage Tank Bottoms
API Std 653	Tank Inspection, Repair, Alteration, and Reconstruction
API Bulletin 939-E	Identification, Repair, and Mitigation of Cracking of Steel Equipment in Fuel Ethanol Service
API Std 2000	Venting Atmospheric and Low-pressure Storage Tanks: Nonrefrigerated and Refrigerated
API RP 350	Overfill Protection for Storage Tanks in Petroleum Facilities

shown in Figure 8.48. Protective coatings are used to control these vertical columns (Figure 8.49).

Impressed current cathodic protection (ICCP) is frequently used to control corrosion in the water-wetted locations at the bottom of these tanks. Figure 8.50 shows ICCP lead wires from a central power source to openings in the tank roof where vertical anodes are strung to near the bottom of a tank.

Exterior corrosion of tank bottoms can also become a problem, and cathodic protection is frequently used to minimize this corrosion. The use of impermeable membranes or sand layers, which have low electrical

Figure 8.48 Corrosion on a vertical column at the bottom of a large AST.

Figure 8.49 Painting for corrosion control on the lower portions of vertical supports on a large AST.

Figure 8.50 Impressed current cathodic protection anode lead wires on the top of a large AST.

Figure 8.51 Corrosion on a floating roof of an AST. *Source*: Photo courtesy of The Hendrix Group, Inc.

conductivity, makes this a difficult proposition in many instances [118].

Procedures have also been developed to insert liners or to coat AST interior tank bottoms. When repairs are necessary, it is sometime possible to jack up the exterior walls and insert new bottom flooring over the degraded floor.

Section C of API 650 discusses floating roofs, which are more common for storing refined products than for water or crude oil [118]. The advantages of floating roofs include reduced vapor losses and reduced likelihood of vapor-related ignitions. The floating roof rises and lowers depending on the volume of liquid stored inside. Flexible seals along the rim are intended to minimize evaporation. Figure 8.51 shows corrosion on a floating roof tank. Note the corrosion of the interior tank wall as well as corrosion on the roof. Problems with drains to remove rainwater and snowmelt are also maintenance considerations for floating roofs. Corrosion of tank roofs, fixed or floating, can also present safety hazards if personnel walk on them for maintenance or inspection.

Heat Exchangers

A wide variety of heat exchangers are used in oilfield processing. Shell and tube heat exchangers predominate in the onshore market, but the reduced weight and processing efficiencies of compact heat exchanger designs lead to their extensive use offshore.

Shell and Tube Heat Exchangers As stated above, shell and tube heat exchangers are the most commonly used designs for onshore oilfield processing [119]. Figure 8.52 is a simplified view of one of these exchangers. It is common for a tubing bundle to have dozens of tubes as well as intermediate support baffles. Most shell

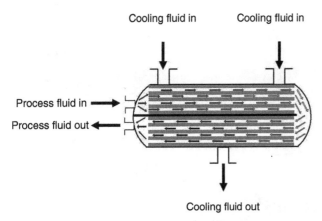

Cooling fluid in   Cooling fluid in

Process fluid in →

Process fluid out ←

Cooling fluid out

Figure 8.52 Schematic of a small shell and tube heat exchanger.

Figure 8.53 Magnesium anodes in the water-cooled header space on a gas cooler.

and tube exchangers operate with the process stream on the inside of the tubing and the heat transfer medium (usually cooling water) on the "shell side" between the tubing and the water jacket or water box. After extensive use it is common for mineral and fouling deposits to form on the outside of the tubing. Thermal expansion and contraction can lead to wear between the tubing and intermediate supports, and this is an area requiring inspection during shutdowns (Figures 5.113 and 7.17). A variety of nondestructive techniques are available for this purpose.

The tubing in a shell and tube heat exchanger is the thinnest component and is subject to erosion corrosion at the inlets, SCC at locations where the tubing has been expanded into the header plates, fretting and fatigue at intermediate support baffle locations, etc. Overheating due to scale buildup, fretting corrosion, and erosion corrosion at tubing inlets are the most common forms of tubing failure.

Header plates are frequently made of slightly less corrosion-resistant materials. They must be galvanically compatible with the tubing, and it is common practice for copper-nickel or NAB headers to be used with both copper-nickel and titanium tubing, although best practice is to use only one alloy family, e.g. copper or titanium for both the tubing, headers, and water box. Crevice corrosion of headers, as shown in Figures 5.41 and 5.42, is a common problem.

Carbon steel water boxes are often used, and it is common for these relatively thick components to have organic protective coatings and galvanic anodes for cathodic protection of the steel. Figure 8.53 shows corroded galvanic anodes in a small onshore heat exchanger.

Compact Heat Exchangers Plate-frame heat exchangers, brazed aluminum heat exchangers (BAHXs), and printed circuit heat exchangers are all classified as compact heat exchangers. While they are also used onshore,

they find their most extensive uses offshore where their weight and size advantages for the same thermal loading lead to significant benefits, offsetting their usually somewhat higher capital costs.

Plate-frame heat exchangers are the most widely used compact heat exchangers. They are almost universally used for offshore seawater cooling [37]. While there are a number of alloys from which the plates can be manufactured, for this service commercially pure titanium (UNS R50400, Grade 2) is the most common, although many organizations specify the palladium-containing grades (UNS R52400 or R52250, Grade 7 or 11 with 0.15 Pd added) for additional elevated-temperature crevice corrosion protection. It should be noted that only Grade 2, commercially pure titanium, and Grade 12, UNS R53400, Ti + 0.3 Mo + 0.8 Ni, are approved by NACE MR0175/ISO 15156 for H_2S service, so the use of Grades 7 or 11 is a departure from many companies' practices of only using H_2S service-approved alloys for any service involving contact with produced fluids.

Problems with availability of titanium have led several organizations to try to qualify other CRAs for this service. Ni–Cr–Mo alloys have been recommended for service up to 50 °C (122 °F) and possibly up to 60 °C (140 °F) [120, 121]. It should be noted that these researchers used commercially pure titanium (Grades 1 and 2) as controls, and this is not a direct comparison with the performance of titanium grades 7 or 11, but the work is an indication that alternatives to titanium may become available, at least at lower seawater temperatures. Figure 8.54 is a schematic of a plate-frame heat exchanger showing how it is assembled from many (often 100 or more) deformed thin plates with channels allowing fluid passage. Figure 8.55 shows crevice corrosion at the contact point between two plates. This is the

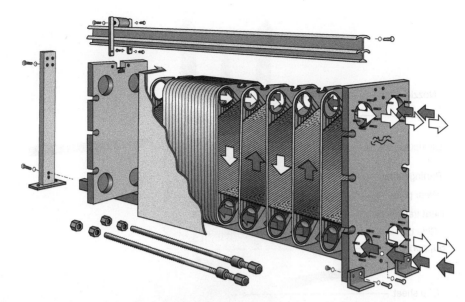

Figure 8.54 Schematic of a plate-frame heat exchanger. *Source*: Image courtesy Alpha Laval, Inc.

Figure 8.55 Crevice corrosion at contact point between two plates in a plate-frame heat exchanger [121]. *Source*: Reproduced with permission of NACE International.

most common form of corrosion on these heat exchangers and is the reason why titanium, with excellent seawater crevice corrosion resistance, has become the standard for this service.

Most failures of these heat exchangers are due to leaking gaskets, which should not be torqued to a specified load; instead they should be tightened to the manufacturer's recommended dimensions.

The principal advantages of these heat exchangers are their relatively light weight. They can also be easily disassembled for cleaning and other maintenance. The disadvantages are relatively limited temperature and pressure ranges dictated by the limitations of the gasket materials.

BAHXs (Figure 8.56) find their primary use in natural gas cooling prior to pipeline transport. The units are fabricated from structural sheets with low-melting aluminum brazing alloys on the surface and corrugated fluid flow channels. The cross section of one of these heat exchangers looks like a corrugated cardboard box. Once the heat exchanger is assembled, it is heated and the low-melting aluminum brazing alloys flow and form a continuous single-piece cooling unit.

Aluminum is not brittle at cryogenic temperatures and is also very light. BAHXs provide approximately 25 times more surface area for the same weight than conventional shell and tube heat exchangers, and this leads to significant savings in structural support requirements, especially offshore [37]. Limitations of BAHXs are their relatively low operating pressure (approximately 120 bar [1750 psi]), inability to be mechanically cleaned, and their susceptibility to mercury attack. Mercury removal processes are necessary prior to natural gas cooling, and these processes are standard for most offshore natural gas processing designs.

Printed circuit heat exchangers are assembled from plates of electrochemically milled metal that are then assembled in stacks and fusion bonded together. Passages are typically 1–2 mm (0.04–0.08 in.) deep. While they can be fabricated from a number of metals, 316 stainless steel (UNS S31600) is the most common alloy used for these exchangers. These heat exchangers are made from stronger materials than BAHXs and can be used at much higher temperatures and at pressures as high as 700 bar (10 000 psi). Unlike aluminum heat exchangers, they are resistant to mercury attack. They have significant weight advantages (approximately 4 : 1) over shell and tube heat exchangers. Figure 8.57 compares the size of an oilfield printed circuit heat exchanger with the size of three shell and tube heat exchangers having the same thermal and fluid processing capacity.

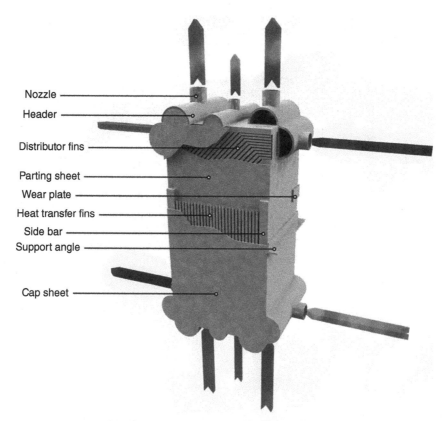

Nozzle

Header

Distributor fins

Parting sheet

Wear plate

Heat transfer fins

Side bar

Support angle

Cap sheet

Figure 8.56 Basic components of a brazed aluminum plate-fin heat exchanger. *Source:* Drawing courtesy Chart Heat Exchangers.

Figure 8.57 Comparable printed circuit (foreground) and shell and tube (background) heat exchangers having the same thermal capacity [122]. *Source:* Figure courtesy Heatric Division of Meggitt (UK) Ltd.

Figure 8.58 Cross section of a diffusion-bonded printed circuit heat exchanger [122]. *Source:* Figure courtesy Heatric Division of Meggitt (UK) Ltd.

Figure 8.58 is a cross section of a diffusion-bonded printed circuit heat exchanger showing the microstructure of the metal and the fluid passage channels.

Failures of these exchangers over the years are mainly due to thermal cycling and subsequent thermal fatigue. The flow passages have sharp edges that can lead to stress concentrations. If temperatures in the exchanger cycle or fluctuate, the differential expansion and contraction of the core material can lead to fatigue cracking [37].

Other problems with these heat exchangers include plugging, usually due to upstream debris. They cannot be mechanically cleaned, and plugging or cracking usually means that the heat exchanger must be replaced.

Other Equipment

Upstream processing of hydrocarbon fluids and associated produced water presents a number of materials challenges. Only a few of the more common problems are addressed in this discussion.

Steam Injection   Steam is injected into oil-bearing formations to heat the formation and enhance productivity by changing the viscosity of the *in situ* oil.

Most steam injection systems use relatively low-quality steam (meaning that the steam may have appreciable water droplets), but the water must be treated to reduce mineral content, or it will result in scaling and plugging of the equipment and the downhole formations. The methods for water treatment prior to steam injection are similar to those used for power boilers. Feedwater with low solids is selected, or suspended matter is removed by filtration or in settling ponds. Zeolite softeners are then used to remove calcium and magnesium ions that would otherwise form scale in the boilers. Dissolved gases are then removed, usually by mechanical deaeration or by simple heating of the water prior to introducing it into the boiler. Steam deaeration, where hot steam is used to heat the water and drive off dissolved gases, is a common method of removing dissolved gases from boiler feedwater [123].

Oxygen scavengers lower the dissolved oxygen in the steam and prevent corrosion of the system. The most common oxygen scavengers are sodium sulfite and hydrazine (N_2H_2), which is toxic, so the use of hydrazine is discouraged in most localities. pH controls are necessary, and boiler water pH is normally kept near 9. Low pHs promote corrosion, and high pHs can lead to both caustic embrittlement cracking (relatively rare in recent years) and scaling.

Process Equipment   There is a wide variety of oilfield process equipment. Likely locations where corrosion may occur, as well as recommended inspection procedures and corrosion control techniques, are summarized in Table 8.10 [26].

Process monitoring can indicate when maintenance problems are likely to occur. Figure 8.59 shows sand accumulation due to a "sand event" that produced much more sand than was expected in the free water knockout vessel on an offshore structure. Table 8.10 warns against corrosion underneath sludge and other deposits on the bottoms of many different vessels.

Bolting, Studs, and Fasteners

The terms bolt, stud, and fastener are often used interchangeably in oilfield practice. This discussion is related to all threaded fasteners. The term "high strength" is used in the threaded fastener industry to refer to bolts that have been quenched and tempered to develop proper strength. These bolts tend to be low-alloy steels with chrome or chrome-molybdenum additions for thick-section hardenability (Figure 4.47).

Bolted connections, flanges, and gaskets are a major source of leaks in oilfield applications. Table 8.11 shows how these problems can account for up to half of all gas leak incidents in a major production area.

Figures 5.10–5.14 showed bolted connections and discussed galvanic corrosion problems associated with bolts. It is important to recognize that bolts are usually the critical component in any assembly and their corrosion or fracture can lead to failure of flanged connections. Most of the critical areas on a bolted connection are hidden from view, so it is important that appropriate bolts be chosen. While it is common oilfield practice to blame failures of bolted connections on workers not following established procedures, problems with inappropriate materials and specifications also exist.

Failure modes for bolts include various forms of weight-loss corrosion, fatigue, and hydrogen embrittlement, which may be due to environmental exposure or may result from manufacturing processes. Manufacturing processes can strongly affect the resistance of bolts to failure during use. As one example, while most threaded connectors have cut (tapped) threads, rolled threads are more resistant to fatigue and other forms of cracking than machined threads. The cold working of rolled threads may require stress relief to meet the hardness requirements of ANSI/NACE MR0175/ISO 15156. The documented improvements in fatigue performance due to cold rolled threads may be offset by increased environmental cracking susceptibility [125].

Materials That Have Been Used for Oilfield Fasteners
Table 8.12 summarizes a number of the bolting materials that have been used in the past in the offshore oil and gas industry.

Many of these alloys perform their intended function in atmospheric exposure but are not considered appropriate for subsea use because of concerns with hydrogen embrittlement from cathodic protection systems.

The most common structural bolts used in oilfield applications meet the requirements of ASTM A 193 Grade B7. These bolts are made from UNS G 41400 or G414420 (AISI 4140 or 4142) alloy steel with Cr and Mo additions for thick-section hardenability (Figure 4.27). The yield strength of B7 bolts varies from 517 to 723 Mpa (75–125 ksi), depending on the size of the bolt, which can be up to 178 mm (7 in.) in diameter. Because these bolts receive their strength from the quenching and tempering process, larger bolts have lower yield strengths, although the overall load capacity does

TABLE 8.10   Internal Corrosion and Corrosion Control of Selected Topside Equipment

System or Equipment (Environment)	Usual Problem Areas	Evaluation Methods (Detection and Monitoring)	Most Common Corrosion Control Methods
Oil Well Produced Fluid Handling			
Oil and gas separators (traps)	Wet gas area (where water condenses)	Failure history	Coatings
	Free water section (bottom)	Inspection: visual, sometimes ultrasonic thickness	Cathodic protection (CP) in the free water
	Water dump valves		Metallurgy of internals and water dump valves
Free water knockout (FWKO)	Free water section, shell, baffles, piping	Failure history	Coatings
	Bottom under deposits	Inspection	Cathodic protection in the free water
	Water dump valves and piping		Periodic flushing and/or bottom cleanout
Heater treaters	Gas section where water condenses	Failure history	Gas section: coatings
	Free water and treating sections; shell and baffles, bottom under deposits	Inspection	Fire tube: cathodic protection and/or scale control chemical
	Fire tube – particularly under scale deposits		Free water and treating sections: coating and cathodic protection
	Water dump valves and waterlines		Routine flushing and cleanout of deposits
			Siphons and waterlines: coatings and plastics
			Water dump valves: metallurgy
Gun barrels (wash tanks, settling tanks)	Free water section	Failure history	Coatings
	Under side decks	Inspection	Cathodic protection in free water
	Gas boot and piping in high H_2S areas		Aluminum decks
	Water dump valves		Plastics and nonmetallics for gas boots and piping
			Water dump valves: metallurgy
			Oxygen exclusion
			Routine flushing and cleanout
Lease tanks	Under side of deck	Failure history	Coatings
	Bottom and lower portion of bottom ring	Inspection	Aluminum deck
			Cathodic protection of bottom area if it maintains a water level
			Oxygen exclusion
			Routine cleanout of bottom
Hydraulic pumping equipment			
Power fluid tanks	Power oil tanks – similar to lease tanks		Tanks – coating where appropriate
Power fluid pumps	Power water tanks – similar to SWD and WI tanks		CP where appropriate
Power fluid lines			Pumps – metallurgy
			Lines – inhibition
			All: oxygen exclusion
			Routine flushing and cleanout
			Note: Power fluids must be clean
Gas-lift systems	Along bottom of line where free water flows or collects	Failure history	Dehydration

TABLE 8.10   (Continued)

System or Equipment (Environment)	Usual Problem Areas	Evaluation Methods (Detection and Monitoring)	Most Common Corrosion Control Methods
	On vessel walls where water condenses	Inspection	Oxygen exclusion
			Inhibition on occasion
Gas Handling Equipment			
Coolers	Where water condenses	Failure history	Metallurgy
		Visual inspection, exchanger tube calipers	Neutralization
Accumulators	Where water collects	Failure history	Coating
		Inspection: visual, ultrasonic	Metallurgy
			Neutralization
			CP in free water
Vessels	Where water collects and in free water portions	Failure history	Coatings
		Inspection: visual, ultrasonic	Neutralization
			Metallurgy
			Inhibition
			CP in free water
Compressors	Valves, cylinders, and bottles	Failure history	Metallurgy
		Inspection: visual	
Glycol dehydrator	Wet glycol lines, contactors: trays and shell	Failure history	Metallurgy
	Regeneration equipment	Inspection	pH control
		Coupons	Oxygen exclusion
		Glycol analysis	Glycol quality
Dry bed dehydrator	Wet gas handling areas, regeneration condensers, etc.	Failure history	Metallurgy
		Inspection	Coatings (where temperature allows)
Water Handling and Injection (Disposal and Flood)			
Vessels and tanks	Shell when water very corrosive	Inspections	Oxygen-free operation (exclusion or removal)
	Bottom under deposits	Failure history	CP
	Vapor space when sour	Coupons, probes	Coatings
		Occasionally iron content and/or bacterial activity	Periodic cleanout
			Occasionally: biocides or inhibitors
			Nonmetals (only for fluids where oxygen exclusion is not important)
Filters	Connections when bimetallic	Same as vessels	Same as vessels
	At filter media (sand) level		
Gathering and injection lines and plant piping	Along bottom under deposits	Same as vessels	Oxygen-free operation
			Coatings and linings (including cement)
			Periodic line pigging
			Occasionally: biocides or inhibitors

(*Continued*)

TABLE 8.10 (Continued)

System or Equipment (Environment)	Usual Problem Areas	Evaluation Methods (Detection and Monitoring)	Most Common Corrosion Control Methods
Injection and transfer pumps	All wetted parts Seals leaking air	Failure history Inspection Galvanic probe (oxygen detector)	Oxygen-free operation Metallurgy
Injection wells	Tubing interior, inhole valves, fittings, etc.	Failure history	Oxygen-free operation
	Wellhead and Christmas tree	Coupons, probes	Coatings and linings (including cement)
	Annulus above packer (depends on packer fluid)	Occasionally iron content and/or bacterial activity Galvanic probes (oxygen detection)	Occasionally: biocides or inhibitors
Miscellaneous Facilities			
Glycol–water heat transfer systems (heating and cooling)	Anywhere in system (particularly where deposits can occur)	Failure history Inspection Coupons Fluid analysis	Inhibition Oxygen exclusion Fluid quality
Boilers and steam systems	Boiler tubes (including creep and swelling) Condensers and condensate return lines	Failure history Inspections Coupons Iron content	Water quality control Deaeration Inhibition
Gas sweetening (MEA and similar amine systems)	Contact tower Reconcentration system	Failure history Inspections Coupons Iron content	CO_2 loading Inhibition (usually inorganic) Oxygen exclusion Fluid quality

Source: Byars [26]. Reproduced with permission of NACE International.

Figure 8.59 Sand accumulation in a free water knockout tank.

TABLE 8.11 Norwegian Continental Shelf Gas Leak Incidents [124]

Year	Total Number of Leakage Incidents	Leaks Caused by Mechanical Connections/Gaskets	
		Number of Incidents	Percentage
1994	194	98	51
1995	117	56	47
1996	171	82	48
1997	177	84	47
1998	248	69	28

Source: Reproduced with permission of NACE International.

TABLE 8.12 Bolting Materials Historically Used in the Offshore Industry [124]

Trade Name/Designation	Nominal Composition	Specification/Standard
AISI 4140	Fe-1Cr-0.2Mo	ASTM A193 Grade B7
Alloy A286 Grade	Fe-26Ni-15Cr-2Ti-2Mn-1M	UNS S66286
17-4PH	Fe-16Cr-4Ni-4Cu	UNS S17400/ASTM A693
254 SMO	Fe-20Cr-18Ni-6Mo	UNS S31254
K-500	Ni-30Cu-3Al	UNS N05500
Alloy 725	Ni-21Cr-8Mo-8Fe-3.5Nb	UNS N07725/ASTM B805
Alloy 625, 625 PH	Ni-21Cr-8Mo-5Fe-3.5Nb	UNS N07725/ASTM B805
Alloy 718	Ni-19Cr-17Fe-5Cb-3Mo	ASTM B637
Ti-6A1-4V ELI	Ti-6Al-4	UNS R56400
MP35N	35Ni-35Co-20Cr-10Mo	UNS R30035

Source: Reproduced with permission of NACE International.

increase with diameter (Figure 4.47). The B8 version of these bolts has higher strength levels due to the fact that these bolts are heat-treated after threading instead of before, but B8 bolts are not common in oilfield practice because their higher yield strength also means higher hardness (HRC 33–39), which causes concerns with hydrogen embrittlement from both H_2S and from cathodic protection.

CRA alloys are also used in the oilfield, primarily in applications involving flanges associated with equipment and piping made from similar alloys. The bolts are "matched" to the larger equipment in an attempt to minimize galvanic corrosion effects. An alternative to this approach is the use of dielectric fittings (Figures 5.13 and 5.14), but pressure capabilities of flanges may be lowered due to creep of the washers in these systems. For subsea applications, these dielectric fittings will also isolate bolts and nuts from cathodic protection, and this can lead to stray current corrosion. It is important that any subsea bolted connection be tied into the cathodic protection system [105].

Stainless and related CRAs that are reported to work for subsea completions include [105]:

- UNS S66286, a precipitation-hardened stainless used for low-pressure units.
- UNS R30035, a nonmagnetic nickel–cobalt–chromium–molybdenum alloy (MP35N®) used at higher pressures.
- UNS N09925, precipitation-hardening nickel–iron–chromium alloy with molybdenum and copper additions.
- UNS N07718, precipitation-hardening nickel–chromium–iron alloy.
- UNS N07725, precipitation-hardening nickel–chromium–iron alloy with higher alloying content than UNS N07718. This alloy has better pitting resistance than the others on this list and is probably most suitable for subsea applications [105].

Several reports indicate that the two precipitation-hardened nickel alloys, UNS N07718 and N07725, are probably the best CRAs for subsea service, although failures of N07718 on a cathodically protected BOP have been reported [105, 126, 127]. The N07718 stud that hydrogen-embrittled in cathodically protected service may have been improperly heat-treated, which resulted in metal having below-specification Charpy impact values. It also had higher than the maximum HRC values suggested by API 6ACRA [25, 126, 128].

Copper-based systems often use aluminum bronze (UNS C63000, ASTM B150) bolts, and most titanium bolts are either UNS R50400, commercially pure titanium, or UNS R56400, titanium with 6% aluminum plus 4% vanadium additions.

While CRA alloy bolts are commercially available, they are seldom used, because low-alloy heat-treatable steel bolts are much stronger in most cases. Electrical isolation practices like those shown in Figures 5.13 and 5.14 are available, although some authorities are of the opinion that they are likely to be overcome through inadvertent grounding in many, if not most, instances.

Recently issued API bolting standards limit CRA bolting to UNS N718, a heat-treatable nickel–chromium alloy, and ASTM A 453, an austenitic stainless steel [129, 130]. It remains to be seen if the limited listing of CRAs in API 20F will be adopted by the worldwide oil and gas industry.

Embrittlement Concerns The high strength nature of most industrial bolts means that they are often made from materials subject to hydrogen embrittlement or environmental cracking. Bolt embrittlement can occur during the manufacturing process, during transportation and storage, or in use. During manufacturing the alloy steels are quenched and tempered to produce the appropriate strength levels. If this is done improperly, it can lead to brittle bolts. Segregation in the alloy can lead to banding (Figure 4.43), the formation of areas in the steel with different chemistries and microstructure, and this is also an embrittlement concern [125].

A more common concern is the effect of corrosion-resistant coatings, usually zinc or sometimes cadmium, which are applied for atmospheric corrosion control. The pickling process (acid cleaning) prior to coating can result in hydrogen entry into the steel. Electroplating processes also introduce hydrogen. This hydrogen entry is accelerated in most electroplating baths because they usually contain cyanides, which help produce quality electroplates but also act as hydrogen-entry poisons in much the same manner as environmental H_2S. The standard means of controlling hydrogen embrittlement in electroplated metal is by using a dissolved-hydrogen bakeout procedure at temperatures from 191 to 218°C (375 to 425°F) for a period of time depending on the size of the part in question [131–139]. Unfortunately, there is no guarantee that all of the hydrogen will be removed from the metal.

Environmental exposure can lead to hydrogen embrittlement of galvanized or electroplated high-strength fasteners at coating holidays, which are inevitable. Hydrogen from atmospheric condensation (typically around pH 5) is enough of a concern that high-strength fasteners are not galvanized because of a concern with hydrogen embrittlement [137, 138]. It should be noted that recent changes to some ASTM fastener standards have removed this restriction, although it remains in force in the recently adopted 2017 version of API 20E, which is intended to provide guidance for all carbon and low-alloy steel fasteners used in the oilfield [130].

Problems with Bolted Connections

In recent years a number of subsea bolted connection failures have occurred that led the US government regulatory organizations to become concerned with safety and environmental hazards. A series of reports have been issued on failures of bolted connections like on the BOP shown in Figure 8.60 [140–145].

The findings of a government panel that reviewed the results from a series of investigations by independent forensic laboratories on failures from a number of different instances of bolt failure were as follows [141]:

- Failures were due to hydrogen-induced SCC.
- Zinc electroplated bolts did not receive post-plating hydrogen bakeout heat treatment.
- No control of maximum hardness of bolts.
- API 16A and other specifications have varying levels of specified maximum hardness.

Figure 8.60 Upper half of a blowout preventer flange assembly where bolt failures led to system failure, with all 36 bolts failing under loading during service. These bolts, 51 mm (2 in.) in diameter, were made of UNS G43400 low-alloy steel, 34-38 HRC, and zinc electroplated [140–145].

The government panel was unable to make conclusive findings about [141]:

- Whether hydrogen charging due to cathodic protection contributed to the failures.
- Use of dissimilar metals.
- Plating requirements for this service class that may not be appropriate for the marine environment.

The key recommendations from the panel include the following [141]:

- Industry needs to develop consistent standards for connections and connection fasteners for offshore subsea:
  ○ Hardness
  ○ Yield strength
  ○ Ultimate tensile strength
- ASTM needs to develop better standards on coatings for marine service.
- Improved industry guidance is needed on cathodic protection voltage limits for use on drillships and similar equipment.

It should be noted that NACE International, the world's largest corrosion society, was not mentioned in the panel's findings and recommendations. This is because ASTM has been the historic leader in developing materials-related standards for many different industries, and NACE materials-specific standards are much fewer, although ANSI/NACE MR0175/ISO 15156 is the most widely accepted international standard on

TABLE 8.13 International Bolting Standards with Hydrogen Embrittlement Warnings

	Hydrogen Embrittlement Warning	HRC Requirement
ASTM		
A143	Safeguarding against embrittlement of hot dip galvanized structural steel products and procedures for detecting embrittlement	>~33
	No warning about EHE cracking	
A325	Structural bolts, steel, heat treated, 120/105 ksi minimum UTS	≤34
	No warning about EHE or IHE cracking	
A354	Quench and tempered alloy steel bolts, studs, and other externally threaded fasteners	≤39 for BD
	Note 4 – Research conducted on bolts of similar material and manufacturing indicates hydrogen stress cracking or stress cracking corrosion may occur on hot-dip galvanized Grade BD bolts	33–36 for BC
B633	Electrodeposited coatings of zinc on iron and steel	
	6.4 All steel parts made having an ultimate tensile strength greater than 1000 Mpa (31 HRC) and…shall be heat treated…to reduce the risk of hydrogen embrittlement	≥31
	6.5 Post-coating treatment of iron and steel for the purpose of reducing the risk of hydrogen embrittlement	≥31
B850	Post-coating treatment of steel for reducing the risk of hydrogen embrittlement	≥31
	4.2 Parts made from steel with actual tensile strength 100 MPa (with corresponding hardness value of 300 HV, 10 kgf, 303 HB, or 31 HRC) and surface-hardened parts may require heat treatment	≥31
F1941	Electrodeposited coatings on mechanical fasteners, inch and metric	≥39
	With normal methods of depositing metallic coatings from aqueous solutions, there is a risk of delayed failure due to hydrogen embrittlement for case hardened fasteners having hardness of 39 HRC or above	
F2329	Zinc coating, hot dip, requirements for application to carbon and low alloy steel bolts, screws, washers, nuts, and special threaded fasteners	≥33
	7.2.2 Effect of hydrogen on the mechanical properties after galvanizing – for high-strength fasteners (having a specified minimum product hardness of 33 HRC), there is a risk of internal hydrogen embrittlement…	
NACE		
MR0175	Standard for avoiding H_2S-related cracking in oilfield environments	
	ASTM A 193 Grade B7M or ASTM A 320 Grade L7M only accepted bolt standards	≤22
API		
17A	Design and operation of subsea production systems – general requirements and recommendations	
	…For high strength martensitic carbon, low alloy, and stainless steels, failure by CP-induced HISC (hydrogen-induced stress cracking) has been encountered	≤35
	6.4 Bolting materials for subsea applications for piping systems and equipment include ASTM A193 Grade B7 for structural applications (hardness to be verified by spot test)	≤32

the effects of H_2S on brittle metal behavior and cracking phenomena [18, 35, 36].

The above-listed recommendations are controversial, and ASTM representatives have suggested that evidence of internal hydrogen embrittlement (IHE) associated with electroplating was not conclusive for these bolt failures [143]. Others have discussed whether or not fatigue or corrosion fatigue could have been involved [142, 144]. A recent publication discussing hydrogen embrittlement of bolts for this and other applications points out that high-strength steel bolting, resistant to environmental hydrogen embrittlement (EHE) cracking in atmospheric service, can fail due to EHE in immersion service [142]. It is also not clear if cathodic

protection can be effective on controlling corrosion or if it will cause hydrogen embrittlement on complicated-geometry structures made of high-strength steel like those involved the BSEE study.

Because of the uncertainties associated with the above findings, the conservative approach is to address all possibilities.

International Bolting Standards

Table 8.13 lists a series of standards with hydrogen embrittlement warnings and a wide variety of hardness limitations. Some of these standards also suggest problems associated with electroplated zinc coatings and the

TABLE 8.14 Bolting Materials Accepted by API Bolting-specific Standards [129, 130]

API 20E Carbon and Low-alloy Bolting Material Grades	API 20F Corrosion-resistant Bolting Material Grades
ASTM A193 Grades B7 and B7M	API 6A718[a]
ASTM A194 Grades 2H, 4, 7, 2HM, and 7M	ASTM A453 Grade 660 Class D[b]
ASTM A320 Grades L7, L7M, and L43	
ASTM A320 Grade L43	
ASTM A540 Grades B22 and B23	

[a] API 6A718 was in force when API 20F was approved. It was replaced in 2015 by API 6A-CRA, which covers other alloys besides UNS N07718, which is a precipitation-hardening nickel-based alloy.
[b] ASTM A453, High-Temperature Bolting, with Expansion Coefficients Comparable to Austenitic Stainless Steels.
Source: Reproduced with permission of NACE International.

associated hydrogen bakeout procedures. New API bolting standards, API 20E, for carbon and low-alloy steels, and API 20F, for CRAs, now provide uniform guidance on bolting of oilfield use. Table 8.14 shows the alloys covered by these two API standards [129, 130]. It is intended that new/revised versions of API standards – e.g. API 17D, "Wellhead and Christmas Tree Equipment," or API 53, "Blowout Prevention Equipment" – will refer to and follow the guidance of the new API bolting standards [22, 104].

API 20E limits acceptable hardness to a maximum of HRC 34 if not otherwise stated at a lower level in the specific international bolting standards for the application in question. It also precludes the use of zinc electroplating for splash zone or subsea service. This combination of listing acceptable hardness standards and avoiding electroplating, with the associated questions of quality control on hydrogen bakeout procedures, is an appropriately conservative approach to the question of what bolts can be used in subsea service, considering the questions that remain:

- Was internal hydrogen embrittlement (IHE) involved in the bolt fractures on subsea equipment? Fastener experts, including representatives from ASTM disagree on this question.
- Did environmental hydrogen embrittlement (EHE-SCC) contribute to the bolt failures?
- What hardness levels can be accepted for quench and tempered low-alloy steel (e.g. UNS G41400 or UNS G43400) bolts?
- How can improper bakeout procedures on electroplated bolts be avoided? Perhaps by not accepting plating for bolts intended for subsea or splash zone service.

API standards API 20E and 20F (Table 8.14) were developed to provide uniform guidance on acceptable bolting for use in oil and gas operations and to address in a conservative manner the concerns and limitations listed above. API 20E for alloy and carbon steel bolting allows the use of both ASTM A 193 Grades B7 and B7M, which is a lower-hardness version of ASTM A 193 Grade B7, the most commonly used structural bolting in the upstream oil and gas industry. This B7M bolting, with a hardness limit of HRC 22, is intended for use in H_2S environments covered by ANSI/NACE MR0175/ISO 15156. The harder B7 bolting is limited to a maximum of HRC 34. This is an attempt to allow stronger bolting for use in applications not covered by ANSI/NACE MR0175/ISO 15156 because they are not exposed to H_2S. This HRC 34 limit is also the same HRC limit suggested by the 2017 version of API 20E. Other international standards with other hardness maxima are likely to be revised in accordance with the 2017 HRC limit of API 20E, which is deliberately conservative.

The very limited listing of CRAs suggested by API 20F and shown in Table 8.14 is likely to change when the standard is revised. The new API 6ACRA expands and replaces the scope of API 6A718, which it replaces. The now-replaced API 6A718 was in force when API 20F was originally developed [146]. It is likely that at least some of the additional alloys listed in the replacement standard, API 6ACRA, will be added to API 6F when it is reviewed and revised, sometime after 2017 when this manuscript was written [128]. API 6CRA, issued in 2015, lists only a small number of nickel-based age-hardened alloys, but it is likely that other alloys, including UNS R30035 (MP35N), which contains more cobalt than nickel, might be added to both API 6CRA and to API 20F in the future. UNS R30035 is already used for some subsea threaded applications and is listed in ANSI/NACE MR0175/UNS 15156 Part 3 for nonbolting applications.

For many years the guidance of ANSI/NACE MR 0175/ISO 15156 was accepted as an overall limitation on the alloys that can be used in oil and gas production [17, 34, 35]. This standard is intended to prevent cracking due to the presence of H_2S in oil and gas production environments, which could mean that it is intended for internal fluid-handling environments and not atmospheric, buried, or immersion environments. The only place that threaded connectors are mentioned in this voluminous standard is in Part 2, "Cracking-resistant carbon and low-alloy steels, and the use of cast irons." Threaded connectors are not mentioned in Part 3, which deals with cracking-resistant CRAs. The only bolting materials listed in this standard are shown in Table 8.15, which is reproduced from Part 2 of the standard.

The same standard also limits carbon steels to a maximum hardness of HRC 22, although it does suggest that

TABLE 8.15   Carbon and Low-alloy Steel Acceptable Bolting Materials IAW ANSI/NACE MR0175/ISO 15156-2 [34]

Bolts	Nuts
ASTM A193 Grade B7M	ASTM A194 Grades 2HM and 7M
ASTM A320 Grade L7M	

Source: Reproduced with permission of NACE International.

higher hardnesses, which are never stated, are acceptable for chrome-molybdenum alloy steels, and most ASTM A 193 Grade B7M bolts are made from AISI 4140 or 4142 chrome-molybdenum steels.

This standard applies to environmental cracking in the presence of H_2S. Bolting that will be directly exposed to sour (H_2S) environments or that will be buried, insulated, equipped with flange protectors, or otherwise denied atmospheric exposure must meet the requirements of this standard, which restricts bolts to the materials shown in Table 8.15. The typical maximum hardness of B7 bolts is approximately HRC 32, which seems to be allowed by Part 2, which states that the normal limitation of HRC 22 for H_2S service can be higher for chromium-molybdenum-containing steels (Paragraph A.2.1.1). Other international standards set hardness maxima at HRC 35.

The two bolt materials listed are in Table 8.15 are the same, but the ASTM Grade L7M is for use at low temperatures and has fracture toughness testing requirements missing from the more commonly used ASTM A 193 Grade B7M standard [139]. The M in the grade designation indicates that the hardness levels for these bolts are held to a maximum of HRC 22, in accordance with the limitations of the NACE/ISO standard. This reduced hardness means that flanges may be derated to lower pressures than would be allowed if standard bolting were used.

Subsea Embrittlement by Cathodic Protection   At one time it was thought that limiting subsea bolting to materials that met the requirements of NACE MR0175 would prevent hydrogen embrittlement of fasteners on cathodically protected subsea assemblies. Unfortunately alloys in the following groups have been found to have hydrogen embrittlement problems when used as fasteners on cathodically protected equipment [105, 147–151]. This listing of CRAs is in addition to the problems with low-alloy steels that caused the introduction of API 20E:

• Martensitic stainless steels
• Ferritic stainless steels
• Duplex stainless steels
• Nickel-based alloys

It is possible that the reported problems were associated with using metals that were too hard, and studies indicate that hydrogen embrittlement should not be a problem for fasteners if the hardness level is kept at HRC 34 or lower [104, 105, 124, 127]. These hardness levels are much higher than the HRC 22 restrictions for H_2S service. The discrepancies between reports that hardness levels in excess of HRC 22 can be used conflict with the requirements of MR0175/ISO 15156 Part 2, which limit hardness levels for any bolts not subjected to atmospheric service to HRC 22 [152].

Bolting Alloys Used for Atmospheric Service   Table 8.12 showed bolting materials that have been used by the offshore industry. While all of these alloys find uses in atmospheric service, several of them have been reported to have hydrogen embrittlement problems when used in subsea applications with cathodic protection. Definitive research on the limitations of most of these alloys is not available, and the introduction of API 20E and 20F is an attempt to address these problems.

Coatings for Bolts   The great majority of oilfield bolting requirements are met by B7 bolts with protective coatings applied for corrosion control. Generic coatings for fasteners are listed in Table 8.16. Most of these systems do not produce adequate performance for use in offshore and other oilfield applications. The most common coating used for corrosion control is zinc. Most organic coatings are intended for lubricity and quick-disconnect properties, because many metallic-coated fasteners cannot be unscrewed and must be removed by cutting after several years of atmospheric or immersion service.

Zinc is the most common protective coating for threaded hardware [154]. It can be applied by electroplating, the sherardizing (mechanical plating) process, or, most commonly, by dipping the steel parts to be coated in molten zinc. While the sherardizing process, which involves vapor deposition of zinc onto steel substrates, is popular in Europe, it is relatively uncommon in other locations. Most precision zinc coatings are thin electroplates, and hot-dipped galvanizing, which produces thicker zinc coatings, typically 1 mil (25 µm) or greater, is used more often for corrosion control in aggressive oilfield environments. The atmospheric corrosion protection provided by zinc coatings is roughly proportional to the thickness of the zinc. Thin electroplated coatings may provide protection for one to two years, while galvanizing may protect for 10 years or more [154].

Embrittlement concerns for zinc coatings are addressed by limiting the strength (hardness) levels of zinc-coated metals. Various standards suggest different maximum hardness levels, but HRC 33–35 hardness levels

TABLE 8.16 Generic Coating Systems for Threaded Fasteners [153]

Genetic Coating System	DFT, mil (μm)	Maximum Temperature °F (°C)	ASTM B 117[a] Exposure Hours	Relative[b] Cost Comparison
Uncoated (bare) steel stud bolt and two nuts[b]	NA	NA	NA	1.0
Aluminum	0.2–0.3 (5–8)	1000 (540)	1200	1.8
Cadmium plating	0.2–0.3 (5–8)	500 (260)	250	1.6
Ceramic metallic	0.8–1.0 (20–35)	1200 (650)	1000	5.0
Hot-dip galvanizing	1.0–2.5 (25–63)	750 (403)	1500	2.0
Inorganic zinc-rich silicate	1.8–2.4 (45–60)	750 (400)	1500	2.0
Manganese phosphate, fluoropolymer	1.0–1.2 (25–30)	500 (260)	750	3.7
Mechanical zinc	0.2–2.5 (5–63)	750 (400)	400	2.0
Zinc aluminum	0.2–0.50 (5–13)	800 (430)	1000	2.0
Zinc phosphate	0.2–0.50 (5–13)	250 (130)	48	1.5
Zinc phosphate and oil	0.3–0.50 (8–13)	250 (130)	250	1.6
Zinc phosphate, fluoropolymer	1.0–1.2 (25–30)	500 (260)	500	3.7
Zinc plating	0.2–0.3 (5–8)	750 (400)	250	1.6
Zinc plating fluoropolymer	1.0–1.2 (25–30)	500 (260)	1000	3.7

[a] ASTM B117 end point is the first sign of red rust.

[b] ASTM A193 bare steel stud bolt and two nuts, size 0.625×4 in. (16×100 mm), are used for cost comparison.

Source: NACE Publication 02107. Reproduced with permission from NACE International.

associated with approximately 150 ksi (100 MPa) yield strengths are generally recommended as the dividing line between fasteners subject to EHE at the inevitable defects in zinc coatings. The new API 20E specification for carbon and alloy steel suggests HRC 34 as the conservative upper limit on hardness. Common bolting standards and whether they can be galvanized (or zinc electroplated) are shown in Table 8.17. Most of the standards in this table are not applicable to oilfield applications, but ASTM A193 Grade B7 is the most common bolting material for oilfield piping. These bolts can be galvanized, and they usually are. Structural bolts are often ASTM A320, which can be galvanized, and ASTM A 490, for which the ASTM standard specifically forbids galvanizing.

One of the problems associated with hot-dipped galvanizing is the thickness of the zinc coating, which can make applying nuts difficult. It is common to overtap the nuts used on galvanized bolts to accommodate oversized threads that result from galvanizing. NACE and ASTM provide guidance on this [154–156].

Cadmium electroplating is sometimes used, but many governments have banned this material due to toxicity questions. It is still acceptable in some locations, and many authorities consider this coating to be superior to electroplated zinc.

It is common to chromate both zinc and cadmium-coated parts to improve the protective qualities of the coating and increase their atmospheric corrosion resistance.

Many oilfield fasteners are covered with fluoropolymer coatings. While they are sometimes marketed for corrosion protection, they are too soft for this purpose and will develop significant holidays during installation. Their proper use is for antigalling purposes so that the fasteners can be removed with wrenches and perhaps even reused. A variety of fluoropolymers are available for this purpose, and they are marketed using trade names such as Teflon®, Viton®, Xylan®, Hyflon®, Kynar®, etc. While the labels imply that they consist of the polymer implied by the trade name, these coatings, which are usually applied as liquid coatings, all include binder resins that determine most of their corrosion resistance. This means that commercial products having the same fluoropolymer additive may have markedly different corrosion and antigalling properties [154, 156]. Test procedures for evaluating these coating systems are available. Fluoropolymer coating suppliers generally recommend that the metal surface be roughened using phosphate tie coats. The acid phosphating baths can dissolve zinc coatings, and this is another reason why thicker hot-dipped coatings are recommended over much thinner electroplated coatings, which have been known to be entirely removed by phosphating processes.

Makeup torque is lowered by the use of fluoropolymer coatings and specifications for bolting must be adjusted accordingly [157].

Additional Comments on Fasteners The above discussion has concentrated on bolting materials and coatings. The most corrosive locations in any bolted connection are locations where the bolt shank is shielded from the overall environment in bolt holes. The use of gel lubricants to fill these holes has been successfully

TABLE 8.17 Recommendations on Galvanizing for Different Bolt Standards [154]

Grade	Can I Galvanize?	Raw Material	Nominal Size (in.)	Minimum Yield Strength	Minimum Tensile Strength	Minimum Hardness
ASTM F1664 Grade 56	Yes	Low-alloy steel	½–4	55	75	—
ASTM A325	Yes	Medium-carbon steel, quenched and tempered	½–1	92	120	C24
			1⅛–1½	81	105	C19
ASTM A449	Yes		¼–1	92	120	C25
			1⅛–1½	81	105	C19
			1⅝–3	58	90	B91
SAE J429 Grade 5	Yes		¼–1	92	120	C25
			1⅛–1½	81	105	C19
ASTM A193 Grade B7	Yes	Medium carbon alloy steel, quenched and tempered	¼–2½	105	125	NA
			2⅝–4	95	115	
ASTM A354 Grade BC	Yes		¼–2½	109	125	C26
			2⅝–4	94	115	C22
ASTM F1554 Grade 105	Yes		½–3	105	125	NA
ASTM A320 Grade L7	Yes		¼–2½	105	125	NA
ASTM A490	No		½–1½	130	150	C33
ASTM A354 Grade BD	No		¼–2½	130	150	C33
			2⅝–4	115	140	C31
SAE J429 Grade 8	No		¼–1½	130	150	C3

Source: Reproduced with permission of Portland Bolt & Mfg. Co.

reported on offshore structures. Filling the cavity with hydrophobic greases greatly prolongs assembly lives [157]. Commercial devices are also available to encapsulate bolt and nut heads and ensure that they are covered with protective greases (Figure 8.61) [158–161]. These protective caps can be filled with water-repelling waxes to keep the metal dry. They can be removed for maintenance. Protector caps that do not engage the bolt/stud threads can be pulled off, and may fall off. The polymer chosen should be UV resistant so they will not degrade in atmospheric service [158].

Most oilfield piping and similar equipment uses ASTM A193M bolting, because this material is approved by MR0175/ISO 15156-2. Galvanized bolting with fluoropolymer antigalling coatings are recommended and used by most major operators.

Continued interest in using the best possible CRA bolting materials for subsea materials seems to indicate that precipitation-hardened nickel alloys may find more extensive use in the future. These alloys must be resistant to hydrogen embrittlement caused by cathodic protection systems. At present the use of these alloys has been restricted by the lack of inclusion of any CRA bolting materials in current versions of NACE MR0176/ISO 15156, which restricts the use of nonlisted alloys for bolts not exposed to atmospheric environments. This means that most subsea assemblies, to include pipelines, must use low-alloy bolts.

The recently published API bolting standards, API 20E and API 20F, should lead to more reliable fastener

Figure 8.61 High-density polyethylene fastener protection cap [158]. *Source*: Reproduced with permission of Deepwater Corrosion Services Inc.

and flanged-connection designs in the future. The reader is cautioned that both of these relatively new standards are likely to be significantly revised in the coming years. It is likely that additional nickel-based alloys and the cobalt-based alloy UNS R30035 (MP35N) will be included in future editions of API 20F.

Flares

Flares are relatively small components of most processing plants, so materials costs are less important, and reliability is emphasized. Table 8.18 shows the materials recommended for flares in ISO 21457 [48]. The API 537 recommendations are similar [162, 163]. Many organizations recommend the hottest parts of flares be made from UNS N08800 iron–nickel–chromium alloy, which is recommended for temperatures up to 600 °C (1100 °F) [164].

Intermediate-temperature alloys having corrosion resistance in the 120–230 °C (250–450 °F) temperature range include:

- UNS N10276, Hastelloy C-276
- UNS N06200, Hastelloy C-2000
- UNS N06686, Inconel 686
- UNS N06059, VDM Alloy 59
- Acid-resistant bricks

Concerns with atmospheric pollution and energy conservation measures mean that many flares are now being used for intermittent service and resistance to corrosion at temperatures below the dew point has become important. This means that alloys must have corrosion

TABLE 8.18 Flare Materials

Equipment	Materials
Relief system, piping, and vessels	Carbon steel, stainless steels – UNS S31603, UNS S31254, UNS N08904
Flare-tip assembly	UNS S31000, UNS S30815, UNS N08810/N08811, UNS N06625

Source: Adapted from Table 5 in ISO Standard 21457.

resistance below the dew point in the presence of CO_2 and H_2S. One alloy that may be chosen is:

UNS N08810, nickel–iron–chromium alloy.

Figure 8.62 shows a typical flare on an offshore production platform. The structural support elements of the flare boom are usually carbon steel with whatever coating system is used elsewhere on the platform. The wind screen or fence shown in Figure 8.62b is intended to prevent high winds from blowing out the flame.

Corrosion Under Insulation

Corrosion underneath insulation (CUI) (Figure 8.63) is a continuing problem in many oilfield processing and steam injection environments. It is common to use stainless steel piping for internal corrosion control, but external corrosion due to moisture leaking through metallic jacketing and insulation problems will affect both stainless steel and carbon steel piping [165–168]. CUI of carbon steel is normally pitting corrosion (Figure 8.64), whereas CUI of austenitic stainless steel is more often chloride-related SCC associated with water-soluble chlorides leached from insulation and concentrated at hot metal surfaces by evaporation (Figure 5.79). NACE SP0198 provides guidance on appropriate standards related to insulation materials that will minimize leaching of chlorides and other problem chemicals [165]. Most CUI problems are associated with aboveground piping (Figure 8.65), but some insulated pipes are buried. Piping systems have more CUI problems than tanks and vessels, and this is due in large part to the complexity of piping systems.

(a)   (b)

Figure 8.62 North Sea flare. (a) Overall view showing support structure. (b) Close-up image of the wind screen/fence.

Figure 8.63 Corrosion underneath insulation on a crude oil piping system.

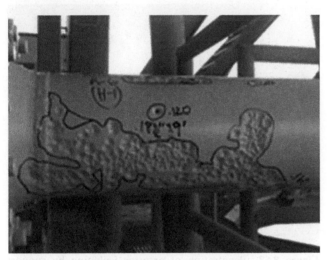

Figure 8.64 Localized pitting corrosion patterns underneath insulation on an offshore piping system.

Figure 8.65 Aboveground insulated piping.

The two most used industry standards that address CUI are NACE SP0198 and API RP583. While NACE SP0198 has been available for many years, the more comprehensive API RP583, first issued in 2014, covers the subject in greater detail and has many suggestions that do not appear in the NACE document [165, 166]. The API standard is based largely on input from the European Federation of Corrosion CUI activities, and EFC Publication 55 has excellent guidance on how to deal with CUI along with descriptions of how industry has evolved in treating this problem [167]. For many years insulated carbon steel piping was not coated, and in recent years the practice has become to use immersion-grade protective coating systems appropriate for the aggressive temperatures associated with the piping service [167].

Most oilfield insulated piping is used outdoors and requires moisture shields (jacketing) to prevent rain, snow, condensation, or water from other sources from wetting the insulation. Moisture lowers the efficiency of insulation, which relies of air spaces to produce low thermal conductivity. It also causes corrosion if liquid water reaches metal surfaces. Moisture shields are placed around the insulation (Figures 8.63 and 8.65), and openings in this jacketing allow water to infiltrate the insulation and cause corrosion at the pipe surfaces.

The metal jacketing shown in Figure 8.63 is galvanized steel, and this is one of several jacketing materials recommended in the latest versions of NACE SP0198 [165]. API RP583 is more restrictive and limits jacketing to 316L (UNS S31603) stainless steel or to aluminum alloy AA3103 (UNS A93103) "or equal" [166]. The NACE document covers all industries where insulated piping is used, to include indoor applications, whereas the API document is intended for use in oil and gas production and processing, where most piping is exposed to outdoor atmospheres.

The spaces between the interior jacketing and the underlying metal surfaces are never full. Allowances are necessary for differences in thermal expansion and contraction, and voids and pockets are inevitable. Figure 8.66 shows voids in flexible foam insulation.

Leaks into insulation systems are inevitable at locations suggested in Figures 3.16–3.18 and Figures 8.67–8.69. The arrows in Figure 8.69 indicate locations where moisture is likely to infiltrate the thermal jacketing system and cause corrosion. Thermal expansion and contraction can cause failures of caulked joints [166]. This is a major problem in geographic locations where the jacketing surfaces are heated by sunlight during the daytime and cool at night [167].

Figure 8.70 shows how the dew point, and the subsequent likelihood of condensation-related CUI, varies in different parts of the world. Cooler humid climates with

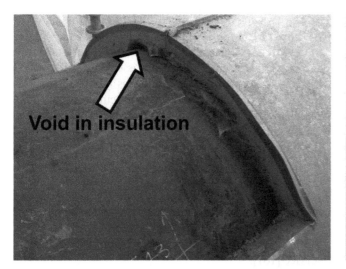

Figure 8.66 Voids in insulation.

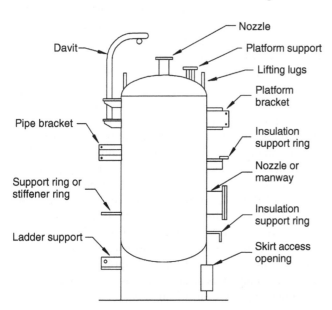

Figure 8.67 Typical vessel attachments where water may bypass insulation [165]. *Source*: NACE SP0198. Reproduced with permission of NACE International.

Figure 8.68 Penetration of exterior jacketing by support structure. Note how the original tight seal between the support and the jacketing has degraded over the years of service due to differences in the relative motion of the piping and the support.

Figure 8.69 Locations where seams in insulation jacketing can lead to moisture ingress and corrosion.

more time below the dew point are more likely to have increased problems with CUI [169]. API RP 571 gives the general temperature range of CUI as [169]:

- 10 °F (−12 °C) and 350 °F (175 °C) for carbon steel.
- 140 °F (60 °C) and 400 °F (205 °C) for austenitic stainless steels.

Chilled waterlines are a common problem. Figure 8.71 shows condensation leading to corrosion on a chilled water system where the insulation was not replaced after a repair. The localized nature of CUI was shown in Figure 8.64 where the insulation was removed and the

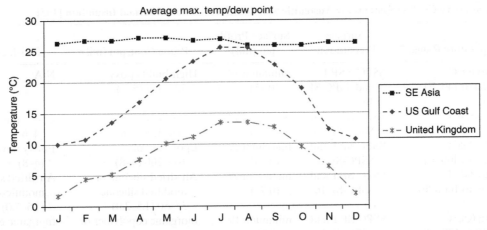

Figure 8.70 Influence of annual variation in temperatures and dew point in different regions on degree of wetness.

Figure 8.71 Condensation on the surface of an insulated chilled waterline where the insulation has not been replaced after repairs.

steel was sandblasted and repainted prior to placing back in service.

Insulated equipment should be coated with immersion-grade protective coating systems rated for the service temperature. Tables 8.19 and 8.20 list NACE-recommended coating systems for both austenitic stainless steel and carbon steel equipment. Similar recommendations are listed in the API CUI recommended practice. Coating systems used for CUI service should be rated for immersion service at operating temperatures.

Many authorities and standards recommend thermal-sprayed aluminum (TSA) for CUI control instead of conventional organic coating systems or the use of inorganic zinc coatings [166].

Aluminum foil is used to wrap around stainless steel to minimize the possibility of SCC [170–172]. API RP583 provides specific guidance, to include illustrations, on how the wrapping should be done [166].

Drainage is very important in controlling CUI, and best practices should include locating jacketing seams near the bottom of the pipe instead of, as is more common, near the top – which is easier for construction workers to install. The longitudinal seams shown in Figures 8.63 and 8.66 are associated with CUI problems might have been minimized if construction workers had been required to install the seams away from the top of the piping.

Because moisture penetration into insulated piping systems seems inevitable, drainage plugs are also available to remove moisture from the bottom of horizontal lines. These plugs are installed at the six o'clock position and are commercially available (Figure 8.72) [17]. Another option is to simply drill holes at the bottom of horizontal jacketing approximately every 3 m (10 ft).

Inspection and detection of CUI is difficult. NACE SP0198 lists several suggestions and approaches, but the emphasis of most authorities is on visual inspection with less reliance on nondestructive inspection.

All operating personnel need to be encouraged to report any indications of moisture infiltration into insulated piping and other equipment [158]. Figure 7.23 shows moisture indications that were routinely observed by operating personnel and not reported. This lack of attention led to the CUI shown in Figure 8.73 on vertical piping over a sensitive marine waterway.

NACE and API documents suggest a variety of NDT techniques [165, 166]. Additional more detailed advice is available:

- When insulation removal is not practical, suitable NDT methods can be used.
- Some of the NDT methods that can be used are Long Range Ultrasonic Testing (LRUT), Pulsed Eddy Current Technique (PEC) and profile radiography.
- Long Range Ultrasonic Testing (LRUT) can be used for pipeline inspection where operating

TABLE 8.19   Protective Coating Systems for Austenitic Stainless Steels Under Thermal Insulation [165]

System Number	Temperature Range[a,b]	Surface Preparation[c]	Surface Profile, μm (mil)[d]	Prime Coat, μm (mil)[e]	Finish Coat, μm (mil)[e]
SS-1	−45 to 60 °C (−50 to 140 °F)	SSPC[(5)]-SP 1[12] and SSPC-SP 16[13]	minimum 19 (0.75)	High-build epoxy, 125–175 (5–7)	N/A
SS-2	−45 to 150 °C (−50 to 300 °F)	SSPC-SP 1 and SSPC-SP 16	minimum 19 (0.75)	Epoxy phenolic, 100–150 (4–6)	Epoxy phenolic, 100–150 (4–6)
SS-3	−45 to 205 °C (−50 to 400 °F)	SSPC-SP 1 and SSPC-SP 16	minimum 19 (0.75)	Epoxy novolac, 100–200 (4–8)	Epoxy novolac, 100–200 (4–8)
SS-4	−45 to 540 °C (−50 to 1000 °F)	SSPC-SP 1 and SSPC-SP 16	minimum 19 (0.75)	Air-dried silicone or modified silicone, 37–50 (1.5–2.0)	Air-dried silicone or modified silicone, 37–50 (1.5–2.0)
SS-5	−45 to 650 °C (−50 to 1200 °F)	SSPC-SP 1 and SSPC-SP 16	minimum 19 (0.75)	Inorganic copolymer or coatings with an inert multipolymeric matrix,[f] 100–150 (4–6)	Inorganic copolymer or coatings with an inert multipolymeric matrix,[f] 100–150 (4–6)
SS-6	−45 to 595 °C (−50 to 1100 °F)	SSPC-SP 1 and SSPC-SP 16	50–100 (2–4)	Thermal-sprayed aluminum (TSA) with minimum of 99% aluminum, 250–375 (10–15)	Optional: sealer with either thinned epoxy-based or silicone coating (depending on max. service temperature) at approximately 40 (1.5) thickness. (not recommended for TSA under insulation)
SS-7	−45 to 540 °C (−50 to 1000 °F)	SSPC-SP 1	N/A	Aluminum foil wrap with min. thickness of 64 (2.5)	N/A

[a] The temperature range shown for a coating system is that over which the coating system is designed to maintain its integrity and capability to perform as specified when correctly applied. However, the owner may determine whether any coating system is required, based on corrosion resistance of austenitic and duplex stainless steels at certain temperatures. Temperature ranges are typical for the coating system; however, specifications and coating manufacturer's recommendations should be followed. SS-4, SS-5, SS-6, and SS-7 may be used under frequent thermal cyclic conditions in accordance with manufacturer's recommendations.

[b] Temperature range refers to the allowable temperature capabilities of the coating system, not service temperatures. An experienced metallurgist should be consulted before exposing duplex stainless steel to temperatures greater than 300 °C (572 °F).

[c] To avoid surface contamination, austenitic and duplex stainless steels shall be blasted with nonmetallic grit such as silicon carbide, garnet, or virgin aluminum oxide. Because there are no specifications for the degree of cleanliness of abrasive blasted austenitic and duplex stainless steels, the owner should state the degree of cleanliness required after abrasive blasting, if applicable, and whether existing coatings are to be totally removed or whether tightly adhering coatings are acceptable.

[d] Typical minimum and maximum surface profile is given for each substrate. Acceptable surface profile range may vary, depending on substrate and type of coating. Coating manufacturer's recommendations should be followed.

[e] Coating thicknesses are typical dry film thickness (DFT) values, but the user should always check the manufacturer's product data sheet for recommended coating thicknesses.

[f] Consult with the coating manufacturer for actual temperature limits of these coatings.

temperature is less than 125 Deg C. A small band of insulation needs to be removed for mounting array of UT transducers band in LRUT technique. It scans the pipeline longitudinally on both sides of transducer ring using guided ultrasonic waves. This technique gives the cross-sectional metal loss of pipelines. TLRUT is suitable for long straight length pipe.

- Pulsed Eddy Current (PEC) may be deployed without removal of insulation on both equipment & pipelines and average metal wall thickness of the location below the insulation can be measured. PEC technique may also be used for inspection of fire proofing, skirts/pipelines. Necessary caution to be taken when PEC is used at projections like nozzle, stiffener ring etc. as the projection also generates additional eddy current.

- Profile radiography may be used for measuring thickness without removing insulation. Cordoning of the area for radiography is the main disadvantage [173].

TABLE 8.20 Protective Coating Systems for Carbon Steels Under Thermal Insulation and Cementitious Fireproofing [165]

System Number	Temperature Range[a,b]	Surface Preparation	Surface Profile, μm (mil)[c]	Prime Coat, μm (mil)[d]	Finish Coat, μm (mil)[d]
CS-1	−45 to 60°C (−50 to 140°F)	NACE No. 2/ SSPC-SP 10[16]	50–75 (2–3)	High-build epoxy, 125 (5)	Epoxy, 125 (5)
CS-2 (shop application only)	−45 to 60°C (−50 to 140°F)	NACE No. 2/ SSPC-SP 10	50–75 (2–3)	N/A	Fusion-bonded epoxy (FBE), 300 (12)
CS-3	−45 to 150°C (−50 to 300°F)	NACE No. 2/ SSPC-SP 10	50–75 (2–3)	Epoxy phenolic, 100–150 (4–6)	Epoxy phenolic, 100–150 (4–6)
CS-4	−45 to 205°C (−50 to 400°F)	NACE No. 2/ SSPC-SP 10	50–75 (2–3)	Epoxy novolac or silicone hybrid, 100–200 (4–8)	Epoxy, novolac or silicone hybrid, 100–200 (4–8)
CS-5	−45 to 595°C (−50 to 1100°F)	NACE No. 1/ SSPC-SP 5[17]	50–100 (2–4)	TSA, 250–375 (10–15) with minimum of 99% aluminum	Optional: Sealer with either a thinned epoxy-based or silicone coating (depending on maximum service temperature) at approximately 40 (1.5) thickness. (not recommended for TSA under insulation)
CS-6	−45 to 650°C (−50 to 1200°F)	NACE No. 2/ SSPC-SP 10	40–65 (1.5–2.5)	Inorganic copolymer or coatings with an inert multi-polymeric matrix, 100–150 (4–6)	Inorganic copolymer or coatings with an inert multipolymeric matrix, 100–150 (4–6)
CS-7	60°C (140°F) maximum	SSPC-SP 2[18] or SSPC-SP 3[19]	N/A	Thin film of petrolatum or petroleum wax primer	Petrolatum or petroleum wax tape, 1–2 (40–80)
CS-8 Bulk or shop-primed pipe, coated with inorganic zinc	−45 to 400°C (−50 to 750°F)	Low-pressure water cleaning to 3,000psi (20MPa) if necessary	N/A	N/A	Epoxy novolac, epoxy phenolic, silicone, modified silicone, inorganic copolymer, or a coating with an inert multipolymeric matrix, is typically applied in the field. Consult coating manufacturer for thickness and service temperature limits.[e]
CS-9 Carbon steel under fireproofing	Ambient	NACE No. 2/ SSPC-SP 10	50–75 (2–3)	Epoxy or epoxy phenolic, 100–150 (4–6)	Epoxy or epoxy phenolic, 100–150 (4–6) A second coat is typically not required under epoxy intumescent fireproofing specifications
CS-10 Galvanized steel under fireproofing	Ambient	Galvanizing; sweep blast with fine, nonmetallic grit	25 (1)	Epoxy or epoxy phenolic (for more information on coatings over galvanizing, see 4.3.3). 100–150 (4–6)	Epoxy or epoxy phenolic, 100–150 (4–6) A second coat is typically not required under epoxy intumescent fireproofing specifications

a The temperature range shown for a coating system (including thermal-cycling within this range) is that over which the coating system is designed to maintain its integrity and capability to perform as specified when correctly applied. However, the owner may determine whether any coating system is required, based on corrosion resistance of carbon steel at certain temperatures. Temperature ranges are typical for the coating system; however, not all coatings in a category are rated for the given minimum/maximum temperature. Specifications and coating manufacturer's recommendations should be followed for a particular coating system.

b Temperature range refers to the allowable temperature capabilities of the coating system, not service temperatures.

c Typical minimum and maximum surface profile is given for each substrate. Acceptable surface profile range may vary, depending on substrate and type of coating. Coating manufacturer's recommendations should be followed.

d Coating thicknesses are typical DFT values, but the user should always check the manufacturer's product data sheet for recommended coating thicknesses.

e If inorganic zinc-rich coating is applied in a shop and topcoat is applied in the field, proper cleaning of the inorganic zinc-rich coating is required. Inorganic zinc-rich coating shall not be used by itself under thermal insulation in the 50 to 175°C (120 to 350°F) service temperature range for long-term or cyclic service (see Paragraph 4.3.5). However, bulk piping is often coated with inorganic zinc-rich coating in the shop and some owners purchase this piping for use under insulation. In these cases, the inorganic zinc-rich coating should be topcoated to extend its life.

Figures 7.21 and 8.74 show NDT images of insulated piping systems. Radiography (Figure 8.73) is commonly used, whereas thermal imaging (Figure 7.21) is not common [173].

Insulation is installed on piping for several reasons including:

- Heat conservation and/or freeze protection
- Process control
- Viscosity control (e.g. the Trans-Alaska Pipeline)
- Sound control
- Condensation control
- Fire protection
- Personnel protection

If the insulation is installed for personnel protection (to keep people away from hot [or dangerously cold] equipment) and does not produce significant advantages associated with the other reasons for insulation, then it can and should be removed. This eliminates CUI concerns on the uninsulated equipment. It can also be argued that insulation should be removed whenever the energy savings or process control are minimal.

API RP583 discusses the idea of personnel protective cages and recommends their use in situations where equipment surfaces are greater than 60 °C (140 °F) [166]. If insulation is removed where it is not needed, then external inspection for corrosion of equipment is much easier and CUI, one of many possible causes of corrosion, is eliminated. Detailed discussions of the advantages of eliminating unnecessary insulation and/or the use of personnel protective cages are available [167, 174]. Commercial versions of personnel protective cages are available, but many organizations choose to fabricate the protective equipment on-site.

Cathodic protection of thermally insulated structures is possible and discussed in a NACE publication [175]. The report discusses problems associated

Figure 8.72 Commercially available drain plug installed in insulated piping. *Source*: Reproduced with kind permission of Temati, Beverwijk, The Netherlands.

CUI suspected locations

Figure 8.74 Radiograph showing full diameter scan of an insulated pipeline.

Figure 8.73 CUI on vertical piping near the locations shown in Figure 7.23a and b where obvious moisture problems could have been noted and reported.

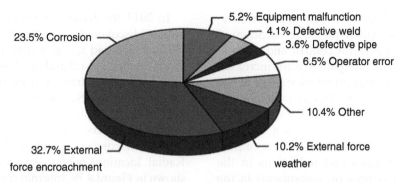

Figure 8.75 Causes of gas pipeline failures. *Source*: From Bruno [177]. Metallurgical Consultants, Inc.

Figure 8.76 Crater caused by a natural gas pipeline explosion.

Figure 8.77 Part of the pipeline shown in Figure 8.76.

with metal-jacketed buried pipelines and suggests that cathodic protection will have limited effects on controlling external corrosion of the carrier pipe if leaks in the outer jacket occur, normally at field joints in prefabricated insulated piping sections. The attack due to moisture ingress into the insulated piping system can be very rapid.

PIPELINES AND FLOWLINES

The terms for pipelines and flowlines are somewhat interchangeable. Many organizations consider flowlines to be piping systems, sometimes buried and sometimes on the surface, that carry fluids from wells to processing equipment. Once fluids are separated or treated in some manner, the term pipeline becomes the generally accepted term, and these pipelines will often extend for many kilometers (miles) [176]. Figure 8.75 shows the causes of onshore gas pipeline failures [177]. The numbers change, but the trends are roughly similar for liquid petroleum pipelines. Corrosion is the second most common form of pipeline failures, exceeded only by

external forces, often termed third-party damage, which is often construction activity related. The percentage numbers differ, but the trends are the same for liquid petroleum pipelines.

Figure 1.1 showed some of the consequences of a fatal pipeline explosion near Carlsbad, New Mexico, in 2000. Up until that time, most corrosion efforts on "sales gas" pipelines were concentrated on external corrosion. One of the lessons learned from this incident is that internal corrosion is also important and deserves attention.

Figures 8.76 and 8.77 show the forces associated with gas pipeline ruptures. Fortunately, the pitting of a wrapped spiral-welded pipeline with external corrosion happened in a remote location and no injuries resulted. The energy stored in compressed gases of all types makes gas pipelines more dangerous than oil pipelines, even though both transport flammable fluids.

Pipeline Problems and Failures

The 2006 Prudhoe Bay crude oil leaks received worldwide attention [178–180]. The spills caused environmental and economic damages that were still being debated years after the incidents. One of the unforeseen consequences of the widely publicized incident at Prudhoe Bay was that corrosion control and maintenance budgets of many oil companies were increased as a result of improved management awareness of the consequences of corrosion of pipelines and other equipment. Unfortunately the turnover and retirement of key personnel in the industry has led to a situation where after a decade of increased concern, new emphases on economical operations and reductions in the prices (and subsequent returns on investment) in the oil and gas industry have caused this increased concern to diminish in recent years.

The 2010 San Bruno, California, pipeline accident was associated with weld seam defects. The pipeline, constructed in the 1950s, did not have the "once in a lifetime" pressure testing after construction that became routine in later decades. The longitudinal seam welds in the failed section had defects that led to the rupture after a series of changes in operating pressure [181]. Vintage pipelines, constructed before approximately 1970, have been found to have inconsistent seam welds and "hard spots" due to rapid quenching when water in the rolling mills rapidly cools the metal surfaces and produces brittle untempered martensite on manufactured pipe surfaces. Modern pipe manufacturing processes have reduced the likelihood of defects of this type [182, 183]. Hard spots can also be produced by welding arc burns in both manufacturing and at girth welds during pipeline construction [184].

Manufacturing defects can become problematic when pressure cycles cause cracks to form at brittle hard spot areas. Figure 8.78 shows the operating pressures on a Canadian pipeline that burst in 2009 in a remote location. The pipeline rupture initiated at a manufacturing-process hard spot [185].

In 2013 the Kashagan oil and gas field, the largest oilfield discovered since Prudhoe Bay, Alaska, in the 1970s, which had recently started production, was shut down due to H_2S-related cracking problems. The entire pipeline system running from offshore production in shallow Caspian Sea waters to the onshore refinery needed to be replaced with CRA-lined piping, which took years to accomplish [186].

Pipeline corrosion can be either external or internal. Radial locations where corrosion is most likely are shown in Figure 8.79. Internal corrosion will occur at the bottom, or six o'clock position, where water and debris are likely to accumulate (Figure 8.80), and near the top of multiphase systems, where condensation creates corrosive conditions in the absence of corrosion inhibitors. External corrosion of buried pipelines is most likely near the four and eight o'clock positions, where the lack of soil compaction is likely to leave air voids that can become wetted with groundwater.

Figure 8.81 shows the uncorroded top interior of a natural gas pipeline and the pitting corrosion associated with water and biofilm collection near the bottom. Note how the liquid phase levels have changed, leading to various depths of pitting attack at the liquid–gas interface.

Construction activities and repairs will also create galvanic cells between the disturbed soil near new pipe and adjacent older pipe that has been buried for longer times (Figure 5.21).

Fatigue failures occur when suspended pipelines encounter vortex shedding due to subsea or river currents [187].

Figure 8.82 shows pipeline cleaning pigs after they have been removed from a pipeline. The main purposes of these pigs are to remove a variety of substances that could lead to blocked fluid flow or corrosion. As

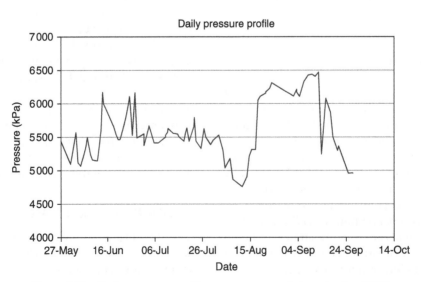

Figure 8.78 Daily pressure profile on a gas transmission pipeline [185].

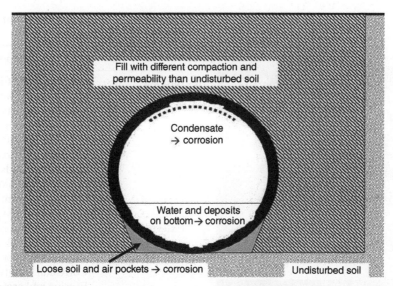

Figure 8.79 Radial locations where corrosion is most likely on buried pipelines.

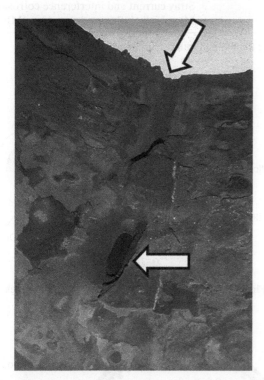

Figure 8.80 Six o'clock corrosion on a crude oil gathering line. Lower arrow shows pitting that produced leaking. Upper arrow highlights the profile of the pipeline.

oilfields age and pipeline fluid flow rates decline, the likelihood of corrosion problems increases due to increased water dropout and accumulation in low spots. Waxy deposits and hydrates (Figure 3.44) are also likely to accumulate. In low-temperature climates the possibility of ice formation, which can interfere with cleaning pig operations, must also be considered [188].

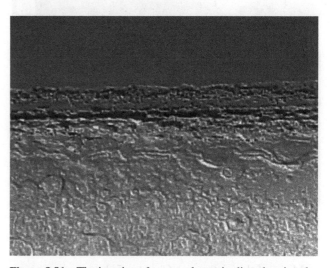

Figure 8.81 The interior of a natural gas pipeline showing the lack of corrosion in the gas phase and extensive pitting corrosion in the water and condensate phase.

Forms of Corrosion Important in Pipelines and Flowlines

Table 8.21 and Figure 8.83 list the forms of corrosion associated with the worldwide operating conditions of a major oil company's pipelines. The list in Table 8.21 does not consider corrosion mechanisms not controlled by operations, e.g. those forms of degradation addressed by proper design and construction practices. These include the various forms of environmental cracking controlled by appropriate materials selection and welding procedures and, in general, not influenced by operating conditions [189], although some forms of SCC might also be characterized as corrosion fatigue, because they are accelerated by differences in operating pressures in gas pipelines [190].

(a)

(b)

Figure 8.82 Crude oil pipeline pigs in maintenance shed after a cleaning run. (a) Before cleaning. (b) After cleaning.

Many of the pictures in Chapter 5 show corrosion on pipelines. The most serious corrosion problems come from pitting corrosion and from environmental cracking. Both forms of corrosion can be due to either internal or external environments and can lead to unexpected pipeline failures. It is not uncommon for corrosion pits to serve as stress risers, leading to SCC failures.

MIC is also a serious concern and has been associated with both the Carlsbad gas pipeline failure and the Prudhoe Bay crude oil pipeline leaks.

A form of corrosion unique to gas and multiphase pipelines is top-of-the-line corrosion. At locations where temperatures and pressures allow condensation, the condensate is frequently a mixture of low-mineral-content water and hydrocarbons, including acetic and formic acid (Figure 8.84). This condensate is corrosive and causes internal corrosion near the top (12 o'clock position) on many pipelines [191]. Corrosion inhibitors are often concentrated in water at the bottom of these same pipelines. Inhibitor pigs, including newly developed

TABLE 8.21   Corrosion Mechanisms Associated with Pipeline Maintenance

	Corrosion Mechanism
Internal corrosion	CO_2/H_2S weight-loss corrosion including preferential weld corrosion, corrosion under deposits, etc.
	Top of line (TOL)
	Microbially influenced corrosion (MIC)
	Erosion corrosion
	Erosion by solids
	Galvanic corrosion and corrosion at insulating joints (external CP)
	Corrosion by oxygen
External corrosion	Atmospheric corrosion
	Corrosion in splash and transition zones, including river crossings
	Corrosion in buried/immersed conditions
	Erosion and erosion corrosion including shore landing and river crossings
	Stray current and interference corrosion
	Microbially influenced corrosion (MIC)

Source: Reproduced with permission of NACE International.

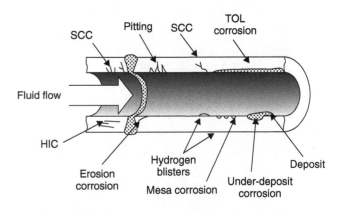

Figure 8.83 Forms of corrosion found on pipelines.

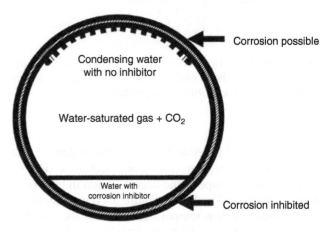

Figure 8.84 Condensation leading to top-of-the-line corrosion in gas and multiphase pipelines.

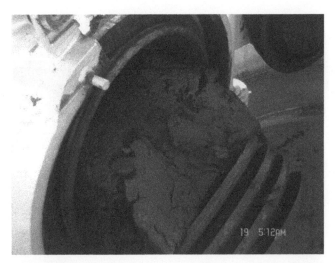

Figure 8.85 Black powder removed from a gas pipeline at a pig trap.

Figure 8.87 Installation of a pipeline repair sleeve over a corroded pipeline leak.

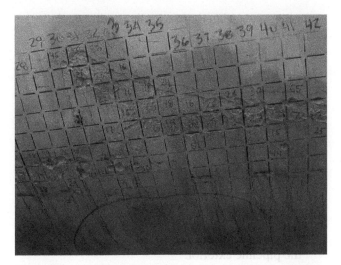

Figure 8.86 Grid pattern marked on the exterior of a pipeline.

spray pigs, have been developed to recycle film-forming corrosion inhibitors from the bottom to the top of pipelines in attempts to control this form of corrosion.

Black powder (Figure 8.85), deposits found in pipelines including iron sulfides and oxides, hydrocarbon solids, and other debris, is a major problem associated with pipelines. This powder can accumulate in low-velocity locations in pipelines and has been known to block flow. It can also cause under-deposit (crevice) corrosion [192].

Repairs and Derating Due to Corrosion

Several guidelines require various corrosion allowances for pipelines, and most pipeline corrosion is localized and can lead to relatively deep penetration or cracking with little average wall loss. Figure 8.86 shows a grid pattern marked on a pipeline exterior. The purpose of

mapping these corrosion pits is to determine if clustered pits are close enough to act as a somewhat larger defect or if the individual pits act independently. Similar calculations, based on defect depth and proximity to other defects, are used to calculate if environmental cracking defects are too close and act as larger defects.

Software programs overestimate the remaining strength and are not conservative enough in some cases when applied to older pipelines, which are considered to be brittle compared with steels produced since approximately 1990. While various software programs have been shown to work quite well based on laboratory and field tests on modern pipeline steels [193–198], they may overestimate the safe operating pressures for steel manufactured by earlier production methods [199]. Pipeline steels were not controlled for brittleness prior to changes in API 5L requirements introduced in the year 2000 revision of the standard. Steels manufactured before that date may be brittle and have not been tested for ductile–brittle behavior.

Pipeline repair methods often employ installation of sleeves over corroded areas (Figures 8.87 and 8.88) [200]. The strength of these sleeves depends on the quality of installation, and manufacturer's recommendations concerning safe operating levels often overestimate safe operating pressures. This is especially true with composite sleeves, which have become popular in recent years, because they do not require welding and can often be applied without depressurization of operating pipelines [201].

Casings for Road and Railway Crossings

Figure 8.89 shows the idea of pipeline casings for road and railway crossings. A strong outer pipe (casing) surrounds the inner carrier pipeline that contains the

Figure 8.88 Repair sleeves installed on corroded crude oil pipeline.

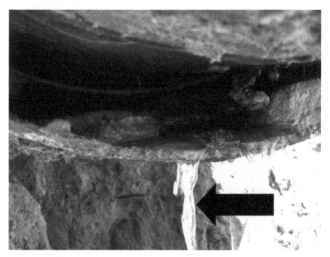

Figure 8.90 Water leaking from a defective pipeline casing annular space. Arrow indicates dripping water.

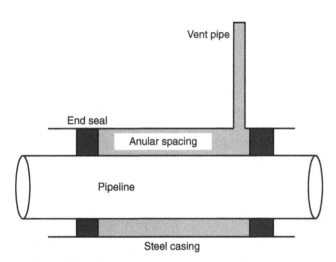

Figure 8.89 Pipeline casing for road or railway crossing.

fluids being transported in the pipeline. Electrical insulators are installed to isolate the two pipes, seals are intended to keep moisture from the annular spacing between the two pipes, and vents are installed so that any moisture or leaking fluid can escape from the casing annulus [202].

Perceived benefits from the use of casings include the following:

- Casings provide structural support and protect carrier pipes from vibrations and surface motion ("live loads").
- Casings protect carrier pipes from the dead weight of the structures above them.
- Vents allow the escape of dangerous material from the right of way.
- Casing systems make leak detection easier.

Problems associated with pipeline casings include the following:

- Seals between casings and carrier pipes can leak and fill the casing with corrosive water. This is shown in Figure 8.90.
- They can also become electrically shorted, invalidating cathodic protection on the carrier pipe [202].

In recent years it has become common to fill the annular spaces with waxes or similar fillers. These fillers several purposes:

- They prevent water from entering the void space.
- They cover the pipeline exterior with a water-repelling hydrophobic layer and avoid water wetting on the pipeline exterior.

Common void space filling practice is to pump the wax in from the bottom of one end of the casing and determine that the voids are filled once wax comes out of the vent pipe on the opposite end.

There are many problems associated with installation and maintenance of pipeline casings, and most authorities recommend against their use, except where required by regulatory agencies. Deeper burial of conductor pipes or the use of thicker steel at crossings is recommended to avoid the need for casings except where required by regulatory agencies, railroad operators, etc. Developments in onshore directional drilling have enabled many pipelines to be installed at greater depths, eliminating requirements for pipeline casings.

Pipeline and Flowline Materials

Most pipelines are constructed from carbon steel in accordance with API 5L or similar standards, although there has been a tendency for subsea pipelines to be

TABLE 8.22 Typical Pipeline Materials Suggested by ISO 21457

Type of Service	Materials
Hydrocarbon production	Carbon steel with or without chemical treatment
	Flexible pipe with carcass in type 316 or duplex stainless steel
	Carbon steel, internally clad/lined with type 316, alloy 825 or alloy 625
	Type 22Cr
	Type 13Cr with low carbon content
Wet hydrocarbon gas (not dehydrated)	Same materials as for unprocessed hydrocarbon production
Dry gas (dehydrated)	Carbon steel
Stabilized or partly stabilized oil or condensate	Carbon steel with or without corrosion inhibitor
Deaerated seawater[a] and produced water for injection	Carbon steel with chemical treatment
	Carbon steel with organic coating or lining
	Flexible pipes with carcass in type 316 or duplex stainless steel

[a] Several failures due to bacterial corrosion have been reported in carbon steel subsea injection flowlines transporting deaerated seawater. The corrosion that occurs in the six o'clock position in the pipe is caused by sulfate-reducing bacteria. The attack is very difficult to control, and even with cleaning pigs and bacterial treatment, corrosion rates in the order of $1\,mm\,yr^{-1}$ have been experienced. Carbon steel with an internal organic coating/lining has been used with success and can be considered for water injection pipelines. Alternatively, controlling injected deaerated seawater chemistry to specified low oxygen levels and including a biocide or nitrate treatment can permit the use of unlined carbon steel. Nitrate improves the corrosion control and, in addition, reduces the reservoir souring and, hence, the H_2S production.
Source: From ISO 21457 [50].

constructed from martensitic stainless steels (13Cr) in recent years because of concerns with internal corrosion [187]. Unfortunately the so-called "Super 13 Chrome" (low-carbon) alloys developed for this purpose are very difficult to weld [203–206], and this practice has been supplanted by the use of other CRAs or by the use of carbon steel pipelines with CRA linings. The 13Cr alloys may also have hydrogen embrittlement problems [207, 208].

Table 8.22 lists typical pipeline materials suggested by ISO 21457 [50]. This suggested standard lists typical materials for a variety of upstream applications, excluding production wells. The intent of the standard is to minimize design efforts by recommending standard materials whenever possible [209].

As of 2017 the worldwide pipeline CRA practice seems to be:

- Super 13Cr alloys (martensitic stainless steels) are used offshore Norway, but not in many other locations. This is in accordance with ISO 21457.

- Duplex and super duplex stainless steels are widely used but have low strength compared with other options. They are more typically used for flowlines, where the pressures are lower and the costs of CRA piping is less than for long-distance pipelines.

- Carbon steel pipe with CRA linings. This is a common approach worldwide. This is the approach that was adopted when the Kashagan field pipeline under the Caspian Sea needed to be replaced due to H_2S-related cracking problems. In these situations the carbon steel is used for pressure containment, and the CRA alloys, which are not as strong, are used in the annealed condition that limits their susceptibility to environmental cracking of all types. Organic coating liners are sometimes used for flow enhancement, but they are not used for corrosion control.

The installed costs of subsea pipelines and flowlines are divided approximately equally between materials (assuming carbon steel) and welding, lay barge, seabed preparation, and insulation and weight coating. Thus doubling the cost of materials by changing from carbon steel to CRAs or carbon steel pipes lined with CRAs is sometimes considered justified. Doubling of the costs for materials and welding only increases the total costs by about 25%. Compared with the costs of lost production and repairs, this is often considered to be a justified expenditure.

All stainless steels and most other CRAs have problems with crevice corrosion, so subsea or buried CRAs must have the same external corrosion control measures – coatings supplemented by cathodic protection – that are used for carbon steel pipelines.

Martensitic stainless steels offer increased corrosion resistance to CO_2 corrosion but only limited resistance for mild H_2S service. For high H_2S applications it is necessary to use much more expensive CRAs. This is why sour gas is usually processed relatively close to the source, whereas CO_2-rich fluids are sometimes transported for long distances in 13Cr multiphase pipelines before onshore or centralized processing.

Untempered martensite (called "hard spots" in pipeline terminology) can occur due to improper thermomechanical processing in pipe fabrication mills or due to improper welding procedures. Magnetic flux leakage (MFL) inspections can identify hard spots. This untempered martensite can occur in carbon steels, 13Cr pipeline steels, and selected other ferrous metals.

Hard spots can be detected by MFL inspection [210]. Unfortunately, it is often necessary to do this after field failures have indicated that entire shipments of questionably processed pipe have been delivered, installed, and placed in service.

The most common grade of API line pipe is X65 with specified minimum yield stresses of 65 ksi. Grades as high as X80 have been accepted and used with minimal

Figure 8.91 Concrete weight coating on the exterior of pipe intended for submerged service.

concerns about EHE, which can come from internal fluids; from H_2S-rich soils, e.g. due to anaerobic bacteria; or from cathodic protection. Steel producers are developing higher-strength pipelines, and the claimed advantages are reduced material shipping costs and lower weights in transport [211]. For most construction the weight of pipes is a secondary concern, and offshore and other submerged pipelines frequently require concrete weight coatings (Figure 8.91) to provide negative buoyancy.

The steels used in pipelines have changed over the years, even though most pipelines are still constructed from carbon steels. In the 1970s hot rolling followed by normalizing was replaced by thermomechanical processing. Continuous casting processing has resulted in an unfortunate tendency for the "cleaner" steels produced in recent years to have more segregation and inclusions near the middle of plate steels. Varying minor alloying additions produce higher-strength fine-grained steels with improved weldability [212, 213]. The results of these improvements are generally better steels, and the introduction of ductility testing requirements to API 5L in 2000 means that newer pipelines are less likely to have some of the brittleness problems (hydrogen embrittlement in its various forms plus low ductile–brittle transition temperatures) associated with older pipelines. Unfortunately, much of the existing pipeline infrastructure was built before these improvements were introduced.

A major problem associated with pipeline steels in the past has been welding. Spiral-welded pipe, used extensively in Canada and Europe, is still considered a lower grade of pipe in many other locations, although major pipelines in the United States were constructed using spiral-welded piping in recent years due to nonavailability of conventional longitudinally welded pipe.

Seamless pipe is only available in relatively small diameters (up to approximately 16 in. or 410 mm). Welding still remains the most likely location for metallurgical defects on pipelines. The practice of orienting joints of pipe so that the longitudinal welds are at 2 o'clock on one joint and 10 o'clock on the next joint (Figure 4.40) is intended to prevent any cracks that do form from running from one joint to the next joint. The 2 and 10 o'clock positions are chosen to avoid the most likely orientations for internal corrosion (6 and 12 o'clock) or external corrosion (4 and 8 o'clock) (Figure 8.79). Causes such as erosion corrosion can also result in deep corrosion in certain locations.

Pipeline Hydrotesting

Hydrotesting is a common means of testing the integrity of pipelines after construction, major alterations, or repairs. The procedure involves filling the pipeline (or other pressure vessels) with water and pressurizing the system to a level higher than the anticipated maximum allowable operating pressure (MAOP), usually 125% of MAOP. This test is intended to identify any defects, either corrosion or, more likely, sharp defects associated with welding or environmental cracking. Hydrotesting may increase resistance to environmental cracking by slowing the growth rates of preexisting cracks, but the primary purpose is to locate and repair defects before the equipment is placed or returned to service [113, 114, 193, 214]. Water used for hydrotesting should be as clean as possible (Table 8.23). While clean water sources are often used for smaller equipment, the large volumes of water needed to pressurize pipelines mean that river, lake, or ocean water is usually used for pipelines.

The main problem associated with hydrotest water is MIC. Biofilms will form on metal surfaces within 48 h to 2 weeks and can cause significant corrosion within a month. Mature biofilms are resistant to biocides, so biocide treatments should accompany initial flooding of the equipment, unless the equipment can be drained

TABLE 8.23   Water Sources Used for Hydrotesting Pipelines and Other Pressurized Equipment [214]

Water Source	
Demineralized water	Most Desirable
High purity steam condensate	
Potable water	
Seawater (clean, more than 50 ft above seabed and 50 ft below sea surface)	↑
River water	
Lake water	
Brackish water	Least desirable

Source: Reproduced with permission of NACE International.

within three to five days of testing [214]. Table 8.24 shows typical concentrations of bacteria in natural waters, and Table 8.25 lists the generic types of biocides commonly used in pipeline hydrotest waters [214].

Dissolved oxygen will normally be consumed in the first days of the test and will not produce appreciable corrosion unless air pockets form. Oxygen scavengers are not recommended, because they will often interfere with biocides, which are usually necessary [214].

Many pipelines cannot be drained efficiently, and biofilm removal using scraper pigs is often advised before placing pipelines in service. MIC problems can be significant for pipelines operating in the 15–45 °C (60–115 °F) range. At temperatures above 80 °C (180 °F), the risks are minimal [214].

Seawater Injection Pipelines/Flowlines

Palmer and King discuss seawater injection pipelines. Oxygen contamination of produced water and the use of surface waters for injection lead to higher corrosion rates for this service than for hydrocarbon pipelines. The higher corrosion rates are usually addressed by a combination of specifying thicker pipe walls (corrosion

allowance) and using oxygen scavengers [187]. This is one area where the use of online electrochemical corrosion rate monitors (electrical resistance probes or LPR probes – see the discussion on these techniques in Chapter 7) is especially useful. Corrosion coupons, which only provide long-term corrosivity trends, are inappropriate for this application, because it is not uncommon for upsets in operating conditions (e.g. air ingress or failing inhibitor pumps) to drastically increase corrosion rates. The causes of upsets and increased corrosion rates need to be identified as soon as possible. With LPR results can be seen in a matter of minutes, and this information is often useful in identifying the source of oxygen contamination. Galvanic probes can offer similar quick-response warnings of oxygen-contamination problems [54].

External Corrosion of Pipelines

External corrosion control of pipelines is considered a well-developed technology, although questions and controversies associated with industrial practices still arise. The use of organic coatings to protect metallic pipelines dates to 1830 in England. Cathodic protection of pipelines was reported around 1906 in Germany and popularized by Kuhn and coworkers in Louisiana. Kuhn suggested the use of a protection potential of –850 mV $Cu/CuSO_4$ in 1933 and led efforts to found NACE in the 1940s [215]. Recent versions of NACE SP0169, the most commonly cited cathodic protection reference, which is heavily oriented toward pipeline corrosion control, have introduced other means of determining if adequate corrosion protection has been provided. The thinking behind these more complicated cathodic protection criteria is that Kuhn's original work was in Louisiana, where current-resistance (IR) drops between the structure surface are minimal, because Kuhn worked with

TABLE 8.24 Typical Concentrations of Bacteria in Natural Waters [214]

Location		Concentration (cells (ml)$^{-1}$)
Seawater	Continental shelf and upper 200m of open ocean	5×10^5
	Deep water (below 200m)	5×10^4
	Deep water (below 320m)	10^2
Freshwaters and saline lakes		10^6
Potable water		10^5

Source: Reproduced with permission of NACE International.

TABLE 8.25 Properties of Biocides Used for Pipeline Hydrotesting [214]

Property		Biocide		
		Quaternary Amine	Glutaraldehyde [27, 28]	THPS [29]
pH stability range		5–9	5–9	2.5–9
Oxygen stability		Excellent	Poor	Poor
Half-life (days)	Biotic anaerobic	100s of days	River water and sediment: <1 Seawater: 32.5	Seawater: 72 d at pH 7 7 d at pH 9
	Biotic aerobic	100s of days	Seawater: <1	Seawater: <1 d to THPO. 25 d to complete degradation
Wetting ability		Good	Improved with surfactant addition	Requires surfactant addition
Treatment of hydrotest water required at end of use?		Yes	Yes	Yes (may not need treatment dependent on discharge water test results)

Source: Reproduced with permission of NACE International.

wet, conductive soils, which is not the case in many other parts of the world [216, 217]. ISO standards, which are somewhat different, are also available and are incorporated by reference into many international standards and other documents for pipelines [218, 219].

The standard way of protecting pipelines from external corrosion is to use protective coatings as the primary means of corrosion control with cathodic protection systems as the secondary, or supplementary, corrosion control technique. Cathodic protection systems are sized and designed to provide sufficient electrical current to protect exposed metal at holidays in new coatings and to provide more electrical current as protective coating systems age and become less effective. While cathodic protection can be applied to uncoated pipelines, reductions in electric current due to protective coatings have been shown to be the preferable approach to external pipeline corrosion control. In current practice, virtually all hydrocarbon pipelines have a combination of protective coatings supplemented by cathodic protection. Gathering lines, which may have shorter intended service lives, are sometimes protected only by protective coatings.

Chapter 6 has extensive discussions of protective coatings and cathodic protection. While protective coatings have many uses and applications, most of the discussions of cathodic protection in Chapter 6, and in the worldwide cathodic protection literature, relate to pipelines and associated equipment.

One of the problems with cathodically protected pipelines is hydrogen embrittlement. This is normally handled by only using pipeline steels that are resistant to hydrogen embrittlement, and this is one reason why most pipelines are limited to steels having yield strengths of 80 ksi (550 MPa). This practice seems to work in most cases, although there is some question about hydrogen charging in concentrated brines in permafrost soils [213].

Protective coating systems for pipelines have changed over the years as new coating systems have been developed, and problems with existing systems have been identified. Current buried and submerged pipeline protective coating practices include the following coating systems:

- Fusion-bonded epoxy.[1]
- Multilayer polyethylene systems – often with epoxy primer and copolymer adhesive layers below the polyethylene surface coating.[2]
- Extruded thermoplastic systems – commonly used on small-diameter pipelines and other piping systems.[3]

[1] Most widely used on pipelines in North America.
[2] More popular in Europe than in North America.
[3] See note 1.

Figure 8.92 Disbonded pipeline coating.

- Multilayer polyurethane – similar to polyethylene and less common.
- Asphalt/coal-tar enamels – health and safety issues limit their use in many locations, and they are sometimes banned from use. Their market share is declining worldwide.

Pipeline tapes and wraps were common at one time, but their market share has been reduced in recent decades due to problems including disbonding and dielectric shielding of cathodic protection. Their use on pipelines is largely restricted to complicated shapes and to repair and rehabilitation projects.

Disbonded (debonded) coatings have been found to shield cathodic protection currents and lead to external pipeline corrosion. This is shown in Figure 8.92. Proper surface preparation is necessary to ensure that coatings adhere to the steel substrate. Disbonding can also be caused by cathodic protection, and ASTM standards have been developed to test coating systems for resistance to coating disbonding at coating holidays [220, 221].

One of the claimed advantages of fusion-bonded pipeline coatings is that if they are damaged and holidays are formed, the holidays do not shield cathodic protection currents from reaching the surface. This is shown in Figures 8.93 and 8.94, which show the exterior of a gas transmission line. The high pH of the water bleeding from the blister indicates that cathodic protection, which causes water to become alkaline or basic near cathodes, has been altering the environment beneath the disbonded coating. The gray surface beneath the blister has been exposed in Figure 8.94 and is not corroded.

One of the problems with hard, damage-resistant coatings is that coating holidays, whether due to application problems or due to mechanical damage in shipping and construction, are hard to repair. The same properties

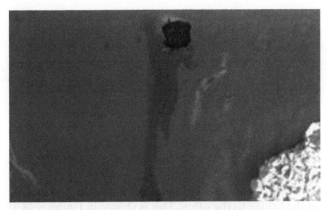

Figure 8.93 Blisters on fusion-bonded epoxy-coated gas transmission pipeline. Note the pH 12 water stains dripping from the broken blister. *Source*: Photo courtesy R. Norsworthy, Polyguard Products, Inc.

Figure 8.94 Underlying metal after nonadherent coating removed in the vicinity of the coating blister shown in Figure 8.93. *Source*: Photo courtesy R. Norsworthy, Polyguard Products, Inc.

Figure 8.95 Application of a field-applied coating at a girth weld on a pipeline. *Source*: Photo courtesy NACE International.

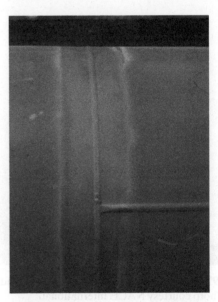

Figure 8.96 Field joint after blast cleaning. The factory-applied coating has been ground to provide a gradual transition for the subsequent field-applied coating. *Source*: Photo courtesy NACE International.

that make holidays hard to repair also mean that field joint coatings, where coatings must be applied to the metal surface and to the adjacent pipeline coating, are also difficult.

Girth (circumferential) welds in pipelines must be coated in the field. This presents difficulties associated with surface preparation and coating application. Figure 8.95 shows the application of a field-applied coating system to a girth weld on a pipeline. Figure 8.96 shows a girth weld field joint after surface preparation and prior to coating application. The edges of the factory-applied coating have been ground smooth to allow for an easy transition from coating the bare metal surface to coating over the factory-applied coating.

The difficulty of applying coatings in the field led to the development and marketing of a variety of shrink

sleeve coatings for pipeline use. These coatings or wraps are applied as flexible solids to the pipeline exterior. Heat is applied to the sleeves, and this causes them to shrink and comply with the underlying surface profile. Properly applied shrink sleeves should be watertight and prevent corrosion of the underlying metal. Unfortunately, disbonded shrink sleeves shield underlying surfaces from cathodic protection. The pitting corrosion in Figure 5.38 was caused by dielectric shielding from a nonadherent shrink sleeve similar to the wrinkled shrink sleeve shown in Figure 8.97. Proponents of shrink sleeves claim that most cases of corrosion beneath nonadherent sleeves were caused by poor surface preparation prior to application of the shrink sleeves. Improvements in shrink sleeve application methods have been reported [222–224].

Most problems with external corrosion of pipelines are due to coating problems. Cathodic protection is intended to supplement protective coatings, and it is effective when dielectric shielding is absent [224].

Pipelines often need to be electrically isolated from other structures. Sometimes this is to ensure that stray currents from the other structures do not cause corrosion on the structure under consideration. Another reason for isolation is to prevent unwanted current drainage to nearby structures, which might mean that the cathodic protection system is not minimizing corrosion on the pipeline segments under consideration. Electrical isolation is often accomplished using isolation fittings like those shown in Figures 5.13 and 5.14 and described in

NACE SP0286 [225]. Isolation fittings work for pipelines carrying nonconductive fluids like crude oil and natural gas. Unfortunately, the conductivity of some produced waters (typically high in ionic salt content) can overcome the insulation and short out the system [226]. Air is an insulator, but wet soil and water (surrounding electrolytes) can also short isolation fittings. This is why electrical isolation flanges are installed in air-exposed locations and are not buried or submerged.

Buried pipelines are often suspended on bridges over waterways and other obstacles. When this happens, it is necessary to provide electrical continuity from one side of the bridge to the other. This is shown in Figure 8.98. If the CP system is designed so that the length of pipeline suspended on the bridge is isolated from the underground sections of the pipeline, a jumper cable is necessary to connect the two buried sections of the pipeline [220].

Internal Corrosion of Pipelines

The Carlsbad, New Mexico, pipeline incident in 2000 prompted increased attention on internal corrosion of pipelines. Internal pipeline coatings are seldom used, because they are hard to apply at field joints, they can disbond due to decompression, and they are subject to mechanical damage from inspection pigs and other sources.

Oil pipelines have relatively few internal corrosion problems, because corrosive water can be entrained in oil–water emulsions and oil-wetted surfaces are not corroded. This approach does not work with lowered flow rates or in other circumstances where water can accumulate. Wicks and Fraser developed a model to explain when oil–water separation was likely to occur, and other models have been developed in recent years [27, 227–229]. Water separation and corrosion at the bottom, six o'clock, position has been reported in crude oil pipelines

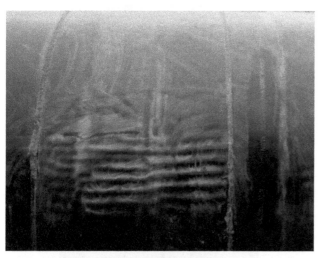

Figure 8.97 Wrinkled shrink sleeve due to soil loading [222]. *Source*: Photo courtesy NACE International.

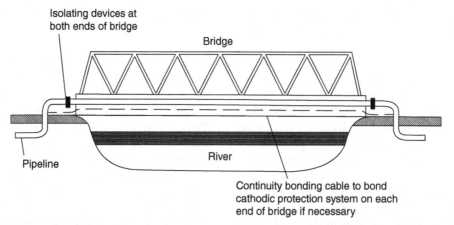

Figure 8.98 Typical use of isolating devices at each end of a bridge for a cathodically protected pipeline. *Source*: Reproduced with permission of NACE International.

Figure 8.99 Galvanic corrosion monitoring probe removed from a crude oil pipeline. Note the paraffin and sludge deposits [231]. *Source*: Photo courtesy NACE International.

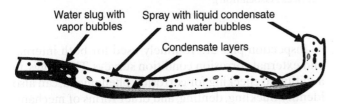

Figure 8.100 Slugging and fluid flow through a multiphase piping system.

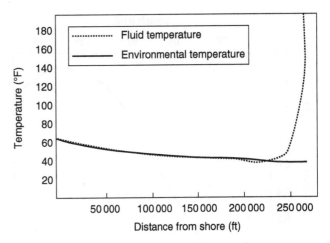

Figure 8.101 Pipeline fluid vs. environmental temperatures for an offshore pipeline.

having as low as ½–2% water, depending on oil viscosity, flow rates, pipe diameters, etc. [152].

Water separation, sludge accumulations, and paraffin deposits that cause corrosion are most likely to occur in locations with low flow rates, water separation from hydrocarbons, and solids accumulation [230]. Sludge and paraffin deposits are shown on the corrosion monitoring probes removed from a crude oil pipeline shown in Figure 8.99 [231]. The 2006 Prudhoe Bay pipeline leaks have been reported to be caused by accumulated corrosion products and biofilms at the bottom of a crude oil pipeline. These leaks, and others, illustrate the need for internal cleaning and inspection operations using specialized pigs.

Condensate and injection water pipelines and flowlines carry corrosive liquids and must be designed accordingly. Internal coatings, corrosion inhibitors, and CRA linings are used for corrosion control in these services [232].

Gas pipelines are generally more corrosive than oil pipelines. This is primarily due to condensation of organic acids that cause any water in the pipeline to become corrosive. Figure 8.100 shows the multiple phases forming in a gas pipeline. The water slug contains most of the corrosion inhibitor, but condensed liquids containing organic acids and newly condensed, inhibitor-free water may collect on the top of the pipeline and cause corrosion. Corrosion inhibitors, dissolved in liquid water, may take up to a month to move through a pipeline, but condensation, which comes from the much faster-moving gas phase, can form at any time along the pipeline at any location where temperature and pressure considerations favor condensation. If the gas phase is not moving fast enough to cause droplets of corrosion inhibitor-containing water from the bottom of the pipe to splash onto the upper portions of the pipe, then top-of-the-line corrosion can occur. Inhibitors should either be continuously injected into these pipelines, or pigging operations, redistributing corrosion inhibitor from the bottom of the pipeline to the top, become necessary. Polymer spheres, gel pigs, and special corrosion inhibitor Venturi spray pigs (Figure 6.70) are all used for this purpose.

Phase-separation prediction models are often used to suggest where corrosion is likely to occur [188].

Figures 8.101 and 8.102 show the effects of distance from the source on fluid temperature for an offshore pipeline (Figure 8.101) and the effects of insulation on the fluid temperature of an onshore pipeline (Figure 8.102). The figures are simplified versions of the illustrations in the original reports [233].

The results of fluid separation and temperature–pressure profile modeling are used to determine where corrosive conditions are most likely to occur. This information is used to determine where corrosion inhibitor injection points and corrosion monitoring probes should be located. They are also used to enable decisions on pigging schedules for corrosion inhibitor distribution, internal pipeline cleaning, and inspection scheduling and methods [227–233].

Dead legs on pipelines, and pipelines that are out of service for extended periods of time, are also subject to

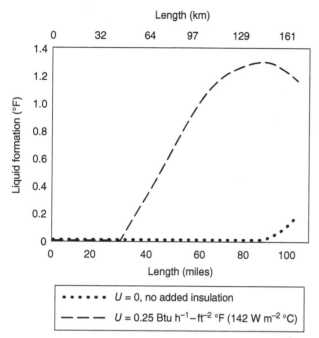

Figure 8.102   Effect of thermal insulation on fluid temperatures in an onshore pipeline.

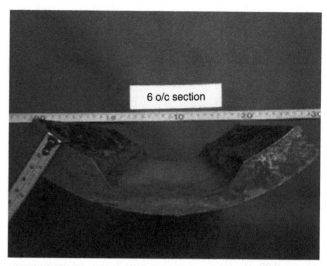

Figure 8.103   Deep erosion corrosion leading to channeling corrosion at the six o'clock position on a subsea water injection line. *Source*: Image courtesy of PPSA (Pigging Products & Services Association).

stagnant water accumulation. This can lead to corrosion, possibly accelerated by microbial activity [113]. The Carlsbad pipeline explosion (Figure 1.1) was attributed to water accumulation and associated MIC in a dead leg [234].

Multiphase pipelines, which carry combinations of crude oil, natural gas, and produced water, have many of the same corrosion problems as wet (water-containing) natural gas pipelines [235].

Inspection, Condition Assessment, and Testing

Pitting corrosion and environmental cracking are the most likely forms of degradation leading to pipeline failure. Unfortunately, the statistical distribution of pipeline pitting corrosion, which has been known and documented for decades, means that small corrosion probes or coupons are unlikely to identify the extent of corrosion-related wall loss [236].

The common methods of pipeline inspection are [187]:

- **Radiography** – normally used during construction, but also used during excavated inspections.
- **Ultrasound** – useful for wall thickness and hydrogen blistering inspections.
- **Magnetic flux leakage** – considered the "smartest" of pig-mounted inspection methods, although it may miss some cracking.
- **Visual inspection** – requires excavation, but can identify important phenomena rapidly.

Inspection pigs are routinely used for both internal and external corrosion condition surveys. These pigs, in addition to checking for corrosion phenomena, can also identify buckling, denting, and other forms of mechanical anomalies. Unfortunately, not all pipelines and gathering lines are piggable. This is especially true at offshore risers, where the transitions from essentially horizontal piping to vertical piping make the systems most vulnerable to all forms of damage, both mechanical and chemical.

External corrosion survey methods have relied on cathodic protection surveys, but problems with dielectric shielding mean that some areas of serious corrosion have been missed by above-surface electrical surveys [222, 223].

It is important to remember that internal corrosion is likely to lead to wall loss at different locations along the pipeline that are associated with water buildup, slugging effects due to elevation changes, changes in diameter, MIC, and erosion patterns [51, 188]. The radial locations can also vary from those shown in Figure 8.79. Figure 8.103 shows channeling corrosion attributed to undesirable oxygen levels in a water injection flowline.

Monitoring and inspection problems have led to the development of internal corrosion direct assessment (ICDA) and external corrosion direct assessment (ECDA) techniques and requirements [116, 217, 237–243]. Direct assessment standards and techniques are relatively new and subject to revision. Updated versions of these standards are likely to have significant changes as field experience provides input. ECDA data elements are discussed in Table 8.26.

TABLE 8.26   ECDA Data Elements

Data Elements	Indirect Inspection Tool Selection	ECDA Region Definition	Use and Interpretation of Results
Pipe related			
Material (steel, cast iron, etc.) and grade	ECDA not appropriate for nonferrous materials	Special considerations should be given to locations where dissimilar metals are joined	Can create local corrosion cells when exposed to the environment
Diameter	May reduce detection capability of indirect inspection tools		Influences CP current flow and interpretation of results
Wall thickness			Affects critical defect size and remaining life predictions
Year manufactured			Older pipe materials typically have lower toughness levels, which reduces critical defect size and remaining life predictions
Seam type		Locations with pre-1970 low-frequency electric resistance welded (ERW) or flash-welded pipe with increased selective seam corrosion susceptibility may require separate ECDA regions	Older pipe typically has lower weld seam toughness that reduces critical defect size. Pre-1970 ERW or flash-welded pipe seams may be subject to higher corrosion rates than the base metal
Bare pipe	Limits ECDA application. Fewer available tools – See NACE SP0207 and NACE Standard TMD109	Segments with bare pipe in coated pipelines should be in separate ECDA regions	Specific ECDA methods provided in NACE SP0207 and NACE Standard TM0109
Construction related			
Year installed			Affects time over which coating degradation may have occurred, defect population estimates, and corrosion rate estimates
Route changes/ modifications		Changes may require separate ECDA regions	
Route maps/aerial photos		Provides general applicability information and ECDA region selection guidance	Typically contain pipeline data that facilitate ECDA
Construction practices		Construction practice differences may require separate ECDA regions	May indicate locations at which construction problems may have occurred (e.g. backfill practices influence probability of coating damage during construction)
Locations of valves, clamps, supports, taps, mechanical couplings, expansion joints, cast iron components, tie-ins, insulating joints, etc.		Significant drains or changes in CP current should be considered separately; special considerations should be given to locations at which dissimilar metals are connected	May affect local current flow and interpretation of results; dissimilar metals may create local corrosion cells at points of contact; coating degradation rates may be different from adjacent regions
Locations of and construction methods used at casings	May preclude use of some indirect inspection tools	Requires separate ECDA regions	May require operator to extrapolate nearby results to inaccessible regions. Additional tools and other assessment activities may be required

(Continued)

TABLE 8.26   (Continued)

Data Elements	Indirect Inspection Tool Selection	ECDA Region Definition	Use and Interpretation of Results
Locations of bends, including miter bends and wrinkle bends		Presence of miter bends and wrinkle bends may influence ECDA region selection	Coating degradation rates may be different from adjacent regions; corrosion on miter and wrinkle bends can be localized, which affects local current flow and interpretation of results
Depth of cover	Restricts the use of some indirect inspection techniques	May require different ECDA regions for different ranges of depths of cover	May affect current flow and interpretation of results
Underwater sections and river crossings	Restricts the use of many indirect inspection techniques	Requires separate ECDA regions	Changes current flow and interpretation of results
Locations of river weights and anchors	Reduces available indirect inspection tools	May require separate ECDA regions	Influences current flow and interpretation of results; corrosion near weights and anchors can be localized, which affects local current flow and interpretation of results
Proximity to other pipelines, structures, high-voltage electric transmission lines, and rail crossings	May preclude use of some indirect inspection methods	Regions where the CP currents are significantly affected by external sources should be treated as separate ECDA regions	Influences local current flow and interpretation of results
Soils/environmental Soil characteristics/ types (refer to appendixes A and C)	Some soil characteristics reduce the accuracy of various indirect inspection techniques	Influences where corrosion is most likely; significant differences generally require separate ECDA regions	Can be useful in interpreting results. Influences corrosion rates and remaining life assessment
Drainage		Influences where corrosion is most likely; significant differences may require separate ECDA regions	Can be useful in interpreting results. Influences corrosion rates and remaining life assessment
Topography	Conditions such as rocky areas can make indirect inspections difficult or impossible		
Land use (current/past)	Paved roads, etc., influence indirect inspection tool selection	Can influence ECDA application and ECDA region selection	
Frozen ground	May affect applicability and effectiveness of some ECDA methods	Frozen areas should be considered separate ECDA regions	Influences current flow and interpretation of results
Corrosion control CP system type (anodes, rectifiers, and locations)	May affect ECDA tool selection		Localized use of sacrificial anodes within impressed current systems may influence indirect inspection. Influences current flow and interpretation of results
Stray current sources/ locations			Influences current flow and interpretation of results
Test point locations (or pipe access points)		May provide input when defining ECDA regions	
CP evaluation criteria			Used in postassessment analysis

TABLE 8.26   (Continued)

Data Elements	Indirect Inspection Tool Selection	ECDA Region Definition	Use and Interpretation of Results
CP maintenance history		Coating condition indicator	Can be useful in interpreting results
Years without CP applied		May make ECDA more difficult to apply	Negatively affects ability to estimate corrosion rates and make remaining life predictions
Coating type (pipe)	ECDA may not be appropriate for disbanded coatings with high dielectric constants, which can cause shielding		Coating type may influence time at which corrosion begins and estimates of corrosion rate based on measured wall loss
Coating type (joints)	ECDA may not be appropriate for coatings that cause shielding		Shielding caused by certain joint coatings may lead to requirements for other assessment activities
Coating condition	ECDA may be difficult to apply with severely degraded coatings		
Current demand			Increasing current demand may indicate areas where coating degradation is leading to more exposed pipe surface area
CP survey data/history			Can be useful in interpreting results
Operational data			
Pipe operating temperature		Significant differences generally require separate ECDA regions	Can locally influence coating degradation rates
Operating stress levels and fluctuations			Affects critical defect size and remaining life predictions
Monitoring programs (coupons, patrol, leak surveys, etc.)		May provide input when defining ECDA regions	May affect repair, remediation, and replacement schedules
Pipe inspection reports (excavation)		May provide input when defining ECDA regions	
Repair history/records (steel/composite repair sleeves, repair locations, etc.)	May affect ECDA tool selection	Prior repair methods, such as anode additions, can create a local difference that may influence ECDA region selection	Provide useful data for postassessment analyses such as interpreting data near repairs
Leak/rupture history (external corrosion)		Can indicate condition of existing pipe	
Evidence of external MIC			MIC may accelerate external corrosion rates
Type/frequency (third-party damage)			High third-party damage areas may have increased indirect inspection coating fault detects
Data from previous over-the-ground or from-the-surface surveys			Essential for preassessment and ECDA region selection
Hydrostatic testing dates/pressures			Influences inspection intervals
Other prior integrity-related activities – CIS, ILI runs, etc.	May affect ECDA tool selection – isolated vs. larger corroded areas		Useful postassessment data

Those items that are shaded are most important for tool selection purposes.
Source: From NACE SP0502 [237].

OFFSHORE AND MARINE APPLICATIONS

Offshore oil and gas production operations are usually self-sustaining and generate their own power and other utility systems. Topside facilities must consider weight, which is one reason why BAHXs, discussed above, are used. Table 8.10 listed corrosion control methods recommended for topside equipment, which are located as shown in Figure 8.37.

The tremendous expenses associated with inspection and replacement of subsea equipment mean that long-term reliability is necessary for any submerged equipment, and this means that subsea

equipment suppliers will often specify more robust designs than might be used onshore.

Offshore Pipelines

Offshore pipelines are usually designed with weight coatings to keep them submerged (Figure 8.91). Most offshore pipelines use galvanic (sacrificial) anodes for cathodic protection because impressed current requires local power supplies. Protective coatings are applied, even on CRA pipelines, because of crevice (underdeposit) corrosion concerns.

Cathodic protection is normally by zinc alloy galvanic bracelet anodes installed at field-applied girth weld locations where field-applied protective coatings are also applied. Figure 8.104 shows field-applied coatings at a submersible pipeline joint, and Figure 8.105 shows a typical bracelet anode installation. It is common to install these bracelet anodes every 300–500 m, depending on the size of the pipeline.

Pipelines are usually laid on the seabed and seldom buried during installation. Shifting bottom conditions can cause burial, and burial is sometimes necessary to protect against third-party damage from shipping and other operations. Subsea scour can produce suspended areas [187].

Zinc galvanic anodes are usually used for this application, because impressed current requires nearby power sources. The weight of zinc is an advantage for marine pipelines, and they also work in the water–mud environment where aluminum might passivate. Magnesium anodes would be consumed too quickly.

Stray current corrosion from nearby platforms is a possibility.

Figure 8.104   Field-applied protective coating at girth weld location between two weight-coated joints.

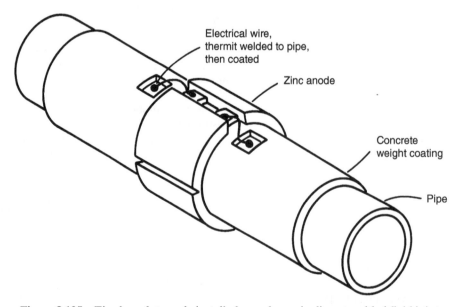

Electrical wire, thermit welded to pipe, then coated

Zinc anode

Concrete weight coating

Pipe

Figure 8.105   Zinc bracelet anode installed on subsea pipeline at welded field joint.

Offshore Structures

Offshore structures are subject to several different corrosion environments. Figure 3.9 shows how corrosion rates vary with elevation on steel piling in seawater. Figure 8.106 shows corrosion on an offshore platform where corrosion control has been lacking for several years.

Figure 8.107 is a simplified drawing of a conventional jacket offshore platform. Below the waterline and in the mud, cathodic protection controls the corrosion rate of the steel platform. Cathodic protection is used to minimize corrosion in the water column and below the mud line. The tidal and splash zones, from mean low tide to above mean high tide, are the hardest locations for corrosion control, so it is common to use thicker metal (greater fatigue and corrosion allowances) at these elevations. Corrosion-resistant sheathing or extra steel is required in the tidal region and splash zone where cathodic protection cannot be effective because of a lack of a continuous electrolyte [244, 245].

Above these lower regions the structural steel of platforms are protected by protective coating systems. Equipment and structural members immediately below the main deck have the highest corrosion and maintenance rates, because they are exposed to moisture and are poorly drained [245].

Figure 8.106 Offshore platform corrosion environments.

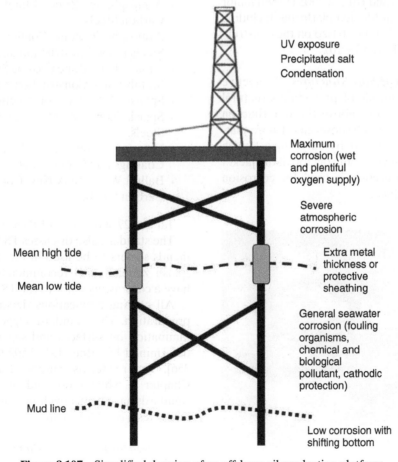

UV exposure
Precipitated salt
Condensation

Maximum corrosion (wet and plentiful oxygen supply)

Severe atmospheric corrosion

Mean high tide

Mean low tide

Extra metal thickness or protective sheathing

General seawater corrosion (fouling organisms, chemical and biological pollutant, cathodic protection)

Mud line

Low corrosion with shifting bottom

Figure 8.107 Simplified drawing of an offshore oil production platform.

While there are a number of other drilling and production platform designs that have been developed in recent years for use in deep waters, the great majority of offshore platforms are steel jackets like shown in Figures 8.106 and 8.107.

Fatigue Loading Fatigue (and corrosion fatigue) is a major consideration in the design of offshore platforms. Topside weight is kept to a minimum. Figure 8.108 shows the Alexander Kielland semisubmersible floating platform, which was involved in a fatal accident due to fatigue failure. The accident was caused by corrosion fatigue that started at a cracked weld defect that should have been detected in the fabrication yard [246–248].

Fatigue is one of many considerations in the design of offshore structures [249–253]. NORSOK N-004 provides detailed guidance on fatigue design to include structural details where "hot spots" (locations with highest stress range) are located. A major emphasis of this document is on where inspection of these hot spots should be concentrated [253]. Welded joints are examples of hot spots, and underwater inspection for corrosion damage concentrates in these areas.

Other platforms have also experienced fatigue problems. Hurricanes Katrina and Rita in 2005 caused major damage to several Gulf of Mexico platforms, including tension leg and other designs not based on the construction shown in Figures 8.106 and 8.107.

Protective Coatings on Offshore Structures Protective coatings are the major means of protecting structural steel and equipment exteriors above the waterline on offshore structures. Figure 8.109 shows an offshore platform jacket ready for transport by barge to a launch location in the Gulf of Mexico. Only the top of the jacket is painted. Cathodic protection is used for corrosion control below the waterline.

Figure 8.108 The capsized Alexander Kielland floating offshore platform.

Figure 8.109 Offshore platform prior to launching in the Gulf of Mexico. Arrows indicate three of the hundreds of 800 lb anodes attached to the structure.

NACE SP0108 covers coatings for offshore platforms from submerged locations to flare stacks [254]. It covers surface preparation and has tables suggesting typical coating systems for various locations and metallic substrates:

- Atmospheric Zone New Construction on Carbon Steels.
- Atmospheric Zone Maintenance Coatings on Carbon Steels.
- Atmospheric Zone Coating Systems on Stainless Steels (New Construction and Maintenance).
- Atmospheric Zone Coating Systems for Nonferrous Metals (New Construction and Maintenance).
- Splash Zone New Construction on Carbon Steels.
- Splash Zone Maintenance Coatings on Carbon Steels.
- Exterior Submerged Zone New Construction Coatings on Carbon Steels.
- Ballast Water Tank New Construction Coatings on Carbon Steels.

Table 8.27 shows one of these tables.

The standard also discusses TSA coatings and recommends sealers to be used in the atmospheric zone and splash zone. It also recommends that sealers should have a contrasting color to the TSA to aid inspection.

All coating applications depend on proper surface preparation. This standard suggests residual salt contamination on surfaces and suggests acceptable levels determined by either ISO 8502-6 or ISO 8502-9 [255, 256]. The reader is cautioned that, as discussed in Chapter 6, while these and other residual salt level standards have been developed, agreement between different tests is seldom achieved.

Submerged Structure Corrosion Control The submerged portions of most offshore structures are carbon steel and are protected from corrosion by cathodic protection.

TABLE 8.27 Typical Atmospheric Zone New Construction Coating Systems on Carbon Steels [249]

Service Category	Coat	Coating System	DFT, μm (mil)	Target DFT, μm (mil)
CN-1 Atmospheric zone at −50 to 120°C (−58 to 248°F) with/ without insulation	1	Zinc-rich primer	50–75 (2–3)	75 (3)
	2	Epoxy	125–175 (5–7)	125 (5)
	3	Polyurethane	50–75 (2–3)	75 (3)
	1	Epoxy primer	125–175 (5–7)	125 (5)
	2	Epoxy	125–175 (5–7)	125 (5)
	3	Polyurethane	50–75 (2–3)	75 (3)
	1	T3A	250–375 (10–15)	250 (10)
	2	Thinned sealer (epoxy)	Do not add to DFT[a]	No additional DFT
	3	Sealer (epoxy)	Do not add to DFT[a]	No additional DFT
CN-2 Atmospheric zone at 120 to 150°C (248 to 302°F) without insulation	1	Inorganic zinc-rich primer	50–75 (2–3)	75 (3)
	2	Silicone acrylic	25–50 (1–2)	50 (2)
	1	TSA	250–375 (10–15)	250 (10)
	2	Thinned sealer (acrylic silicone or epoxy phenolic)	Do not add to DFT[a]	No additional DFT
	3	Sealer (acrylic silicone or epoxy phenolic)	Do not add to DFT[a]	No additional DFT
CN-3 Atmospheric zone at 120 to 150°C (248 to 302°F) with insulation	1	Epoxy phenolic	100–125 (4–5)	125 (5)
	2	Epoxy phenolic	100–125 (4–5)	125 (5)
	1	TSA	250–375 (10–15)	250 (10)
	2	Thinned sealer (silicone acrylic or epoxy phenolic)	Do not add to DFT[a]	No additional DFT
	3	Sealer (silicone acrylic or epoxy phenolic)	Do not add to DFT[a]	No additional DFT
CN-4 Atmospheric zone at 150 to 450°C (302 to 842°F) with/ without insulation	1	TSA	250–375 (10–15)	250 (10)
	2	Thinned sealer (silicone)	Do not add to DFT[a]	No additional DFT
	3	Sealer (silicone)	Do not add to DFT[a]	No additional DFT
	1	Inorganic zinc-rich primer	50–75 (2–3)	75 (3)
	2	Silicone	25–50 (1–2)	50 (2)
	3	Silicone	25–50 (1–2)	50 (2)
CN-5 Decks and floors – light and normal duty	1	Zinc-rich primer	50–75 (2–3)	75 (3)
	2	High-solids epoxy	125–175 (5–7)	125 (5)
	3	Antiskid epoxy[b]	125–175 (5–7)[c]	125 (5)[c]
	4	Polyurethane	50–75 (2–3)	75 (3)
	1	Epoxy primer	125–175 (5–7)	125 (5)
	2	High-solids epoxy	125–175 (5–7)	125 (5)
	3	Antiskid epoxy[b]	125–175 (5–7)[c]	125 (5)[c]
	4	Polyurethane	50–75 (2–3)	75 (3)
	1	TSA	250–375 (10–15)	250 (10)
	2	Sealer (polyurethane)	Do not add to DFT[a]	No additional DFT
	1	Antiskid high-build (HB) epoxy	Vendor specification	Vendor specification
CN-6 Decks and floors—heavy duty and helidecks	1	Zinc-rich primer	50–75 (2–3)	75 (3)
	2	High-solids epoxy	200–300 (8–12)	250 (10)
	3	Antiskid epoxy[b]	200–300 (8–12)[c]	250 (10)[c]
	4	Polyurethane safety marking	50–75 (2–3)	75 (3)
	1	Epoxy primer	125–175 (5–7)	125 (5)
	2	High-solids epoxy	200–300 (8–12)	250 (10)
	3	Antiskid epoxy[b]	200–300 (8–12)[c]	250 (10)[c]
	4	Polyurethane safety marking	50–75 (2–3)	75 (3)
	1	Prealloyed aluminum/aluminum oxide TSA[d]	300–400 (12–16)	300 (12)
	2	Sealer (polyurethane)	Do not add to DFT[a]	No additional DFT
	1	Antiskid (HB) epoxy[b]	Vendor specification	Vendor specification

[a] The sealers seal the porosity of the TSA coating and should not add DFT to the existing TSA coating. Allow thinned sealer to dry >30 minutes before application of next sealer coat.

[b] Antiskid grits should be mixed in the liquid coating prior to application to obtain good wetting of the grits. Fine grits should be used with the application of antiskid epoxy.

[c] DFT of the applied coating shall be calculated prior to the addition of antiskid grits.

[d] TSA gun parameters and gun hardware should be adjusted so that the finished TSA coating has an antiskid profile set at a desired coarseness specification. Although TSA coatings inherently contain hard, wear-resistant aluminum oxide particles that are part of the TSA matrix, prealloyed TSA wire, which is 90% aluminum/10% aluminum oxide, or its equivalent with even greater amounts of aluminum oxide, should be used.

Source: Reproduced with permission of NACE International.

NORSOK suggests an additional corrosion allowance (thicker metal) and the use of thick-film protective coatings in the splash zone for both carbon steel and martensitic (13Cr) stainless steels [59]. These guidelines are reflected in Figures 8.107 and 8.109.

Most offshore jackets are not painted below the waterline, because the cathodic protection systems produce a pH shift in the water and cause calcareous deposits (mostly $CaCO_3$ but sometimes other minerals) to be deposited on the surface. These mineral deposits and the accumulation of marine growth serve to shield the surface from and reduce the need for cathodic protection (Figure 8.110) [257, 258]. Deposit formation depends on the electrical current density and water temperature. Calcite, the predominant mineral in these deposits, is less likely to form in colder waters or at depth. That is the reason why both the NACE and DNV standards suggest relatively high current densities for new platforms, while the deposits are building, and in colder waters, where calcite minerals are more soluble and less likely to deposit.

Table 8.28 shows the NACE suggested current densities for various locations around the world. Note how the current density recommendations for deep water in the Gulf of Mexico are higher than for the warmer surface waters. As offshore production moved into deeper, and colder, water, several platforms were found to be underprotected. Cook Inlet in Alaska has the highest recommended current densities. This is because of the cold water combined with high tidal currents that keep the water oxygen saturated (and therefore corrosive) all the way to the bottom of the water column [265]. Figure 1.4 showed how turbulent the water is during tidal flows in Cook Inlet.

The DNV standard for offshore cathodic protection is newer than the NACE standard and organizes some information in a different manner. Whereas the NACE standard provides suggested current densities for uncoated submerged steel based on geographic locations (Table 8.28), the DNV standard suggests current densities based on water depth and surface water temperature (Table 8.29). It should be noted that neither standard addresses lake water (examples are North American Great Lakes offshore natural gas production and Lake Maracaibo in Venezuela) nor low-salinity seawater locations such as the Caspian Sea, where the salinity and mineral contents of the water are lower than the locations identified in these documents.

The DNV standard also lists suggestions for the following:

- Tubular standoff anodes for steel substructures and on subsea templates.
- Drag forces in surface waters may be significant.
- Bracelet anodes for pipelines.
  - Mounted on surfaces where standoff anodes would produce too much drag
- Flush-mounted anodes should have protective coating on the surface facing the protected object to avoid buildup of calcareous deposits, which can cause distortion and eventual fracture of fastening devices.

Both NACE and DNV have recommendations for enclosed flooded spaces. If the flooded spaces are airtight, minimal corrosion will occur until the dissolved oxygen is used up and then corrosion will become minimal [257].

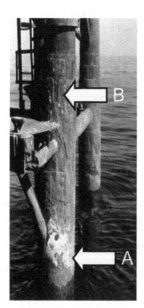
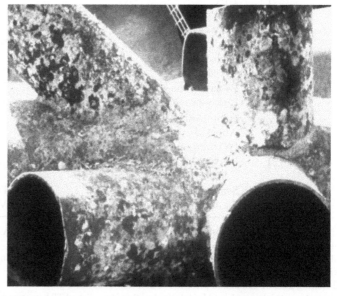

Figure 8.110 Calcareous deposits on offshore platforms. A – Arrows indicate the deposits form in the tidal region as well as on continuously immersed portions of the platform. B – Deposits on a platform node removed from service.

TABLE 8.28   NACE Design Criteria for Offshore Marine Structures [257]

Production Area	Water Resistivity[a] (ohm-m)	Water Temperature (°C)	Environmental Factors[b] Turbulence Factor (Wave Action)	Lateral Water Flow	Typical Design Current Density[c] mA m^{-2} (mA ft^{-2}) Initial[d]	Mean[e]	Final[f]
Gulf of Mexico	0.20	22	Moderate	Moderate	110 (10)	54 (5)	75 (7)
Deepwater GOM	0.29	4–12	Low	Varies	194 (18)	75 (7)	86 (8)
US West Coast	0.24	15	Moderate	Moderate	150 (14)	86 (8)	100 (9)
Cook Inlet	0.50	2	Low	High	430 (40)	380 (35)	380 (35)
Northern	0.26–0.33	0–12	High	Moderate	180 (17)	86 (8)	120 (11)
North Sea[g] Southern	0.26–0.33	0–12	High	Moderate	150 (14)	86 (8)	100 (9)
North Sea[g] Arabian Gulf	0.15	30	Moderate	Low	130 (12)	65 (6)	86 (8)
Southeast Australia	0.23–0.30	12–18	High	Moderate	130 (12)	86 (8)	86 (8)
Northwest Australia					120 (11)	65 (6)	65 (6)
Brazil	0.20	15–20	Moderate	High	180 (17)	65 (6)	86 (8)
West Africa	0.20–0.30	5–21	Low	Low	130 (12)	65 (6)	86 (8)
Indonesia	0.19	24	Moderate	Moderate	110 (10)	54 (5)	75 (7)
South China Sea	0.18	30	Low	Low	100 (9)	32 (3)	32 (3)

[a] Water resistivities are a function of both chlorinity and temperature.

[b] Typical values and ratings based on average conditions, remote from river discharge.

[c] In ordinary seawater, a current density less than the design value suffices to hold the structure at protective potential once polarization has been accomplished and calcareous coatings are built up by the design current density. CAUTION: Depolarization can result from storm action.

[d] Conditions in the North Sea can vary greatly from the northern to the southern area, from winter to summer, and during storm periods.

[e] Initial and final current densities are calculated using Ohm's Law and a resistance equation such as Dwight's or McCoy's equation with the original dimensions of the anode.

[f] Mean current densities are used to calculate the total weight of anodes required to maintain the protective current to the structure over the design life.

[g] Final current densities are calculated in a manner similar to the initial current density, except that the depleted anode dimensions are used.

Source: Reproduced with permission of NACE International.

TABLE 8.29   DNV Cathodic Protection Design Current Density Recommendations for Bare Steel Production Jackets [253]

Recommended initial and final design current densities (A m^{-2}) for seawater exposed bare metal surfaces as a function of depth and "climatic region" based on surface water temperature

Depth (m)	Tropical (>20 °C) Initial	Final	Subtropical (12–20 °C) Initial	Final	Temperate (7–11 °C) Initial	Final	Arctic (<7 °C) Initial	Final
0–30	0.150	0.100	0.170	0.110	0.200	0.130	0.250	0.170
>30–100	0.120	0.080	0.140	0.090	0.170	0.110	0.200	0.130
>100–300	0.140	0.090	0.160	0.110	0.190	0.140	0.220	0.170
>300	0.180	0.130	0.200	0.150	0.220	0.170	0.220	0.170

Recommended mean design current densities (A m^{-2}) for seawater exposed bare metal surfaces as a function of depth and "climatic region" based on surface water temperature

Depth (m)	Tropical (>20 °C)	Subtropical (12–20 °C)	Temperate (7–11 °C)	Arctic (<7 °C)
0–30	0.070	0.080	0.100	0.120
>30–100	0.060	0.070	0.080	0.100
>100–300	0.070	0.080	0.090	0.110
>300	0.090	0.100	0.110	0.110

Source: From NORSOK Standard N-004, Design of steel structures.

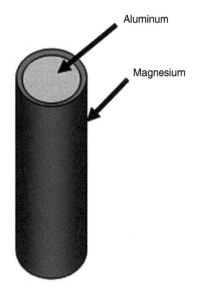

Figure 8.111   Composite anode with magnesium outer layer and aluminum core.

Most platforms use galvanic anodes to provide the necessary cathodic protection current. Aluminum anodes are most common for platforms, because they provide the best combination of predictable long life and have better current density/weight characteristics than zinc. Magnesium anodes were sometimes used in early offshore developments, but they do not last long enough and are seldom used.

Composite anodes with magnesium outer layers and aluminum cores (Figure 8.111) have been used in some applications. The idea behind these anodes is that the magnesium will corrode quickly, providing the high current densities that promote calcareous deposits. Once these deposits, and marine growth, lower, the aluminum anodes will provide sufficient current to maintain protection for the design life of the system.

The tremendous weight of galvanic anodes has caused some operators to switch to ICCP for deepwater structures. Most organizations prefer the simplicity of galvanic anode systems to the higher training levels and increased inspection requirements of impressed current systems.

Impressed current systems are also more likely to produce stray current corrosion on nearby equipment, but galvanic anode (sacrificial anode) systems can also cause stray current corrosion.

REFERENCES

1   Lyons, W.C. and Plisga, G.J. (2005). *Standard Handbook of Petroleum & Natural Gas Engineering*. Houston, TX: Gulf Professional Publishing.

2   Chillingar, G., Mourhatch, R., and Al-Qahtani, G. (2008). *The Fundamentals of Corrosion and Scaling for Petroleum and Environmental Engineers*. Houston, TX: Gulf Publishing.

3   Vaisberg, O., Vincke, O., Perrin, G. et al. (2002). Fatigue of drillstring: state of the art. *Oil & Gas Science and Technology* 57 (1): 7–37.

4   Brondel, D., Edwards, R., Hayman, A. et al. (1994). Corrosion in the oil industry. *Oilfield Review* 6 (2): 4–18.

5   Craig, B.D. (2004). *Oilfield Metallurgy and Corrosion*, 3. Denver, CO: MettCorr.

6   Papavinasam, S. (2013). *Corrosion Control in the Oil and Gas Industry*. Waltham, MA: Gulf Publishing, Elsevier.

7   Lv, S.L., Yuan, P.B., Luo, F.Q. et al. (2010). Sulfide stress cracking of a S135 drill pipe. *Materials Performance* 49 (3): 66–69.

8   Grant, R.S. and Texter, H.G. (1941). *Causes and Prevention of Drill Pipe and Tool Joint Troubles*, 9–43. New York: API Drilling and Production Practice.

9   API 6D. *Drill Pipe*. Washington, DC: API.

10   ISO 11961. *Petroleum and Natural Gas Industries – Steel Drill Pipe*. Geneva: ISO.

11   Joosten, M.W., Shute, J., and Ferguson, R.A. (1985). New study shows how to predict accumulated drill pipe fatigue. *World Oil* (October): 6–69.

12   Azar, J.J. (1979). How O_2, CO_2 and chlorides affect drill-pipe fatigue. *Petroleum Engineer International* (March): 73–78.

13   Bush, H.E. (1973). Controlling corrosion in petroleum drilling and in packer fluids. In: *Corrosion Inhibitors* (ed. C.C. Nathan), 102–113. Houston, TX: NACE International.

14   ISO/CD 10407-1. *Petroleum and Natural Gas Industries – Drilling and Production Equipment – Part 1: Drill Stem Design and Operating Limits*. Geneva: ISO.

15   API RP13B-1. *Annex E, Field Testing of Drilling Fluids-Drill Pipe Corrosion Ring Coupon*. Washington, DC: API.

16   NACE 1D177. *Monitoring Techniques and Corrosion Control for Drill Pipe, Casing, and Other Steel Components in Contact with Drilling Fluids*. Houston, TX: NACE International.

17   NACE RP0775. *Preparation, Installation, Analysis, and Interpretation of Corrosion Coupons in Oilfield Operations*. Houston, TX: NACE International.

18   ANSI/NACE MR0175/ISO 15156-1. *Petroleum and Natural Gas Industries – Materials for Use in H_2S-Containing Environments in Oil and Gas Production – Part 1: General Principles for Selection of Cracking-Resistant Materials*. Houston, TX: NACE International.

19   Jellison, M.J. (2003). Drill pipe materials defy H_2S. *Hart's E&P* 76 (8): 6–7.

20   Spoerker, H.F., Havlik, W., and Jellison, M.J. (2008). JIP investigates: what really happens to high-strength drill pipe after exposure to sour gas environment? *Drilling Contractor* (May/June): 110–115.

21   Enform IRP (2015). Section 1.8, Drill string design and metallurgy. In: *Critical Sour Drilling*, vol. Volume 1. Calgary, AB: Enform Canada.

22   API 53. *Blowout Prevention Equipment Systems for Drilling Wells*. Washington, DC: API.

23 Iannuzzi, M. (2011). Chapter 15: Environmentally-assisted cracking in oil and gas production. In: *Stress Corrosion Cracking: Theory and Practice* (ed. V. Raja and T. Shoji), 570–607. Cambridge, UK: Woodhead Publishing, Ltd.

24 Kane, R. (2006). Corrosion in petroleum production operations. In: *Metals Handbook, Volume 13C – Corrosion: Corrosion in Specific Industries*, 922–966. Materials Park, OH: ASM International.

25 API Spec 6A. *Wellhead and Christmas Tree Equipment*. Washington, DC: API.

26 Byars, H. (1999). *Corrosion Control in Petroleum Production*, 2. Houston, TX: NACE International.

27 Kolts, J., Joosten, M., and Singh, P. (2006). *An Engineering Approach to Corrosion/Erosion Prediction*, NACE 06560. Houston, TX: NACE International.

28 Martinez-Niembro, A., Palmer, J.C.W., and Shouly, W.S. (2008). Corrosion mitigation advances at Abu Dhabi Marine Operating Company: lessons learned in 45 years. *Proceedings, 12th Middle East Corrosion Conference and Exhibition*, Paper No. 08085.

29 Martin, R. (2003). Corrosion inhibitors for oil and gas production. In: *ASM Handbook, Volume 13A, Corrosion Fundamentals, Testing, and Protection*, 878–886. Materials Park, OH: ASM International.

30 Nyborg, R. (September 2009). *Guidelines for Prediction of CO$_2$ Corrosion in Oil and Gas Production Systems*. Norway, Report IFE-KR/E-2009/003: Institute for Energy Technology.

31 Oberndorfer, M., Thayer, K., and Havlik, W. (2007). *Corrosion Control in the Oil and Gas Production – 5 Successful Case Histories*, NACE 07317. Houston, TX: NACE International.

32 Place, M.C. (1979). Corrosion control – deep sour gas production. *54th Annual Fall Technical Conference and Exhibition of SPE-AIME*, Las Vegas, NV, SPE (September 1979).

33 Crolet, J.L. and Bonis, M.R. (2010). *Algorithm of the Protectiveness of Corrosion Layers 1 – Protectiveness Mechanisms and CO$_2$ Corrosion Prediction*, NACE 10363. Houston, TX: NACE International.

34 Crolet, J.L. and Bonis, M.R. (2010). *Algorithm of the Protectiveness of Corrosion Layers 2 – Protectiveness Mechanisms and H$_2$S Corrosion Prediction*, NACE 10365. Houston, TX: NACE International.

35 ANSI/NACE MR0175/ISO 15156-2. *Petroleum and Natural Gas Industries – Materials for Use in H$_2$S-Containing Environments in Oil and Gas Production – Part 2: Cracking-Resistant Carbon and Low-Alloy Steels, and the Use of Cast Irons*. Houston, TX: NACE International.

36 ANSI/NACE MR0175/ISO 15156-3. *Petroleum and Natural Gas Industries – Materials for Use in H$_2$S-Containing Environments in Oil and Gas Production – Part 3: Cracking-Resistant CRAs (Corrosion-Resistant Alloys) and Other Alloys*. Houston, TX: NACE International.

37 Campbell, J.M., Lilly, L.L., and Maddox, R.N. (2004). *Gas Conditioning and Processing, Volume 2: The Equipment Modules*, 8 (ed. R. Hubbard). Norman, OK: John M. Campbell Co.

38 Bonis, M.R. and Crolet, J.L. (1989). Basics of the prediction of the risks of CO$_2$ corrosion in oil and gas wells. *NACE Corrosion 89*, Paper No. 466. Houston, TX: NACE International.

39 Garber, J.D., Perkins, R.S., Jangama, V.R., and Alapati, R.R. (1996). *Calculation of Downhole pH and Delta pH in the Presence of CO$_2$ and Organic Acids*, NACE 96176. Houston, TX: NACE International.

40 Dougherty, J.A. (2004). *A Review of the Effect of Organic Acids on CO$_2$ Corrosion*, NACE 04376. Houston, TX: NACE International.

41 Garsany, Y., Pletcher, D., and Hedges, B. (2002). *The Role of Acetate in CO$_2$ Corrosion of Carbon Steel: Has the Chemistry Been Forgotten?*, NACE 02273. Houston, TX: NACE International.

42 Garber, J.D., Knierim, K., Patil, V.B., and Hebert, J. (2010). *Role of Acetates on Pitting Corrosion in a CO$_2$ System*, NACE 10185. Houston, TX: NACE International.

43 Hedges, B. and McVeigh, L. (1999). *The Role of Acetate in CO$_2$ Corrosion: The Double Whammy*, NACE 99021. Houston, TX: NACE International.

44 Joosten, M.W., Kolts, J., Hembree, J.W., and Achour, M. (2002). *Organic Acid Corrosion in Oil and Gas Production*, NACE 02294. Houston, TX: NACE International.

45 Burke, P.A. (1988). *CO$_2$ Corrosion Behavior of Carbon and Alloy Steels in High Concentration Brine Solutions*, NACE 88212. Houston, TX: NACE International.

46 Kane, R.D. (1996). *Corrosion in High Density Packer Fluids at High Temperatures*, NACE 96066. Houston, TX: NACE International.

47 Ke, M., Stevens, R.F., and Qu, Q. *Novel Corrosion Inhibitor for High Density ZnBr$_2$ Completion Brines at High Temperatures*, NACE 08630. Houston, TX: NACE International.

48 OLGA (2002). *Transient Multiphase Flow Model*. Scandpower Petroleum Technology AS.

49 Srinivasan, S., Lagad, V., and Kane, R.D. (2005). *Automating Evaluation and Selection of Corrosion Resistant Alloys for Oil and Gas Production*, NACE 05054. Houston, TX: NACE International.

50 ISO 21457. *Materials Selection and Corrosion Control for Oil and Gas Production Systems*. Geneva: ISO.

51 Heidersbach, K. and van Roodselaar, A. (2012). *Understanding, Preventing, and Identification of Microbial Induced Erosion-Corrosion (Channeling) in Water Injection Pipelines*, NACE 2012-1221. Houston, TX: NACE International.

52 Patton, C. and Foster, A. (2007). *Applied Water Technology*, 3. Norman, OK: John M. Campbell & Company.

53 NACE SP0590. *Prevention, Detection, and Correction of Deaerator Cracking*. Houston, TX: NACE International.

54 Hedges, B., Chen, H.J., Bieri, T.H., and Sprague, K. (2006). *A Review of Monitoring and Inspection Techniques for CO$_2$ and H$_2$S Corrosion in Oil and Gas Production Facilities: Location, Location, Location*, NACE 06120. Houston, TX: NACE International.

55 Ciaraldi, S., Ghazel, H., Abou Shadey, T., and El-Leil, H. (1999). *Progress in Combating Microbiologically Induced*

Corrosion in Oil Production, NACE 99181. Houston, TX: NACE International.

56 Coulter, A., Hendrickson, A., and Martinez, S. (1987). Chapter 54: Acidizing. In: *Petroleum Engineering Handbook*. Richardson, TX: Society of Petroleum Engineers.

57 Singh, A. and Quraishi, M. (2015). Acidizing corrosion inhibitors: a review. *Journal of Materials and Environmental Science* 6 (1): 224–235.

58 M. Yadav, S. Kumar, and P. Yadav, Corrosion inhibition of tubing steel during acidization of oil and gas wells, *Journal of Petroleum Engineering*, 2013 (2013) Article ID 354630, 9. https://www.hindawi.com/journals/jpe/2013/354630/ (accessed 19 March 2017).

59 NORSOK Standard M-001. *Materials Selection*. Lysaker, Norway: Standards Norway.

60 Singh, T.B., Dey, A.K., Gaur, B., and Singh, D.D.N. (1995). Are corrosion, hydrogen absorption and mechanical strength of steel exposed to inhibited acid solution inter-related? *Anti-Corrosion Methods and Materials* 42 (6): 19–22.

61 Riggs, O.L. and Hurd, R.M. (1968). Effect of inhibitors on scale removal in HCl pickling solutions. *Corrosion* 24 (2): 45–49.

62 El-Meligi, A.A., Turgoose, S., Ismail, A.A., and Sanad, S.H. (2000). Technical note effect of corrosion inhibitors on scale removal during pickling of mild steel. *British Corrosion Journal* 35 (1): 75–77.

63 Hill, D. and Romijn, H. (2000). *Reduction of Risk to the Marine Environment from Oilfield Chemicals – Environmentally Improved Acid Corrosion Inhibition for Well Stimulation*, NACE 2000-00342. Houston, TX: NACE International.

64 API 5CT. *Steel Pipes for Use as Casing or Tubing for Wells*. Washington, DC: API.

65 Papavinasam, S. (2013). *Corrosion Control in the Oil and Gas Industry*. New York: Gulf Professional Publishing – Elsevier.

66 DNV Report 2006-3496 (2006). *Material Risk – Ageing Offshore Installations API 5L, Line Pipe*. Oslo, Norway: DNVGL.

67 Holtsbaum, W.B. (2006). Well casing external corrosion and cathodic protection. In: *Metals Handbook, Volume 13C – Corrosion: Corrosion in Specific Industries*, 97–106. Materials Park, OH: ASM International.

68 Prinz, W. and Leutner, B. (1997). Cathodic protection of well casings. In: *Handbook of Cathodic Protection* (ed. W. Baeckmann, W. Schwenk and W. Prinz), 414–426. Houston, TX: Gulf Publishing.

69 Klechka, E.W., Koch, G.H., Kowalski, A.R., and Al-Mithin, A.W. (2006). *Cathodic Protection of Well Casings Using E-log I Criteria*, NACE 06071. Houston, TX: NACE International.

70 NACE SP0186. *Application of Cathodic Protection for External Surfaces of Steel Well Casings*. Houston, TX: NACE International.

71 Hausler, R. (1985). Corrosion inhibition in the presence of corrosion product layers. *Proceedings of the Sixth European Conference on Corrosion Inhibitors*, Ferrara, pp. 41–65.

72 Simon Thomas, M.J.J. (2000). *Corrosion Inhibitor Selection – Feedback from the Field*, NACE 00056. Houston, TX: NACE International.

73 Fang, C.S. and Speed, C.F. (1990). *Corrosion Film in Batch Treatment of Gas Wells*, NACE 90038. Houston, TX: NACE International.

74 Poetker, R. and Stone, J. (1956). Squeeze inhibitor into formation. *Petroleum Engineering* 58 (5): B29–B34.

75 Poetker, R. (1960). Case histories show value of corrosion inhibitor squeeze treatment. *Journal of Petroleum Technology* 12 (6): SPE-1129-G.

76 Samad, S., Al Sawadi, O., Odeh, N., and Afzal, M. (2010). *Robust Implementation of Corrosion Inhibition Squeeze (CIS) Program Extends the Well Completion Lifetime (Case Study)*, SPE-137622-MS. Society of Petroleum Engineers.

77 Alzahrani, M., Senusi, K., Lewis, D. et al. (2014). Effective wells corrosion mitigation in two major Middle East fields: a case study. *IPRC 2014: International Petroleum Technology Conference* (19 January 2014).

78 Graham, G., Bowering, D., MacKinnon, K. et al. (2014). *Corrosion Inhibitors Squeeze Treatments – Misconceptions, Concepts and Potential Benefits*, SPE-1696040MS. Society of Petroleum Engineers.

79 NACE Report 1G286. (2002). *Oilfield Corrosion Inhibitors and Their Effects on Elastomeric Seals*. Houston, TX: NACE International.

80 Eckert, R. and Amend, B. (2017). MIC and materials selection. In: *Microbiologically Influenced Corrosion in the Upstream Oil and Gas, Industry* (ed. T. Skovhus, D. Enning and J. Lee), 36–56. Boca Raton, FL: CRC Press.

81 Nelson, J. and Davis, R. (2000). *Internal Tubular Coatings Used to Maximize Hydraulic Efficiency*, NACE 00173. Houston, TX: NACE International.

82 Garverick, L. (1994). *Corrosion in the Petrochemical Industry*, 268. Materials Park, OH: ASM International.

83 NACE SP0181. *Liquid-Applied Internal Protective Coatings for Oilfield Production Equipment*. Houston, TX: NACE International.

84 NACE SP0188. *Discontinuity (Holiday) Testing of New Protective Coatings on Conductive Substrates*. Houston, TX: NACE International.

85 NACE SP0191. *Application of Internal Plastic Coatings for Oilfield Tubular Goods and Accessories*. Houston, TX: NACE International.

86 NACE RP 0291. *Care, Handling, and Installation of Internally Plastic-Coated Oilfield Tubular Goods and Accessories*. Houston, TX: NACE International.

87 NACE TM0185. *Evaluation of Internal Plastic Coatings for Corrosion Control of Tubular Goods by Autoclave Testing*. Houston, TX: NACE International.

88 NACE TM187. *Evaluating Elastomeric Materials in Sour Gas Environments*. Houston, TX: NACE International.

89 NACE TM 0192. *Evaluating Elastomeric Materials in Carbon Dioxide Decompression Environments*. Houston, TX: NACE International.

90 Wood, R.J.K., Symonds, N., and Mellor, B. *Wireline Wear Resistance of Polymeric Corrosion Barrier Coatings for Downhole Applications*, NACE 00170. Houston, TX: NACE International.

91 Afghoul, A.C., Amaravadi, S., Boumali, A. et al. (2004). Coiled tubing: the next generation. *Oilfield Review* Spring: 38–57.

92 Van Amam, W.D., McCoy, T., Cassidy, J., and Rosine, R. (2000). The effect of corrosion in coiled tubing and its prevention. *SPE/ICoTA Coiled Tubing Roundtable* (5–6 April 2000), Houston, TX.

93 Nasar-El-Din, H.A. and Metcalf, A.S. (2008). Workovers in sour environments: how do we avoid coiled tubing (CT) failures? *SPE Production & Operations* 23 (2): 112–118.

94 Crabtree, A.R. and Gavin, W. (2005). Coiled tubing in sour environments: theory and practice. *SPE Drilling & Completion* 20 (1): 71–80.

95 API Spec 11B. *Sucker Rods*. Washington, DC: API.

96 ANSI/API RP 11BR. *Care and Handling of Sucker Rods*. Washington, DC: API.

97 NACE – TM0275. *Performance Testing of Steel and Reinforced Plastic Sucker Rods by the Mixed-String, Alternate-Rod Method*. Houston, TX: NACE International.

98 API RP 11L. *Design Calculations for Sucker Rod Pumping Systems*. Washington, DC: API.

99 API Bulletin 11L3 (1970). *Sucker Rod Pumping System Design Book*. Washington, DC: API.

100 API 11C. *Reinforced Plastic Sucker Rods*. Washington, DC: API.

101 Hara, T., Asahi, H., and Kaneta, H. (1996). *Galvanic Corrosion in Oil and Gas Environments*, NACE 96063. Houston, TX: NACE International.

102 ISO 6072. *Rubber – Compatibility Between Hydraulic Fluids and Standard Elastomeric Materials*. Geneva: ISO.

103 ISO 16010. *Elastomeric Seals – Material Requirements for Seals Used in Pipes and Fittings Carrying Gaseous Fuels and Hydrocarbon Fluids*. Geneva: ISO.

104 API Spec 17D. *Subsea Wellhead and Christmas Tree Equipment*. Washington, DC: API.

105 Maligas, M.N. and Skogsberg, L.A. (2001). *Material Selection for Deep Water Wellhead Applications*, NACE 01001. Houston, TX: NACE International.

106 API RP 14E. *Offshore Production Platform Piping Systems*. Washington, DC: API.

107 Salama, M. and Venkatesh, E. (1983). Evaluation of API RP 14EW erosional velocity limitations for offshore gas wells, OTC-4485-MS. *Proceedings, Offshore Technology Conference*.

108 ISO 10423. *Drilling and Production Equipment – Wellhead and Christmas Tree Equipment*. Geneva: ISO.

109 Mack, R. and Norton, S. (2001). *Brittle Fracture in an Upper Tree Connector System at Mensa*, NACE 01015. Houston, TX: NACE International.

110 NACE Report 1F192. *Use of Corrosion-Resistant Alloys in Oilfield Environments*. Houston, TX: NACE International.

111 Bradford, S.A. (2001). *Corrosion Control*, 2. Edmonton, Alberta: CASTI Publishing Inc.

112 API RP14G. *Fire Prevention and Control*. Washington, DC: API.

113 Skovhus, T., Enning, D., and Lee, J. (eds.) (2017). *Microbiologically Influenced Corrosion in the Upstream Oil and Gas, Industry*. Boca Raton, FL: CRC Press.

114 Javaherdashti, R. (2017). *Microbiologically Influenced Corrosion: An Engineering Insight*, 2. London: Springer-Verlag.

115 Penkala, J., Fichter, J., and Ramachandran, S. (2010). *Protection Against Microbiologically Influenced Corrosion by Effective Treatment and Monitoring During Hydrotest Shut-In*, NACE 2010-10404. Houston, TX: NACE International.

116 NACE SP0499. *Corrosion Control and Monitoring in Seawater Injection Systems*. Houston, TX: NACE International.

117 API 653. *Tank Inspection, Repair, Alteration, and Reconstruction*. Washington, DC: API.

118 API 650. *Welded Tanks for Oil Storage*. Washington, DC: API.

119 ISO 16812. *Shell-and-Tube Heat Exchangers*. Geneva: ISO.

120 Crum, J.R., Hazeldine, P., Shoemaker, L.E., and Peschel, J. (2002). *Evaluation of Materials for Seawater Plate Heat Exchanger Applications*, NACE 02748. Houston, TX: NACE International.

121 Cassagne, T., Houlle, P., Zuili, D. et al. (2010). *Replacing Titanium in Sea Water Plate Heat Exchangers*, NACE 10391. Houston, TX: NACE International.

122 Southall, D., Le Pierres, R., and Dewson, S.J. (2008). Design considerations for compact heat exchangers. *Proceedings of ICAPP-08*, Anaheim, CA, USA (9–12 June 2008), Paper 9009.

123 Revie, R.W. and Uhlig, H.H. (2008). *Corrosion and Corrosion Control*, 326–328. Hoboken, NJ: Wiley-Interscience.

124 Fischer, K.P. (2003). *A Review of Offshore Experiences with Bolts and Fasteners*, NACE 03017. Houston, TX: NACE International.

125 Craig, B.D. (2015). On the contradiction of applying rolled threads to bolting exposed to hydrogen-bearing environments. *Oil and Gas Facilities* 4: 66–71.

126 Jones, R. and Buehler, W. (2010). *Examination of Three Failed Inconel 718 Studs*, Report No. 0091-19492R (1 March). Houston, TX: Stork Testing & Metallurgical Consulting, Inc.

127 D. Bush, NACE MR0175 and the valve industry, *Valve Magazine*, Spring 1998, No. 2, pages 2–8, http://www.documentation.emersonprocess.com/groups/public/documents/articles_articlesreprints/d350807x012.pdf (accessed 20 April 2010).

128 API 6ACRA (2015). *Age-Hardened Nickel-Based Alloys for Oil and Gas Drilling and Production Equipment*. Washington, DC: API.

129 API 20F (June 2015). *Corrosion Resistant Bolting for Use in the Petroleum and Natural Gas Industries*. Washington, DC: API.

130 API 20E (February 2017). *Alloy and Carbon Steel Bolting for Use in the Petroleum and Natural Gas Industries*. Washington, DC: API.

131 Elliott, P. (2003). Materials selection for corrosion control. In: *ASM Handbook, Volume 13A, Corrosion: Fundamentals, Testing, and Protection* (ed. S.D. Cramer and B.S. Covino), 909–928. Materials Park, OH: ASM International.

132 ASTM B850. *Post-Coating Treatments of Steel for Reducing the Risk of Hydrogen Embrittlement*. West Conshohocken, PA: ASTM International.

133 BS ISO 9588. *Metallic and Other Inorganic Coatings. Post-coating Treatments of Iron or Steel to Reduce the Risk of Hydrogen Embrittlement*. Geneva: ISO.

134 ASTM B633. *Electrodeposited Coatings of Zinc on Iron and Steel*. West Conshohocken, PA: ASTM International.

135 ASTM F1136. *Zinc/Aluminum Corrosion Protective Coatings for Fasteners*. West Conshohocken, PA: ASTM International.

136 ASTM F2339. *Zinc Coating, Hot-Dip, Requirements for Application to Carbon and Alloy Steel Bolts, Screws, Washers, Nuts, and Special Threaded Fasteners*. West Conshohocken, PA: ASTM International.

137 Townsend, H. (1975). Effects of zinc coatings on the stress corrosion cracking and hydrogen embrittlement of low-alloy steel. *Metallurgical Transactions A* 6A: 877–883.

138 Raymond, L. (1998). Chapter 39: The susceptibility of fasteners to hydrogen embrittlement and stress corrosion cracking. In: *Handbook of Bolts and Bolted Joints* (ed. J.H. Bickford and S. Nassar), 723–756. New York: M. Dekker.

139 ASTM A193/A193M. *Alloy-Steel and Stainless Steel Bolting Materials for High Temperature or High Pressure Service and Other Special Purpose Applications*. West Conshohocken, PA: ASTM International.

140 *Lower Marine Riser Package Connector Failure*, Safety Alert No. 303 (29 January 2013). US Department of the Interior, Bureau of Safety and Environmental Enforcement

141 *QC-FIT Evaluation of Connector and Bolt Failures Summary of Findings*, QC-FIT Report #2014-01 (August 2014). US Department of the Interior, Bureau of Safety and Environmental Enforcement. https://www.bsee.gov/sites/bsee_prod.opengov.ibmcloud.com/files/reports/drilling/bolt-report-final-8-4-14.pdf (accessed 19 April 2018).

142 Chung, Y. and Fulton, L.K. (2017). Environmental hydrogen embrittlement of G41400 and G43400 steel bolting in atmospheric versus immersion services. *Journal of Failure Analysis and Prevention* 17: 330–339.

143 Greenslade, J. (2015). ASTM F16 fastener committee info. Presented at the 2015 BSEE Domestic and International Standards Workshop. https://www.bsee.gov/sites/bsee.gov/files/technical-presentations/standards/greenslade-bsee-astm-f16-may-8-presentation-1.pdf (accessed 19 April 2018).

144 Chaudhury, G. (2017). Pragmatic new solution improves reliability of bolts used subsea. *Proceeding of the Offshore Technology Conference* (May 2017), OTC 27591-MS.

145 Zheng, H. (2016). Subsea bolts performance and critical drill-through equipment fastener study. https://www.bsee.gov/sites/bsee.gov/files/bsee_bolt_forum_lbnl_presentation_8292016_0.pdf (accessed 19 April 2018).

146 API 6A718 (2006). *Nickel Base Alloy 718 (UNS N07718) for Oil and Gas Drilling and Production Equipment*. Washington, DC: API.

147 Efird, K.D. (1985). Failure of Monel Cu-Ni-Al alloy K-500 bolts in seawater. *Materials Performance* 23 (4): 37–40.

148 Hendrix, D.E. (1997). Hydrogen embrittlement of high strength steel fasteners in atmospheric service. *Materials Performance* 36 (12): 54–56.

149 Wolfe, L.H., Burnette, C.C., and Joosten, M.W. (1993). *Hydrogen Embrittlement of Cathodically Protected Subsea Bolting Alloys*, NACE 93288. Houston, TX: NACE International.

150 Wolfe, L.H. and Joosten, M.W. (1988). Failures of nickel/copper bolts in subsea applications. *SPE Production Engineering* 3 (3): 382–386.

151 Pfeifer, A.R. (2006). *Fastener Coating Trends*, NACE 00619. Houston, TX: NACE International.

152 API RP 17A/ISO 13628. *Design and Operation of Subsea Production Systems*. Washington, DC: API.

153 NACE Publication 02107 (2002). *Coatings for Protection of Threaded Fasteners Used with Structural Steel, Piping, and Equipment*. Houston, TX: NACE International.

154 McKinnon, D. (2018). Galvanizing high strength bolts. http://www.portlandbolt.com/faqs/galvanizing-high-strength-bolts (accessed 19 April 2018).

155 ASTM A563. *Carbon Steel and Alloy Steel Nuts*. West Conshohocken, PA: ASTM International.

156 Mills, B.M. (2006). *Fluoropolymer Fastener Coating Systems*, NACE 00618. Houston, TX: NACE International.

157 Bauman, J. (2004). *Materials Selection and Corrosion Issues on an Offshore Producing Platform*, NACE 04126. Houston, TX: NACE International.

158 Britton, J. (1998). *New Paint Preservation Technologies for Offshore & Marine Equipment*. Houston, TX: Deepwater Corrosion. http://stoprust.com/technical-papers/30-marine-paint-preservation (accessed 19 April 2018).

159 Davison, T., Joosten, M., and Matheson, I. (2011). *Bolt Corrosion Prevented by Corrosion-Inhibiting Spray-On Thermoplastic*, NACE C2011-11017. Houston, TX: NACE International.

160 Stevens, K. (2000). *Corrosion Protection for Nuts and Bolts*, NACE C2000-00626. Houston, TX: NACE International.

161 Lomasney, C., Zaharias, C., Lomasney, S., and Joosten, M. (2017). *Zinc–Nickel Nanolaminate – Advanced Coating for Bolt Corrosion Control*, NACE C2017-9220. Houston, TX: NACE International.

162 API 537. *Flare Details for General Refinery and Petrochemical Service*. Washington, DC: API.

163 Kodssi, I. (2017). An overview of flare systems for the oil and gas industry. *Inspectioneering* (March/April). https://inspectioneering.com/journal/2017-04-27/6417/an-overview-of-flare-systems-for-the-oil-and-gas-industry (accessed 10 May 2017).

164 IPS-E-TP-740 (March 2010). *Corrosion Consideration in Material Selection, Iranian Petroleum Standards, First Revision.* http://igs.nigc.ir/STANDS/IPS/e-tp-740.PDF (accessed 14 March 2017).

165 NACE SP0198. *Control of Corrosion Under Thermal Insulation and Fireproofing Materials – A Systems Approach.* Houston, TX: NACE International.

166 API RP583 (2014). *Corrosion Under Insulation and Fireproofing.* Washington, DC: API.

167 Winnik, S. (2016). *Corrosion Under Insulation (CUI) Guidelines*, 2, European Federation of Corrosion Publications Number 55e. Cambridge: Woodhead Publishing-Elsevier.

168 CINI Manual. *Insulation for Industries*, updates issued annually. Spijkenisse, The Netherlands: CINI Foundation (Committee Industrial Insulation).

169 API RP 571. *Damage Mechanisms Affecting Fixed Equipment in the Refining Industry.* Washington, DC: API.

170 Richardson, J. and Fitzsimmons, T. (1985). Use of aluminum foil for preventing stress corrosion cracking of austenitic stainless steel under thermal insulation. In: *ASTM STP880, Corrosion of Metals Under Insulation* (ed. W. Pollock and J. Barnhart). West Conshohocken, PA: ASTM International.

171 Sentjens, J. (2011). CUI and coatings. *Minutes of EFC WP15, Corrosion in the Refining Industry* (7 September 2011). http://efcweb.org/efcweb_media/-p-644.pdf?rewrite_engine=id (accessed 12 May 2017).

172 Soman, A. (2015). A systematic inspection plan for corrosion under insulation (CUI) in process plants. https://www.corrosionpedia.com/2/4763/corrosion-under-insulation-cui/cui-detection-techniques-for-process-pipelines-part-1 (accessed 19 April 2018).

173 Soman, A. (2016). CUI detection techniques for process pipelines (Part 2). https://www.corrosionpedia.com/2/5364/corrosion-under-insulation-cui/cui-detection-techniques-for-process-pipelines-part-2 (accessed 12 May 2017).

174 Sentjens, J. (2018). Corrosion under insulation: the challenge and need for insulation. https://www.corrosionpedia.com/2/1372/corrosion/corrosion-under-insulation-the-challenge-and-need-for-insulation (accessed 19 April 2018).

175 NACE Publication 10A392 (2006). *Effectiveness of Cathodic Protection on Thermally Insulated Underground Metallic Structures.* Houston, TX: NACE International.

176 Revie, R.W. (ed.) (2015). *Oil and Gas Pipelines: Integrity and Safety Handbook.* Hoboken, NJ: Wiley.

177 Bruno, T.V. (1997). *Causes and Prevention of Pipeline Failures.* Houston, TX: Metallurgical Consultants, Inc.

178 Jacobsen, G.A. (2007). Corrosion at Prudhoe Bay – a lesson on the line. *Materials Performance* 46 (8): 27–32, 34.

179 USATODAY.com (2006). Oil giant BP already under scrutiny for allowing Alaska pipeline to crumble. http://www.usatoday.com/news/washington/2006-08-09-oil-field_x.htm (accessed 17 May 2017).

180 Wikipedia (2006). Prudhoe Bay oil spill. http://en.wikipedia.org/wiki/2006_Alaskan_oil_spill (accessed 17 May 2017).

181 National Transportation Safety Board (2011). *Pacific Gas and Electric Company Natural Gas Transmission Pipeline Rupture and Fire, San Bruno, California, 9 September 2010*, NTSB/PAR-11/-1, PB2011-916501. Washington, DC: National Transportation Safety Board.

182 Clark, E., Leis, B., and Eiber, R. (2004). *Integrity Characteristics of Vintage Pipelines.* Columbus, OH: Battelle Memorial Institute.

183 Pollard, L., Belanger, A., and Clarke, T. (2004). *Managing HIC Affected Pipelines Utilizing MFL Hard Spot Technology*, NACE C2004-04172. Houston, TX: NACE International.

184 Belanger, A. and Barker, T. (2014). Multiple data inspection of hard spots and cracking. *10th International Pipeline Conference, Volume 2, Pipeline Integrity Management*, Calgary, Alberta, Canada, Paper No. IPC2014-33060. New York: ASME.

185 Transportation Safety Board of Canada (2015). *Natural Gas Pipeline Rupture, TransCanada PipeLine Limited, Line 100-1, 762-Millmetre-Diamter Pipeline, Main Line Valve 111A-1, from Kilometres 11.12 to 11.16, Near Marten River, Ontario, 26 September 2009*, Report P09H0083. Gatineau, Canada: Transportation Safety Board of Canada.

186 Nurshayeva, R. (2014). Update 1 – new pipelines to cost Kashagan oil project up to $3.6 bn. *Reuters* (10 October). http://www.reuters.com/article/oil-kashagan-idUSL6N0S52P420141010 (accessed 17 May 2017).

187 Palmer, A.C. and King, R.A. (2008). *Subsea Pipeline Engineering*, 2. Tulsa, OK: Pennwell.

188 Alyeska Pipeline Service Company (2011). Low Flow Impact Study. Final Report (15 June 2011).

189 Chevrot, T.C., Bonis, M., and Stroe, M. (2009). *Corrosion Assessment Methodology for an Existing Pipeline Network*, NACE 09113. Houston, TX: NACE International.

190 W. Zheng, R. Surtherby, R. W. Revie, W. R. Tyson, and G. Shen. (2002). Stress corrosion cracking of linepipe steels in near-neutral pH environment: issues related to the effects of stress. *Corrosioneering, The Corrosion Journal for the Online Community.* http://www.corrosionsource.com/corrosioneering/journal/Jul02_Revie/Jul02_Revie_1.htm, (accessed 24 April 2010).

191 Nyborg, R. and Dugstad, A. (2007). *Top of Line Corrosion and Water Condensation Rates in Wet Gas Pipelines*, NACE C2007-07555. Houston, TX: NACE International.

192 Smart, J. (2007). Determining the velocity required to keep solids moving in pipelines. Presented at the 19th Pipeline Pigging and Integrity Management Conference, Houston, TX (February 2007). Published in *The Journal of Pipeline Engineering* 6 (1) (1st Qtr, 2007).

193 Cosham, A., Hopkins, P., and Macdonald, K.A. (2007). Best practice for the assessment of defects in

pipelines – corrosion. *Engineering Failure Analysis* 14 (7): 1254–1265.

194 *Manual for Determining the Remaining Strength of Corroded Pipelines: A Supplement to ASME B31G Code for Pressure Piping*, ASME B31G. New York: ASME.

195 Kiefner, J.F. and Vieth, P.H. (1990). A modified criterion for evaluating the remaining strength of corroded pipe. *Proceedings of the American Gas Association, Distribution/Transmission Conference*, Los Angeles (May 1990).

196 Kiefner, J.F. and Maxey, W.A. (2000). Hydrostatic testing – conclusion – model helps prevent failures. *Oil and Gas Journal* 98 (32): 54–58.

197 Vieth, P.H. and Kiefner, J.F. (1993). *RSTRENG User's Manual*. American Gas Association, Pipeline Research Committee, Catalog No. L51688 (31 March).

198 DNV RP-F-101. *Corroded Pipelines*. Oslo, Norway: DNVGL.

199 Chauhan, V., Swankie, T.D., Espiner, R., and Wood, I. (2009). *Developments in Methods for Assessing the Remaining Strength of Corroded Pipelines*, NACE 09115. Houston, TX: NACE International.

200 Kiefner, J.F. (2005). Welding criteria permit safe and effective pipeline repair. In: *Pipeline Rules of Thumb Handbook*, 6 (ed. E.W. McCallister), 70–76. Gulf Professional Publishing.

201 Mateer, M.W. and Chang, J. (2005). *Performance of Pipeline Composite Sleeves*, NACE 05139. Houston, TX: NACE International.

202 NACE SP0200. *Steel-Cased Pipeline Practices*. Houston, TX: NACE International.

203 Wilms, M.E., Huizinga, S., de Jong, J.G. et al. (2006). *Susceptibility of Weldable Martensitic Stainless Steel (13Cr) Pipelines to Internal Cracking*, NACE 06493. Houston, TX: NACE International.

204 van Gestel, W. (2004). *Girth Weld Failures in 13Cr Sweet Wet Gas Flowlines*, NACE 04114. Houston, TX: NACE International.

205 Barnett, T. (2004). Minimizing hydrogen cracking critical for 13Cr pipeline welds. *Offshore* 64 (2): 4. 67–68.

206 Enerhaug, J., Olsen, S., Rorvik, G. et al. (2005). *Robustness of Supermartensitic Stainless Steel Girth Welds – Experiences from the Kristin Field Development Project*, NACE C2005-05097. Houston, TX: NACE International.

207 Hesjekik, S., Olsen, S., and Rorvik, G. (2004). *Hydrogen Embrittlement from Cathodic Protection on Supermartensitic Stainless Steels – Case History*, NACE C2004-0456. Houston, TX: NACE International.

208 Olsen, S. and Hesjevik, S. (2004). *Hydrogen Embrittlement from CP on Supermartensitic Stainless Steels – Case History*, NACE C2004-0456. Houston, TX: NACE International.

209 Olsen, S. (2008). ISO 21457 materials selection and corrosion control for oil and gas production systems. Presentation to WP13 (10 September).

210 Crump, H.M. and Papenfuss, S. (1991). Use of magnetic flux leakage inspection pigs for hard spot detection and repair. *Materials Performance* 30 (6): 26–28.

211 Graf, M.K., Hillenbrand, H.-G., Heckmann, C.J., and Niederhoff, K.A. (2004). High-strength large-diameter pipe for long-distance high-pressure gas pipelines. *International Journal of Offshore and Polar Engineering* 14 (1): 68.

212 Hillenbrand, H.G., Heckman, C.J., and Niederhoff, K.A. (2002). X80 line pipe for large-diameter high strength pipelines. *APIA 2002 Annual Conference, X80 Pipeline Workshop*, Hobart, Australia (October 2002).

213 Xi, J., Yang, L., Worthingham, B., and King, F. (2009). *Hydrogen Effects on High Strength Pipeline Steels*, NACE 09120. Houston, TX: NACE International.

214 Darwin, A., Annadorai, K., and Heidersbach, K. (2010). *Prevention of Corrosion in Carbon Steel Pipelines Containing Hydrotest Water – An Overview*, NACE 10401. Houston, TX: NACE International.

215 Toncre, A.C. (1981). The relationship of coatings and cathodic protection for underground corrosion control. In: *Underground Corrosion, ASTM STP 741* (ed. E. Escalante), 166–181. West Conshohocken, PA: ASTM International.

216 Bianchetti, R. (ed.) (2018). *Peabody's Control of Pipeline Corrosion*, 3. Houston, TX: NACE International.

217 NACE SP0169. *Control of External Corrosion on Underground or Submerged Metallic Piping Systems*. Houston, TX: NACE International.

218 ISO 15589-1. *Cathodic Protection of Pipeline Systems – Part 1: On-land Pipelines*. Geneva: ISO.

219 ISO 15589-2. *Cathodic Protection of Pipeline Transportation Systems – Part 2: Offshore Pipelines*. Geneva: ISO.

220 ASTM G8. *Cathodic Disbonding of Pipeline Coatings*. West Conshohocken, PA: ASTM International.

221 ASTM G42. *Cathodic Disbonding of Pipeline Coatings Exposed to Elevated Temperatures*. West Conshohocken, PA: ASTM International.

222 Norman, D. and Argent, C. (2005). *Pipeline Coatings, External Corrosion and Direct Assessment*, NACE 05034. Houston, TX: NACE International.

223 Norsworthy, R. (2011). Chapter 68: Selection and use of coatings for underground or submersion service. In: *Uhlig's Corrosion Handbook*, 3 (ed. R.W. Revie). Hoboken, NJ: Wiley.

224 Norsworthy, R. (2009). *Cases of External Corrosion on Coated and Cathodically Protected Pipelines*, NACE 09116. Houston, TX: NACE International.

225 NACE SP0286. *Electrical Isolation of Cathodically Protected Pipelines*. Houston, TX: NACE International.

226 Canto-Ibanez, J., Martínez de la Escalera, L.M., Cervantes, J. et al. (2013). *Cathodic Protection in High Consequence Areas: Challenges and Solutions in Eastern Ecuador Oil Production Fields*, NACE C2013-2408. Houston, TX: NACE International.

227 Wicks, M. and Fraser, J.P. (1975). Entrainment of water by flowing oil. *Materials Performance* 14 (5): 9–12.

228 Lagad, V.V., Srinivasan, S., and Kane, R.D. (2008). *Facilitating Internal Corrosion Direct Assessment Using Advanced Flow and Corrosion Prediction Models*, NACE 08131. Houston, TX: NACE International.

229 Dragomelo, N. and Soltani, S. (2010) Quality assurance in corrosion prediction of multiphase lines. *Pipeline and Gas Journal* 237(3). https://pgjonline.com/magazine/2010/march-2010-vol-237-no-3/features/quality-assurance-in-corrosion-prediction-of-multiphase-lines (accessed April 2018).

230 Garber, J., Farshad, F., and Tadepally, V. (2005). A comprehensive model for predicting internal corrosion rates in flowlines and pipelines. *1st International Symposium on Oilfield Corrosion*, Aberdeen, UK (28 May 2004), SPE 87556. Richardson, TX: Society of Petroleum Engineers.

231 Pickthall, T., Carlile, A., and Duncan, T. (2005). *Predictive Internal Corrosion Monitoring on a Crude Oil Pipeline: A Case Study*, NACE 05163. Houston, TX: NACE International.

232 Olsen, S., Lunde, O., and Dugstad, A. (1999). *pH Stabilization in the Troll Gas Condensate Pipelines*, NACE-99019. Houston, TX: NACE International.

233 Moshfeghian, M. (2006). *The Impact of Insulation on Pipeline Performance, April 2006, Gas Processing Tip of the Month*. Norman, OK: JM Campbell & Co. http://www.jmcampbell.com/tip-of-the-month/2006/04/the-impact-of-insulation-on-pipeline-performance (accessed 18 May 2017).

234 National Transportation Safety Board (2003). *Pipeline Accident Report, Natural Gas Pipeline Rupture and Fire Near Carlsbad, New Mexico, 19 August 2000*. NTSB/PAR-03/01 (11 February 2003). Washington, DC: National Transportation Safety Board.

235 NACE SP0116. *Multiphase Flow Internal Corrosion Direct Assessment (MP-ICDA) Methodology for Pipelines*. Houston, TX: NACE International.

236 Scott, G.N. (1933). Exhibit "A", report of A.P.I. Research Associate to Committee of Corrosion of Pipe Lines, Part I – adjustment of soil-corrosion pit-depth measurements for size of sample, and Part II – a preliminary study of the rate of pitting of iron pipe in soils. *American Petroleum Industry Bulletin* 204–220.

237 NACE SP0502. *Pipeline External Corrosion Direct Assessment Methodology*. Houston, TX: NACE International.

238 NACE SP0206. *Internal Corrosion Direct Assessment Methodology for Pipelines Carrying Normally Dry Natural Gas (DG-ICDA)*. Houston, TX: NACE International.

239 ANSI/NACE SP0208. *Internal Corrosion Direct Assessment Methodology for Liquid Petroleum Pipelines*. Houston, TX: NACE International.

240 NACE SP0110. *Wet Gas Internal Corrosion Direct Assessment Methodology for Pipelines*. Houston, TX: NACE International.

241 NACE/PODS SP0507. *External Corrosion Direct Assessment (ECDA) Integrity Data Exchange (IDX)*. Houston, TX: NACE International.

242 NACE SP0207. *Performing Close-Interval Potential Surveys on Buried of Submerged Pipelines*. Houston, TX: NACE International.

243 NACE TM0109. *Above Ground Survey Techniques for the Evaluation of Underground Coating Condition*. Houston, TX: NACE International.

244 Smart, J.S. and Heidersbach, R. (1987). (Marine) organic coatings. In: *Metals Handbook, Volume 13, Corrosion*, 912–919. Materials Park, OH: ASM International.

245 Heidersbach, R., Brandt, J., Johnson, D., and Smart, J. (2006). Marine cathodic protection. In: *Metals Handbook, Volume 13C – Corrosion: Environments and Industries*, 73–78. Materials Park, OH: ASM International.

246 Almar-Naess, A. et al. (1984). Investigation of the Alexander L. Kielland failure – metallurgical and fracture analysis. *Journal of Energy Resources Technology* 106 (1): 24–32.

247 Moan, T. (2010). *The Alexander L. Kielland Accident – 30 Years Later*. http://www.psa.no/getfile.php/1312419/PDF/Konstruksjonsseminar%20aug2010/Alexander%20L.%20Kielland%20ulykken%20%E2%80%93%2030%20%C3%A5r%20etter%20-%20%20Torgeir%20Moan%20%28NTNU%29.pdf (accessed 20 April 2018).

248 Moan, T. (1985). The progressive structural failure of the Alexander L. Kielland platform. In: *Case Histories in Offshore Engineering* (ed. G. Maeir), 1–42. Berlin: Springer-Verlag.

249 API RP2A-WSD. *Planning, Designing, and Constructing Fixed Offshore Platforms – Working Stress Design*. Washington, DC: API.

250 API 2A-LRFD. *Planning, Designing, and Constructing Fixed Offshore Platforms – Load and Resistance Factor Design*. Washington, DC: API.

251 ISO 19902. *Fixed Steel Offshore Structures*. Geneva: ISO.

252 Yang, L. and Moczulski, M. (2012). *DNV Fatigue Calculations for Existing Gulf of Mexico Fixed Structures*, BOEMRE TR 675. US Bureau of Ocean Energy Management, Regulations and Enforcement.

253 NORSOK Standard N-004. *Design of Steel Structures*. Lysaker, Norway: NORSOK.

254 NACE SP0108. *Corrosion Control of Offshore Structures by Protective Coatings*. Houston, TX: NACE International.

255 ISO 8502-6. *Preparation of Steel Substrates Before Application of Paints and Related Products – Tests for the Assessment of Surface Cleanliness – Part 6: Extraction of Soluble Contaminants for Analysis – The Bresle Method*. Geneva: ISO.

256 ISO 8502-9. *Preparation of Steel Substrates Before Application of Paints and Related Products – Tests for the Assessment of Surface Cleanliness – Part 9: Field Method for the Conductometric Determination of Water-Soluble Salts*. Geneva: ISO.

257 NACE SP0176. *Corrosion Control of Submerged Areas of Permanently Installed Steel Offshore Structures Associated with Petroleum Production*. Houston, TX: NACE International.

258 DNV-RP-B401. *Cathodic Protection Design*. Oslo, Norway: DNVGL.

INDEX

Metallurgy and Corrosion Control in Oil and Gas Production, Second Edition. Robert Heidersbach.
© 2018 John Wiley & Sons, Inc. Published 2018 by John Wiley & Sons, Inc.